Job #: 111368

Author Name: Conlon

Title of Book: Differentiable Manifolds

ISBN #: 9780817647667

Modern Birkhäuser Classics

Many of the original research and survey monographs in pure and applied mathematics published by Birkhäuser in recent decades have been groundbreaking and have come to be regarded as foundational to the subject. Through the MBC Series, a select number of these modern classics, entirely uncorrected, are being re-released in paperback (and as eBooks) to ensure that these treasures remain accessible to new generations of students, scholars, and researchers.

Differentiable Manifolds
Second Edition

Lawrence Conlon

Reprint of the 2001 Second Edition

Birkhäuser
Boston • Basel • Berlin

Lawrence Conlon
Department of Mathematics
Washington University
St. Louis, MO 63130-4899
U.S.A.

Originally published in the series *Birkhäuser Advanced Texts*

ISBN-13: 978-0-8176-4766-7 e-ISBN-13: 978-0-8176-4767-4
DOI: 10.1007/978-0-8176-4767-4

Library of Congress Control Number: 2007940493

Mathematics Subject Classification (2000): 57R19, 57R22, 57R25, 57R30, 57R45, 57R35, 57R55, 53A05, 53B05, 53B20, 53C05, 53C10, 53C15, 53C22, 53C29, 22E15

Cover design by Alex Gerasev.

Printed on acid-free paper.

9 8 7 6 5 4 3 2 1

www.birkhauser.com

Lawrence Conlon

Differentiable Manifolds
Second Edition

Birkhäuser
Boston • Basel • Berlin

Lawrence Conlon
Department of Mathematics
Washington University
St. Louis, MO 63130-4899
U.S.A.

Library of Congress Cataloging-in-Publication Data

Conlon, Lawrence, 1933-
Differentiable manifolds / Lawrence Conlon.–2nd ed.
p. cm.– (Birkhäuser advanced texts)
Includes bibliographical references and index.
ISBN 0-8176-4134-3 (alk. paper)–ISBN 3-7643-4134-3 (alk. paper)
1. Differentiable manifolds. I. Title. II. Series.
QA614.3.C66 2001
516.'6–dc21

2001025140

AMS Subject Classifications: 57R19, 57R22, 57R25, 57R30, 57R35, 57R45, 57R50, 57R55, 53A05, 53B05, 53B20, 53C05, 53C10, 53C15, 53C22, 53C29, 22E15

Printed on acid-free paper.

Birkhäuser 

ISBN 0-8176-4134-3 SPIN 10722989
ISBN 3-7643-4134-3

Typeset by the author in LaTeX.
Printed and bound by Hamilton Printing Company, Rensselaer, NY.
Printed in the United States of America.

9 8 7 6 5 4 3 2 1

This book is dedicated to my wife Jackie, with much love

Contents

Preface to the Second Edition

In revising this book for a second edition, I have added a significant amount of new material, dropping the subtitle "A first course". It is hoped that this will make the book more useful as a reference while still allowing it to be used as the basis of a first course on differentiable manifolds. In such a course, one should omit some or all of the material marked with an asterisk. More information about these optional topics will be given below.

Presupposed is a good grounding in general topology and modern algebra, especially linear algebra and the analogous theory of modules over a commutative, unitary ring. Mastery of the central topics of this book should prepare students for advanced courses and seminars in differential topology and geometry.

There are certain basic themes of which the student should be aware. The first concerns the role of differentiation as a process of linear approximation of nonlinear problems. The well-understood methods of linear algebra are then applied to the resulting linear problem and, where possible, the results are reinterpreted in terms of the original nonlinear problem. The process of solving differential equations (*i.e.*, integration) is the reverse of differentiation. It reassembles an infinite array of linear approximations, resulting from differentiation, into the original nonlinear data. This is the principal tool for the reinterpretation of the linear algebra results referred to above.

It is expected that the student has been exposed to the above processes in the setting of Euclidean spaces, at least in low dimensions. This is what we will refer to as *local calculus*, characterized by explicit computations in a fixed coordinate system. The concept of a "differentiable manifold" provides the setting for *global calculus*, characterized (where possible) by coordinate-free procedures. Where (as is often the case) coordinate-free procedures are not feasible, we will be forced to use local coordinates that vary from region to region of the manifold. When theorems are proven in this way, it becomes necessary to show independence of the choice of coordinates. The way in which these local reference frames fit together globally can be extremely complicated, giving rise to problems of a *topological* nature. In the global theory, geometric topology and, sometimes, algebraic topology become essential features.

These themes of *linearization*, (*re*)*integration*, and *global versus local* will be emphasized repeatedly.

Although a certain familiarity with the local theory is presupposed, we will try to reformulate that theory in a more organized and conceptual way that will make it easier to treat the global theory. Thus, this book will incorporate a modern treatment of the elements of multivariable calculus.

Fundamental to the global theory of differentiable manifolds is the concept of a *vector bundle*. As the global theory is developed, the tangent bundle, the cotangent bundle and various tensor bundles will play increasingly important roles, as will the related notions of infinitesimal G-structures and integrable G-structures.

For conceptual simplicity, all manifolds, functions, bundles, vector fields, Lie groups, homogeneous spaces, *etc.*, will be smooth of class C^∞. It is possible to adapt the treatment to smoothness of class C^k, $1 \le k < \infty$, but the technical problems that arise are distracting and the usefulness of this level of generality is limited. On the other hand, in much of the literature, the study of Lie groups and homogeneous spaces is carried out in the real analytic (C^ω) category. In these treatments, it is customary to note that C^∞ groups can be proven to be analytic, hence that no generality is lost. It seems to the author, however, that nothing would be gained by this approach and that the ideal of keeping this book as self-contained as possible would be compromised.

The optional topics (sections, subsections and one chapter, with titles terminating in an asterisk) can safely be omitted without creating serious gaps in the overall presentation. One topic that is new to this edition, covering spaces and the fundamental group, is not starred and should not be omitted unless the students have seen it in some prior course.

Some of the optional topics fall into subgroupings, any one of which can be included without dependence on the others. Thus, Subsection 2.9.B and Sections 3.10 and 4.2 constitute a brief introduction to Morse theory, one of the most useful tools in differential topology. Similarly, Sections 3.9, 6.5, and 8.7 constitute an introduction to degree theory, together with some classical topological applications, but in this case any one of these three sections can be treated without serious logical dependence on the others. Apart from minor revisions, this treatment of degree theory is not new to this edition. In Subsection 1.6.B, we classify 1-manifolds. This intuitively plausible result needed is only in the optional Section 3.9. Also easily omitted is the brief Subsection 1.6.A, this being an extended remark on cobordism theory.

New to this edition is an optional introductory treatment of Whitney's imbedding theorems (Subsection 3.7.C). We prove only the "easy" Whitney theorem, while stating carefully the general theorem. Imbeddings of manifolds in Euclidean space will be used only in treating some other optional topics, namely, the smoothing of continuous maps and homotopies (Subsection 3.8.B) and the existence of Morse functions (Section 3.10).

In Chapter 5, an introduction to Lie theory, adequate for a first course on manifolds, requires only the first two sections. Accordingly, Sections 5.3 (the closed subgroup theorem and related topics) and 5.4 (homogeneous spaces) are optional.

Certain topics in de Rham theory, Sections 8.8 (Poincaré duality) and 8.9 (a version of the de Rham theorem), can be omitted, as can the treatment of foliations defined by closed 1-forms (Section 9.3). Also easily omitted is the brief treatment of Riemannian homogeneous and symmetric spaces (Section 10.7). Finally, Chapter 11, on principal bundles and their role in geometry, gathers together and slightly expands on topics treated in various parts of the first edition and can be reserved to introduce a more advanced course or seminar.

There are some significant changes in the appendices also. The original Appendix A has been replaced by one that gives the construction of the universal covering space. The former Appendix D (Sard's theorem) has been moved to the main body of the text. The current Appendix D (formerly Appendix E) has been expanded to include a proof of the de Rham theorem for singular as well as Čech cohomology.

Acknowledgments

I am grateful to the late Robby Gardner and his students at Chapel Hill who "beta tested" eight chapters of a preliminary version of the first edition of my book in an intensive, one-semester graduate course. Their many suggestions were most helpful in the final revisions. Others whose input was helpful include Geoffrey Mess, Gary Jensen, Alberto Candel, Nicola Arcozzi and Tony Nielsen. I particularly want to thank Filippo De Mari, whose beautiful class notes, written when he was one of my students in an earlier version of this course, were immensely useful in subsequent revisions and first suggested to me the idea of writing a book. Finally, my students in the academic year 1999–2000 have offered much helpful input toward the final version of this edition.

CHAPTER 1

Topological Manifolds

This chapter pertains to the global theory of manifolds. See also [**3**, Chapter I] and [**41**, Chapter 1].

1.1. Locally Euclidean Spaces

Classical analysis is carried out in Euclidean space, the operations being defined by local formulas. One might hope, therefore, to extend this classical theory to all topological spaces that are *locally* Euclidean. While this is not generally possible without further restrictions on the spaces, the locally Euclidean condition is fundamental.

Definition 1.1.1. A topological space X is locally Euclidean if, for every $x \in X$, $\exists n \geq 0$ (an integer), an open neighborhood $U \subseteq X$ of x, an open subset $W \subseteq \mathbb{R}^n$ and a homeomorphism $\varphi : U \to W$.

If we can show that n is uniquely determined by x, we will write $n = d(x)$ and call this the *local dimension* of X at x.

Example 1.1.2. Any open subset $X \subseteq \mathbb{R}^n$ is a locally Euclidean space that is also Hausdorff and 2nd countable. We will see that the local dimension is n at every $x \in X$.

Example 1.1.3. Let $X = \mathbb{R} \sqcup \{*\}$, where $*$ is a single point and *sqcup* denotes disjoint union. Topologize this set so that a basis of open subsets $V \subset X$ consists of the following:

- If $* \notin V$, then V is open as a subset of $\mathbb{R}$.
- If $* \in V$, then $0 \notin V$ and there is an open neighborhood $W \subset \mathbb{R}$ of 0 such that $V = (W \smallsetminus \{0\}) \cup \{*\}$.

This space is locally Euclidean and 2nd countable. It is *not* Hausdorff since every open neighborhood of $*$ meets every open neighborhood of 0. In this case, the local dimension is everywhere 1, even at $*$ and at 0.

Example 1.1.4. In [**41**, Appendix A], there is described a bizarre space called the *long line*. It is connected, Hausdorff, and locally Euclidean with $d(x) \equiv 1$, but it is not 2nd countable. In fact, the long line contains an uncountable family of disjoint, open intervals.

Exercise 1.1.5. Prove that each connected component of a locally Euclidean space X is an open subset of X.

Exercise 1.1.6. Prove that a connected, locally Euclidean space X is path connected.

Exercise 1.1.7. Give an example of a connected, 2nd countable, Hausdorff space that is not path connected.

Recall that a regular space X is one in which any proper closed subset $C \subset X$ and point $x \in X \smallsetminus C$ can be separated by disjoint, open neighborhoods of each.

Exercise 1.1.8. Prove that a locally compact (in particular, a locally Euclidean), Hausdorff space must be regular.

By a theorem of Urysohn [**8**, p. 195], 2nd countable regular spaces are metrizable. Thus, manifolds are metrizable and questions of continuity, closure, compactness *etc.* involving manifolds can be reduced to corresponding questions of sequential continuity, sequential closure, sequential compactness, *etc.*

The following difficult result, known as L. E. J. Brouwer's theorem on invariance of domain, is needed in order to show that local dimension is always well defined on locally Euclidean spaces. The proof will not be given. It is best carried out by the methods of algebraic topology [**10**, p. 303], [**39**, p. 199], [**13**, p. 110]. For a more classical proof, see [**20**, pp. 95–96]. In the theory of smooth manifolds, differential calculus reduces the appropriate analogue of this theorem to elementary linear algebra.

Theorem 1.1.9 (Invariance of domain). *If $U \subseteq \mathbb{R}^n$ is open and $f : U \to \mathbb{R}^n$ is continuous and one-to-one, then $f(U)$ is open in $\mathbb{R}^n$.*

Corollary 1.1.10. *If $U \subseteq \mathbb{R}^n$ and $V \subseteq \mathbb{R}^m$ are open subsets such that U is homeomorphic to V, then $n = m$.*

Proof. Assume that $m \neq n$, say, $m < n$. Define $i : \mathbb{R}^m \hookrightarrow \mathbb{R}^n$ by

$$i(x^1, \dots, x^m) = (x^1, \dots, x^m, \underbrace{0, \dots, 0}_{n-m}).$$

This map is continuous and one-to-one and $i(\mathbb{R}^m)$ is *not* open in $\mathbb{R}^n$ and does not even contain a subset that is open in $\mathbb{R}^n$. By assumption, there is a homeomorphism $\varphi : U \to V$, so the composition

$$f : U \xrightarrow{\varphi} V \xrightarrow{i|V} \mathbb{R}^n$$

is continuous and one-to-one. Also, while $U \subseteq \mathbb{R}^n$ is open, we see that $f(U) = i(V) \subseteq i(\mathbb{R}^m)$ cannot be open in $\mathbb{R}^n$. This contradicts Theorem 1.1.9. □

Corollary 1.1.11. *If X is locally Euclidean, then the local dimension is a well-defined, locally constant function $d : X \to \mathbb{Z}^+$.*

Proof. Let $x \in X$ and suppose that there are open neighborhoods U and V of x in X, together with open subsets $\widetilde{U} \subseteq \mathbb{R}^n$, $\widetilde{V} \subseteq \mathbb{R}^m$, and homeomorphisms

$$\varphi : U \to \widetilde{U}$$
$$\psi : V \to \widetilde{V}.$$

Since $V \cap U$ is open in X, it follows that

$$\varphi(V \cap U) \subseteq \widetilde{U} \subseteq \mathbb{R}^n$$

are inclusions of open subsets and, similarly, that $\psi(V \cap U)$ is open in $\mathbb{R}^m$. But

$$\varphi \circ \psi^{-1} : \psi(U \cap V) \to \varphi(U \cap V)$$

is a homeomorphism, so $m = n$ by the previous corollary. All assertions follow. □

Corollary 1.1.12. *If X is a connected, locally Euclidean space, then the local dimension $d : X \to \mathbb{Z}^+$ is a constant called the dimension of X.*

Exercise 1.1.13. Let X and Y be connected, locally Euclidean spaces of the same dimension. If $f : X \to Y$ is bijective and continuous, prove that f is a homeomorphism.

1.2. Topological Manifolds

Some authors designate by the term "manifold" an arbitrary locally Euclidean space. It is more common, however, to require more.

Definition 1.2.1. A topological space X is a manifold of dimension n (an n-manifold) if

(1) X is locally Euclidean and $d(x) \equiv n = \dim X$;
(2) X is Hausdorff;
(3) X is 2nd countable.

Of the three examples in the previous section, only the open subsets of $\mathbb{R}^n$ were manifolds.

Lemma 1.2.2. *If X is a compact, connected, metrizable space that is locally Euclidean, then X is an n-manifold, for some $n \in \mathbb{Z}^+$.*

Proof. Indeed, X is Hausdorff because it is metrizable. It is 2nd countable because it is locally Euclidean and compact. Being locally Euclidean and connected, X has constant local dimension. □

Here are some examples of manifolds.

Example 1.2.3. The n-sphere $S^n = \{v \in \mathbb{R}^{n+1} \mid \|v\| = 1\}$ is an n-manifold. One way to see that it is locally Euclidean is by *stereographic projection*. Let

$$p_+ = (0, \dots, 0, 1),$$
$$p_- = (0, \dots, 0, -1)$$

be the north and south poles of S^n, respectively. Then the stereographic projections

$$\pi_+ : S^n \smallsetminus \{p_+\} \to \mathbb{R}^n$$
$$\pi_- : S^n \smallsetminus \{p_-\} \to \mathbb{R}^n$$

onto the subspace

$$\mathbb{R}^n = \{(x^1, \dots, x^n, 0)\}$$

are homeomorphisms and $\{S^n \smallsetminus \{p_-\}, S^n \smallsetminus \{p_+\}\}$ is an open cover of S^n. (For a pictorial definition of π_+, see Figure 1.2.1.) Since S^n is compact and metrizable, it is an n-manifold.

Exercise 1.2.4. In Example 1.2.3, write down formulas for the stereographic projections $\pi_\pm$ and prove carefully that they are homeomorphisms.

Example 1.2.5. If N is an n-manifold and M is an m-manifold, then $N \times M$ is an $(n + m)$-manifold. Indeed, if $(x, y) \in N \times M$, let U be a neighborhood of x in N homeomorphic to an open subset of $\mathbb{R}^n$ and V a neighborhood of y in M homeomorphic to an open subset of $\mathbb{R}^m$. Then, the neighborhood $U \times V \subseteq N \times M$ of (x, y) is homeomorphic to an open subset of $\mathbb{R}^n \times \mathbb{R}^m = \mathbb{R}^{n+m}$. Since M and N are Hausdorff and 2nd countable, so is $N \times M$.

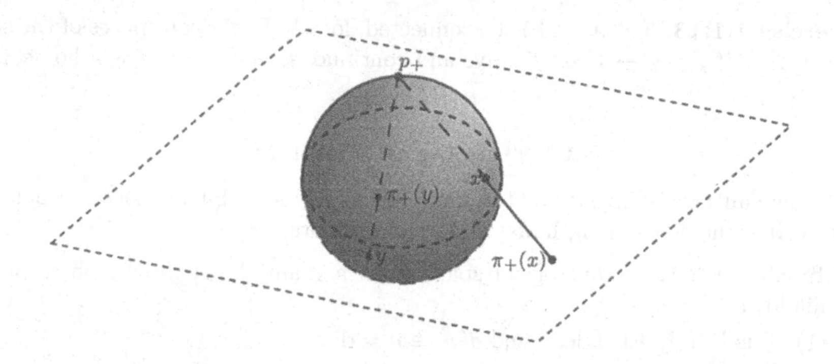

Figure 1.2.1. Stereographic projection from p_+

Example 1.2.6. The n-torus

$$T^n = \underbrace{S^1 \times S^1 \times \cdots \times S^1}_{n \text{ factors}}$$

is an n-dimensional manifold. Indeed, by Example 1.2.3, S^1 is a 1-manifold and Example 1.2.5, applied successively, then implies that T^n is an n-manifold.

Example 1.2.7. A vector $w \in \mathbb{R}^{n+1}$ is defined to be tangent to S^n at $v \in S^n$ if $w \perp v$. This conforms to naive geometric intuition and can be seen to conform to the general definition of tangent vectors to differentiable manifolds that we will give later. In order to keep track of the point of tangency, we will denote this tangent vector by $(v, w) \in \mathbb{R}^{n+1} \times \mathbb{R}^{n+1}$. Thus, the set of all tangent vectors to S^n is

$$T(S^n) = \{(v, w) \in \mathbb{R}^{n+1} \times \mathbb{R}^{n+1} \mid \|v\| = 1, w \perp v\}.$$

This space is topologized as a subspace of $\mathbb{R}^{n+1} \times \mathbb{R}^{n+1}$. We also define the continuous map $p : T(S^n) \to S^n$ by $p(v, w) = v$. Thus, p assigns to each tangent vector its point of tangency. For each $v_0 \in S^n$, consider

$$T_{v_0}(S^n) = \{(v_0, w) \in T(S^n)\} = p^{-1}(v_0),$$

the set of all vectors tangent to S^n at v_0. This is an n-dimensional vector space under the operations

$$r \cdot (v_0, w) = (v_0, r \cdot w)$$
$$(v_0, w_1) + (v_0, w_2) = (v_0, w_1 + w_2).$$

This structure, $p : T(S^n) \to S^n$, is called the *tangent bundle* of S^n. The space $T(S^n)$ is the *total space* of the bundle, the space S^n is the *base space* of the bundle, and p is the *bundle projection.* By a common abuse of terminology, the total space is often referred to as the tangent bundle itself. In Exercises 1.2.10 and 1.2.11, you are going to prove that $T(S^n)$ is a $2n$-manifold. A couple of definitions are needed first.

Definition 1.2.8. If $U \subseteq S^n$ is an open subset, then $T(S^n)|U = T(U)$ is the space $p^{-1}(U)$. The tangent bundle of U is given by $p_U : T(U) \to U$, where p_U denotes the restriction $p|T(U)$.

Definition 1.2.9. If $U \subseteq S^n$ is open, a vector field on U is a continuous map $s : U \to T(U)$ such that $p_U \circ s = \mathrm{id}_U$.

Exercise 1.2.10. Given $v_0 \in S^n$, show that there is an open neighborhood $U \subset S^n$ of v_0 and vector fields $s_i : U \to T(U)$, $1 \le i \le n$, such that

$$\{s_1(v), s_2(v), \dots, s_n(v)\}$$

is a basis of the vector space $T_v(S^n)$, $\forall\, v \in U$.

Exercise 1.2.11. Let $U \subset S^n$ be as in the previous exercise. Using that exercise, construct a continuous bijection $\varphi : U \times \mathbb{R}^n \to T(U)$ and prove that φ is a homeomorphism. (This is not very deep. You do not, for instance, need Theorem 1.1.9.) Using this, prove that $T(S^n)$ is a $2n$-manifold. Prove also that, for each $v \in U$, the formula $\varphi_v(w) = \varphi(v, w)$ defines an isomorphism $\varphi_v : \mathbb{R}^n \to T_v(S^n)$ of vector spaces.

A thorough understanding of the tangent bundle of S^n eluded topologists for several decades. For instance, it was long unknown what is the maximum number $r(n)$ of vector fields $s_i : S^n \to T(S^n)$, $1 \le i \le r(n)$, that are everywhere linearly independent. That is, we require that, for each $v \in S^n$, the vectors $\{s_1(v), \dots, s_{r(n)}(v)\}$ be linearly independent in $T_v(S^n)$ and that no set of $r(n) + 1$ fields has this property. The problem of computing $r(n)$ was known as the "vector field problem for spheres".

Definition 1.2.12. The sphere S^n is parallelizable if $r(n) = n$.

This brings us to a striking example of global *versus* local properties. If S^n is parallelizable, Exercise 1.2.11 implies that $T(S^n) \cong S^n \times \mathbb{R}^n$. For general n, this same exercise implies that the tangent bundle $T(S^n)$ is *locally* a Cartesian product of an open set $U \subset S^n$ with $\mathbb{R}^n$, but it is only *globally* such a product when S^n is parallelizable.

Not every sphere is parallelizable. For instance, it has long been known that $r(2n) = 0$. This means that every vector field on S^{2n} is somewhere zero. In the case of S^2, this is sometimes stated facetiously as "you can't comb the hair on a coconut". It was also known for some time that S^1, S^3, and S^7 are parallelizable. The following was finally proven in the late 1950s [**4**], [**27**].

Theorem 1.2.13 (R. Bott and J. Milnor, M. Kervaire). *The sphere S^n is parallelizable if and only if $n = 0, 1, 3,$ or 7.*

The case $n = 0$ is the trivial fact that the 0-sphere $S^0 = \{\pm 1\} \subset \mathbb{R}$ admits 0 independent fields.

There is an interesting relationship between Theorem 1.2.13 and the problem of defining a bilinear multiplication on $\mathbb{R}^n$ without divisors of zero. Such a multiplication is a bilinear map

$$\mu : \mathbb{R}^n \times \mathbb{R}^n \to \mathbb{R}^n,$$

written $\mu(v, w) = vw$, such that $vw = 0 \Leftrightarrow v = 0$ or $w = 0$.

Theorem 1.2.14. *There is a multiplication on $\mathbb{R}^{n+1}$ without divisors of zero if and only if $n = 0, 1, 3,$ or 7.*

Indeed, $\mathbb{R}^1 = \mathbb{R}$, $\mathbb{R}^2 = \mathbb{C}$, and $\mathbb{R}^4 = \mathbb{H}$ (the quaternions). The multiplication on $\mathbb{R}^8$ is given by the *Cayley numbers,* a nonassociative division algebra whose elements are ordered pairs (x, y) of quaternions [**43**, pp. 108–109]. This proves the "if" in Theorem 1.2.14.

Exercise 1.2.15. If $\mathbb{R}^{n+1}$ admits a multiplication without divisors of zero, prove that S^n is parallelizable. In light of Theorem 1.2.13, this gives the "only if" part of Theorem 1.2.14.

The full solution to the vector field problem for spheres was given by F. Adams in the early 1960s [**1**], culminating a long history of research on that problem by several algebraic topologists. We state Adams' result. Define the function $\rho(n)$, $n \geq 1$, by requiring that S^{n-1} admit $\rho(n) - 1$ everywhere linearly independent vector fields, but not $\rho(n)$ such fields (thus, $r(n) = \rho(n+1) - 1$). Write each natural number n uniquely as

$$n = (2r+1)2^{c+4d},$$

where r, c, d are nonnegative integers and $c \leq 3$. This uses the unique factorization theorem and the division algorithm mod 4.

Theorem 1.2.16 (F. Adams). $\rho(n) = 2^c + 8d$.

Remark that $\rho(\text{odd}) = 1$, since $c = d = 0$. This gives the classical result that every vector field on an even dimensional sphere is somewhere zero.

The easier part of Adams' theorem is that S^{n-1} does admit at least $2^c + 8d - 1$ independent vector fields. The original proof, using only linear algebra, was given by Radon and Hurwitz and, in 1942, an improved proof was given by Eckmann [**9**]. The harder part of the theorem, that there are *at most* $2^c + 8d - 1$ such fields, is much more advanced.

Exercise 1.2.17. Using Theorem 1.2.16, show that $\rho(n) = n$ if and only if $n = 1, 2, 4,$ or 8. This gives back Theorem 1.2.13.

Exercise 1.2.18. Verify that $\rho(2) = 2$ without using Theorem 1.2.16.

1.3. Quotient Constructions and 2-Manifolds

We continue the project of constructing manifolds. In this section, examples will be constructed using the *quotient topology*. Before giving a careful definition of quotient spaces, we look at some intuitive examples of 2-manifolds constructed in this way.

Consider the square $D = [0,1] \times [0,1]$. In Figure 1.3.1, the arrows indicate that the opposite sides of D are to be glued together so that the vertical edges are glued bottom to top and the horizontal edges from left to right. When the first pair of edges are glued, the result is a cylinder. When the second pair are glued, the cylinder becomes the 2-torus $T^2 = S^1 \times S^1$.

In Figure 1.3.2, the horizontal edges of D are identified in the same sense, but the left and right are glued together in opposite senses. If the left and right are identified first, the result is a "Möbius strip". It is not very easy, then, to picture the rest of the identification. If the top and bottom are identified first, the result is a cylinder whose boundary circles are then to be identified with an orientation-reversing "flip". The resulting 2-manifold is called the "Klein bottle" K^2 and can be pictured only in $\mathbb{R}^3$ if one allows the surface to intersect itself. Figure 1.3.2 is an attempt at such a picture, the self-intersection occurring along a circle. This circle of intersection corresponds to the vertical line ℓ and the circle c in Figure 1.3.3.

In order to view K^2 without self intersection, it is necessary to situate it in a 4-dimensional framework. This is not as psychologically hopeless as it might seem.

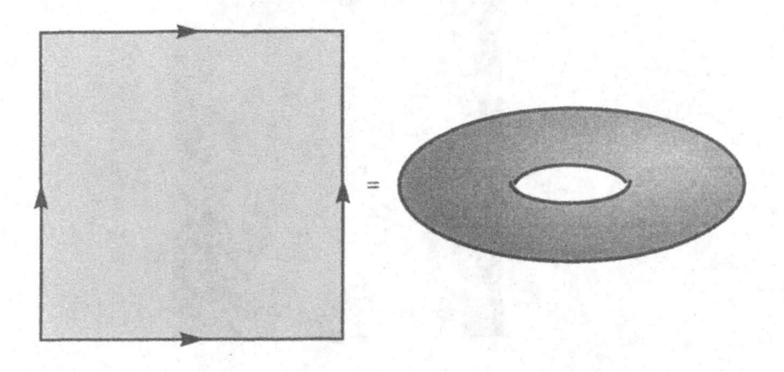

Figure 1.3.1. The 2-torus

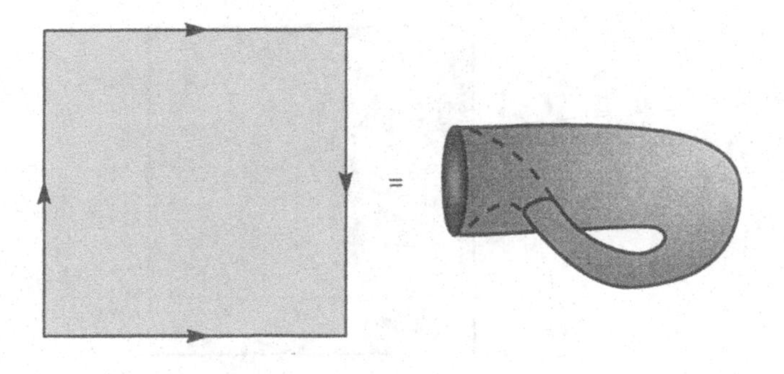

Figure 1.3.2. The Klein bottle with self-intersection

One can, for example, color K^2 by shades of grey, varying continuously over the Klein bottle, in such a way that the circle c in Figure 1.3.3 has no points with the same shade as any point in the line ℓ. By continuously assigning numbers from 0 to 1 (lightest to darkest) to these shades, one introduces a fourth "dimension" and the shaded Klein bottle has no self intersections. What was the circle of intersection is now two *disjoint* shaded circles ℓ and c. One can think of the shaded Klein bottle as a topologically imbedded Klein bottle in $\mathbb{R}^4$.

In Figure 1.3.4, we identify each pair of opposite sides of D with a reverse of orientation. The resulting 2-manifold is called the projective plane P^2 and, once again, it cannot be imbedded in $\mathbb{R}^3$. The topologist (but not the geometer) can view D as the unit disk $D^2 = \{v \in \mathbb{R}^2 \mid \|v\| \leq 1\}$ in such a way that the gluing

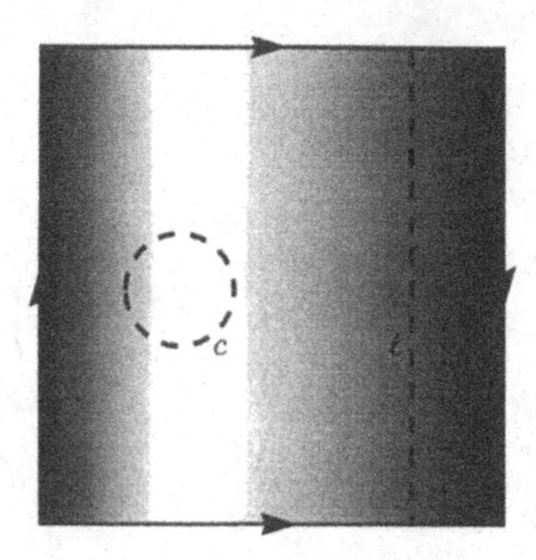

Figure 1.3.3. Using variable shading to imbed K^2 in $\mathbb{R}^4$

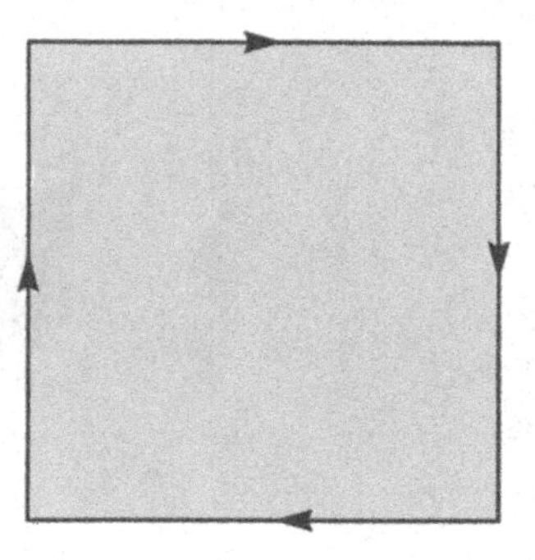

Figure 1.3.4. The projective plane

identifies antipodal pairs of boundary points. That is, if $\|v\| = 1$, then v and $-v$ are identified. Another way to think of P^2 is to start with $S^2 \subset \mathbb{R}^3$ and to identify the antipodal pairs $\{w, -w\}$. To see that this also yields P^2, first carry out the identification for the antipodal pairs not lying on the equatorial circle $z = 0$. The resulting disk has the equatorial circle as boundary and the remaining identifications give the previous description of P^2.

One also considers surfaces with edges. These are not manifolds in the sense we have defined, but they are manifolds with boundary in the sense to be defined in Section 1.6. For example, the disk D^2 is such a surface, as is the cylinder (also called the *annulus*) $S^1 \times [0, 1]$. Of particular interest is the *Möbius strip* $\mathfrak{M}^2$, already referred to in the construction of the Klein bottle. It is the result of gluing one pair

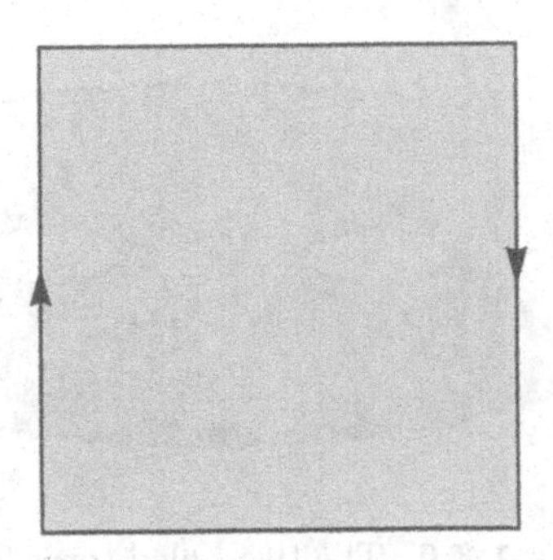

Figure 1.3.5. The Möbius strip $\mathfrak{M}^2$

of opposite edges of a rectangle with an orientation-reversing flip, but leaving the other two edges alone (Figure 1.3.5). This *nonorientable* surface has just one edge (a circle).

Definition 1.3.1. Let M_1 and M_2 be 2-manifolds and let $D_i \subset M_i$ be imbedded disks, $i = 1, 2$. If M_i has edges, require that D_i be disjoint from the edges. Let $M_i' = M_i \smallsetminus \operatorname{int}(D_i)$, $i = 1, 2$, and glue these together by a homeomorphism of $\partial M_1' = \partial D_1$ to $\partial M_2' = \partial D_2$. The resulting 2-manifold $M_1 \# M_2$ is called the connected sum of M_1 and M_2.

When this definition has been put on a rigorous footing, it can be shown that the connected sum is well defined up to homeomorphism. Thus, strictly speaking, the symbols M_1, M_2, and $M_1 \# M_2$ should denote *homeomorphism classes* of 2-manifolds. With this understanding, the connected sum can be seen to be commutative,

$$M_1 \# M_2 = M_2 \# M_1,$$

associative,

$$(M_1 \# M_2) \# M_3 = M_1 \# (M_2 \# M_3),$$

(hence parentheses can be dropped) and to admit (the homeomorphism class of) S^2 as a 2-sided identity

$$M_1 \# S^2 = M_1 = S^2 \# M_1.$$

In this way, the set of (homeomorphism classes of) surfaces becomes an abelian semigroup. Of particular interest is the subsemigroup of compact, connected surfaces.

Exercise 1.3.2. You are to give an intuitively clear "proof" that the Klein bottle is the connected sum of two projective planes. Proceed as follows.

(1) Remove the interior of an imbedded disk from P^2 and show that the resulting surface M is the Möbius strip. (Hint. Use the description of P^2 as the result of antipodal identifications on S^2.) Thus, $P^2 \# P^2$ is obtained by gluing

Figure 1.3.6. The two-holed torus $T^2 \# T^2$

together two copies of the Möbius strip by a homeomorphism between their boundary circles.

(2) Consider the square with edge identifications in Figure 1.3.2 and use two horizontal lines to divide this square into three congruent rectangles. Show that one of these is really a Möbius strip, as is the union of the other two, showing that K^2 is also obtained by gluing together two copies of the Möbius strip by a homeomorphism between their boundary circles.

Example 1.3.3. The connected sum $T^2 \# T^2$ of two copies of the torus is the *two-holed torus* pictured in Figure 1.3.6. It is possible to view this surface as the result of gluing together pairs of edges of an octagon, as we now sketch. Consider the irregular pentagon, as pictured in Figure 1.3.7, and glue together each pair of parallel edges as indicated there. Remark that this identifies all five vertices to a single point and that the resulting surface is just the complement in T^2 of the interior of an imbedded disk. Now glue two copies of this surface together as indicated in Figure 1.3.8 to obtain $T^2 \# T^2$. A topologically equivalent picture in Figure 1.3.9 shows that the two-holed torus is obtained from an octagon with edges identified pairwise as indicated there. By an inductive procedure, one can continue this process, showing that the *g-holed* torus, produced as a connected sum of g copies of T^2, is obtained by pairwise identifications of edges of a $4g$-gon.

Exercise 1.3.4. Give an intuitively plausible proof that

$$P^2 \# T^2 = P^2 \# K^2 = P^2 \# P^2 \# P^2.$$

Of course, the second equality is by Exercise 1.3.2. For the first, proceed as follows.

(1) Remove the interiors of two disjoint, imbedded disks in a rectangle $[0, 1] \times [0, 1]$. There are essentially two distinct ways to attach a cylinder $S^1 \times [0, 1]$, identifying its boundary circles respectively to the two boundary circles left by the excised disks. Show that one of these is homeomorphic to the surface obtained by removing the interior of a disk from T^2, the other to the surface obtained by the corresponding operation on K^2.

(2) Show that, if one identifies one pair of opposite edges of the square with an orientation-reversing flip, both of the surfaces obtained in (1) become homeomorphic.

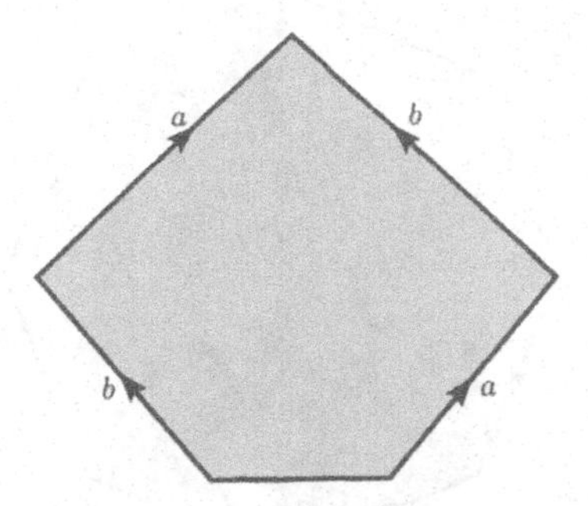

Figure 1.3.7. Identify the directed edges with the same labels.

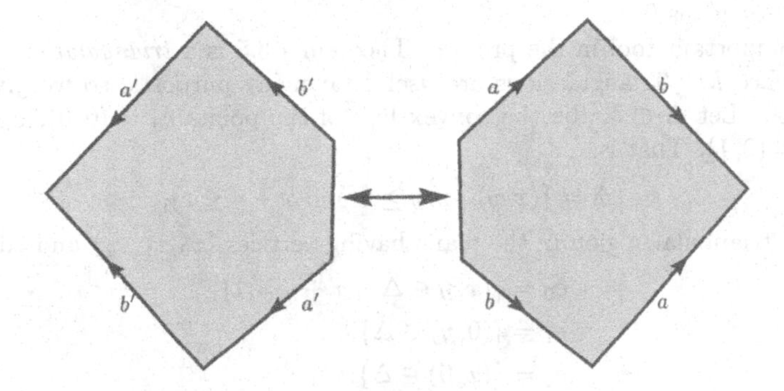

Figure 1.3.8. Glue two copies together along their free edges

(3) Show that the surface obtained in (2) can be viewed both as the connected sum $\mathfrak{M}^2 \# T^2$ and as $\mathfrak{M}^2 \# K^2$.
(4) Since the Möbius strip is obtained by deleting an open disk from P^2 (part (1) of Exercise 1.3.2), conclude to the desired homeomorphism.

Remark. Exercise 1.3.4 shows, in particular, that the semigroup of compact surfaces does not have cancellation. It is not a group. This exercise and constructions as in Example 1.3.3 are involved in the proof of the following classification theorem for compact surfaces. A proof of this theorem, in the spirit of our current discussion, can be found in [**26**, Chapter 1].

Theorem 1.3.5. *The semigroup of compact, connected 2-manifolds with no edges is generated by T^2 and P^2. Indeed, every element other than the identity S^2 can be uniquely written as a connected sum of finitely many copies of T^2 or of finitely many copies of P^2.*

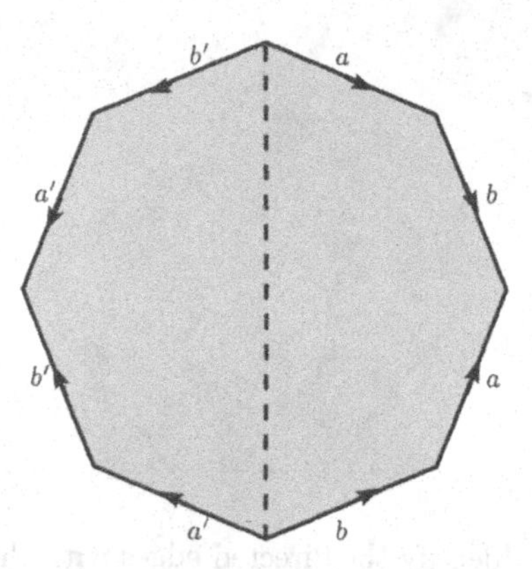

Figure 1.3.9. Identify the directed edges with the same labels.

The number g of copies of T^2 or of P^2 in the unique connected sum representation of a compact, connected surface $M \neq S^2$ is called the *genus* of M. We agree that S^2 has genus 0.

An important tool in the proof of Theorem 1.3.5 is a *triangulation* of a compact surface M. Triangulations are useful for many purposes, so we give a brief discussion. Let $\Delta \subset \mathbb{R}^2$ be the convex hull of the points $v_0 = (0,0)$, $v_1 = (1,0)$, and $v_2 = (0,1)$. That is,

$$\Delta = \{(x,y) \mid x,y \geq 0 \text{ and } x+y \leq 1\},$$

a closed, triangular region in the plane having vertices $\{v_0, v_1, v_2\}$ and edges

$$e_0 = \{(x,y \in \Delta) \mid x+y=1\},$$
$$e_1 = \{(0,y) \in \Delta\},$$
$$e_2 = \{(x,0) \in \Delta\}.$$

The triangle $\Delta \subset \mathbb{R}^2$ is called the *standard 2-simplex*. One decomposes M into pieces $\Delta_1, \Delta_2, \dots, \Delta_r$, together with homeomorphisms

$$\varphi_i : \Delta \to \Delta_i, \quad 1 \leq i \leq r,$$

in such a way that any two of the triangles Δ_i, Δ_j are either disjoint, have in common just one vertex $\varphi_i(v_\ell) = \varphi_j(v_k)$, or have in common just one edge $\varphi_i(e_\ell) = \varphi_j(e_k)$. Such a decomposition is called a triangulation of M.

Theorem 1.3.6 (T. Radó). *Every compact surface M admits a triangulation. Equivalently, M can be constructed, up to homeomorphism, by taking finitely many copies of the standard 2-simplex Δ and gluing them together appropriately along edges.*

This theorem is intuitive but nontrivial [**31**, pages 58–64]. The first proof was given by T. Radó [**36**].

The standard 2-simplex is *oriented* by a choice of ordering of its three vertices. Two orderings give equivalent orientations if they differ by an even permutation. Thus, up to equivalence, there are two orientations of Δ. Such an orientation

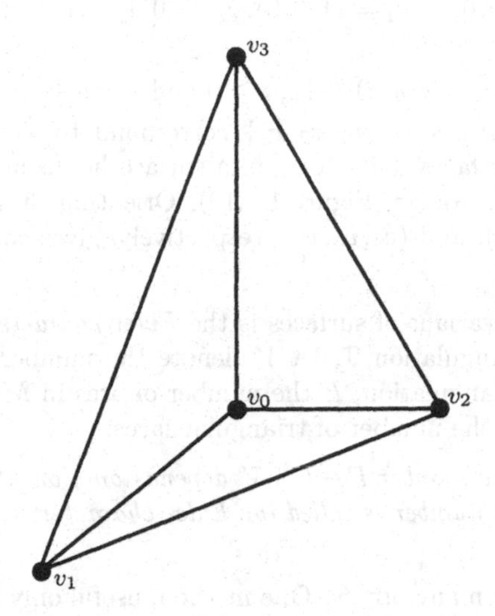

Figure 1.3.10. A triangulation of S^2

induces a direction along the edges of Δ, leading to the terminology "clockwise orientation" and "counterclockwise" orientation. The standard orientation of Δ, given by the ordering (v_0, v_1, v_2), is the counterclockwise orientation. Note that the homeomorphism $\alpha : \Delta \to \Delta$, defined by $\alpha(x, y) = (y, x)$ is orientation reversing.

Triangulations give us a way of defining the notion of "orientability" for a compact surface. The idea is that a triangle $\Delta_i = \varphi_i(\Delta)$ has the standard orientation $(\varphi_i(v_0), \varphi_i(v_1), \varphi_i(v_2))$. Two triangles Δ_i and Δ_j that have a common edge e are *coherently oriented* if their standard orientations induce *opposite* directions along e. A little thought should convince the reader that this is the natural notion if we are to picture the orientations of both Δ_i and Δ_j as counterclockwise. If Δ_i and Δ_j have no edge in common, they are also said to be coherently oriented.

Given a triangulation of a connected surface M, one can attempt to make all the orientations coherent as follows. Starting with one triangle, say Δ_1, look at any triangle, say Δ_2, having an edge in common with Δ_1. If they are coherently oriented, well and good. If not, replace φ_2 with $\varphi_2 \circ \alpha$, making them coherently oriented. Continuing in this way either orients the triangulation or leads to conflicting orientations on at least one simplex. In the first case, the triangulation is said to be *orientable* and, in the second, to be nonorientable. If M has more than one component, each must be treated separately and the triangulation of M is orientable if and only if the triangulation of each component is orientable. It can be proven that some triangulation of M is orientable if and only if every triangulation of M is orientable.

Definition 1.3.7. We say that M is orientable if some, hence every, triangulation of M is orientable.

Example 1.3.8. The sphere S^2 is orientable. Indeed, define the standard 3-simplex $\Delta^3 \subset \mathbb{R}^3$ to be the convex hull of the set of points

$$\{v_0 = (0,0,0), v_1 = (1,0,0), v_2 = (0,1,0), v_3 = (0,0,1)\}.$$

That is,

$$\Delta^3 = \{(x,y,z) \mid x,y,z \geq 0 \text{ and } x+y+z \leq 1\}.$$

Any three of the vertices $\{v_0, v_1, v_2, v_3\}$ correspond to a triangular face of Δ^3. These four triangular faces unite to form a surface homeomorphic to S^2, defining thereby a triangulation of S^2 (Figure 1.3.10). Orienting these faces by (v_1, v_2, v_3), (v_0, v_2, v_1), (v_3, v_2, v_0), and (v_3, v_0, v_1), respectively, gives coherent orientations to all the triangles.

An interesting invariant of surfaces is the *Euler characteristic*. If the compact surface M has a triangulation $\mathcal{T}$, let V denote the number of points of M that are vertices of the triangulation, E the number of arcs in M that are edges of the triangulation, and F the number of triangular faces.

Theorem 1.3.9. *The number $F - E + V$ depends only on M, not on the choice of triangulation $\mathcal{T}$. This number is called the Euler characteristic of M and is denoted $\chi(M)$.*

This theorem has many proofs. One method, useful only for *smoothly imbedded* surfaces in $\mathbb{R}^3$, is to use the Gauss–Bonnet theorem, which asserts that the integral of the Gauss curvature over M is $2\pi\chi(M)$ ([**15**, p. 111], [**34**, pp. 380–382], [**44**, pp. 237–239]).

Example 1.3.10. In the triangulation of S^2 given in Figure 1.3.10, we have $V = 4$, $E = 6$, and $F = 4$, so $\chi(S^2) = 2$.

Exercise 1.3.11. Give an intuitive but completely convincing proof that $V - E + F = 2$ for every triangulation of S^2.

Exercise 1.3.12. Consider a subdivision of S^2 into n-gons, where $n \geq 3$ is fixed, such that any two of these n-gons meet, if at all, only along a common edge or a common vertex. Assume further that there is a fixed integer $m \geq 3$ such that exactly m edges meet at each vertex. Using the fact that $\chi(S^2) = 2$, prove that the number of n-gons in this subdivision must be 4, 6, 8, 12 or 20. This proves the result, known from antiquity, that there are only five regular polyhedra (the "Platonic solids"). What are the values of n and m for these regular polyhedra?

Exercise 1.3.13. Produce triangulations of the surfaces T^2, K^2, and P^2. Use these to prove that T^2 is orientable and that K^2 and P^2 are not. Use Theorem 1.3.5 to determine all orientable, compact, connected surfaces and all nonorientable ones.

Exercise 1.3.14. For compact, connected surfaces M_1 and M_2, prove that

$$\chi(M_1 \# M_2) = \chi(M_1) + \chi(M_2) - 2.$$

Use this equation, Theorem 1.3.5 and Exercise 1.3.13 to compute the Euler characteristics of all compact, connected surfaces without edges. All compact, connected surfaces with edges are obtained from those without edges by removing the interiors of finitely many disjoint, imbedded disks. The resulting surface, obtained (say) from M, is said to have the same genus as M and is orientable if and only if M is orientable. Give a formula for the Euler characteristic of this surface in terms of $\chi(M)$ and the number of disks removed.

Our discussion of the topology of 2-manifolds belongs to the "cut and paste" brand of topology. To put such constructions on a rigorous footing, it is necessary to introduce quotient spaces. Let X be a topological space and let $\sim$ be an equivalence relation on X. That is, the relation satisfies the following three conditions:

- $x \sim x$, $\forall x \in X$;
- $x \sim y \Rightarrow y \sim x$;
- $x \sim y$ and $y \sim z \Rightarrow x \sim z$.

Such a relation partitions X into a collection $\{X_\alpha\}_{\alpha\in\mathfrak{A}}$ of disjoint subspaces, the equivalence classes of $\sim$.

Definition 1.3.15. The set $\{X_\alpha\}_{\alpha\in\mathfrak{A}}$ of equivalence classes of X is called the quotient space of X modulo $\sim$ and is denoted by $X/\!\sim$. The surjection

$$\pi : X \to X/\!\sim$$

is the map that assigns to each $x \in X$ its equivalence class $\pi(x) \in X/\!\sim$. A subset $U \subseteq X/\!\sim$ is said to be open if and only if $\pi^{-1}(U)$ is open in X.

It is trivial to check that these open sets constitute a topology on the quotient space $X/\!\sim$. This is the quotient topology, characterized as the largest (*i.e.*, the finest) topology on $X/\!\sim$ relative to which the canonical projection

$$\pi : X \to X/\!\sim$$

is continuous.

Remark. Here and throughout this book, the term "canonical" will appear. While it has a precise meaning in category theory, our use will be less formal. It always indicates some sort of independence of arbitrary choices, the word "canon" being the Latin word for "law". For the theologically inclined, the term might be read as "God-given".

Example 1.3.16. Let $X = \mathbb{R} \times \{0,1\}$, topologized as the disjoint union of two copies of $\mathbb{R}$. Define an equivalence relation on X by setting

$$(x,\alpha) \sim (y,\beta)$$

if and only if either $\alpha = \beta$ and $x = y$, or $\alpha \neq \beta$ and $x = y \neq 0$. Thus $\{(0,0)\}$ and $\{(0,1)\}$ each constitute a distinct equivalence class, but all other classes are pairs $\{(x,0),(x,1)\}$ where $x \neq 0$. The quotient space $X/\!\sim$ is the non-Hausdorff, locally Euclidean space given in Example 1.1.3.

As the above example shows, even if X is Hausdorff, the quotient $X/\!\sim$ may fail to be Hausdorff.

Lemma 1.3.17. *If the space X is compact, so is $X/\!\sim$.*

Proof. The continuous image $\pi(X) = X/\!\sim$ of a compact space X is compact. □

Lemma 1.3.18. *If the space X is connected, so is $X/\!\sim$.*

Proof. The continuous image $\pi(X) = X/\!\sim$ of a connected space X is connected. □

Definition 1.3.19. A map $f : X \to Y$ respects an equivalence relation $\sim$ on X if $x \sim y \Rightarrow f(x) = f(y)$. In this case, the induced map

$$\bar{f} : X/\!\sim \to Y$$

is well defined by

$$\bar{f}(\pi(x)) = f(x).$$

Lemma 1.3.20. *Let X and Y be topological spaces, let $\sim$ be an equivalence relation on X, and let $f : X \to Y$ be a map respecting this equivalence relation. Then f is continuous if and only if $\bar{f}$ is continuous.*

Proof. Consider the commutative diagram

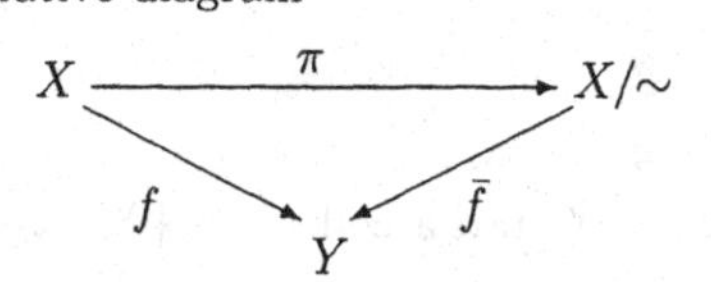

Since $f = \bar{f} \circ \pi$, continuity of $\bar{f}$ implies continuity of f. For the converse, assume that f is continuous, hence that $f^{-1}(U)$ is open in X whenever U is open in Y. But this implies, via the commutative diagram, that

$$\pi^{-1}(\bar{f}^{-1}(U)) = f^{-1}(U)$$

is an open subset of X. By the definition of the quotient topology, $\bar{f}^{-1}(U)$ is open in $X/\sim$, so $\bar{f}$ is continuous. □

Example 1.3.21. Let $X = [0,1]$ and let $Y = S^1 \subset \mathbb{C}$. Define

$$f : X \to Y$$

by $f(t) = e^{2\pi it}$, a continuous surjection. It is not quite one-to-one since $f(1) = f(0)$. On X, define the equivalence relation $x \sim y$ by requiring either that $x = y$ or that $\{x, y\} = \{0, 1\}$. Clearly, f respects this relation. Denote $X/\sim$ by $[0,1]/\{0,1\}$ and remark that

$$\bar{f} : [0,1]/\{0,1\} \to S^1$$

is bijective. By Lemma 1.3.20, $\bar{f}$ is also continuous. But $[0,1]/\{0,1\}$ is compact by Lemma 1.3.17 and S^1 is Hausdorff. A one-to-one, continuous map from a compact space onto a Hausdorff space is a homeomorphism, so $\bar{f}$ gives a canonical homeomorphism $[0,1]/\{0,1\} \cong S^1$. Intuitively, we have glued together the two ends of the interval $[0,1]$ to obtain a circle.

Example 1.3.22. Consider the map

$$p : S^1 \times [0,1] \to D^2$$

defined by viewing $S^1 \subset D^2 \subset \mathbb{C}$ and writing

$$p(z,t) = (1-t)z.$$

This is one-to-one on $S^1 \times [0,1)$ and collapses $S^1 \times \{1\}$ to the single point $0 \in \mathbb{C}$. Arguing as in the previous example, we see that the quotient space $(S^1 \times [0,1])/(S^1 \times \{0\})$ is canonically homeomorphic to D^2. Intuitively, we have collapsed the top of the cylinder $S^1 \times [0,1]$ to a point, obtaining a cone that can then be flattened to a disk.

Generally, if $A \subseteq X$, one can define the equivalence relation $\sim_A$ by writing $x \sim_A y$ if and only if either $x = y$ or $x, y \in A$. The quotient space $X/\sim_A$ is thought of as the result of crushing A to a single point in X and will be denoted by X/A. Some care should be taken in using this notation. If $X = G$ is a topological group and $H \subseteq G$ is a subgroup, then G/H denotes the space of left cosets of H, *not* the space obtained by collapsing H alone to a point. The context should make clear which interpretation is intended. Remark that G/H is also a set of equivalence classes, the relation being $g_1 \sim g_2$ if and only if $g_1^{-1}g_2 \in H$. Thus, G/H can be given the quotient topology and the continuous map $\pi : G \to G/H$ is $\pi(g) = gH$.

Definition 1.3.23. If G is a topological group and $H \subseteq G$ is a subgroup, the quotient space G/H is called the (left) coset space of G mod H.

Many important topological groups are also manifolds. These are called *Lie groups* and are enormously important throughout mathematics and in mathematical physics.

We close this section with some exercises using quotient constructions to produce manifolds of higher dimension.

Exercise 1.3.24. Let $D^n = \{v \in \mathbb{R}^n \mid \|v\| \leq 1\}$, the unit n-disk (also called the closed n-ball) with boundary $\partial D^n = S^{n-1}$. Prove that D^n/S^{n-1} is homeomorphic to S^n. (We proved the case $n = 1$ in Example 1.3.21.)

Exercise 1.3.25. View $\mathbb{R}^n$ as an abelian group under vector addition. This is a topological group. The integer lattice

$$\mathbb{Z}^n = \{(m_1, m_2, \dots, m_n) \mid m_i \in \mathbb{Z}, \quad 1 \leq i \leq n\}$$

is a (normal) subgroup of $\mathbb{R}^n$. Let $T^n = \mathbb{R}^n/\mathbb{Z}^n$ be the coset space (actually, a Lie group). Prove that this space is homeomorphic to

$$\underbrace{S^1 \times S^1 \times \cdots \times S^1}_{n \text{ factors}},$$

our definition of the n-torus in Example 1.2.6. Use this to show that the surface constructed in Figure 1.3.1 is, indeed, T^2.

Exercise 1.3.26. Define an equivalence relation on $S^n \subset \mathbb{R}^{n+1}$ by writing $v \sim w$ if and only if $v = \pm w$. The quotient space $P^n = S^n/{\sim}$ is called projective n-space. (This is one of the ways that we defined the projective plane P^2.) The canonical projection $\pi : S^n \to P^n$ is just $\pi(v) = \{\pm v\}$. Define $U_i \subset P^n$, $1 \leq i \leq n+1$, by setting

$$U_i = \{\pi(x^1, \dots, x^{n+1}) \mid x^i \neq 0\}.$$

Prove

(1) U_i is open in P^n.
(2) $\{U_1, \dots, U_{n+1}\}$ covers P^n.
(3) There is a homeomorphism $\varphi_i : U_i \to \mathbb{R}^n$.
(4) P^n is compact, connected, and Hausdorff, hence is an n-manifold.

1.4. Partitions of Unity

Partitions of unity play a crucial role in manifold theory. Their existence is a consequence of the fact that manifolds are paracompact. In this section, we establish these facts. For further information about paracompact spaces, the reader is referred to [**8**, pp. 162–169].

Definition 1.4.1. A family $\mathcal{C} = \{C_\alpha\}_{\alpha \in \mathfrak{A}}$ of subsets of X is locally finite if each $x \in X$ admits an open neighborhood W_x such that $W_x \cap C_\alpha \neq \emptyset$ for only finitely many indices $\alpha \in \mathfrak{A}$.

The following exercise will be useful.

Exercise 1.4.2. If $\mathcal{C} = \{C_\alpha\}_{\alpha\in\mathfrak{A}}$ is a locally finite family of closed subsets of X, prove that $\bigcup_{\alpha\in\mathfrak{A}} C_\alpha$ is a closed subset of X.

Definition 1.4.3. Let $\mathcal{U} = \{U_\alpha\}_{\alpha\in\mathfrak{A}}$ and $\mathcal{V} = \{V_\beta\}_{\beta\in\mathfrak{B}}$ be open covers of a space X. We say that $\mathcal{V}$ is a refinement of $\mathcal{U}$ if there is a function $i : \mathfrak{B} \to \mathfrak{A}$ such that $V_\beta \subseteq U_{i(\beta)}$, $\forall\, \beta \in \mathfrak{B}$.

Definition 1.4.4. A Hausdorff space X is paracompact if it is regular and if every open cover of X admits a locally finite refinement.

Actually, it is redundant to require regularity, but it will shorten some arguments.

Theorem 1.4.5. *Every locally compact, 2nd countable Hausdorff space X is paracompact.*

Proof. By Exercise 1.1.8, X is regular. Since the theorem is evident for compact Hausdorff spaces, we assume that X is not compact. Consequently, there is a countable, increasing nest

$$K_1 \subset K_2 \subset \cdots \subset K_r \subset \cdots$$

of compact subsets of X such that

$$K_r \subset \operatorname{int}(K_{r+1}), \quad 1 \le r < \infty,$$

$$X = \bigcup_{r=1}^{\infty} \operatorname{int}(K_r).$$

Indeed, let $\{W_i\}_{i=1}^\infty$ be a countable base of the topology of X such that each $\overline{W}_i$ is compact. We set $K_1 = \overline{W}_1$ and, assuming inductively that K_j has been defined, $1 \le j \le r$, we let ℓ denote the least integer such that

$$K_r \subseteq \bigcup_{i=1}^{\ell} W_i$$

and set

$$K_{r+1} = \bigcup_{i=1}^{\ell+r} \overline{W}_i.$$

The required properties are easily checked.

Let $\mathcal{U} = \{U_\alpha\}_{\alpha\in\mathfrak{A}}$ be an open cover of X. We select a refinement as follows. We can choose finitely many $V_i = U_{\alpha_i} \in \mathcal{U}$, $1 \le i \le \ell_1$, that cover the compact set K_1. Extend this by $\{U_{\alpha_i}\}_{i=\ell_1+1}^{\ell_2}$ to an open cover of K_2. Since X is Hausdorff, the compact set K_1 is closed, so $V_i = U_{\alpha_i} \smallsetminus K_1$ is open, $\ell_1+1 \le i \le \ell_2$, and $\{V_i\}_{i=1}^{\ell_2}$ is an open cover of K_2. We have arranged that K_1 does not meet V_i, $i > \ell_1$. Proceeding inductively, we obtain a refinement $\mathcal{V} = \{V_i\}_{i=1}^\infty$ of $\mathcal{U}$ with the property that K_r meets only finitely many elements of $\mathcal{V}$, $\forall\, r \ge 1$. Given $x \in X$, choose $r \ge 1$ such that $x \in \operatorname{int}(K_r)$, a neighborhood of x that meets only finitely many elements of $\mathcal{V}$. □

Corollary 1.4.6. *Every manifold is paracompact.*

This corollary is the main reason that we required manifolds to be 2nd countable.

Definition 1.4.7. Let $\mathfrak{U} = \{U_\alpha\}_{\alpha\in\mathfrak{A}}$ be an open cover of a space X. A partition of unity, subordinate to $\mathfrak{U}$, is a collection $\lambda = \{\lambda_\alpha\}_{\alpha\in\mathfrak{A}}$ of continuous functions $\lambda_\alpha : X \to [0,1]$ such that

(1) $\operatorname{supp}(\lambda_\alpha) \subset U_\alpha$, $\forall\,\alpha \in \mathfrak{A}$ (where the support $\operatorname{supp}(\lambda_\alpha)$ is the closure of the subset of X on which $\lambda_\alpha \neq 0$);
(2) for each $x \in X$, there is a neighborhood W_x of x such that $\lambda_\alpha|W_x \not\equiv 0$ for only finitely many indices $\alpha \in \mathfrak{A}$;
(3) the sum $\sum_{\alpha\in\mathfrak{A}} \lambda_\alpha$, well defined and continuous by the above, is the constant function 1.

Our goal is to prove that open covers of manifolds admit subordinate partitions of unity. In fact, we will prove this for all paracompact spaces.

Lemma 1.4.8. *If X is paracompact and $\mathfrak{U} = \{U_\alpha\}_{\alpha\in\mathfrak{A}}$ is an open cover of X, there is a locally finite refinement $\mathcal{V} = \{V_\alpha\}_{\alpha\in\mathfrak{A}}$, indexed on the same set $\mathfrak{A}$, such that $\overline{V}_\alpha \subset U_\alpha$, $\forall\,\alpha \in \mathfrak{A}$. This will be called a* precise refinement.

Proof. Indeed, paracompact spaces are regular, so it is possible to find a refinement $\mathcal{W} = \{W_\kappa\}_{\kappa\in\mathfrak{K}}$ with $j : \mathfrak{K} \to \mathfrak{A}$ such that $\overline{W}_\kappa \subset U_{j(\kappa)}$, $\forall\,\kappa \in \mathfrak{K}$. Passing to a locally finite refinement of $\mathcal{W}$ gives a locally finite refinement $\mathcal{V}' = \{V'_\beta\}_{\beta\in\mathfrak{B}}$ of $\mathfrak{U}$ with associated map $i : \mathfrak{B} \to \mathfrak{A}$ such that $\overline{V}'_\beta \subset U_{i(\beta)}$, $\forall\,\beta \in \mathfrak{B}$. Remark that $\{\overline{V}'_\beta\}_{\beta\in\mathfrak{B}}$ is a locally finite family of closed subsets of X. For each $\alpha \in \mathfrak{A}$, let $\mathfrak{B}_\alpha = i^{-1}(\alpha)$, a possibly empty subset of indices in $\mathfrak{B}$. Remark that $\mathfrak{B}_\alpha \cap \mathfrak{B}_\gamma = \emptyset$ if $\alpha \neq \gamma$. Setting $V_\alpha = \bigcup_{\beta\in\mathfrak{B}_\alpha} V'_\beta$ (possibly empty), $\forall\,\alpha \in \mathfrak{A}$, gives the desired refinement $\mathcal{V}$. Indeed, the local finiteness of $\mathcal{V}$ follows from that of $\mathcal{V}'$ and

$$\overline{V}_\alpha = \bigcup_{\beta\in\mathfrak{B}_\alpha} \overline{V}'_\beta \subset U_\alpha$$

by Exercise 1.4.2. □

Recall that a topological space X is said to be *normal* if, whenever $A, B \subset X$ are closed, disjoint subsets, there is an open set $U \supset A$ such that $\overline{U} \cap B = \emptyset$.

Lemma 1.4.9. *If X is paracompact, it is normal.*

Proof. Let $A, B \subset X$ be closed, disjoint subsets. The space X being regular, there is a family $\{U_\alpha\}_{\alpha\in\mathfrak{A}}$ of open subsets of X, covering A and such that $\overline{U}_\alpha \subset X \smallsetminus B$, $\forall\,\alpha \in \mathfrak{A}$. The space X being paracompact, the open cover of X, obtained by adjoining $X \smallsetminus A$ to $\{U_\alpha\}_{\alpha\in\mathfrak{A}}$, has a locally finite refinement. Thus, we lose no generality in assuming that $\{U_\alpha\}_{\alpha\in\mathfrak{A}}$ is a locally finite family of open sets, hence that $\{\overline{U}_\alpha\}_{\alpha\in\mathfrak{A}}$ is also locally finite. Set $U = \bigcup_{\alpha\in\mathfrak{A}} U_\alpha$, an open neighborhood of A. By Exercise 1.4.2, $\overline{U} = \bigcup_{\alpha\in\mathfrak{A}} \overline{U}_\alpha$ and this set does not meet B. □

In the construction of partitions of unity, we will use the following well-known property of normal spaces [**8**, pp. 146–147]. We do not give the proof here since, ultimately, our interest is in the C^∞ version that we will prove later (Corollary 3.5.5).

Theorem 1.4.10 (Urysohn's lemma). *If X is a normal space and if A and B are closed, disjoint subsets, then there is a continuous function*

$$f : X \to [0,1]$$

such that $f|A \equiv 1$ and $\operatorname{supp}(f) \subset X \smallsetminus B$.

The existence of partitions of unity can now be established.

Theorem 1.4.11. *If X is a paracompact space and $\mathfrak{U} = \{U_\alpha\}_{\alpha\in\mathfrak{A}}$ is an open cover of X, then there exists a partition of unity subordinate to $\mathfrak{U}$.*

Proof. Use Lemma 1.4.8 to choose a precise refinement $\mathcal{V} = \{V_\alpha\}_{\alpha\in\mathfrak{A}}$ of $\mathfrak{U}$ and a precise refinement $\mathcal{W} = \{W_\alpha\}_{\alpha\in\mathfrak{A}}$ of $\mathcal{V}$. For each $\alpha \in \mathfrak{A}$, use Theorem 1.4.10 to define a continuous function $\gamma_\alpha : X \to [0,1]$ such that

$$\gamma_\alpha|W_\alpha \equiv 1,$$
$$\operatorname{supp}(\gamma_\alpha) \subset \overline{V}_\alpha \subset U_\alpha.$$

The local finiteness of $\mathcal{V}$ implies that $\{\operatorname{supp}(\gamma_\alpha)\}_{\alpha\in\mathfrak{A}}$ is also locally finite and $\{\gamma_\alpha\}_{\alpha\in\mathfrak{A}}$ satisfies properties (1) and (2) in Definition 1.4.7. It is also clear that

$$\gamma = \sum_{\alpha\in\mathfrak{A}} \gamma_\alpha < \infty$$

is continuous and nowhere 0. Therefore,

$$\left\{\lambda_\alpha = \frac{\gamma_\alpha}{\gamma}\right\}_{\alpha\in\mathfrak{A}}$$

is a partition of unity subordinate to $\mathfrak{U}$. □

Corollary 1.4.12. *Every open cover of a manifold admits a subordinate partition of unity.*

Partitions of unity will be needed for Riemann integration and Riemannian geometry on (smooth) manifolds. For these and similar applications we will need *smooth* partitions of unity, a notion that we are not yet ready to define. In the next section, we will use the existence of continuous partitions of unity to prove a topological imbedding theorem for manifolds.

Exercise 1.4.13. Let M be a manifold, $U \subseteq M$ an open subset, $K \subset U$ a set that is closed in M, and let $f : U \to \mathbb{R}$ be continuous. Prove that the restriction $f|K$ extends to a continuous function $f : M \to \mathbb{R}$.

Exercise 1.4.14. Let $K \subset S^n$ be a closed subset, $U \supset K$ an open neighborhood of K, v a vector field defined on U. Prove that $v|K$ extends to a vector field on all of S^n.

1.5. Imbeddings and Immersions

We will prove that compact manifolds can always be imbedded in Euclidean spaces of suitably large dimensions.

Definition 1.5.1. Let N and M be topological manifolds of respective dimensions $n \le m$. A topological imbedding of N in M is a continuous map $i : N \to M$ that carries N homeomorphically onto its image $i(N)$.

Definition 1.5.2. If N and M are as above, a topological immersion of N into M is a continuous map $i : N \to M$ such that, for each $x \in N$, there is an open neighborhood W of x in N such that $i|W : W \to M$ is a topological imbedding.

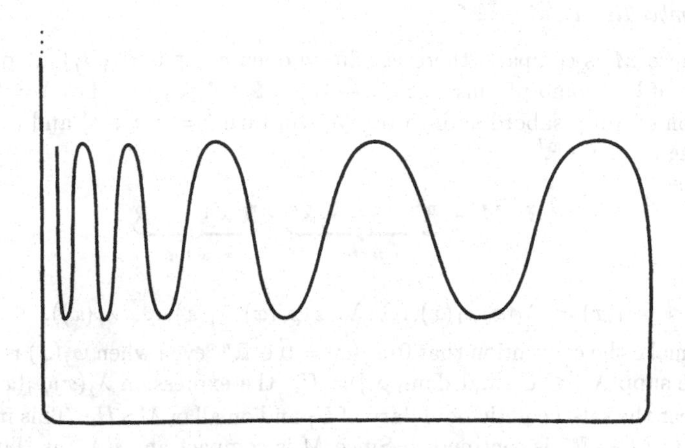

Figure 1.5.1. The topologist's sine curve

For example, Figure 1.3.2 depicts an immersion of the Klein bottle K^2 into $\mathbb{R}^3$. Every imbedding is also an immersion, but even one-to-one immersions can fail to be imbeddings. The immersion of $\mathbb{R}$ in $\mathbb{R}^2$, pictured in Figure 1.5.1, is one-to-one, but is not a homeomorphism onto its image.

Definition 1.5.3. If M is a manifold and $X \subseteq M$ is a subspace, we say that X is a submanifold if there is a manifold N and an imbedding $i : N \hookrightarrow M$ such that $X = i(N)$.

Definition 1.5.4. The image of a one-to-one immersion $i : N \to M$ is called an immersed submanifold of M.

Some authors use the term "submanifold" to include immersed submanifolds. From the point of view of a topologist, this seems dangerously misleading.

Exercise 1.5.5. Define $f : S^2 \to \mathbb{R}^4$ by the formula

$$f(x, y, z) = (yz, xz, xy, x^2 + 2y^2 + 3z^2).$$

Prove that f passes to a well-defined, topological imbedding

$$\bar{f} : P^2 \hookrightarrow \mathbb{R}^4.$$

(It is known that P^2 cannot be imbedded in $\mathbb{R}^3$.)

Exercise 1.5.6. Let $g : S^2 \to \mathbb{R}^3$ be defined by

$$g(x, y, z) = (yz, xz, xy).$$

Find six points $p_1, \ldots, p_6 \in P^2$ such that

$$\bar{g} : P^2 \smallsetminus \{p_1, \ldots, p_6\} \to \mathbb{R}^3$$

is a topological immersion. The mapping $\bar{g}$ of P^2 itself into $\mathbb{R}^3$ is known as Steiner's surface. It is simply the imbedding $\bar{f}$ into $\mathbb{R}^4$ followed by projection onto a three-dimensional subspace of $\mathbb{R}^4$. Prove that $\bar{g}$ does not restrict to an imbedding of any neighborhood of p_i, $1 \leq i \leq 6$. (It is known that $p \circ \bar{f}$ cannot be an immersion for any linear surjection $p : \mathbb{R}^4 \to \mathbb{R}^3$.)

Theorem 1.5.7. *If M is a compact n-manifold, then there is an integer $k > n$ and an imbedding $i : M \hookrightarrow \mathbb{R}^k$.*

Proof. Since M is compact, there is a finite open cover $\mathcal{U} = \{U_j\}_{j=1}^r$ of M and a collection of homeomorphisms $\varphi_j : U_j \to W_j \subseteq \mathbb{R}^n$, $1 \le j \le r$. Let $\lambda = \{\lambda_j\}_{j=1}^r$ be a partition of unity subordinate to $\mathcal{U}$. We will take $k = r(n+1)$ and construct an imbedding $i : M \hookrightarrow \mathbb{R}^k$.

Define

$$i : M \to \underbrace{\mathbb{R}^n \times \cdots \times \mathbb{R}^n}_{r \text{ factors}} \times \underbrace{\mathbb{R} \times \cdots \times \mathbb{R}}_{r \text{ factors}}$$

by

$$i(x) = (\lambda_1(x)\varphi_1(x), \dots, \lambda_r(x)\varphi_r(x), \lambda_1(x), \dots, \lambda_r(x)).$$

Here we make the convention that $0 \cdot \varphi_j(x) = \vec{0} \in \mathbb{R}^n$, even when $\varphi_j(x)$ is undefined.

Since $\operatorname{supp}(\lambda_j) \subset U_j$ and $\operatorname{dom}(\varphi_j) = U_j$, the expression $\lambda_j(x)\varphi_j(x)$ is identically $\vec{0}$ near the set-theoretic boundary of U_j and on all of $M \smallsetminus U_j$. This implies that the map $i : M \to \mathbb{R}^k$ is continuous. Since M is compact and $i(M)$ is Hausdorff, we only need to prove that i is one-to-one in order to prove that i is a homeomorphism onto its image.

Let $x, y \in M$ and suppose that $i(x) = i(y)$. Since λ is a partition of unity, there is a value of j such that $\lambda_j(x) \neq 0$. But the $(nr+j)$th coordinates of $i(x)$ and $i(y)$ are $\lambda_j(x) = \lambda_j(y)$, so $x, y \in \operatorname{supp}(\lambda_j) \subset U_j$. Also, $\lambda_j(x)\varphi_j(x) = \lambda_j(y)\varphi_j(y)$, so $\varphi_j(x) = \varphi_j(y)$. Since $\varphi_j : U_j \to \mathbb{R}^n$ is one-to-one, it follows that $x = y$. □

The imbedding dimension $k = r(n+1)$ given by this theorem for compact n-manifolds is often much too generous. For example, Exercise 1.3.26 gives a covering of P^2 by $r = 3$ open sets homeomorphic to $\mathbb{R}^2$, so the theorem guarantees only that P^2 can be imbedded in $\mathbb{R}^9$. In fact, it is possible to imbed P^2 into $\mathbb{R}^4$, as you showed in Exercise 1.5.5. Generally, if an n-manifold is *differentiable* (Definition 3.1.6), it can be proven that M imbeds in $\mathbb{R}^{2n+1}$ (Theorem 3.7.12). This result, due to H. Whitney, is best possible in the sense that there exist n-manifolds that cannot be imbedded in $\mathbb{R}^{2n}$.

1.6. Manifolds with Boundary

Manifolds are modeled locally on Euclidean n-space. Something like the closed n-ball $D^n = \{v \in \mathbb{R}^n \mid \|v\| \le 1\}$ fails to be a manifold because a point on the boundary $\partial D^n = S^{n-1}$ does not have a neighborhood homeomorphic to an open subset of $\mathbb{R}^n$. It does, however, have a neighborhood homeomorphic to an open subset of Euclidean *half-space.*

Definition 1.6.1. The Euclidean half-space of dimension n is

$$\mathbb{H}^n = \{(x^1, \dots, x^n) \in \mathbb{R}^n \mid x^1 \le 0\}$$

and the boundary of $\mathbb{H}^n$ is

$$\partial\mathbb{H}^n = \{(x^1, \dots, x^n) \in \mathbb{R}^n \mid x^1 = 0\}.$$

The interior of $\mathbb{H}^n$ is

$$\operatorname{int}(\mathbb{H}^n) = \mathbb{H}^n \smallsetminus \partial\mathbb{H}^n = \{(x^1, \dots, x^n) \in \mathbb{R}^n \mid x^1 < 0\}.$$

Remark that $\partial\mathbb{H}^n$ is canonically identified with $\mathbb{R}^{n-1}$ by suppressing the coordinate $x^1 = 0$ and renumbering the remaining coordinates as $y^i = x^{i+1}$, $1 \le i \le n-1$.

Definition 1.6.2. A topological space X is an n-manifold with boundary if

(1) for each $x \in X$, there is an open neighborhood U_x of x in X, an open subset $W_x \subset \mathbb{H}^n$, and a homeomorphism $\varphi : U_x \to W_x$;
(2) X is Hausdorff;
(3) X is 2nd countable.

Definition 1.6.3. Let X be a manifold with boundary. We say that $x \in X$ is a boundary point if a suitable choice of $\varphi : U_x \to W_x$, as above, carries x to a point $\varphi(x) \in \partial\mathbb{H}^n$. The interior points $x \in X$ are those such that a suitable homeomorphism $\varphi : U_x \to W_x$ carries x to a point $\varphi(x) \in \operatorname{int}(\mathbb{H}^n)$. The boundary ∂X is the set of all the boundary points of X and the interior $\operatorname{int}(X)$ is the set of all the interior points of X.

It is clear from the definition that every point of X is either a boundary point or an interior point. That is, $X = \partial X \cup \operatorname{int}(X)$. It is not immediately evident, however, that a point cannot be both an interior point and a boundary point.

Exercise 1.6.4. Use Theorem 1.1.9 to prove that $\partial X \cap \operatorname{int}(X) = \emptyset$. Also prove that ∂X is an $(n-1)$-manifold and that $\operatorname{int}(X)$ is an n-manifold.

Example 1.6.5. The closed, unit n-ball D^n is a manifold with boundary $\partial D^n = S^{n-1}$. The interior is the open ball $B^n = \{v \in \mathbb{R}^n \mid \|v\| < 1\}$. To see this rigorously, recall that the orthogonal group $\mathrm{O}(n)$ is the group of all $n \times n$ matrices A over $\mathbb{R}$ such that $A^{\mathrm{T}} = A^{-1}$ and that the standard matrix action of $\mathrm{O}(n)$ on $\mathbb{R}^n$ restricts to a group of homeomorphisms of D^n onto itself. If $e_1 \in S^{n-1}$ is the column vector with first entry 1 and remaining entries 0, then Ae_1 is the first column of the matrix $A \in \mathrm{O}(n)$. Since *every* unit vector $v \in S^{n-1}$ occurs as the first column of some matrix $A \in \mathrm{O}(n)$, it follows that $v \in S^{n-1}$ is carried by $A^{-1} \in \mathrm{O}(n)$ to e_1. Thus, an arbitrary point of S^{n-1} will be a boundary point of D^n if and only if e_1 is a boundary point. Let $U \subset D^n$ be the open neighborhood of e_1 defined by the condition $x^1 > 0$. Then $\varphi : U \to \mathbb{H}^n$, defined by

$$\varphi(x^1, x^2, \dots, x^n) = \left(x^1 - \sqrt{1 - (x^2)^2 - \cdots - (x^n)^2}, x^2, \dots, x^n\right),$$

carries the unit vectors in U into $\partial\mathbb{H}^n$ and the other points of U into $\operatorname{int}(\mathbb{H}^n)$. Furthermore, φ has continuous inverse ψ given by the formula

$$\psi(y^1, y^2, \dots, y^n) = \left(y^1 + \sqrt{1 - (y^2)^2 - \cdots - (y^n)^2}, y^2, \dots, y^n\right).$$

We have proven that $S^{n-1} \subseteq \partial D^n$. Since the open ball $B^n = D^n \setminus S^{n-1}$ is an open subset of $\mathbb{R}^n$, it is clear that $B^n \subseteq \operatorname{int}(D^n)$. By Exercise 1.6.4, all assertions follow.

Exercise 1.6.6. If $\partial M = \emptyset$ and $\partial N \neq \emptyset$, show that $N \times M$ is a manifold with boundary

$$\partial(N \times M) = \partial N \times M.$$

Example 1.6.7. By Exercise 1.6.6, the n-torus T^n is the boundary of the *solid* torus $D^2 \times T^{n-1}$, a compact $(n+1)$-manifold with boundary.

Example 1.6.8. If W_1 and W_2 are compact 3-manifolds with nonempty, connected boundaries, let $D_i \subset \partial W_i$ be a closed, imbedded 2-ball (a disk), $i = 1, 2$. By fixing a homeomorphism $f : D_1 \to D_2$, we set up an equivalence relation $\sim_f$ on the disjoint union $W_1 \sqcup W_2$ by defining $x \sim_f y$ if and only if either $x = y$, $y = f(x)$, or or $x = f(y)$. The quotient space is denoted by $W_1 \cup_f W_2$ and can be proven to be a compact 3-manifold with boundary the connected sum $\partial W_1 \# \partial W_2$. This is intuitively obvious, and we will not attempt a rigorous proof. It follows that, if M and N are compact, connected 2-manifolds that bound suitable compact 3-manifolds, then the connected sum $M \# N$ is also the boundary of some compact 3-manifold. By Examples 1.6.5 and 1.6.7, S^2 and T^2 are such boundaries, hence the classification of compact surfaces (Theorem 1.3.5) implies that every compact, connected, *orientable* surface bounds a compact 3-manifold. What about the compact, connected, *nonorientable* surfaces? It is known that the Klein bottle is such a boundary $K^2 = \partial W$. Intuitively, we construct W from the solid cylinder $D^2 \times [0,1]$ by gluing $D^2 \times \{1\}$ to $D^2 \times \{0\}$ by an orientation-reversing flip about a diameter. Since $P^2 \# P^2 = K^2$, it follows that the connected sum of an *even* number of projective planes is the boundary of a compact 3-manifold. By Theorem 1.3.5, every compact, connected, nonorientable surface is a connected sum of finitely many copies of P^2. Thus, the only compact, connected surfaces that might fail to bound a compact 3-manifold are connected sums of an odd number of projective planes. In fact, these do fail to bound. We will not prove this but, in Exercise 1.6.12, you will reduce the assertion to the special case that P^2 does not bound.

1.6.A. Cobordism*. Example 1.6.8 suggests an interesting topological problem. Given a compact n-manifold (connected or not), is it the boundary of some compact $(n+1)$-manifold? A closely related problem is that of classifying compact n-manifolds up to *cobordism,* an equivalence relation that was first defined by H. Poincaré [**35**, Section 5].

Definition 1.6.9. Let M and N be compact n-manifolds with empty boundary. We say that M and N are cobordant, and write $M \sim_\partial N$, if there is a compact $(n+1)$-manifold W such that ∂W is homeomorphic to the disjoint union $M \sqcup N$. The cobordism class of M is denoted by $[M]_\partial$. The set of cobordism classes of compact n-manifolds is denoted by $\mathfrak{N}_n$.

The fact that cobordism is an equivalence relation is elementary (Exercise 1.6.10) as is the fact that homeomorphic manifolds are cobordant. If we make the standard convention that the empty set is a manifold of *every* dimension, the cobordism class $[\emptyset]_\partial \in \mathfrak{N}_n$ is precisely the set of compact n-manifolds that bound.

In Exercise 1.6.11, you will show that disjoint union well defines an abelian group structure on $\mathfrak{N}_n$ under which every element is its own inverse. It is a nontrivial fact that this group is finitely generated, hence $\mathfrak{N}_n \cong \mathbb{Z}_2 \oplus \mathbb{Z}_2 \oplus \cdots \oplus \mathbb{Z}_2$. It is rather obvious that $\mathfrak{N}_0 = \mathbb{Z}_2$, the nontrivial element being $[\text{point}]_\partial$. We are about to see that, up to homeomorphism, the only compact, connected, boundaryless 1-manifold, is $S^1 = \partial D^2$, so $\mathfrak{N}_1 = 0$. In Exercise 1.6.12, you will show that $\mathfrak{N}_2 = \mathbb{Z}_2$, generated by $[P^2]_\partial$.

Exercise 1.6.10. Prove that cobordism is an equivalence relation on the set of all compact n-manifolds (ignore Zermelo–Frankel set-theoretic scruples about the meaning of "set of all compact n-manifolds"). Prove that homeomorphic manifolds are cobordant.

Exercise 1.6.11. Show that the operation $[M]_\partial + [N]_\partial = [M \sqcup N]_\partial$ is a well-defined binary operation on $\mathfrak{N}_n$. Prove that this operation makes $\mathfrak{N}_n$ into an abelian group with 0 element the class $[\emptyset]_\partial$ and $-[N]_\partial = [N]_\partial$, $\forall\, [N] \in \mathfrak{N}_n$.

Exercise 1.6.12. Using Example 1.6.8 and assuming that P^2 is not a boundary, prove that $\mathfrak{N}_2 = \mathbb{Z}_2$, the nontrivial element being $[P^2]_\partial$. If M and N are compact, connected surfaces, prove that $[M]_\partial + [N]_\partial = [M \# N]_\partial$.

1.6.B. Classification of 1-manifolds*. We sketch the classification of compact 1-manifolds, possibly with boundary. It is enough to classify the connected ones. The result is "intuitively evident" and will be needed only in the proof of Lemma 3.9.4, so this subsection can be omitted in a first reading.

Theorem 1.6.13. *If N is a compact, connected 1-manifold, then N is homeomorphic either to S^1 or to $[0,1]$.*

The following exercise is a critical step in the proof of this theorem.

Exercise 1.6.14. Let N be as in Theorem 1.6.13. We say that $V \subseteq N$ is an open interval in the 1-manifold N if it is an open subset that is homeomorphic to an open interval in $\mathbb{R}$. You are to prove that there is a *maximal* open interval in N, proceeding as follows.

(1) Let $\mathfrak{I}$ be the family of open intervals in N, partially ordered by inclusion. Let $\mathcal{V} = \{V_\alpha\}_{\alpha \in \mathfrak{A}}$ be an infinite, linearly ordered subset of $\mathfrak{I}$. Prove that there is a sequential nest
$$V_{\alpha_1} \subset V_{\alpha_2} \subset \cdots \subset V_{\alpha_k} \subset \cdots$$
in $\mathcal{V}$ such that, for each $\alpha \in \mathfrak{A}$, there is an integer $k > 0$ with $V_\alpha \subset V_{\alpha_k}$. (Hint: N is 2nd countable.)
(2) For $\{V_{\alpha_k}\}_{k=1}^\infty$ as in part (1), prove that it is possible to choose the homeomorphisms h_k of V_{α_k} to open intervals in $\mathbb{R}$ so that $h_{k+1}|V_{\alpha_k} = h_k$, $\forall\, k \geq 1$. Show that these assemble to define a homeomorphism h of $\bigcup_{k=1}^\infty V_{\alpha_k}$ onto an open interval in $\mathbb{R}$.
(3) Using the above, show that the partially ordered set $\mathfrak{I}$ is inductive, hence contains a maximal element (Zorn's lemma).

Proof of Theorem 1.6.13. Let $U \subset N$ be a maximal open interval and fix a homeomorphism $f : (0,1) \to U$. Here we use the well-known fact that any two open intervals in $\mathbb{R}$ are homeomorphic. Since N is a compact Hausdorff space, there is a strictly increasing subsequence $\{a_k\}_{k=1}^\infty \subseteq (0,1)$ such that $a_k \uparrow 1$ and
$$\lim_{k\to\infty} f(a_k) = x_+ \in N \smallsetminus U$$
exists. It follows easily that, for *every* monotonic sequence $x_k \uparrow 1$ in $(0,1)$,
$$\lim_{k\to\infty} f(x_k) = x_+.$$
Similarly, there is a unique $x_- \in N \smallsetminus U$ with
$$\lim_{k\to\infty} f(y_k) = x_-$$
whenever $y_k \downarrow 0$. Therefore, $\overline{U} = U \cup \{x_+, x_-\}$ is the closure of U in N, a compact, connected subset. Define $g : [0,1] \to \overline{U}$ to be the extension of f by
$$g(0) = x_- \text{ and } g(1) = x_+,$$
clearly a continuous map. We consider the cases $x_+ = x_-$ and $x_+ \neq x_-$.

If $x_+ = x_-$, g induces a one-to-one, continuous map
$$\overline{g} : [0,1]/\{0,1\} \to \overline{U}.$$

Since $S^1 \cong [0,1]/\{0,1\}$ is compact and N is Hausdorff, this defines an imbedding $S^1 \hookrightarrow N$ of a compact, connected, boundaryless 1-manifold into a connected 1-manifold. By an application of Theorem 1.1.9, the image of this imbedding is an open subset of N. Being compact, this subset is also closed in N and, being nonempty, it is all of N. This proves that N is homeomorphic to S^1.

If $x_+ \neq x_-$, it follows that $g : [0,1] \to N$ is one-to-one, hence is a topological imbedding. If $x_+ \in \partial N$, let $W \subseteq N$ be an open neighborhood of x_+ and find $0 < \delta < 1$ such that $g(1-\delta, 1] = J \subseteq W$. We can assume that there is a homeomorphism $h : W \to (-\infty, 0]$ such that $h(x_+) = 0$, so $h(J)$ is a connected subset of $(-\infty, 0]$ containing 0 and having more than one point. Using the standard fact that the only connected subsets of $\mathbb{R}$ are the intervals, we conclude that $h(J)$ is a nondegenerate interval. By an application of Theorem 1.1.9, one sees that $h(J)$ has the form $(-\epsilon, 0]$ for suitable $\epsilon > 0$, from which it follows that J is open in the topology of N. By Theorem 1.1.9, $g(0,1)$ is also open in N, so $g(0,1]$ is open. Similarly, if $x_- \in \partial N$, $g[0,1)$ is an open subset of N. Thus, if x_+ and x_- both belong to ∂N, the image of g is open and, being compact, is also closed in N. By connectivity, this image is all of N, proving that N is homeomorphic to $[0,1]$.

Suppose, therefore, that $x_+ \in \mathrm{int}(N)$ and deduce a contradiction. The same argument will show that $x_- \notin \mathrm{int}(N)$. Let $h : V \to \mathbb{R}$ be a homeomorphism of an open neighborhood V of x_+ in N onto an open subset of $\mathbb{R}$ and let J denote the connected component of $g(0,1] \cap V$ containing x_+. Since x_+ is the limit of a sequence in $g(0,1)$, J does not degenerate to a single point, hence $h(J)$ is a half-open interval with $h(x_+)$ as its endpoint. Rechoosing h, if necessary, assume that $h(J) = (-1, 0]$. For a small value of $\epsilon > 0$, there is an open set $W \subset V$ such that $h(W) = (-1, \epsilon)$ and $J = g(0,1] \cap W$. It follows easily that $W \cup g(0,1) = W \cup U$ is homeomorphic to an open interval in $\mathbb{R}$, contradicting the maximality of U. □

The following innocuous corollary will be quite important for our treatment of mod 2 degree theory in Section 3.9.

Corollary 1.6.15. *Every compact, 1-dimensional manifold M has ∂M equal to a finite set of points with an even number of elements.*

1.7. Covering Spaces and the Fundamental Group

Covering spaces play a fundamental role, not only in manifold theory, but throughout topology. In this section, there will be no reason to restrict our attention to manifolds, everything being true for a quite large class of topological spaces. Accordingly, we fix only the following hypothesis for the entire section.

Hypothesis. All spaces are locally path-connected.

Note that local path-connectedness implies that connected spaces are path-connected. Note also that we do not require the Hausdorff property. There are actually useful applications of covering space theory to non-Hausdorff 1-manifolds.

1.7.A. The basics of covering spaces.

Definition 1.7.1. Let $p : Y \to X$ be a continuous map. An open, connected subspace $U \subseteq X$ is said to be *evenly covered* by p, if each connected component of $p^{-1}(U)$ is carried homeomorphically by p onto U.

Definition 1.7.2. A continuous map $p : Y \to X$ is a *covering map* if X is connected and each point $x \in X$ has a connected neighborhood that is evenly covered by p. The triple (Y, p, X) is called a covering space of X.

In practice, one usually abuses this terminology, referring to Y itself as the covering space. Note that we do not require Y to be connected. However, we will be mostly interested in connected covering spaces.

Of considerable importance are *automorphisms* of covering spaces, defined precisely as follows.

Definition 1.7.3. Let $p : Y \to X$ be a covering map. A *covering transformation*, also known as a *deck transformation* or an automorphism, is a homeomorphism $h : Y \to Y$ such that $p \circ h = p$. That is, the following diagram commutes:

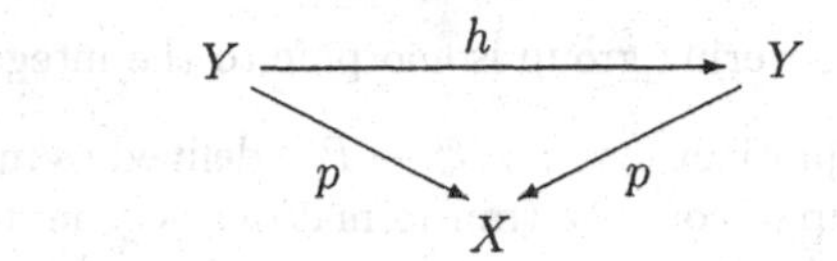

Lemma 1.7.4. *The set Γ of covering transformations associated to a covering map $p : Y \to X$ forms a group under composition, called the covering group.*

Proof. Indeed, if h_1 and h_2 are covering transformations,

$$p \circ (h_1 \circ h_2) = (p \circ h_1) \circ h_2 = p \circ h_2 = p.$$

Furthermore,

$$p \circ h = p \Rightarrow p = (p \circ h) \circ h^{-1} = p \circ h^{-1}.$$

Finally, it is clear that id_Y is a covering transformation, so Γ is a group. □

Example 1.7.5. The map $p : \mathbb{R} \to S^1$, defined by

$$p(t) = e^{2\pi i t},$$

is a covering map. Indeed, if $z_0 = e^{2\pi i t_0} \in S^1$, p carries the compact interval $[t_0 - 1/4, t_0 + 1/4]$ one-to-one, hence homeomorphically, onto a compact arc in S^1 containing z_0 in its interior U. Then $p^{-1}(U)$ is the disjoint union of open intervals $(t_0 + n - 1/4, t_0 + n + 1/4)$, as n ranges over the set $\mathbb{Z}$ of all integers. Evidently, each of these intervals is a connected component of $p^{-1}(U)$ and is carried by p homeomorphically onto U. Finally, $p(t) = p(s)$ if and only if $s = t + m$, for some $m \in \mathbb{Z}$. Thus, $h : \mathbb{R} \to \mathbb{R}$ is a covering transformation if and only if $h(t) = t + m_t$ where $m_t \in \mathbb{Z}$, $-\infty < t < \infty$. By continuity, m_t depends continuously on t. But $\mathbb{Z}$ is a discrete space and $\mathbb{R}$ is connected, so $m_t \equiv m$ is constant. The group of covering transformations is the group of translations by integers, hence is canonically isomorphic to the additive group $\mathbb{Z}$.

Exercise 1.7.6. Let G be a connected, locally path-connected, topological group. Let $H \subset G$ be a closed, discrete subgroup. (Recall that a subspace is discrete if, in the relative topology, each of its points is open.) Prove that the coset projection

$$p : G \to G/H$$

is a covering space (where G/H has the quotient topology). Show that the group Γ of covering transformations consists of *right* translations

$$g \in G \mapsto gh$$

by elements $h \in H$. More precisely, prove that

$$\varphi : h \in H \to \varphi_h \in \Gamma,$$
$$\varphi_h(g) = gh^{-1},$$

defines an isomorphism of the group H to the group Γ. (The need for h^{-1} rather than h is due to the possible noncommutativity of these groups. With this definition, one has that $\varphi_{h_1 h_2} = \varphi_{h_1} \circ \varphi_{h_2}$. For commutative groups, left and right translation are equivalent and the inverse could be omitted.)

Note that Example 1.7.5 is a special case of Exercise 1.7.6. More generally, the projection

$$p : \mathbb{R}^n \to \mathbb{R}^n/\mathbb{Z}^n = T^n$$

is a covering map with covering group isomorphic to the integer lattice $\mathbb{Z}^n$.

Example 1.7.7. The quotient map $p : S^n \to P^n$, defined as in Exercise 1.3.26, is a covering map. The group of covering transformations is generated by the antipodal interchange map, hence is $\mathbb{Z}_2$.

Definition 1.7.8. If

$$\begin{array}{ccc} Y' & \xrightarrow{\widetilde{f}} & Y \\ {\scriptstyle p'}\downarrow & & \downarrow{\scriptstyle p} \\ X' & \xrightarrow[f]{} & X \end{array}$$

is a commutative diagram of continuous maps, where p' and p are covering maps, we say that $\widetilde{f}$ is a *lift* of f to the covering spaces. In the case that $X' = X$ and $f = \mathrm{id}_X$, such a lift is called a *homomorphism* of covering spaces. A homomorphism of covering spaces that is also a homeomorphism is called an isomorphism of covering spaces.

Of course, an automorphism of covering spaces, as defined earlier, is an isomorphism.

Lemma 1.7.9. *If $\widetilde{f}$ is a lift of f, as in the preceding definition, and if the covering space Y' is connected, then $\widetilde{f}$ is completely determined by f and by the value of $\widetilde{f}$ at a single point.*

Proof. Let $\widetilde{f}$ and $\widehat{f}$ be lifts of f which agree at some point $y \in Y'$. Let $x = p'(y)$ and note that $f(x) = p(\widetilde{f}(y)) = p(\widehat{f}(y))$. Denote this point by z. Let U' be an evenly covered neighborhood of x and let U be an evenly covered neighborhood of z such that $f(U') \subseteq U$. This is possible since f is continuous. Let V' be the component of $(p')^{-1}(U')$ that contains y and V the component of $p^{-1}(U)$ that contains $\widetilde{f}(y) = \widehat{f}(y)$. By the definition of "lift", the diagrams

$$\begin{array}{ccc} V' & \xrightarrow{\widetilde{f}} & V \\ {\scriptstyle p'}\downarrow & & \downarrow{\scriptstyle p} \\ U' & \xrightarrow[f]{} & U \end{array}$$

and

$$\begin{array}{ccc} V' & \xrightarrow{\widehat{f}} & V \\ {\scriptstyle p'}\big\downarrow & & \big\downarrow{\scriptstyle p} \\ U' & \xrightarrow[f]{} & U \end{array}$$

exist and commute. But p' and p are one-to-one on the components V' and V, so the maps $\widetilde{f}$ and $\widehat{f}$ must agree on the open subset $V' \subseteq Y'$. This proves that the set of points on which $\widehat{f} = \widetilde{f}$ is open in Y'. On the other hand, the set of points on which $\widetilde{f}$ and $\widehat{f}$ do not agree is also open. Indeed, if $\widetilde{f}(y) \neq \widehat{f}(y)$, then the component V_1 of $p^{-1}(U)$ containing $\widetilde{f}(y)$ is disjoint from the component V_2 containing $\widehat{f}(y)$. Since V' is connected and $y \in V'$, we conclude that $\widetilde{f}(V') \subset V_1$ and $\widehat{f}(V') \subset V_2$. That is, $\widetilde{f}$ and $\widehat{f}$ disagree at every point of the open set V'. The fact that Y' is connected and that the functions agree at some point implies that $\widetilde{f} \equiv \widehat{f}$. ☐

Corollary 1.7.10. *Let $p : Y \to X$ be a covering map, assume that Y is connected, and let $y \in Y$. Then a covering transformation h is uniquely determined by the point $h(y)$.*

Indeed, a covering transformation is a lift of $\mathrm{id} : X \to X$, where we take $Y' = Y$ and $p' = p$.

Another important type of lift is one for which the covering $p' : Y' \to X'$ is the trivial covering $\mathrm{id} : X' \to X'$. In this case, the continuous lift fits into a commutative triangle

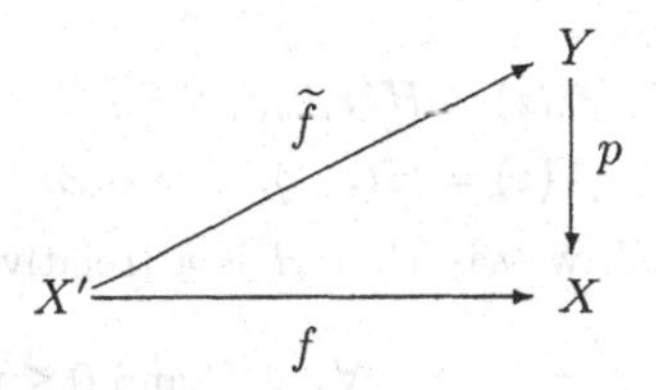

Lemma 1.7.11 (Path-lifting property). *Let $p : Y \to X$ be a covering space, let $x \in X$ and $y \in p^{-1}(x)$, and let $\sigma : [a, b] \to X$ be a continuous path with $\sigma(a) = x$. Then there is a unique lift $\widetilde{\sigma} : [a, b] \to Y$ such that $\widetilde{\sigma}(a) = y$.*

Proof. For each $t \in [0, 1]$, let U_t denote an evenly covered neighborhood of $\sigma(t)$. Then $\{\sigma^{-1}(U_t)\}_{0 \le t \le 1}$ is an open cover of $[0, 1]$. Let J_t denote the component of $\sigma^{-1}(U_t)$ containing t, a (relatively) open interval in $[0, 1]$, remarking that $\{J_t\}_{0 \le t \le 1}$ is again an open cover. Pass to a finite subcover $\{J_i\}_{i=0}^r$, indexed so that $0 \in J_0$, $J_i \cap J_{i+1} \neq \emptyset$, $0 \le i < r$, and $1 \in J_r$. We can choose $0 = t_0 < t_1 < \cdots < t_{r+1} = 1$ so that $[t_i, t_{i+1}] \subset J_i$, $0 \le i \le r$. The segment $\sigma_i = \sigma|[t_i, t_{i+1}]$, $1 \le i \le r$, has its image in an evenly covered neighborhood, call it U_i, and we can define the desired lift by finite induction on i. Indeed, there is a unique component $\widetilde{U}_0$ of $p^{-1}(U_0)$ that contains y and, since $p : \widetilde{U}_0 \to U_0$ is a homeomorphism, there is a unique lift $\widetilde{\sigma}_0$ of σ_0 such that $\widetilde{\sigma}_0(0) = y$. Then there is a unique component $\widetilde{U}_1$ of $p^{-1}(U_1)$ containing $\widetilde{\sigma}_0(t_1)$ and one lifts σ_1 uniquely to a path $\widetilde{\sigma}_1$ in $\widetilde{U}_1$ starting at $\widetilde{\sigma}_0(t_1)$. Thus, these lifts fit together to form a continuous lift of $\sigma|[0, t_2]$. Iterating this procedure finitely often creates the desired lift of σ. ☐

Since each $x \in X$ has an evenly-covered neighborhood U, it is clear that distinct points of the "fiber" $p^{-1}(x)$ lie in distinct components of $p^{-1}(U)$, so the fiber is a discrete space. Covering transformations must map $p^{-1}(x)$ one-to-one onto itself, hence Corollary 1.7.10 requires that the covering group Γ permute this set in such a way that only the identity element of Γ has a fixed point. A permutation group with this property is said to act "simply". The most useful covering spaces are those in which Γ must also be *transitive.* (This means that, for each pair of points $y, z \in p^{-1}(x)$, there is $h \in \Gamma$ such that $h(y) = z$.) In this case, we say that Γ permutes the fiber "simply transitively".

Exercise 1.7.12. Use the path-lifting property to prove that the group Γ of covering transformations is transitive on one fiber if and only if it is transitive on every fiber. (You only need X to be connected, not Y.)

Definition 1.7.13. A covering space $p : Y \to X$ is said to be *regular* if the group Γ of covering transformations permutes each fiber $p^{-1}(x)$ simply transitively.

Thus, for regular coverings, Γ can be put in one-to-one correspondence with $p^{-1}(x)$ by selecting a "basepoint" $y_0 \in p^{-1}(x)$ and setting up the correspondence

$$h \in \Gamma \leftrightarrow h(y_0) \in p^{-1}(x).$$

Another important lifting property for covering spaces is the *homotopy-lifting property.* We need a definition.

Definition 1.7.14. Let f_0 and f_1 be continuous maps of a space Z into a space W. A homotopy between these maps is a continuous map

$$H : Z \times [0,1] \to W$$

such that

$$f_0(z) = H(z,0), \quad \forall z \in Z,$$
$$f_1(z) = H(z,1), \quad \forall z \in Z.$$

If $C \subseteq Z$ and $f_0|C \equiv f_1|C$, we say that H is a (relative) homotopy mod C if, in addition,

$$H(z,t) = f_0(z), \quad \forall z \in C \text{ and } 0 \le t \le 1.$$

If there is a homotopy between f_0 and f_1, we say that f_0 is homotopic to f_1 and write $f_0 \sim f_1$. If this is a homotopy mod C, we write $f_0 \sim_C f_1$.

We think of a homotopy H as a continuous deformation of the map f_0 to the map f_1 through intervening maps

$$f_t(z) = H(z,t), \quad 0 \le t \le 1.$$

Exercise 1.7.15. Prove that homotopy and homotopy mod C are equivalence relations. Also prove that, if $f \sim g$ (respectively, $f \sim_C g$), then $u \circ f \sim u \circ g$ (respectively, $u \circ f \sim_C u \circ g$), whenever these compositions are defined. The equivalence classes under this relation are called homotopy classes.

Exercise 1.7.16. Consider a commutative triangle

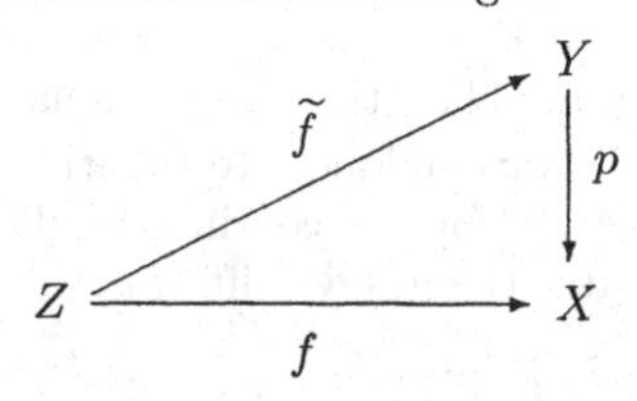

of continuous maps, where p is a covering map. If $H : Z \times [0,1] \to X$ is a homotopy such that $H|Z \times \{0\} = f$, prove that there is a unique lift $\widetilde{H} : Z \times [0,1] \to Y$ such that $\widetilde{H}|Z \times \{0\} = \widetilde{f}$. (Hint. $H|\{z\} \times [0,1]$ is a path, for each $z \in Z$.) This is the homotopy-lifting property for covering spaces.

A particularly important case of relative homotopy will be that in which the maps are paths

$$\sigma_i : [a,b] \to W, \quad i = 0,1,$$

and $C = \partial[a,b] = \{a,b\}$. Then the curves have the same endpoints

$$\sigma_0(a) = \sigma_1(a) = x,$$
$$\sigma_0(b) = \sigma_1(b) = y$$

and the homotopy mod $\{a,b\}$ deforms the one curve to the other while keeping the endpoints x and y fixed. Since this situation arises so often, we will use the notation $\sigma_0 \sim_\partial \sigma_1$ for "homotopy mod the endpoints". In this case, the homotopy-lifting property implies the following.

Lemma 1.7.17. *If $p : Y \to X$ is a covering map, if $\sigma_i : [a,b] \to X$ are paths with the same endpoints, $i = 0,1$, and if $\sigma_0 \sim_\partial \sigma_1$, then lifts of these paths starting at the same point must also terminate at the same point.*

Proof. Indeed, let $\sigma_i(a) = w$, $\sigma_i(b) = v$, let H be a homotopy mod the endpoints, and let $\widetilde{\sigma}_i$ be lifts starting at a point $w' \in p^{-1}(w)$, $i = 0,1$. These exist by the path-lifting property. If $v' \in p^{-1}(v)$ is the terminal point of $\widetilde{\sigma}_0$, we must prove that it is also the terminal point of $\widetilde{\sigma}_1$. By the homotopy-lifting property, there is a unique lift $\widetilde{H}$ of the homotopy that agrees with $\widetilde{\sigma}_0$ along $[a,b] \times \{0\}$. Being a lift, $\widetilde{H}$ must carry the interval $\{a\} \times [0,1]$ into $p^{-1}(w)$ and $\{b\} \times [0,1]$ into $p^{-1}(v)$. Since these fibers are discrete and $\widetilde{H}$ is continuous, the images of these intervals must be the respective singletons $\{w'\}$ and $\{v'\}$. Thus, $\widetilde{H}$ restricts to $[a,b] \times \{1\}$ to define a lift of σ_1 starting at w', hence equal to $\widetilde{\sigma}_1$. This lift terminates at the point v'. □

Usually, we will parametrize paths on the unit interval $[0,1]$. If σ and τ are two such paths such that $\sigma(1) = \tau(0)$, we can join them at this common point to produce a path $\sigma \cdot \tau$ joining $\sigma(0)$ to $\tau(1)$ and parametrized on $[0,1]$ as follows:

$$(\sigma \cdot \tau)(t) = \begin{cases} \sigma(2t), & 0 \le t \le \frac{1}{2}, \\ \tau(2t-1), & \frac{1}{2} \le t \le 1. \end{cases}$$

Intuitively, one runs along σ at twice the original speed, then along τ at twice the original speed. Since $\sigma(1) = \tau(0)$, the resulting path is continuous.

Exercise 1.7.18. Suppose that $\sigma(1) = \tau(0)$, as above, and that σ' and τ' are similar paths such that $\sigma \sim_\partial \sigma'$ and $\tau \sim_\partial \tau'$. Prove that $\sigma \cdot \tau \sim_\partial \sigma' \cdot \tau'$.

1.7.B. Simple connectivity. Of particular interest are *loops* at a point $x_0 \in X$, these being paths

$$\sigma : [0,1] \to X$$

such that $\sigma(0) = \sigma(1) = x_0$. We will also let x_0 denote the constant loop $\sigma \equiv x_0$.

Definition 1.7.19. A topological space X is simply connected if the following conditions hold:

(1) X is path-connected;

(2) there is a point $x_0 \in X$ such that every loop σ at x_0 satisfies $\sigma \sim_\partial x_0$.

A (not necessarily connected) space X is locally simply connected if each neighborhood of each point $x \in X$ contains a simply connected open neighborhood of x.

Example 1.7.20. The open unit ball

$$B^n = \{x \in \mathbb{R}^n \mid \|x\| < 1\}$$

and the closed unit disk

$$D^n = \{x \in \mathbb{R}^n \mid \|x\| \leq 1\}$$

are both simply connected. Indeed, every convex subset $C \subseteq \mathbb{R}^n$ is simply connected. To see this, choose an arbitrary point $x_0 \in C$ and consider a loop σ based at x_0. For each fixed $t \in [0,1]$, consider the line segment $\{sx_0 + (1-s)\sigma(t)\}$, $0 \leq s \leq 1$. By convexity, this segment lies in C. Also, if $\sigma(t) = x_0$, the segment is the constant path x_0. Thus, we define a continuous homotopy $H : [0,1] \times [0,1] \to C$ by

$$H(t,s) = sx_0 + (1-s)\sigma(t),$$

noting that this is a homotopy $\sigma \sim_\partial x_0$. Since manifolds are locally homeomorphic to B^n, manifolds are locally simply connected.

It will be convenient to have various equivalent characterizations of simple connectivity.

Lemma 1.7.21. *The following properties of a path-connected space X are equivalent:*

(1) *Every continuous map $f : S^1 \to X$ is homotopic to a constant map.*
(2) *Every continuous map $f : S^1 \to X$ extends to a continuous map $F : D^2 \to X$.*
(3) *If σ and τ are paths in X having the same initial point and the same terminal point, then $\sigma \sim_\partial \tau$.*
(4) *If σ is a loop at an arbitrary point $x \in X$, then $\sigma \sim_\partial x$.*
(5) *X is simply connected.*

Proof. By Example 1.3.22, it is immediate that (1) $\Rightarrow$ (2). For the implication (2) $\Rightarrow$ (3), one needs the fact that $[0,1] \times [0,1]$ is homeomorphic to D^2. All closed rectangular domains in $\mathbb{R}^2$ are homeomorphic via suitable affine transformations. Thus, we can replace the rectangle $[0,1] \times [0,1]$ with $[-1,1] \times [-1,1]$. The unit disk is inscribed in this rectangle and a suitable radial transformation gives a homeomorphism. The elementary details are left to the reader. Thus, if σ and τ are paths parametrized on $[0,1]$, both joining x_0 to x_1, we obtain a map on the boundary of $[0,1] \times [0,1]$ by using σ along the bottom edge, τ along the top, the constant map x_0 along the left edge and x_1 along the right. By (2), this map extends over $[0,1] \times [0,1]$ and this extension is the desired homotopy $\sigma \sim_\partial \tau$. Evidently, (4) is just a special case of (3), so (3) $\Rightarrow$ (4). Likewise, (4) $\Rightarrow$ (5) is immediate. For the implication (5) $\Rightarrow$ (1), we need the path-connectivity of X. We assume that there is a point $x_0 \in X$ such that every loop at x_0 is homotopic mod the boundary to the constant loop x_0. Let $f : S^1 \to X$ be continuous. In Exercise 1.7.22, you will be asked to show that path-connectivity implies that $f \sim f_0$, where $f_0(1) = x_0$. By example 1.3.21, this map can be viewed as a loop at x_0, hence, by (5), it is homotopic mod the boundary to x_0. This homotopy is constantly equal to x_0 along

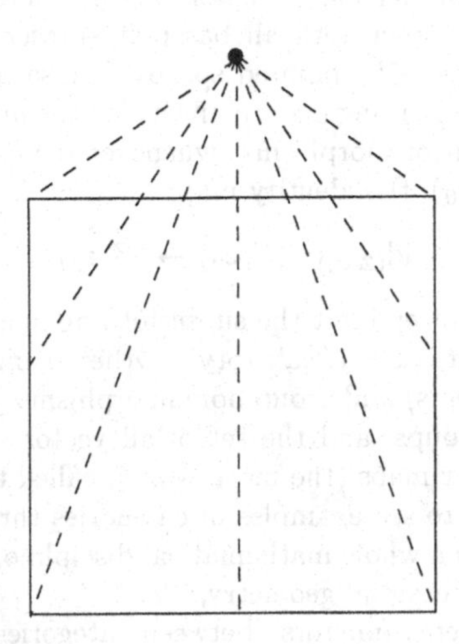

Figure 1.7.1. Projection of the square onto the union of three boundary segments

the two vertical edges and the top edge of the boundary of $[0,1] \times [0,1]$. Identifying $(0,t)$ with $(1,t)$, $0 \le t \le 1$, we view the homotopy as a map on $S^1 \times [0,1]$. Then, appealing to Example 1.3.22, we view this as a map of the disk, obtaining the desired extension of f. □

Exercise 1.7.22. Given a continuous map $f : S^1 \to X$ and assuming that X is path-connected, let $x_0 \in X$ and construct a homotopy $f \sim f_0$ such that $f_0(1) = x_0$. (Hint. Think about the projection of $[0,1] \times [0,1]$ onto part of its boundary indicated in Figure 1.7.1.)

1.7.C. The universal covering. Reasonably nice spaces admit a particularly useful type of covering space, defined by a "universal" property.

Definition 1.7.23. A covering map $\pi : \widetilde{X} \to X$, is said to be universal if $\widetilde{X}$ is connected and, for any covering map $p: Y \to X$ such that Y is connected, there is a continuous map $\widetilde{\pi} : \widetilde{X} \to Y$ that is a lift of π. The covering space $\widetilde{X}$ is called a universal covering space of X.

Exercise 1.7.24. Prove that $\widetilde{\pi}$ is a covering map, hence that every connected covering space is intermediate between X and the universal covering space $\widetilde{X}$.

At this point it will be useful to start working in the category of *pointed spaces.*

Definition 1.7.25. A pointed space is a pair (Z, z_0), where $z_0 \in Z$ is a fixed choice of *basepoint* and Z is path-connected. A map

$$f : (Z, z_0) \to (W, w_0)$$

is a continuous map of Z into W such that $f(z_0) = w_0$. These are called *basepoint-preserving maps.*

Remark. The set of all pointed spaces (ignore Zermelo–Frankel scruples about sets that are "too large"), together with all basepoint-preserving maps, is called the *category* of pointed spaces. The pointed spaces themselves are the *objects* of the category, the basepoint-preserving maps being called the *morphisms* of the category. Remark that composition of morphisms, whenever defined, is again a morphism and, for each object (Z, z_0), the identity map

$$\mathrm{id}_{(Z,z_0)} : (Z, z_0) \to (Z, z_0)$$

is a morphism. (We will often omit the subscript and simply write id.) These are the defining properties of the term "category". Other examples of categories are the set of all groups (the objects) and group homomorphisms (the morphisms), usually called the category of groups, and the set of all vector spaces over a given field K (the objects) and linear maps (the morphisms), called the category of K-vector spaces. We will continue to see examples of categories throughout this book. One can think of a category as a whole mathematical discipline, such as topology, group theory, linear algebra, differential geometry, *etc.*

Typically, one considers "functors" between categories, these being maps that take the objects of one category to the objects of the other and take the morphisms of the first to morphisms of the second, preserving the category structure. Thus, if $F : \mathcal{A} \to \mathcal{B}$ is a functor, we must have

(1) $F(\mathrm{id}_A) = \mathrm{id}_{F(A)}$, for each object $A \in \mathcal{A}$;
(2) $F(f \circ g) = F(f) \circ F(g)$, whenever f and g are morphisms of $\mathcal{A}$ for which $f \circ g$ is defined.

Functors are used to transform problems in one category to analogous problems in another. Shortly, we will introduce a functor, the fundamental group, from the category of pointed spaces to the category of groups. A proof of the Brouwer fixed point theorem will be given to illustrate the usefulness of this functor.

Actually, what we have just defined is called a *covariant* functor. Later in the book, we will have occasion to consider *contravariant* functors. These reverse all morphism arrows and, consequently, satisfy

$$F(f \circ g) = F(g) \circ F(f).$$

In the category of pointed spaces, notions such as "covering map" and "covering space" have their obvious meanings. Note, however, that covering transformations are not basepoint-preserving. In this category, the definition of "universal covering map" $\pi : (\widetilde{X}, \widetilde{x}_0) \to (X, x_0)$ requires that, for any covering map $p : (Y, y_0) \to (X, x_0)$, there be a commutative diagram

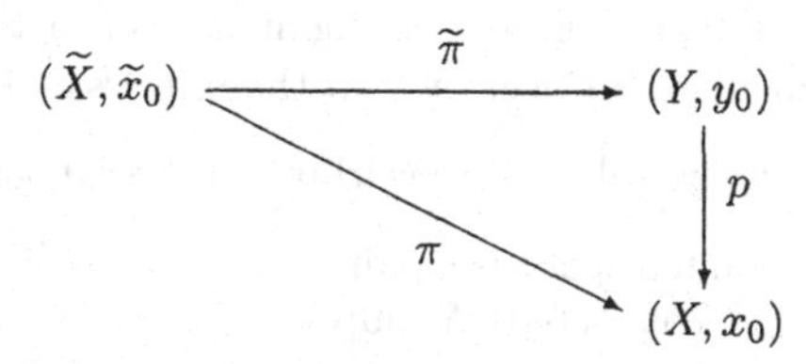

According to Lemma 1.7.9, the lift $\widetilde{\pi}$ is *uniquely* determined by the requirement, built into the language of pointed spaces, that $\widetilde{\pi}(\widetilde{x}_0) = y_0$.

Lemma 1.7.26. *If $\pi : (\widetilde{X}, \widetilde{x}_0) \to (X, x_0)$ and $\widehat{\pi} : (\widehat{X}, \widehat{x}_0) \to (X, x_0)$ are both universal covering maps, then there is a unique homeomorphism φ making the diagram*

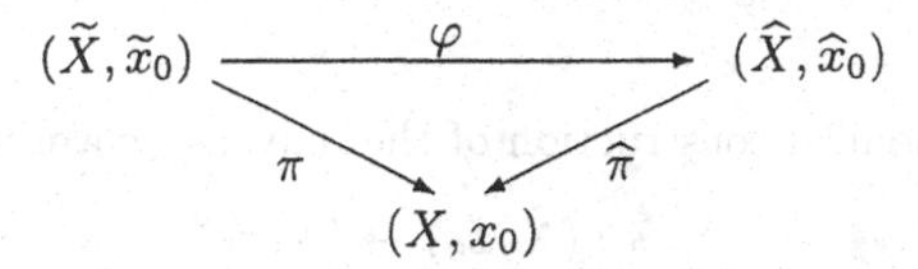

commutative.

Proof. Indeed, the existence of a continuous map φ, making the diagram commute, is guaranteed by the fact that $\widehat{\pi}$ is a covering map and π is a universal covering map. As remarked above, uniqueness is by Lemma 1.7.9. Interchanging the roles of $\widehat{\pi}$ and π, we get a unique commutative triangle

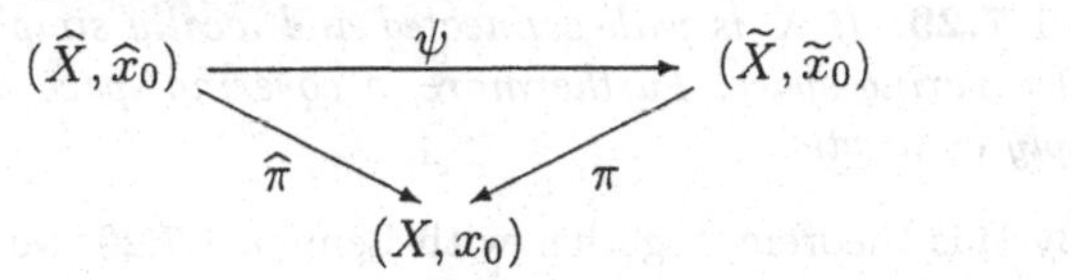

and the composition $\varphi \circ \psi$ is the unique continuous map making the triangle

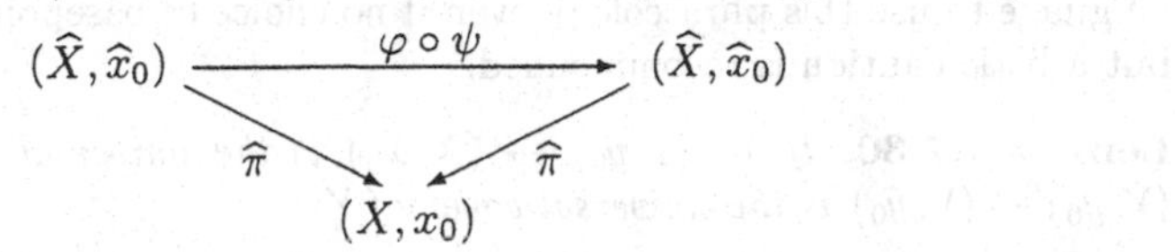

commutative. But the identity map $\mathrm{id}_{(\widehat{X}, \widehat{x}_0)}$ would also make the triangle commutative, so

$$\varphi \circ \psi = \mathrm{id}_{(\widehat{X}, \widehat{x}_0)} .$$

Similarly, $\psi \circ \varphi$ is the identity map on $(\widetilde{X}, \widetilde{x}_0)$, so ψ and φ are mutually inverse homeomorphisms. □

By this lemma, we can speak of "the" universal covering space of (X, x_0), provided that it exists.

Corollary 1.7.27. *If it exists, the universal covering space is a regular covering.*

Proof. Indeed, if $\widetilde{x}_0$ and $\widetilde{y}_0$ lie in the fiber $\pi^{-1}(x_0)$, we can take $\widehat{X} = \widetilde{X}$, $\widehat{\pi} = \pi$ and $\widehat{x}_0 = \widetilde{y}_0$ in Lemma 1.7.26. The map φ, interpreted as a homeomorphism of $\widetilde{X}$ to itself, is obviously a covering transformation taking $\widetilde{x}_0$ to $\widetilde{y}_0$. □

Lemma 1.7.26 is the uniqueness lemma for universal covering spaces. Existence is harder and requires that we further restrict the space X.

Hypothesis. From now on, all spaces will be locally simply connected.

By Example 1.7.20, this hypothesis includes all manifolds. In fact, the hypothesis can be weakened to require only that spaces be "semi-locally simply connected" (see [**26**, Theorem 10.2 on page 175]). This means that the homotopy $\sigma \sim_\partial x_0$ of loops based at x_0 in a neighborhood of x_0 is not itself required to stay in the neighborhood.

Exercise 1.7.28. If $p : (Y, y_0) \to (X, x_0)$ is a covering, prove that any simply connected, open subset of X is evenly covered. Conclude that the composition of covering maps is a covering map and that $\widetilde{\pi}$ in Definition 1.7.23 is also the universal covering map.

There is a canonical construction of the universal covering space

$$\pi : (\widetilde{X}, \widetilde{x}_0) \to (X, x_0)$$

in which the points of $\widetilde{X}$ are the $\sim_\partial$ homotopy classes $[\sigma]$ of paths σ in X with $\sigma(0) = x_0$. The basepoint $\widetilde{x}_0$ is taken to be $[x_0]$, the class of the constant loop, and the projection map is defined by $\pi[\sigma] = \sigma(1)$. The details are a bit involved and could be distracting, so we relegate them to Appendix A, where the following important theorem will be proven.

Theorem 1.7.29. *If X is path-connected and locally simply connected, it admits a universal covering space. Furthermore, a covering space is universal if and only if it is simply connected.*

Remark. By this theorem, together with Lemma 1.7.26, we are now fully justified in speaking of "the" universal covering space of (X, x_0) in the category of (path-connected and locally simply connected) pointed spaces. It is a common abuse of language to use this phraseology even if no choice of basepoints has been specified, but a little caution is recommended.

Lemma 1.7.30. *If $p : (Y, y_0) \to (X, x_0)$ is the universal cover of X, then* id : *$(Y, y_0) \to (Y, y_0)$ is the universal cover of Y*

Proof. Indeed, by Exercise 1.7.28, commutativity of the diagram

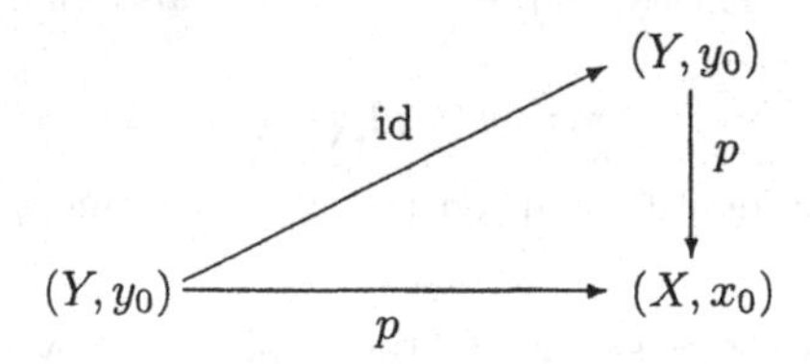

implies the assertion. □

1.7.D. The fundamental group. The group Γ of covering transformations of the universal covering space $\widetilde{X}$ of X is isomorphic to a group $\pi_1(X, x_0)$, called the *fundamental group* of (X, x_0), that plays an important role in *algebraic topology.* Although they are abstractly isomorphic, these groups are subtly different. Since covering transformations, other than the identity, fix no point, they do not really coexist comfortably with the category of pointed spaces. On the other hand, the basepoint x_0 plays an essential role in the study of the fundamental group. The choice of this basepoint is also essential for specifying an isomorphism $\Gamma \cong \pi_1(X, x_0)$.

Let $\mathcal{P}(X, x_0)$ denote the space of paths $\sigma : [0,1] \to X$ issuing from the basepoint x_0 and let $\Omega(X, x_0) \subset \mathcal{P}(X, x_0)$ denote the subset of *loops* in X based at x_0. These are the paths issuing from x_0 at time $t = 0$ and returning to x_0 at time $t = 1$. The homotopy relation $\sim_\partial$ restricts to an equivalence relation on $\Omega(X, x_0)$ and we set

$$\pi_1(X, x_0) = \Omega(X, x_0)/\sim_\partial .$$

If $\sigma, \tau \in \Omega(X, x_0)$, it should be clear that $\sigma \cdot \tau \in \Omega(X, x_0)$. By Exercise 1.7.18, this passes to a well-defined "multiplication" on $\pi_1(X, x_0)$:

$$[\sigma][\tau] = [\sigma \cdot \tau].$$

This is our candidate for the group operation. Our candidate for the identity element is $[x_0]$, and our candidate for the inverse of $[\sigma] \in \pi_1(X, x_0)$ will be the homotopy class of the loop σ^{-1} obtained by traversing σ backwards. Formally,

$$\sigma^{-1}(t) = \sigma(1-t), \quad 0 \leq t \leq 1.$$

Theorem 1.7.31. *With the above operations, $\pi_1(X, x_0)$ is a group that is isomorphic to the group Γ of covering transformations of the universal covering space $\pi : \widetilde{X} \to X$. This isomorphism depends only on the choices of basepoints $x_0 \in X$ and $\widetilde{x}_0 \in \pi^{-1}(x_0)$.*

Indeed, let $\pi : (\widetilde{X}, \widetilde{x}_0) \to (X, x_0)$ be the universal covering space, and let Γ be the group of covering transformations of $\widetilde{X}$. Given $\sigma \in \Omega(X, x_0)$, let $\widetilde{\sigma}$ be the unique lift of σ to a path in $\widetilde{X}$ starting at $\widetilde{x}_0$. By Lemma 1.7.17, $\widetilde{\sigma}(1)$ depends only on $[\sigma] \in \pi_1(X, x_0)$. Furthermore, if the lifts $\widetilde{\tau}$ and $\widetilde{\sigma}$ of $\tau, \sigma \in \Omega(X, x_0)$ have $\widetilde{\tau}(1) = \widetilde{\sigma}(1)$, the simple connectivity of $\widetilde{X}$ implies that $\widetilde{\tau} \sim_\partial \widetilde{\sigma}$. Thus, by Exercise 1.7.15,

$$[\sigma] = [\pi \circ \widetilde{\sigma}] = [\pi \circ \widetilde{\tau}] = [\tau]$$

and we have set up a one-to-one correspondence between the set $\pi_1(X, x_0)$ and the set $\pi^{-1}(x_0)$. Given $[\sigma] \in \pi_1(X, x_0)$, let $\varphi_\sigma \in \Gamma$ be the unique covering transformation that satisfies $\varphi_\sigma(\widetilde{x}_0) = \widetilde{\sigma}(1)$. Thus, $[\sigma] \leftrightarrow \varphi_\sigma$ is a one-to-one correspondence between $\pi_1(X, x_0)$ and Γ, canonically determined by the choice of basepoints. Since Γ is a group, we obtain a group structure on $\pi_1(X, x_0)$. That this is the "correct" group structure, as described above, is the content of the next lemma.

Lemma 1.7.32. *The correspondence $\sigma \mapsto \varphi_\sigma$ has the following properties:*

(1) $\varphi_{\sigma \cdot \tau} = \varphi_\sigma \circ \varphi_\tau$;
(2) $\varphi_{\sigma^{-1}} = \varphi_\sigma^{-1}$;
(3) $\varphi_{x_0} = \mathrm{id}$.

Proof. Let $\sigma, \tau \in \Omega(X, x_0)$ and, as usual, denote the respective lifts to paths issuing from $\widetilde{x}_0$ by $\widetilde{\sigma}$ and $\widetilde{\tau}$. As $\varphi_\sigma \circ \widetilde{\tau}$ is the unique lift of τ to a path issuing from $\widetilde{\sigma}(1)$, it is evident that the lift of $\sigma \cdot \tau$ issuing from $\widetilde{x}_0$ is

$$\widetilde{\sigma \cdot \tau} = \widetilde{\sigma} \cdot (\varphi_\sigma \circ \widetilde{\tau}).$$

Thus,

$$\varphi_{\sigma \cdot \tau}(\widetilde{x}_0) = \widetilde{\sigma \cdot \tau}(1) = \varphi_\sigma(\widetilde{\tau}(1)) = \varphi_\sigma(\varphi_\tau(\widetilde{x}_0)).$$

Since a covering transformation is uniquely determined by its value at one point, $\varphi_{\sigma \cdot \tau} = \varphi_\sigma \circ \varphi_\tau$, proving (1). For (2), remark that the unique lift of σ^{-1} to a path starting at $\widetilde{\sigma}(1)$ will be a path ending at $\widetilde{x}_0$. Applying the above argument to $\tau = \sigma^{-1}$, we see that $\varphi_{\sigma \cdot \sigma^{-1}}(\widetilde{x}_0) = \widetilde{x}_0$, implying that $\varphi_\sigma \circ \varphi_{\sigma^{-1}} = \mathrm{id}$, as desired. Finally, it is evident that $\varphi_{x_0}(\widetilde{x}_0) = \widetilde{x}_0$, so $\varphi_{x_0} = \mathrm{id}$ and (3) is proven. □

The proof of Theorem 1.7.31 is complete. Remark that, if we construct $\widetilde{X}$ as indicated prior to the statement of Theorem 1.7.29, there is a canonical choice of the basepoint $\widetilde{x}_0$, so the identification $\Gamma \cong \pi_1(X, x_0)$ can be said to depend only on the choice of basepoint x_0. There are situations, however, in which one prefers not to think of $\widetilde{X}$ in this way and in which the choice of basepoint $\widetilde{x}_0$ should be free.

Example 1.7.33. By Example 1.7.5 and the fact that simply connected covering spaces are universal (Theorem 1.7.29), the universal covering space of the circle is $p : \mathbb{R} \to S^1$ ($\mathbb{R}$, being convex, is simply connected) and the group of covering transformations is isomorphic to the infinite cyclic group $\mathbb{Z}$. Thus, $\pi_1(S^1, x_0) \cong \mathbb{Z}$, where $x_0 = p(\mathbb{Z})$. Remark that a loop on S^1, based at x_0 and generating $\pi_1(S^1, x_0)$, is given by the restriction $\sigma = p|[0,1]$.

Exercise 1.7.34. Verify directly, without using covering spaces, that the operation $[\sigma][\tau] = [\sigma \cdot \tau]$ makes $\pi_1(X, x_0)$ into a group, the inverses and identity element being as described above. This requires constructing a number of homotopies. For example, you need to construct a homotopy

$$\sigma \cdot (\tau \cdot \gamma) \sim_\partial (\sigma \cdot \tau) \cdot \gamma.$$

Exercise 1.7.35. If X is (path-) connected and $x_0, x_1 \in X$, show that a choice of path σ from x_0 to x_1 can be used to define a group homomorphism $\theta_\sigma : \pi_1(X, x_1) \to \pi_1(X, x_0)$ that depends only on $[\sigma]$. Show that $\theta_{\sigma^{-1}}$ is a two-sided inverse to θ_σ, hence that the two fundamental groups are isomorphic. Since this isomorphism generally depends on the choice of $[\sigma]$, it is not canonical. At any rate, triviality of the fundamental group at one basepoint implies its triviality at all basepoints.

By the definition of the fundamental group, we obtain the following addition to the list of equivalent properties in Lemma 1.7.21

Lemma 1.7.36. *If X is path-connected, it is simply connected if and only if $\pi_1(X, x_0)$ is trivial, for some (hence every) basepoint $x_0 \in X$.*

Consider a continuous, basepoint-preserving map

$$f : (X, x_0) \to (Y, y_0)$$

between pointed spaces. If $\sigma \in \Omega(X, x_0)$, then, since f is basepoint-preserving, $f \circ \sigma \in \Omega(Y, y_0)$. If $\sigma \sim_\partial \tau$, then $f \circ \sigma \sim_\partial f \circ \tau$ by Exercise 1.7.15. This enables us to define an *induced* map

$$f_* : \pi_1(X, x_0) \to \pi_1(Y, y_0)$$
$$f_*([\sigma]) = [f \circ \sigma].$$

We use the notation $f \sim_{x_0} g$ for homotopy mod the singleton $\{x_0\}$. This is a homotopy through basepoint-preserving maps. The following is a very routine exercise, but important.

Exercise 1.7.37. Prove that f_* is well defined and is a group homomorphism. Show that, whenever

$$(X, x_0) \xrightarrow{g} (Y, y_0) \xrightarrow{f} (Z, z_0),$$

then

$$(f \circ g)_* = f_* \circ g_*,$$

and that $\mathrm{id}_* = \mathrm{id}$ (where we use "id" for identity maps on any suitable domain). Finally, if $f \sim_{x_0} g$, prove that $f_* = g_*$.

Remark. These properties are summed up by saying that the fundamental group defines a homotopy-invariant, covariant functor from the category of pointed spaces and continuous, basepoint-preserving maps to the category of groups and group homomorphisms. This makes it possible to "paint algebraic pictures" of difficult topological problems. As with all pictures, a great deal of detail is lost (for example, homotopic maps become indistinguishable), but the algebraic problem that appears on this "canvas" is often more manageable. After solving this problem, one then tries to interpret the solution in terms of the original topological problem. The following example is a good case in point.

Example 1.7.38. We indicate how to use properties in Exercise 1.7.37 to prove the *Brouwer fixed point theorem.* This theorem asserts that every continuous map $f : D^2 \to D^2$ has a fixed point. One proceeds by assuming that f has no fixed point and deriving a contradiction. For each $x \in D^2$, we assume that $x \neq f(x)$. Out of the point $f(x)$, draw the unique ray R_x that passes through x and let $h(x)$ denote the point of intersection $R_x \cap \partial D^2$. This defines a map

$$h : D^2 \to \partial D^2 = S^1$$

and it is geometrically plausible that h is continuous. In fact, with a little care, one can write down an explicit formula for h that makes the continuity evident. Fix a point $x_0 \in \partial D^2 \subset D^2$ to serve as basepoint for both of these spaces. Remark that, if $x \in \partial D^2 = S^1$, it is immediate from our definition that $h(x) = x$. In particular, h is basepoint-preserving, as is the inclusion map $i : S^1 \hookrightarrow D^2$. It should be clear that the diagram

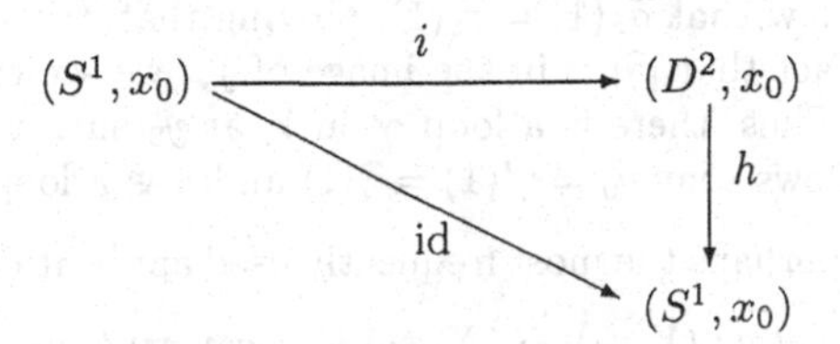

is commutative. Then, by Exercise 1.7.37, so is the diagram

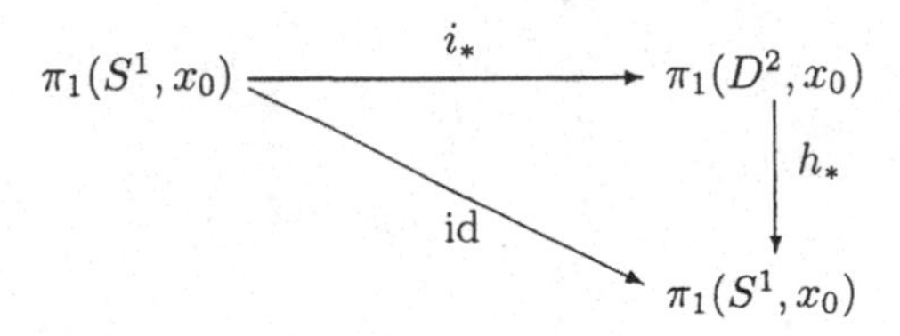

By Example 1.7.33, $\pi_1(S^1, x_0) = \mathbb{Z}$. Also, $\pi_1(D^2, x_0) = 0$, since $D^2 \subset \mathbb{R}^2$ is convex, hence simply connected. This gives the commutative diagram

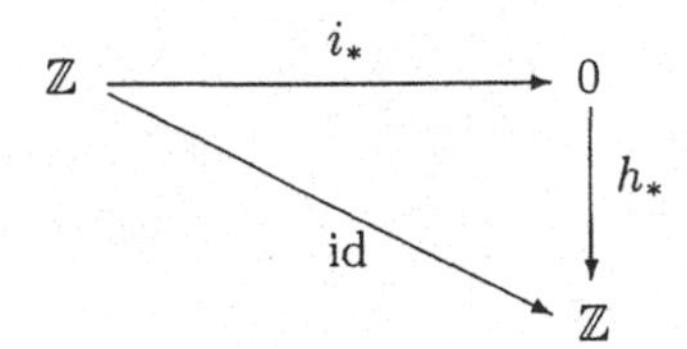

and this is transparently absurd. From this contradiction, we conclude that f had a fixed point.

Finally, using the fundamental group, we formulate a very important necessary and sufficient condition for the existence of lifts.

Theorem 1.7.39. *Let $p : (Y, y_0) \to (X, x_0)$ be a covering map and let $f : (Z, z_0) \to (X, x_0)$ be a continuous, basepoint-preserving map. Then there is a lift $\widetilde{f} : (Z, z_0) \to (Y, y_0)$ if and only if*

$$f_*(\pi_1(Z, z_0)) \subseteq p_*(\pi_1(Y, y_0)).$$

Proof. If the lift exists, then

$$f_*(\pi_1(Z, z_0)) = p_* \circ \widetilde{f}_*(\pi_1(Z, z_0)) \subseteq p_*(\pi_1(Y, y_0)).$$

For the converse, we assume that $f_*(\pi_1(Z, z_0)) \subseteq p_*(\pi_1(Y, y_0))$ and construct the lift using the path-lifting property.

First, we fix some notation. If $\sigma : [0, 1] \to Z$ is an arbitrary path with $\sigma(0) = z_0$, we will denote by $\widehat{\sigma}$ the path $f \circ \sigma$ in X, starting at x_0, and by $\widetilde{\sigma}$ the lift of $\widehat{\sigma}$ to a path in Y, starting at y_0.

For each $z \in \mathbb{Z}$, choose a path σ_z from z_0 to z. We attempt to define $\widetilde{f}(z) = \widetilde{\sigma}_z(1)$. If $\widetilde{f}$ is well defined, it will be continuous by an easy argument that can be left to the reader. Evidently, $p \circ \widetilde{f} = f$ and $\widetilde{f}(z_0) = y_0$, so we will have constructed the required lift. In order to prove that $\widetilde{f}$ is well defined, let τ_z be another path from z_0 to z. Then $\gamma = \sigma_z \cdot \tau_z^{-1}$ is a loop determining $[\gamma] \in \pi_1(Z, z_0)$, $\widehat{\gamma}$ is a loop at x_0, and $f_*[\gamma] = [\widehat{\gamma}]$. The lift of $\widehat{\gamma}$ is $\widetilde{\gamma} = \widetilde{\sigma}_z \cdot (\widetilde{\tau}_z)^{-1}$. If we can prove that $\widetilde{\gamma}$ is a loop at y_0, it will follow that $\widetilde{\sigma}_z(1) = \widetilde{\tau}_z(1)$, proving that $\widetilde{f}$ is well defined. By our hypothesis and the fact that $[\widehat{\gamma}]$ is in the image of f_*, we see that this class is also in the image of p_*. Thus, there is a loop γ' in Y at y_0 such that $p \circ \gamma' \sim_\partial \widehat{\gamma}$. By Lemma 1.7.17, it follows that $y_0 = \gamma'(1) = \widetilde{\gamma}(1)$ and $\widetilde{\gamma}$ is a loop at y_0. □

The following is perhaps the most frequently used application of Theorem 1.7.39.

Corollary 1.7.40. *Let $p : (Y, y_0) \to (X, x_0)$ be a covering map and let Z be simply connected. Then every continuous, basepoint-preserving map $f : (Z, z_0) \to (X, x_0)$ has a unique lift $\widetilde{f} : (Z, z_0) \to (Y, y_0)$.*

CHAPTER 2

The Local Theory of Smooth Functions

In this chapter, we treat the fundamentals of differential calculus in open subsets of Euclidean spaces. Everything will be set up so as to extend naturally to global differential calculus on smooth manifolds.

Notation. Elements of $\mathbb{R}^n$, when thought of as vectors, will be written as column n-tuples. When thought of as points, they will be written as rows.

2.1. Differentiability Classes

Let $U \subseteq \mathbb{R}^n$ be an open subset. Let $x = (x^1, \dots, x^n)$ denote the general (variable) point of U and let $p = (p^1, \dots, p^n)$ be a fixed but arbitrary point of U. Let $f : U \to \mathbb{R}$ be a function and let $L_p : U \to \mathbb{R}$ be an affine (*i.e.*, inhomogeneous linear) map

$$L_p(x) = c + \sum_{i=1}^{n} b_i x^i$$

such that

$$L_p(p) = f(p).$$

Definition 2.1.1. If f and L_p are as above and if

$$\lim_{x \to p} \frac{f(x) - L_p(x)}{\|x - p\|} = 0,$$

then L_p is called a derivative of f at p. If f admits a derivative at p, then f is said to be differentiable at p.

We think of a derivative L_p as a linear approximation of f near p. By the definition, the error involved in replacing $f(x)$ by $L_p(x)$ is negligible compared to the distance of x from p, provided that this distance is sufficiently small. It follows from the definition that an affine map is a derivative of itself.

The above definition of "derivative" as a linear approximation embodies the real philosophy of differential calculus. As it stands, however, this definition is a bit unsatisfying. The use of the indefinite article (*a* derivative) raises the issue of uniqueness, while the relationship of the notion of a derivative to the familiar operation of differentiation is also unclear. The following exercise resolves these doubts.

Exercise 2.1.2. If $L_p(x) = c + \sum_{i=1}^{n} b_i x^i$ is a derivative of f at p, then

$$b_i = \frac{\partial f}{\partial x^i}(p),$$

$1 \leq i \leq n$. In particular, if f is differentiable at p, these partial derivatives exist and the derivative L_p is unique.

Having seen that derivatives are given by partial derivatives, we center our attention on these more familiar operators.

Definition 2.1.3. The class of continuous functions $f : U \to \mathbb{R}$ is denoted by $C^0(U)$. If $r \geq 1$, the class $C^r(U)$ of functions $f : U \to \mathbb{R}$ that are smooth of order r is specified inductively by requiring that $\partial f / \partial x^i$ exist and belong to $C^{r-1}(U)$, $1 \leq i \leq n$. The functions that are smooth of order r are also called C^r-smooth.

Exercise 2.1.4. Inductively, prove that

$$C^0(U) \supseteq C^1(U) \supseteq \cdots \supseteq C^{r-1}(U) \supseteq C^r(U) \supseteq \cdots.$$

Examples show that these inclusions are all *proper.*

Definition 2.1.5. The set of infinitely smooth functions on U is

$$C^\infty(U) = \bigcap_{r \geq 0} C^r(U).$$

It is not uncommon simply to call C^∞ functions "differentiable" or "smooth". We will be concerned primarily with such functions and will usually refer to them as "smooth". Remark that the coordinates in U are themselves smooth functions $x^i : U \to \mathbb{R}$. Thus, $q \in U$ has coordinates $x^i(q)$, $1 \leq i \leq n$, and we can write $q = (x^1(q), \ldots, x^n(q))$.

Exercise 2.1.6. If $\dim U = 1$, prove that the derivative L_p exists if and only if $f'(p)$ exists. If, however, $\dim U = 2$ and $p = (0, 0)$, find a function $f : U \to \mathbb{R}$ such that both partial derivatives exist at every point of U, but such that the derivative $L_{(0,0)}$ does *not* exist.

Exercise 2.1.7. Let

$$D(U) = \{f : U \to \mathbb{R} \mid f \text{ is differentiable at } x, \ \forall x \in U\}.$$

Show that $C^0(U) \supseteq D(U) \supseteq C^1(U)$. Produce examples to show that both of these inclusions are *proper.* (Hint: First do this for $\dim U = 1$ and then extend to arbitrary dimensions.)

Exercise 2.1.8. Let $U \subseteq \mathbb{R}^n$ be open, let $f \in C^r(U)$, where $1 \leq r \leq \infty$, and let $g : \mathbb{R} \to \mathbb{R}$ be C^r-smooth also. Prove that the composition $g \circ f$ belongs to $C^r(U)$.

2.2. Tangent Vectors

We continue to let $U \subseteq \mathbb{R}^n$ be a fixed but arbitrary open set. We fix $p \in U$ and describe the tangent space $T_p(U)$ of U at p. In calculus, it is customary to translate a tangent vector $\vec{a}$ at p to the origin $0 \in \mathbb{R}^n$, thereby identifying $\vec{a}$ canonically with an element of $\mathbb{R}^n$. That is, we set $T_p(U) = \mathbb{R}^n$. This will not do for our purposes since we are trying to set up a local calculus that will make sense on manifolds where, generally, there will be no preferred coordinate system and translation will be meaningless. Also, the custom of representing vectors as directed straight line segments in $\mathbb{R}^n$ will not do, since a straight line segment in one coordinate system may look like a curved line segment in another. Instead of these naive definitions, we will view a tangent vector as a certain type of *operator* on functions. This definition will have no dependence on the choice of coordinates.

In standard calculus, the vector

$$\vec{a} = \begin{bmatrix} a^1 \\ \vdots \\ a^n \end{bmatrix}$$

defines a *directional derivative* $D_{\vec{a}}$ at p by the formula

$$D_{\vec{a}}(f) = \lim_{h \to 0} \frac{f(p + h\vec{a}) - f(p)}{h} = \sum_{i=1}^{n} a^i \frac{\partial f}{\partial x^i}(p),$$

where f is an arbitrary smooth function defined on an open neighborhood of p. In the notation of differential operators,

$$D_{\vec{a}} = \sum_{i=1}^{n} a^i \left. \frac{\partial}{\partial x^i} \right|_p .$$

Applying this operator to the coordinate functions x^i gives

$$D_{\vec{a}}(x^i) = a^i.$$

So the vector $\vec{a}$ is uniquely determined by its associated directional derivative.

Another way to obtain this directional derivative is to consider a curve

$$s : (-\delta, \epsilon) \to U$$

(where $\epsilon, \delta > 0$), written

$$s(t) = (x^1(t), \ldots, x^n(t)),$$

such that each $x^i(t)$ is of class at least C^1 and

$$s(0) = p,$$
$$\dot{s}(0) = \vec{a}.$$

That is,

$$x^i(0) = p^i,$$
$$\dot{x}^i(0) = a^i,$$

for $1 \leq i \leq n$. By standard calculus,

$$D_{\vec{a}}(f) = \lim_{h \to 0} \frac{f(s(h)) - f(p)}{h},$$

an equation that makes sense without explicit reference to coordinates. In other words, although the directional derivative was defined as differentiation at $t = 0$ along a straight line curve

$$\ell(t) = p + t\vec{a}, \quad -\infty < t < \infty,$$

with constant velocity $\vec{a}$, it could just as well have been defined via *any* C^1 curve out of p with initial velocity $\vec{a}$. While the notion of "straight line" will not have meaning on general manifolds, the notion of "C^1 curve" will.

Definition 2.2.1. Given $p \in U$, $C^\infty(U, p)$ will denote the set of smooth, real valued functions f with $\text{dom}(f)$ an open subset of U and $p \in \text{dom}(f)$.

Definition 2.2.2. The set of all C^1 curves $s : (-\delta, \epsilon) \to U$ (where the numbers $\delta > 0$ and $\epsilon > 0$ depend on s) such that $s(0) = p$ is denoted by $S(U, p)$.

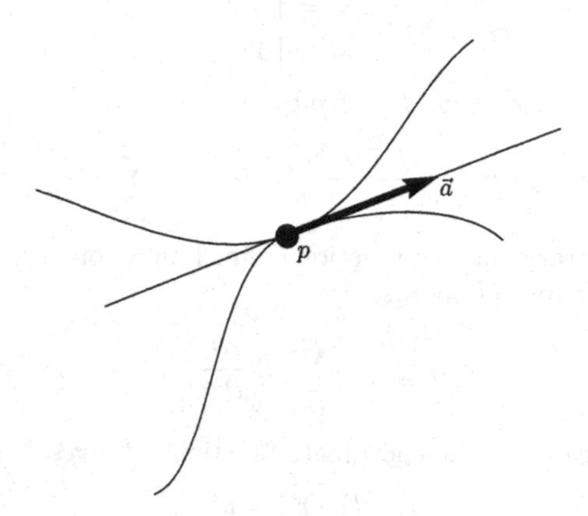

Figure 2.2.1. Some infinitesimally equivalent curves at p

Definition 2.2.3. If $s_1, s_2 \in S(U,p)$, we say that s_1 and s_2 are infinitesimally equivalent at p and we write $s_1 \approx_p s_2$ if and only if

$$\frac{d}{dt}f(s_1(t))\bigg|_{t=0} = \frac{d}{dt}f(s_2(t))\bigg|_{t=0},$$

for all $f \in C^\infty(U,p)$.

It is easy to check that $\approx_p$ is an equivalence relation on the set $S(U,p)$. Following Isaac Newton, we think of each equivalence class as an "infinitely short curve", but not as a single point. In fact, we are simply lumping together all curves sharing the same position p and velocity vector $\vec{a}$ at time $t = 0$ (see Figure 2.2.1). But, while the notion of "velocity vector" will not have an obvious meaning for curves in a manifold, the definition of "infinitesimal equivalence" will be meaningful in that context, allowing us to define the velocity vector as an "infinitesimal curve".

Definition 2.2.4. The infinitesimal equivalence class of s in $S(U,p)$ is denoted by $\langle s \rangle_p$ and is called an infinitesimal curve at p. An infinitesimal curve at p is also called a tangent vector to U at p and the set

$$T_p(U) = S(U,p)/{\approx_p}$$

of all tangent vectors at p is called the tangent space to U at p.

Remark. Once we are sufficiently familiar with these notions, we will replace the notation $\langle s \rangle_p$ with the more standard $\dot{s}(0)$ and call this the velocity vector of s at time $t = 0$.

The following is immediate by the definition of infinitesimal equivalence.

Lemma 2.2.5. *For each $\langle s \rangle_p \in T_p(U)$, the operator*

$$D_{\langle s \rangle_p} : C^\infty(U,p) \to \mathbb{R}$$

is well-defined by choosing any representative $s \in \langle s \rangle_p$ and setting

$$D_{\langle s \rangle_p}(f) = \left. \frac{d}{dt} f(s(t)) \right|_{t=0},$$

for all $f \in C^\infty(U,p)$. Conversely, $\langle s \rangle_p$ is uniquely determined by the operator $D_{\langle s \rangle_p}$.

Note that $D_{\langle s \rangle_p}$ is a *linear* operator. That is,

$$D_{\langle s \rangle_p}(af + bg) = aD_{\langle s \rangle_p}(f) + bD_{\langle s \rangle_p}(g),$$

$\forall f, g \in C^\infty(U,p)$, $\forall a, b \in \mathbb{R}$.

Since the whole reason for introducing tangent vectors is to produce linear approximations to nonlinear problems, it will be necessary to exhibit a natural vector space structure on $T_p(U)$. In order to carry this structure over to manifolds, we do not want it to be dependent on the coordinates of $\mathbb{R}^n$. The key lemma for this follows.

Lemma 2.2.6. *Let $\langle s_1 \rangle_p, \langle s_2 \rangle_p \in T_p(U)$ and $a, b \in \mathbb{R}$. Then there is a unique infinitesimal curve $\langle s \rangle_p$ such that the associated operators on $C^\infty(U,p)$ satisfy*

$$D_{\langle s \rangle_p} = aD_{\langle s_1 \rangle_p} + bD_{\langle s_2 \rangle_p}.$$

Proof. We are allowed to use the coordinates of $\mathbb{R}^n$ to prove this assertion. The important point is that the assertion itself is coordinate-free. It is clear, then, that

$$s(t) = as_1(t) + bs_2(t) - (a + b - 1)p,$$

defined by coordinatewise operations for all values of t sufficiently near 0, is a C^1 curve in U with $s(0) = p$, and that this curve represents $\langle s \rangle_p \in T_p(U)$ such that $D_{\langle s \rangle_p}$ is the desired operator. By Lemma 2.2.5, $\langle s \rangle_p$ is uniquely determined by $D_{\langle s \rangle_p}$. □

Definition 2.2.7. Let $\langle s_1 \rangle_p, \langle s_2 \rangle_p \in T_p(U)$ and $a, b \in \mathbb{R}$. Then

$$a \langle s_1 \rangle_p + b \langle s_2 \rangle_p \in T_p(U)$$

is defined to be the unique infinitesimal curve $\langle s \rangle_p$ given by Lemma 2.2.6.

Exercise 2.2.8. Prove that the operation of linear combination, as in Definition 2.2.7, makes $T_p(U)$ into an n-dimensional vector space over $\mathbb{R}$. The zero vector is the infinitesimal curve represented by the constant p. If $\langle s \rangle_p \in T_p(U)$, then $- \langle s \rangle_p = \langle s^- \rangle_p$ where $s^-(t) = s(-t)$, defined for all sufficiently small values of t.

The operator $D_{\langle s \rangle_p}$, described above, does not "see" all of $f \in C^\infty(U,p)$, only the behavior of f in arbitrarily small neighborhoods of p. The proper way to say this is that $D_{\langle s \rangle_p}(f)$ depends only on the "germ" of f at p. There is some usefulness in formalizing this.

Definition 2.2.9. We say that the elements $f, g \in C^\infty(U,p)$ are germinally equivalent at p, and write $f \equiv_p g$, if there is an open neighborhood W of p in U such that $W \subseteq \operatorname{dom}(f) \cap \operatorname{dom}(g)$ and $f|W = g|W$.

It is clear that $\equiv_p$ is an equivalence relation on $C^\infty(U,p)$.

Definition 2.2.10. The germinal equivalence class $[f]_p$ of $f \in C^\infty(U,p)$ is called the germ of f at p. The set $C^\infty(U,p)/{\equiv_p}$ of germs at p is denoted by $\mathfrak{G}_p$.

Definition 2.2.11. For each $s \in S(U,p)$ the operator

$$D_{\langle s \rangle_p} : \mathfrak{G}_p \to \mathbb{R}$$

is defined by

$$D_{\langle s \rangle_p}[f]_p = \frac{d}{dt} f(s(t))|_{t=0}.$$

Remark. The discussion so far would have worked equally well if we had fixed an integer $k \geq 1$, replaced $C^\infty(U,p)$ with the set $C^k(U,p)$ of C^k functions defined in neighborhoods of p, and taken $\mathfrak{G}_p = \mathfrak{G}_p^k$ to be the germs of these functions. This remark is crucial if one wants to formulate the theory of C^k manifolds.

There is a purely algebraic characterization of $T_p(U)$ that, though admittedly more formalistic than the one we have given, has its charms. This definition of the tangent space is valid only for the C^∞ case (the default).

First, recall that an *algebra* $\mathfrak{K}$ over $\mathbb{R}$ is a vector space over $\mathbb{R}$, together with a bilinear map

$$\mathfrak{K} \times \mathfrak{K} \to \mathfrak{K},$$

$$(\xi, \zeta) \mapsto \xi\zeta,$$

called multiplication. If $\xi\zeta = \zeta\xi$, for all $\xi, \zeta \in \mathfrak{K}$, the algebra is said to be *commutative.* If, for all $\xi, \zeta, \chi \in \mathfrak{K}$, $(\xi\zeta)\chi = \xi(\zeta\chi)$, the algebra is *associative.* If there is $\iota \in \mathfrak{K}$ such that $\iota\xi = \xi\iota = \xi$, for all $\xi \in \mathfrak{K}$, then ι is called a *unity.*

Now, define algebraic operations on germs as follows.

- Scalar multiplication: $t[f]_p = [tf]_p$, $\forall t \in \mathbb{R}$, $\forall [f]_p \in \mathfrak{G}_p$.
- Addition: $[f]_p + [g]_p = [f|W + g|W]_p$, $\forall [f]_p, [g]_p \in \mathfrak{G}_p$, where W is an open neighborhood of p in $\operatorname{dom}(f) \cap \operatorname{dom}(g)$.
- Multiplication: $[f]_p[g]_p = [(f|W)(g|W)]_p$, $\forall [f]_p, [g]_p \in \mathfrak{G}_p$, where W is again as above.

Lemma 2.2.12. *The above operations are well defined and make $\mathfrak{G}_p$ a commutative and associative algebra over $\mathbb{R}$ with unity.*

The elementary proof of Lemma 2.2.12 is left to the reader. The unique unity, of course, is the germ of the constant function 1.

Definition 2.2.13. The evaluation map $e_p : \mathfrak{G}_p \to \mathbb{R}$ is defined by

$$e_p[f]_p = f(p).$$

The following is immediate.

Lemma 2.2.14. *The evaluation map $e_p : \mathfrak{G}_p \to \mathbb{R}$ is a well-defined homomorphism of algebras.*

Definition 2.2.15. A derivative operator (or, simply, a derivative) on $\mathfrak{G}_p$ is an $\mathbb{R}$-linear map $D : \mathfrak{G}_p \to \mathbb{R}$ such that

$$D(ab) = D(a)e_p(b) + e_p(a)D(b),$$

for all $a, b \in \mathfrak{G}_p$.

Temporarily, we will denote the set of all derivatives on $\mathfrak{G}_p$ by $T(\mathfrak{G}_p)$. However, we will see shortly that it is a vector space that is canonically isomorphic to $T_p(U)$.

We define algebraic operations on $T(\mathfrak{G}_p)$.

- scalar multiplication: $(tD)(a) = t(D(a))$, $\forall t \in \mathbb{R}$ and $\forall D \in T(\mathfrak{G}_p)$, $\forall a \in \mathfrak{G}_p$;
- addition: $(D_1 + D_2)(a) = D_1(a) + D_2(a)$, $\forall D_1, D_2 \in T(\mathfrak{G}_p)$, $\forall a \in \mathfrak{G}_p$.

Lemma 2.2.16. *The space $T(\mathfrak{G}_p)$ is a vector space over $\mathbb{R}$ under the above operations.*

Again, the proof will be left to the reader.

Example 2.2.17. Define $D_{i,p} : \mathfrak{G}_p \to \mathbb{R}$ by

$$D_{i,p}[f]_p = \frac{\partial f}{\partial x^i}(p).$$

This is a well-defined, $\mathbb{R}$-linear map. Furthermore, by the Leibniz rule for partial derivatives,

$$\begin{aligned} D_{i,p}([f]_p[g]_p) &= \frac{\partial}{\partial x^i}(fg)(p) \\ &= \frac{\partial f}{\partial x^i}(p)g(p) + f(p)\frac{\partial g}{\partial x^i}(p) \\ &= D_{i,p}([f]_p)e_p([g]_p) + e_p([f]_p)D_{i,p}([g]_p). \end{aligned}$$

Thus $D_{i,p}$ is a derivative, $1 \leq i \leq n$.

Example 2.2.18. If $\langle s \rangle_p$ is an infinitesimal curve, then $D_{\langle s \rangle_p} : \mathfrak{G}_p \to \mathbb{R}$ is a derivative. Indeed,

$$D_{\langle s \rangle_p}[f]_p = \sum_{i=1}^{n} a^i D_{i,p}[f]_p,$$

where

$$\dot{s}(0) = \begin{bmatrix} a^1 \\ \vdots \\ a^n \end{bmatrix};$$

so the assertion follows from the previous example. It is obvious that

$$\langle s \rangle_p \mapsto D_{\langle s \rangle_p}$$

defines a linear map from the space of infinitesimal curves into the space of derivatives of $\mathfrak{G}_p$. It is also clear that this linear map is injective. The fact that it is an isomorphism of vector spaces (Corollary 2.2.22) requires proof.

Lemma 2.2.19. *If c is a constant function on U and $D \in T(\mathfrak{G}_p)$, then*

$$D[c]_p = 0.$$

Proof. Consider first the case $c = 1$. Then

$$\begin{aligned} D[1]_p &= D([1]_p[1]_p) \\ &= D([1]_p)e_p([1]_p) + e_p([1]_p)D([1]_p) \\ &= 2D[1]_p, \end{aligned}$$

from which it follows that $D[1]_p = 0$. For an arbitrary constant c,

$$D[c]_p = cD[1]_p = 0,$$

by linearity. □

In order to get more information on $T(\mathfrak{G}_p)$, we need a technical lemma. Let $x = (x^1, \dots, x^n)$ and $p = (x^1(p), \dots, x^n(p))$.

Lemma 2.2.20. *Let $f \in C^\infty(U, p)$. Then there exist functions*

$$g_1, \dots, g_n \in C^\infty(U, p)$$

and a neighborhood $W \subset \operatorname{dom}(f) \cap \operatorname{dom}(g_1) \cap \cdots \cap \operatorname{dom}(g_n)$ of p such that

(1) $f(x) = f(p) + \sum_{i=1}^n (x^i - x^i(p)) g_i(x)$, $\forall x \in W$;
(2) $g_i(p) = \frac{\partial f}{\partial x^i}(p)$, $1 \le i \le n$.

Proof. Define

$$g_i(x) = \int_0^1 \frac{\partial f}{\partial x^i}(t(x-p)+p)\,dt.$$

This is clearly a smooth function defined at all points x sufficiently near p. In order to prove (2), consider

$$\begin{aligned} g_i(p) &= \int_0^1 \frac{\partial f}{\partial x^i}(p)\,dt \\ &= \frac{\partial f}{\partial x^i}(p) \int_0^1 dt \\ &= \frac{\partial f}{\partial x^i}(p). \end{aligned}$$

In order to prove (1), consider

$$\begin{aligned} f(x) - f(p) &= \int_0^1 \frac{d}{dt}(f(t(x-p)+p)\,dt \\ &= \int_0^1 \left\{ \sum_{i=1}^n \frac{\partial f}{\partial x^i}(t(x-p)+p)(x^i - x^i(p)) \right\} dt \\ &= \sum_{i=1}^n \left\{ \int_0^1 \frac{\partial f}{\partial x^i}(t(x-p)+p)\,dt \right\} (x^i - x^i(p)) \\ &= \sum_{i=1}^n g_i(x)(x^i - x^i(p)). \end{aligned}$$

□

Theorem 2.2.21. *The set $\{D_{1,p}, \dots, D_{n,p}\}$ is a basis of the vector space $T(\mathfrak{G}_p)$.*

Proof. Suppose that

$$\sum_{i=1}^n a^i D_{i,p} = 0.$$

For the coordinate functions x^j, $1 \le j \le n$,

$$D_{i,p}[x^j]_p = \frac{\partial x^j}{\partial x^i}(p) = \delta_{ij},$$

the Kronecker delta. Thus,

$$0 = \left(\sum_{i=1}^n a^i D_{i,p} \right) [x^j]_p = a^j,$$

for $1 \le j \le n$. This proves that $\{D_{1,p}, \dots, D_{n,p}\}$ is a linearly independent subset of $T(\mathfrak{G}_p)$. We must prove that it is also a spanning set.

Let $D \in T(\mathfrak{G}_p)$. Set $a^i = D[x^i]_p$, $1 \le i \le n$. Given an arbitrary $[f]_p \in \mathfrak{G}_p$, write

$$f(x) = f(p) + \sum_{i=1}^{n}(x^i - x^i(p))g_i(x)$$

as in Lemma 2.2.20. Then,

$$\begin{aligned}
D[f]_p &= D[f(p) + \sum_{i=1}^{n}(x^i - x^i(p))g_i(x)]_p \\
&= \sum_{i=1}^{n} D[(x^i - x^i(p))g_i]_p \\
&= \sum_{i=1}^{n}\{D[x^i]_p g_i(p) + (x^i(p) - x^i(p))D[g_i]_p\} \\
&= \sum_{i=1}^{n} a^i \frac{\partial f}{\partial x^i}(p) \\
&= \left(\sum_{i=1}^{n} a^i D_{i,p}\right)[f]_p.
\end{aligned}$$

Since $[f]_p \in \mathfrak{G}_p$ is *arbitrary*, it follows that $D = \sum_{i=1}^{n} a^i D_{i,p}$. □

Remark. The above proof would not work for derivatives of the algebra $\mathfrak{G}_p^k$ of germs of C^k functions, $k < \infty$. The problem is that $g_i \in C^{k-1}$, $1 \le i \le n$; so $D[g_i]_p$ is not even defined. In fact, for $0 < k < \infty$, the space of derivatives of $\mathfrak{G}_p^k$ is infinite-dimensional [**33**].

Corollary 2.2.22. *The spaces $T_p(U)$ and $T(\mathfrak{G}_p)$ are canonically isomorphic vector spaces.*

Proof. Indeed, the linear injection, defined in Example 2.2.18, between the two vector spaces must be an isomorphism since both are n-dimensional. □

We write $T_p(U)$ and $T(\mathfrak{G}_p)$ interchangeably, usually preferring $T_p(U)$.

Remark. We can identify this vector space *canonically* with $\mathbb{R}^n$ via

$$\sum_{i=1}^{n} a^i D_{i,p} \leftrightarrow \begin{bmatrix} a^1 \\ \vdots \\ a^n \end{bmatrix}.$$

On the other hand, one should be wary since $T_p(U)$ should not be thought of as identical with $T_q(U)$ when $p \ne q$. There will be no such canonical identification on manifolds.

Let $T(U) = \bigsqcup_{x \in U} T_x(U)$, a disjoint union. There is a one-to-one correspondence $T(U) \leftrightarrow U \times \mathbb{R}^n$ given by

$$\sum_{i=1}^{n} a^i D_{i,x} \leftrightarrow \left(x, \begin{bmatrix} a^1 \\ \vdots \\ a^n \end{bmatrix}\right).$$

We use this to transfer the topology of $U \times \mathbb{R}^n$ to $T(U)$.

Remark. This method of topologizing $T(U)$ does seem to use the coordinates. We will see, however, that the topology on $T(U)$ is actually independent of the choice of coordinates.

Definition 2.2.23. The tangent bundle $\pi : T(U) \to U$ is defined by

$$\pi\left(\sum_{i=1}^{n} a^i D_{i,p}\right) = p.$$

Via the canonical identification $T(U) = U \times \mathbb{R}^n$, π is just the standard projection onto the first factor. For each $x \in U$, $T_x(U)$ should be thought of as the linear approximation of U at x. This is going to enable us to approximate smooth maps between open subsets of Euclidean spaces by linear maps.

Exercise 2.2.24. Let $\mathfrak{G}_p^* \subset \mathfrak{G}_p$ be the kernel of the evaluation map e_p and let $\mathfrak{G}_p^{**} \subset \mathfrak{G}_p^*$ be the vector subspace spanned by the germs of functions gf, where $g, f \in C^\infty(U,p)$ and $g(p) = 0 = f(p)$. Prove that the quotient space $\mathfrak{G}_p^*/\mathfrak{G}_p^{**}$ is canonically isomorphic to the vector space dual of $T_p(U)$. In particular, this quotient space is n-dimensional.

2.3. Smooth Maps and their Differentials

Let $U \subseteq \mathbb{R}^n$ and $V \subseteq \mathbb{R}^m$ be open subsets. Consider functions $\Phi : U \to V$ and their coordinate representations

$$\Phi = \left(\Phi^1, \Phi^2, \ldots, \Phi^m\right),$$

where each $\Phi^i : U \to \mathbb{R}$.

Definition 2.3.1. We say that $\Phi : U \to V$ is a map of class C^k (where $0 \le k \le \infty$) if $\Phi^i \in C^k(U)$, $1 \le i \le m$. If Φ is of class C^∞, it is called a smooth map.

The following lemma is clear by the standard chain rule.

Lemma 2.3.2. *Wherever defined, compositions of smooth maps are also smooth.*

Lemma 2.3.3. *Let $\Phi : U \to V$ be smooth and let $p \in U$. If $s \in S(U,p)$, then the infinitesimal equivalence class of $\Phi \circ s$ at $\Phi(p)$ depends only on the infinitesimal equivalence class of s at p*

Proof. Indeed, let $f \in C^\infty(V, \Phi(p))$ be arbitrary and note that

$$\left.\frac{d}{dt} f(\Phi(s(t)))\right|_{t=0}$$

can be interpreted as the derivative of $f \circ \Phi$ along s at $t = 0$. This depends only on $\langle s \rangle_p$. □

Definition 2.3.4. If $\Phi : U \to V$ is smooth and if $p \in U$, let

$$d\Phi_p = \Phi_{*p} : T_p(U) \to T_{\Phi(p)}(V)$$

be defined by

$$\Phi_{*p} \langle s \rangle_p = \langle \Phi \circ s \rangle_{\Phi(p)},$$

for arbitrary $\langle s \rangle_p \in T_p(U)$. This is called the *differential* of Φ at p.

Under the identification of an infinitesimal curve $\langle s\rangle_p$ with its associated derivative operator $D_{\langle s\rangle_p}$, we can write

$$d\Phi_p(D_{\langle s\rangle_p}) = \Phi_{*p}(D_{\langle s\rangle_p}) = D_{\langle \Phi\circ s\rangle_{\Phi(p)}}.$$

If, for simplicity of notation, we let $f \in C^\infty(V, \Phi(p))$ stand in for its germ $[f]_{\Phi(p)}$, this operator has the form

$$\Phi_{*p}(D_{\langle s\rangle_p})(f) = D_{\langle \Phi\circ s\rangle_p}(f) = \frac{d}{dt} f(\Phi(s(t)))\bigg|_{t=0} = D_{\langle s\rangle_p}(f\circ\Phi).$$

That is, to differentiate f by $\Phi_{*p}(D_{\langle s\rangle_p})$, one "pulls back" f to the function $f\circ\Phi \in C^\infty(U,p)$ and differentiates that function by $D_{\langle s\rangle_p}$.

This "pullback" of functions is denoted by $\Phi^*(f) = f\circ\Phi$. It passes to a well-defined map

$$\Phi_p^* : \mathfrak{G}_{\Phi(p)} \to \mathfrak{G}_p,$$
$$\Phi_p^*([f]_{\Phi(p)}) = [f\circ\Phi]_p.$$

It is almost immediate that this is a homomorphism of algebras and one obtains the following.

Lemma 2.3.5. *Under the identifications $T(\mathfrak{G}_p) = T_p(U)$ and $T(\mathfrak{G}_{\Phi(p)}) = T_{\Phi(p)}(V)$, the formula for the differential $\Phi_{*p} : T(\mathfrak{G}_p) \to T(\mathfrak{G}_{\Phi(p)})$ becomes*

$$\Phi_{*p}(D) = D\circ\Phi_p^*, \quad \forall D \in T(\mathfrak{G}_p).$$

In particular, Φ_{*p} is a linear map.

Exercise 2.3.6. Relative to the respective bases $\{D_{i,p}\}_{i=1}^n$ of $T_p(U)$ and $\{D_{j,\Phi(p)}\}_{j=1}^m$ of $T_{\Phi(p)}(V)$, show that the matrix of the linear map Φ_{*p} is the Jacobian matrix

$$J\Phi(p) = \begin{bmatrix} \frac{\partial\Phi^1}{\partial x^1}(p) & \frac{\partial\Phi^1}{\partial x^2}(p) & \cdots & \frac{\partial\Phi^1}{\partial x^n}(p) \\ \frac{\partial\Phi^2}{\partial x^1}(p) & \frac{\partial\Phi^2}{\partial x^2}(p) & \cdots & \frac{\partial\Phi^2}{\partial x^n}(p) \\ \vdots & \vdots & & \vdots \\ \frac{\partial\Phi^m}{\partial x^1}(p) & \frac{\partial\Phi^m}{\partial x^2}(p) & \cdots & \frac{\partial\Phi^m}{\partial x^n}(p) \end{bmatrix}$$

of Φ at p. In particular, if $\Phi : U \to \mathbb{R}$ and $D_p \in T_p(U)$, conclude that, under the canonical identification $T_{\Phi(p)}(\mathbb{R}) = \mathbb{R}$,

$$\Phi_{*p}(D_p) = D_p(f).$$

Remark. The differentials Φ_{*x}, computed at all points $x \in U$, assemble to a mapping

$$\Phi_* = d\Phi : T(U) \to T(V),$$

called the differential of Φ on U and given by

$$\Phi_*\left(x, \begin{bmatrix} a^1 \\ \vdots \\ a^n \end{bmatrix}\right) = \left(\Phi(x), J\Phi(x)\begin{bmatrix} a^1 \\ \vdots \\ a^n \end{bmatrix}\right).$$

Here, we have identified $T(U)$ with $U \times \mathbb{R}^n$ and $T(V)$ with $V \times \mathbb{R}^m$.

Corollary 2.3.7. *Relative to the identifications $T(U) = U \times \mathbb{R}^n$ and $T(V) = V \times \mathbb{R}^m$, the differential $d\Phi = \Phi_* : T(U) \to T(V)$ is a smooth map from an open subset of $\mathbb{R}^{2n}$ to an open subset of $\mathbb{R}^{2m}$.*

Corollary 2.3.8. *Let $\Phi : U \to V$ be of the form*

$$\Phi(x) = L(x) + y_0,$$

where $L : \mathbb{R}^n \to \mathbb{R}^m$ is linear and $y_0 \in \mathbb{R}^m$ is fixed. Then, denoting by $\widehat{L}$ the matrix of L relative to the standard coordinates of $\mathbb{R}^n$ and $\mathbb{R}^m$, we have that

$$J\Phi(p) = \widehat{L}, \ \forall p \in U.$$

In this case,

$$d\Phi = (\Phi, \widehat{L}) : U \times \mathbb{R}^n \to V \times \mathbb{R}^m.$$

In case $\Phi = L$ is itself linear, $dL = (L, \widehat{L})$. If we identify $T_p(U) = \mathbb{R}^n$ and $T_{\Phi(p)}(V) = \mathbb{R}^m$, we can write $dL_p = L$.

Theorem 2.3.9 (The general chain rule). *If $U \subseteq \mathbb{R}^n$, $V \subseteq \mathbb{R}^m$, and $W \subseteq \mathbb{R}^q$ are open subsets and if $\Phi : U \to V$ and $\Psi : V \to W$ are smooth, then*

$$d(\Psi \circ \Phi)_p = d\Psi_{\Phi(p)} \circ d\Phi_p.$$

Proof. Indeed,

$$d(\Psi \circ \Phi)_p \langle s \rangle_p = \langle \Psi \circ \Phi \circ s \rangle_{\Psi(\Phi(p))} = d\Psi_{\Phi(p)} \langle \Phi \circ s \rangle_{\Phi(p)} = d\Psi_{\Phi(p)}(d\Phi_p \langle s \rangle_p).$$

□

Remark. In terms of Jacobian matrices, the general chain rule can be written

$$J(\Psi \circ \Phi)(p) = J\Psi(\Phi(p)) \cdot J\Phi(p).$$

One can verify this directly by applying the less general chain rule for real-valued functions and the formulas for matrix multiplication. This is the usual proof in multivariable calculus, but the proof via infinitesimal curves is more elegant and more intuitive.

Exercise 2.3.10. Viewing Φ_{*p} in terms of derivatives of germs, give a direct proof of the chain rule (again avoiding Jacobian matrices).

Definition 2.3.11. If $U \subseteq \mathbb{R}^n$ and $V \subseteq \mathbb{R}^m$ are open, a map $\Phi : U \to V$ is a diffeomorphism if it is smooth and bijective and if $\Phi^{-1} : V \to U$ is also smooth.

Proposition 2.3.12. *If $\Phi : U \to V$ is a diffeomorphism of an open subset of $\mathbb{R}^n$ onto an open subset of $\mathbb{R}^m$ and if $p \in U$, then*

$$d\Phi_p : T_p(U) \to T_{\Phi(p)}(V)$$

is a linear isomorphism. In particular, $n = m$.

Proof. Since $\Phi^{-1} \circ \Phi = \mathrm{id}_U$ is the restriction to U of $\mathrm{id}_{\mathbb{R}^n}$, a linear map, Corollary 2.3.8 shows that $d(\Phi^{-1} \circ \Phi)_p = \mathrm{id}_{T_p(U)}$, for each $p \in U$. By the general chain rule, it follows that the linear map $d\Phi_p$ is invertible with inverse $d(\Phi^{-1})_{\Phi(p)}$. □

Remark. Thus the dimension n of an open subset $U \subseteq \mathbb{R}^n$ is a diffeomorphism invariant. We saw earlier that the Brouwer theorem of invariance of domain implies the equality of dimensions, even if Φ were only a *homeomorphism.* That theorem was very deep, while the proof of Proposition 2.3.12 is quite elementary. This is an example of the technique of reducing nonlinear problems to linear ones via derivatives.

Example 2.3.13. Let $\mathfrak{M}(n)$ denote the set of $n \times n$ matrices with real entries. This is a vector space over $\mathbb{R}$ and, by suitably ordering its entries, we can fix an identification of $\mathfrak{M}(n)$ with $\mathbb{R}^{n^2}$. An important subset $\mathrm{Gl}(n)$ of $\mathfrak{M}(n)$ is the set of *nonsingular* matrices. These form a group under matrix multiplication, called the *general linear group.* This is an open subset of $\mathfrak{M}(n)$. Indeed, the determinant function $\det : \mathfrak{M}(n) \to \mathbb{R}$ is a polynomial, hence is smooth. The set $\mathbb{R}^*$ of nonzero reals is open in $\mathbb{R}$, hence the general linear group $\mathrm{Gl}(n) = \det^{-1}(\mathbb{R}^*)$ is open.

If $P \in \mathfrak{M}(n)$, the left multiple map

$$L_P : \mathfrak{M}(n) \to \mathfrak{M}(n)$$

is given by $L_P(Y) = PY$, $\forall\, Y \in \mathfrak{M}(n)$. Similarly, the right multiple map is given by $R_P(Y) = YP$, $\forall\, Y \in \mathfrak{M}(n)$. Clearly, both L_P and R_P are linear transformations. Thus, by Corollary 2.3.8 and the subsequent remark,

$$\begin{aligned} d(L_P)_Y &: T_Y(\mathfrak{M}(n)) \to T_{PY}(\mathfrak{M}(n)), \\ d(R_P)_Y &: T_Y(\mathfrak{M}(n)) \to T_{YP}(\mathfrak{M}(n)) \end{aligned}$$

are given, via the natural identifications of these tangent spaces with the underlying Euclidean space $\mathfrak{M}(n)$, by

$$\begin{aligned} d(L_P)_Y(A) &= PA, \\ d(R_P)_Y(A) &= AP, \end{aligned}$$

$\forall\, A \in \mathfrak{M}(n)$.

If $P \in \mathrm{Gl}(n)$, then the restrictions of L_P and R_P to $\mathrm{Gl}(n)$ are diffeomorphisms of this open set onto itself. Indeed, the respective inverses are $L_{P^{-1}}$ and $R_{P^{-1}}$. These diffeomorphisms $L_P, R_P : \mathrm{Gl}(n) \to \mathrm{Gl}(n)$ are called, respectively, the left and right translations by P.

The precise sense in which the differential $d\Phi_p$ is a linear approximation of Φ near p is given by the following theorem, in which $d\Phi_y$ is interpreted as a linear map of $\mathbb{R}^n \to \mathbb{R}^m$, $\forall\, y \in \mathrm{dom}(\Phi)$.

Theorem 2.3.14. *If $U \subseteq \mathbb{R}^n$ is open and $\Phi : U \to \mathbb{R}^m$ is smooth, then, for each $p \in U$,*

$$\lim_{(x,y)\to(p,p)} \frac{\Phi(x) - \Phi(y) - d\Phi_y(x-y)}{\|x-y\|} = 0.$$

Proof. Using the coordinate representation

$$\Phi = \left(\Phi^1, \Phi^2, \dots, \Phi^m\right),$$

we see that it is enough to prove the assertion for maps $\Phi^j = f : U \to \mathbb{R}$. In Lemma 2.2.20, let p be a variable point y and write

$$\begin{aligned} g_i(x,y) &= \int_0^1 \frac{\partial f}{\partial x^i}(t(x-y)+y)\,dt \\ g_i(y,y) &= \frac{\partial f}{\partial x^i}(y) \\ f(x) - f(y) &= \sum_{i=1}^n (x^i - y^i) g_i(x,y). \end{aligned}$$

Thus,

$$f(x) - f(y) = \|x - y\| \underbrace{\sum_{i=1}^{n} \frac{x^i - y^i}{\|x - y\|}(g_i(x,y) - g_i(y,y))}_{R(x,y)} + \underbrace{\sum_{i=1}^{n} (x^i - y^i) \frac{\partial f}{\partial x^i}(y)}_{df_y(x-y)}.$$

Since $(x^i - y^i)/\|x - y\|$ is bounded, $1 \le i \le n$, it is clear that

$$\lim_{(x,y)\to(p,p)} R(x,y) = 0$$

and the assertion follows. □

Exercise 2.3.15. If $f \in C^\infty(U,p)$ and L_p is the derivative of f at p (Definition 2.1.1), use Theorem 2.3.14 to express L_p in terms of df_p.

Exercise 2.3.16. Let $p \in S^n \subset \mathbb{R}^{n+1}$ and define

$$T_p(S^n) = \{\langle s \rangle_p \in T_p(\mathbb{R}^{n+1}) \mid s : (-\epsilon, \epsilon) \to \mathbb{R}^{n+1} \text{ has } \operatorname{im}(s) \subset S^n\}.$$

Prove that $T_p(S^n)$ is the linear subspace of $\mathbb{R}^{n+1} = T_p(\mathbb{R}^{n+1})$ consisting of all $v \perp p$. This is what we earlier called the tangent space of S^n at p (Example 1.2.7).

In Exercises 2.3.18 and 2.3.19, you will need the following definition.

Definition 2.3.17. If $A \subseteq \mathbb{R}^n$ is an arbitrary subset, define $C^\infty(A)$ to be the set of all functions $f : A \to \mathbb{R}$ such that $f = \tilde{f}|A$, where $\tilde{f} : U \to \mathbb{R}$ is a smooth function defined on some open neighborhood U of A.

Exercise 2.3.18. If $A = [0,1] \times [0,1]$ and $f \in C^\infty(A)$, let $\tilde{f} : U \to \mathbb{R}$ be a smooth extension as in Definition 2.3.17. Prove that $d\tilde{f}_{(0,0)}$ depends only on f, not on the choice of $\tilde{f}$.

Exercise 2.3.19. If $A = S^n$, $f \in C^\infty(S^n)$, $p \in S^n$, and $\tilde{f}$ is as in Definition 2.3.17, show by an example that $d\tilde{f}_p$ may well depend on the choice of extension $\tilde{f}$, but prove that $d\tilde{f}_p|T_p(S^n)$ depends only on f.

2.4. Diffeomorphisms and Maps of Constant Rank

If $\Phi : U \to V$ is a diffeomorphism between open subsets of $\mathbb{R}^n$, then the Jacobian matrix $J\Phi(p)$ is nonsingular, $\forall p \in U$. While the converse is not exactly true, a strong version of the converse is true *locally.*

Theorem 2.4.1 (Inverse function theorem). *Let $\Phi : U \to V$ be smooth, where $U, V \subseteq \mathbb{R}^n$ are open subsets, and let $p \in U$. If*

$$d\Phi_p : T_p(U) \to T_{\Phi(p)}(V)$$

is a linear isomorphism, then there is an open neighborhood W_p of p in U such that $\Phi|W_p$ is a diffeomorphism of W_p onto an open neighborhood $\Phi(W_p)$ of $\Phi(p)$ in V.

This is a remarkable and fundamental result. From a single piece of linear information at one point, it concludes to information in a whole neighborhood of that point. This theorem is often proven in courses in advanced calculus. We will give a proof in Appendix B that works for C^k maps, $1 \le k \le \infty$, and even works for maps between open subsets of a Banach space.

There is a generalization of Theorem 2.4.1, called the "constant rank theorem", that is actually equivalent to the inverse function theorem. Our main goal in this section is to prove the constant rank theorem using the inverse function theorem. The statement and proof are greatly facilitated by "smooth changes of coordinates".

2.4.A. Diffeomorphisms as coordinate changes. Let $\widetilde{W}$ and W be open subsets of $\mathbb{R}^n$ and let $F : W \to \widetilde{W}$ be a diffeomorphism. If we denote the coordinates of W by $x = (x^1, \dots x^n)$ and those of $\widetilde{W}$ by $w = (w^1, \dots, w^n)$, then the coefficient functions F^i of F can be denoted by $w^i(x)$. That is, F is given by a system of smooth equations

$$w^i = w^i(x^1, \dots, x^n), \quad 1 \le i \le n.$$

The existence of a smooth inverse F^{-1} can be interpreted as the existence of smooth solutions

$$x^i = x^i(w^1, \dots, w^n), \quad 1 \le i \le n.$$

These systems can be viewed as defining changes of coordinates.

Suppose that $\Phi : W \to Z$ is a smooth map between open subsets $W \subseteq \mathbb{R}^n$ and $Z \subseteq \mathbb{R}^m$ and that $F : W \to \widetilde{W}$ and $G : Z \to \widetilde{Z}$ are diffeomorphisms and set $\widetilde{\Phi} = G \circ \Phi \circ F^{-1}$. This gives a commutative diagram

$$\begin{array}{ccc} W & \xrightarrow{F} & \widetilde{W} \\ {\scriptstyle \Phi}\downarrow & & \downarrow{\scriptstyle \widetilde{\Phi}} \\ Z & \xrightarrow[G]{} & \widetilde{Z} \end{array}$$

and we can interpret $\widetilde{\Phi}$ as a new formula for the map Φ relative to the respective coordinate changes F and G in the domain and range of Φ. One looks for coordinate changes that make the formula for Φ simpler.

Example 2.4.2. Let $\Phi : W \to Z$ be smooth and let $p \in W$. If we take $F : \mathbb{R}^n \to \mathbb{R}^n$ to be translation by $-p$ and $G : \mathbb{R}^m \to \mathbb{R}^m$ to be translation by $-\Phi(p)$, we can set $\widetilde{W} = F(W)$, $\widetilde{Z} = G(Z)$, and view F and G as coordinate changes. In the new coordinates, p is replaced by $0 \in \mathbb{R}^n$, $\Phi(p)$ is replaced by $0 \in \mathbb{R}^m$ and the new formula for Φ satisfies

$$\widetilde{\Phi}(0) = 0.$$

We say that, "by suitable translations in the range and domain of Φ, we lose no generality in assuming that $p = 0$ and $\Phi(p) = 0$".

Example 2.4.3. Linear changes of coordinates are frequently useful. We note particularly those linear changes that simply permute the order of the coordinates. More precisely, suppose that σ is a permutation of $\{1, 2, \dots, n\}$, that τ is a permutation of $\{1, 2, \dots, m\}$ and that the coordinate changes $F : W \to \widetilde{W}$ and $G : Z \to \widetilde{Z}$ are given by

$$F^{-1}(w^1, \dots, w^n) = (w^{\sigma(1)}, \dots, w^{\sigma(n)}),$$
$$G(y^1, \dots, y^m) = (y^{\tau(1)}, \dots, y^{\tau(m)}).$$

Thus, if $\Phi : W \to Z$ has coordinate functions Φ^i, its new formula $\widetilde{\Phi}$ has coordinate functions

$$\widetilde{\Phi}^i(w^1, \dots, w^n) = \Phi^{\tau(i)}(w^{\sigma(1)}, \dots, w^{\sigma(n)}), \quad 1 \le i \le m.$$

It follows that $J\widetilde{\Phi}$ is obtained from $J\Phi$ by permuting the rows by τ and the columns by σ. Thus, for example, if the matrix $J\Phi_p$ has rank k, we can assume, after suitable permutations of the coordinates in the domain and range, that $J\Phi_p$ has as its upper left $k \times k$ block a nonsingular matrix.

2.4.B. The constant rank theorem.

Definition 2.4.4. A smooth map $\Phi : U \to V$, between open subsets of Euclidean spaces of possibly different dimensions, has constant rank k if the rank of the linear map $d\Phi_x : T_x(U) \to T_{\Phi(x)}(V)$ is k at every point of U. Equivalently, the Jacobian matrix $J\Phi$ has constant rank k on U.

Example 2.4.5. Consider the composition

$$\mathbb{R}^k \times \mathbb{R}^{n-k} \xrightarrow{\pi} \mathbb{R}^k \xrightarrow{i} \mathbb{R}^m,$$

where $k < n$, $k < m$, and

$$\begin{aligned} \pi(x^1, \dots, x^k, y^1, \dots, y^{n-k}) &= (x^1, \dots, x^k) \\ i(x^1, \dots, x^k) &= (x^1, \dots, x^k, \underbrace{0, \dots, 0}_{m-k}). \end{aligned}$$

The Jacobian of $i \circ \pi$ is constantly the $m \times n$ matrix having I_k as its upper left $k \times k$ corner and 0's elsewhere. The rank is constantly k.

The constant rank theorem asserts that, in a certain precise sense, maps of constant rank k locally "look like" the above example.

Theorem 2.4.6 (Constant rank theorem). *Let $U \subseteq \mathbb{R}^n$ and $V \subseteq \mathbb{R}^m$ be open and let $\Phi : U \to V$ be smooth. Let $p \in U$ and suppose that, in some neighborhood of p, Φ has constant rank k. Then there are open neighborhoods W of p in U and $Z \supseteq \Phi(W)$ of $\Phi(p)$ in V, together with smooth changes of coordinates*

$$F : W \to \widetilde{W},$$

$$G : Z \to \widetilde{Z},$$

such that, throughout the neighborhood $\widetilde{W}$ of $F(p)$, the new formula $\widetilde{\Phi}$ for Φ is

$$\widetilde{\Phi}(w^1, \dots, w^n) = \left(w^1, \dots, w^k, 0, \dots, 0\right).$$

Proof. By Example 2.4.2, we make preliminary changes of coordinates so as to assume that $p = 0 \in \mathbb{R}^n$ and $\Phi(p) = 0 \in \mathbb{R}^m$. Similarly, by Example 2.4.3, we assume that the upper left $k \times k$ block

$$\frac{\partial(\Phi^1, \dots, \Phi^k)}{\partial(x^1, \dots, x^k)} = \begin{bmatrix} \frac{\partial\Phi^1}{\partial x^1} & \cdots & \frac{\partial\Phi^1}{\partial x^k} \\ \vdots & & \vdots \\ \frac{\partial\Phi^k}{\partial x^1} & \cdots & \frac{\partial\Phi^k}{\partial x^k} \end{bmatrix}$$

of $J\Phi$ is nonsingular at $p = 0$.

Let $x = (x^1, \dots, x^n)$ and define $F : U \to \mathbb{R}^n$ by

$$F(x) = \left(\Phi^1(x), \dots, \Phi^k(x), x^{k+1}, \dots, x^n\right).$$

Then $F(0) = 0$ and

$$JF = \left[\begin{array}{c|c} \frac{\partial(\Phi^1,\dots,\Phi^k)}{\partial(x^1,\dots,x^k)} & * \\ \hline 0 & I_{n-k} \end{array}\right]$$

is a matrix that is nonsingular at $p = 0$. By Theorem 2.4.1, there is a neighborhood W of 0 on which F is a diffeomorphism onto an open set $\widetilde{W} \subseteq \mathbb{R}^n$. Let $w =$

$(w^1, \ldots, w^n)$ denote the coordinates of $\widetilde{W}$. Then, for suitable smooth functions $\varphi^{k+1}, \ldots, \varphi^m$, we get the formula

$$\Phi \circ F^{-1}(w) = \left(w^1, \ldots, w^k, \varphi^{k+1}(w), \ldots, \varphi^m(w)\right).$$

Since $F^{-1}(0) = 0$ and $\Phi(0) = 0$, we note that the functions φ^j all vanish at the origin. We also note that

$$J\Phi \cdot JF^{-1} = J(\Phi \circ F^{-1}) = \left[\begin{array}{c|ccc} I_k & & 0 & \\ \hline & \frac{\partial \varphi^{k+1}}{\partial w^{k+1}} & \cdots & \frac{\partial \varphi^{k+1}}{\partial w^n} \\ * & \vdots & & \vdots \\ & \frac{\partial \varphi^m}{\partial w^{k+1}} & \cdots & \frac{\partial \varphi^m}{\partial w^n} \end{array}\right].$$

Since $J\Phi$ has rank k in a neighborhood of 0 and JF^{-1} is nonsingular on the neighborhood $\widetilde{W} = F(W)$ of 0, we can choose W smaller, if necessary, so as to assume that the above matrix has rank k at every point of $\widetilde{W}$. It follows that the lower right block must consist entirely of 0s, hence that the functions φ^j depend locally only on $(w^1, \ldots, w^k)$. Thus, choosing W smaller if necessary, we can write

$$\varphi^j(w) = \varphi^j(w^1, \ldots, w^k), \quad k+1 \leq j \leq m.$$

Let $y = (y^1, \ldots, y^m)$ and define

$$G(y) = \left(y^1, \ldots, y^k, y^{k+1} - \varphi^{k+1}(y^1, \ldots, y^k), \ldots, y^m - \varphi^m(y^1, \ldots, y^k)\right).$$

This is defined on a suitably small neighborhood of 0 in $\mathbb{R}^m$. It is clear that

$$JG = \left[\begin{array}{c|c} I_k & 0 \\ \hline * & I_{m-k} \end{array}\right]$$

is a nonsingular matrix, hence Theorem 2.4.1 implies that G is a diffeomorphism of a small enough neighborhood Z of 0 in $\mathbb{R}^m$ onto a neighborhood $\widetilde{Z} = G(Z)$ of 0. Taking W smaller, if necessary, we can assume that $\Phi(W) \subseteq Z$. From the formulas, it is clear that

$$G \circ \Phi \circ F^{-1}(w^1, \ldots, w^n) = \left(w^1, \ldots, w^k, 0, \ldots, 0\right)$$

on $\widetilde{W}$. □

Exercise 2.4.7. Deduce Theorem 2.4.1 from Theorem 2.4.6. Since we deduced Theorem 2.4.6 from Theorem 2.4.1, the two theorems are equivalent.

There are two particularly important cases of Theorem 2.4.6, the immersion theorem and the submersion theorem.

Definition 2.4.8. Let $U \subseteq \mathbb{R}^n$ and $V \subseteq \mathbb{R}^m$ be open subsets. A smooth map $\Phi : U \to V$ is a submersion if it has constant rank m on U. It is an immersion if it has constant rank n on U.

Remark that, if Φ is a submersion, then $n \geq m$. If it is an immersion, then $m \geq n$. If it is both a submersion and an immersion, then $n = m$ and Φ is *locally* a diffeomorphism by Theorem 2.4.1. The next two corollaries are immediate applications of Theorem 2.4.6.

Corollary 2.4.9 (Submersion theorem). *Let $\Phi : U \to V$ be a submersion and let $p \in U$. Then there are open neighborhoods W of p in U and $Z \supseteq \Phi(W)$ of $\Phi(p)$ in V, together with smooth coordinate changes*

$$F : W \to \widetilde{W},$$
$$G : Z \to \widetilde{Z},$$

such that the new formula $\widetilde{\Phi}$ for Φ on $\widetilde{W}$ is

$$\widetilde{\Phi}(w^1, \dots, w^n) = (w^1, \dots, w^m).$$

Corollary 2.4.10 (Immersion theorem). *Let $\Phi : U \to V$ be an immersion and let $p \in U$. Then there are open neighborhoods W of p in U and $Z \supseteq \Phi(W)$ of $\Phi(p)$ in V, together with smooth coordinate changes*

$$F : W \to \widetilde{W},$$
$$G : Z \to \widetilde{Z},$$

such that the new formula $\widetilde{\Phi}$ for Φ on $\widetilde{W}$ is

$$\widetilde{\Phi}(w^1, \dots, w^n) = (w^1, \dots, w^n, 0, \dots, 0).$$

Thus, submersions look locally like projections onto the first m coordinates and immersions look locally like the canonical imbeddings

$$\mathbb{R}^n = \mathbb{R}^n \times \{(0, \dots, 0)\} \hookrightarrow \mathbb{R}^m.$$

Corollary 2.4.11 (Implicit function theorem). *Let $U \subseteq \mathbb{R}^n$ be open and let $p \in U$. Let $f : U \to \mathbb{R}$ be smooth with $f(p) = a$. If*

$$\frac{\partial f}{\partial x^k}(p) \neq 0,$$

then, on some open neighborhood W of p in U, the set of solutions to the equation $f(x) = a$ is the graph of a smooth function

$$x^k = g(x^1, \dots, x^{k-1}, x^{k+1}, \dots, x^n).$$

Exercise 2.4.12. Use the proof of Theorem 2.4.6 to prove the implicit function theorem.

Exercise 2.4.13. Use Corollary 2.4.11 to prove that the unit sphere S^n is a topological submanifold of $\mathbb{R}^{n+1}$ of dimension n.

2.5. Smooth Submanifolds of Euclidean Space

We already know what is meant by a topological submanifold of $\mathbb{R}^n$ (Definition 1.5.3). We extend this notion to the smooth category. The model will be the standard imbedding $\mathbb{R}^r \hookrightarrow \mathbb{R}^n$, $r \leq n$, given by

$$(x^1, \dots, x^r) \mapsto (x^1, \dots, x^r, \underbrace{0, \dots, 0}_{n-r}).$$

Whenever we view $\mathbb{R}^r \subseteq \mathbb{R}^n$, we understand $\mathbb{R}^r$ to be the image of this imbedding.

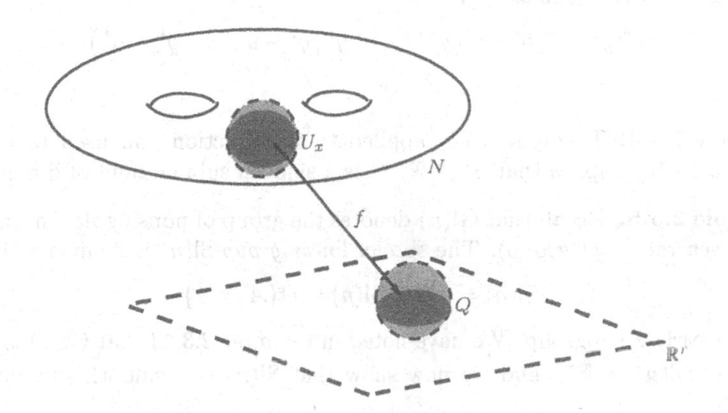

Figure 2.5.1. A submanifold is locally flat

Definition 2.5.1. Let $U \subseteq \mathbb{R}^n$ be open. A topological subspace $N \subseteq U$ is said to be a smooth submanifold of U of dimension $r \leq n$ if, for each $x \in N$, $\exists U_x \subseteq U$, an open neighborhood of x, and a diffeomorphism $f : U_x \to Q$ onto an open subset $Q \subseteq \mathbb{R}^n$ such that $f(N \cap U_x) = Q \cap \mathbb{R}^r$. The empty set $\emptyset \subset U$ is a smooth submanifold of every dimension $r \leq n$.

That is, if we view f as a local change of coordinates, we see that N looks locally like the flat imbedding of $\mathbb{R}^r$ in $\mathbb{R}^n$. This is illustrated in Figure 2.5.1.

Lemma 2.5.2. *If $N \subseteq U$ is a smooth submanifold of dimension r, then N is also a topological submanifold of dimension r of U.*

Proof. Let $x \in N$ and use the notation of Definition 2.5.1. Then $N \cap U_x$ is an open neighborhood of x in the relative topology of N in U. Since f carries $N \cap U_x$ homeomorphically onto the open subset $Q \cap \mathbb{R}^r$ of $\mathbb{R}^r$, it follows that N, with the relative topology, is locally Euclidean of dimension r, the inclusion map $i : N \hookrightarrow U$ being a topological imbedding. As a topological subspace of Euclidean space, N is Hausdorff and second countable. □

Theorem 2.5.3. *Let $U \subseteq \mathbb{R}^n$ and $V \subseteq \mathbb{R}^m$ be open and let $\Phi : U \to V$ be a smooth map of constant rank k. Let $q \in V$. Then $\Phi^{-1}(q)$ is a smooth submanifold of U of dimension $n - k$.*

Proof. If $\Phi^{-1}(q) = \emptyset$, the assertion is true by convention. Assume that this set is nonempty and let x be one of its points. Choose U_x to be the neighborhood W as in Theorem 2.4.6. Without loss of generality, we can replace W with $\widetilde{W}$ and $\Phi|W$ with $G \circ \Phi \circ F^{-1}$ on $\widetilde{W}$, all as in that theorem. That is, on U_x we assume that

$$\Phi(y^1, \dots, y^n) = (y^1, \dots, y^k, 0, \dots, 0)\,.$$

Thus

$$q = (a^1, \dots, a^k, 0, \dots, 0)$$

and $U_x \cap \Phi^{-1}(q)$ is the set of all points in U_x of the form

$$(a^1, \dots, a^k, y^{k+1}, \dots, y^n).$$

The desired diffeomorphism f will be

$$f(y^1, \dots, y^n) = \left(y^{k+1}, \dots, y^n, y^1 - a^1, \dots, y^k - a^k\right).$$

□

Example 2.5.4. Theorem 2.5.3, applied to the function you used to carry out Exercise 2.4.13, implies that $S^n \subset \mathbb{R}^{n+1}$ is a smooth submanifold of dimension n.

Example 2.5.5. Recall that $\mathrm{Gl}(n)$ denotes the group of nonsingular matrices over $\mathbb{R}$ (the general linear group). The *special linear group* $\mathrm{Sl}(n)$ is defined to be

$$\mathrm{Sl}(n) = \{A \in \mathrm{Gl}(n) \mid \det(A) = 1\}.$$

This is clearly a subgroup. We have noted in Example 2.3.13 that $\mathrm{Gl}(n)$ is an open subset of $\mathfrak{M}(n) = \mathbb{R}^{n^2}$ and we now show that $\mathrm{Sl}(n)$ is a smooth submanifold of $\mathrm{Gl}(n)$.

The determinant function $\det : \mathfrak{M}(n) \to \mathbb{R}$ is a polynomial, hence is smooth. We claim that $\det : \mathrm{Gl}(n) \to \mathbb{R}$ has constant rank 1. To prove this, we need to show that, for arbitrary $A \in \mathrm{Gl}(n)$, the linear map

$$d(\det)_A : T_A(\mathrm{Gl}(n)) \to T_{\det(A)}(\mathbb{R})$$

has rank 1. For this, we only need to find an infinitesimal curve $\langle s \rangle_A$ such that $\det_{*A} \langle s \rangle_A = \langle \det \circ s \rangle_{\det(A)} \neq 0$. Define $s(t)$ to be the matrix obtained by multiplying the first row of A by $1 + t$. Then $s(0) = A$ and the fact that $\mathrm{Gl}(n)$ is open in $\mathbb{R}^{n^2}$ implies that $s(t) \in \mathrm{Gl}(n)$ for $|t|$ small enough. Since $\det(s(t)) = (1 + t)\det(A)$ and $\det(A) \neq 0$, it is clear that $\langle \det \circ s \rangle_{\det(A)} \neq 0$. By the above remarks and Theorem 2.5.3, $\mathrm{Sl}(n) = \det^{-1}(1)$ is a smooth submanifold of $\mathrm{Gl}(n)$ of dimension $n^2 - 1$.

Definition 2.5.6. If $N \subseteq U$ is an r-dimensional, smooth submanifold of the open set $U \subseteq \mathbb{R}^n$, and if $x \in N$, a vector $v \in T_x(U)$ is tangent to N at x if, as an infinitesimal curve, $v = \langle s \rangle_x$ has a representative $s : (-\epsilon, \epsilon) \to U$ such that $s(t) \in N$, $-\epsilon < t < \epsilon$. The subset $T_x(N) \subseteq T_x(U)$, consisting of all vectors tangent to N at x, is called the tangent space to N at x.

Lemma 2.5.7. *If $N \subseteq U$ is an r-dimensional, smooth submanifold of the open set $U \subseteq \mathbb{R}^n$, and if $x \in N$, the tangent space $T_x(N)$ is an r-dimensional vector subspace of $T_x(U)$.*

Proof. For the "model" case $N = \mathbb{R}^r \subseteq \mathbb{R}^n$, the assertion is evident. Let all notation be as in Definition 2.5.1. If the smooth path $s : (-\epsilon, \epsilon) \to U$ has image in N and $s(0) = x$, then the diffeomorphism $f : U_x \to Q$ sends s to a smooth path $f \circ s$ in $\mathbb{R}^r$ through $f(x)$. Thus, the linear isomorphism

$$df_x : T_x(U_x) \to T_{f(x)}(Q)$$

carries $T_x(N)$ into the vector space $T_{f(x)}(Q \cap \mathbb{R}^r) = \mathbb{R}^r$. But $Q \cap \mathbb{R}^r$ is mapped onto $U_x \cap N$ by f^{-1} and the same argument shows that the inverse isomorphism $d(f^{-1})_{f(x)} = (df_x)^{-1}$ carries $T_{f(x)}(Q \cap \mathbb{R}^r)$ into $T_x(N)$. The assertion follows. □

Example 2.5.8. The subspace $T_I(\mathrm{Sl}(n)) \subset \mathfrak{M}(n)$ is the space of matrices of trace 0. To prove this, we first show that this tangent space is the kernel of $\det_{*I}$ and then that $\det_{*I} = \mathrm{tr} : \mathfrak{M}(n) \to \mathbb{R}$.

Both $T_I(\mathrm{Sl}(n))$ and $\ker(\det_{*I})$ have dimension $n^2 - 1$; so equality will follow if we show that the first is a subspace of the second. If $v \in T_I(\mathrm{Sl}(n))$ is thought of as an infinitesimal curve, then $v = \langle s \rangle_I$ where $s : (-\epsilon, \epsilon) \to \mathrm{Sl}(n)$, $s(0) = I$. Thus, $\det \circ s \equiv 1$ is a constant curve, and so

$$\det_{*I}(v) = \langle \det \circ s \rangle_1 = 0.$$

The linear functionals $\det_{*I}$ and tr on $\mathfrak{M}(n)$ will be equal if they agree on a basis. The matrices E_{ij}, $1 \le i, j \le n$, having 1 in the (i, j) position and 0s elsewhere, form a basis. As an infinitesimal curve at I, $E_{ij} = \langle s_{ij} \rangle_I$ where $s_{ij}(t) = I + tE_{ij}$. But

$$\det(s_{ij}(t)) = \begin{cases} 1 + t, & i = j, \\ 1 & i \neq j, \end{cases}$$

from which it follows that

$$\langle \det \circ s_{ij} \rangle_1 = \begin{cases} 1 = \mathrm{tr}(E_{ij}), & i = j, \\ 0 = \mathrm{tr}(E_{ij}), & i \neq j. \end{cases}$$

Another interesting example is the subgroup $\mathrm{O}(n) \subset \mathrm{Gl}(n)$ of orthogonal matrices. These are the matrices having as columns an orthonormal basis of $\mathbb{R}^n$. Equivalently, $A^{\mathrm{T}} = A^{-1}$, for each $A \in \mathrm{O}(n)$. Evidently, this is a subgroup, but less evident is the fact that it is a compact submanifold. The following exercise leads you through a proof of this fact and the determination of $T_I(\mathrm{O}(n)) \subset \mathfrak{M}(n)$.

Exercise 2.5.9. The map $\Phi : \mathrm{Gl}(n) \to \mathrm{Gl}(n)$, defined by $\Phi(Y) = Y^{\mathrm{T}} Y$, is smooth since its coordinate functions are quadratic polynomials. Prove the following:

(1) Relative to the standard identification $T_I(\mathrm{Gl}(n)) = \mathfrak{M}(n)$, the differential $d\Phi_I : T_I(\mathrm{Gl}(n)) \to T_I(\mathrm{Gl}(n))$ has the formula
$$d\Phi_I(A) = A^{\mathrm{T}} + A.$$
(2) The map Φ has constant rank $n(n+1)/2$.
(3) Using the above, conclude that the orthogonal group $\mathrm{O}(n) \subset \mathrm{Gl}(n)$ is a smooth, compact submanifold of dimension $n(n-1)/2$.
(4) Show that the vector subspace $T_I(\mathrm{O}(n)) \subset \mathfrak{M}(n)$ is the space of skew symmetric matrices.

Let U be an open subset of $\mathbb{R}^n$ and let $N \subseteq U$ be a smooth r-dimensional submanifold. Exactly as we did for S^n in Section 1.2, we can define the topological subspace $T(N) \subseteq \mathbb{R}^n \times \mathbb{R}^n = \mathbb{R}^{2n}$ to be the set of pairs (x, v), where $x \in N$ and $v \in T_x(N)$. We define

$$p : T(N) \to N$$

by $p(x, v) = x$. The fiber $p^{-1}(x) = T_x(N)$ is a vector space, $\forall\, x \in N$.

Definition 2.5.10. The structure $p : T(N) \to N$ is called the tangent bundle of the submanifold $N \subseteq U$. The total space of the bundle is $T(N)$, the base space is N, and p is called the bundle projection.

In the case of the model submanifold $\mathbb{R}^r \subseteq \mathbb{R}^n$,

$$T(\mathbb{R}^r) = \mathbb{R}^r \times \mathbb{R}^r \subseteq \mathbb{R}^n \times \mathbb{R}^n.$$

We will say that this bundle is *trivial* (*cf.* Definition 2.5.11).

If $Y \subseteq N$ is an open subset, then Y is also a smooth submanifold and $T(Y) = p^{-1}(Y)$ is the total space of the tangent bundle $p : T(Y) \to Y$.

In the following definition, the term "diffeomorphism" refers to a bijection between subsets of Euclidean spaces that, together with its inverse, is smooth in the sense of Definition 2.3.17.

Definition 2.5.11. The tangent bundle $p : T(N) \to N$ is trivial if there is a commutative diagram

$$\begin{array}{ccc} T(N) & \xrightarrow{\varphi} & N \times \mathbb{R}^r \\ {\scriptstyle p}\downarrow & & \downarrow{\scriptstyle p_N} \\ N & \xrightarrow[\text{id}]{} & N \end{array}$$

where p_N is projection onto the factor N and φ is a diffeomorphism with the property that, $\forall\, y \in N$, $\varphi|T_y(N) \to \{y\} \times \mathbb{R}^r$ is a linear isomorphism.

Definition 2.5.12. If the tangent bundle $p : T(N) \to N$ is trivial, the submanifold N is said to be parallelizable.

Definition 2.5.13. If $N \subseteq \mathbb{R}^n$ is a smooth submanifold as above, a smooth vector field on N is a map $X : N \to T(N) \subseteq \mathbb{R}^{2n}$ that is smooth as a map from the subset $N \subseteq \mathbb{R}^n$ into $\mathbb{R}^{2n}$ and satisfies $p \circ X = \text{id}_N$.

Exercise 2.5.14. Show that the r-dimensional submanifold N is parallelizable if and only if there are r smooth vector fields on N that, at each point of N, are linearly independent. Using this, show that the submanifolds $\mathrm{Sl}(n) \subset \mathbb{R}^{n^2}$ and $\mathrm{O}(n) \subset \mathbb{R}^{n^2}$ are parallelizable.

Exercise 2.5.15. Let $N \subseteq \mathbb{R}^n$ be a smooth submanifold of dimension r and let $x \in N$. Prove the following:

(1) There is an open neighborhood $Y \subset N$ of x such that

$$p : T(Y) \to Y$$

is trivial. (We say that the tangent bundle of N is *locally trivial.*)

(2) The subspace $T(N) \subseteq \mathbb{R}^{2n}$ is a smooth submanifold of dim $2r$.

Remark. When we get to the global theory, smooth manifolds will be defined without need of an ambient Euclidean space and their tangent bundles will be defined in an intrinsic way. A key property of these bundles will be local triviality.

2.6. Constructions of Smooth Functions

The main goal in this section is the proof of the following special case of the C^∞ Urysohn Lemma.

Theorem 2.6.1. *Let $K \subseteq \mathbb{R}^n$ be compact and let $U \subseteq \mathbb{R}^n$ be an open neighborhood of K. Then, there is a smooth map $f : \mathbb{R}^n \to [0,1]$ such that $f|K \equiv 1$ and* $\mathrm{supp}(f) \subset U$.

Define $h : \mathbb{R} \to [0,1)$ by

$$h(t) = \begin{cases} e^{-1/t^2}, & t \neq 0, \\ 0, & t = 0. \end{cases}$$

Exercise 2.6.2. Prove that the function h is C^∞, even at $t = 0$, where the derivatives are

$$\frac{d^n h}{dt^n}(0) = 0,$$

for all integers $n \geq 1$. (One says that h is C^∞-flat at $t = 0$.)

The graph of h is depicted in Figure 2.6.1.

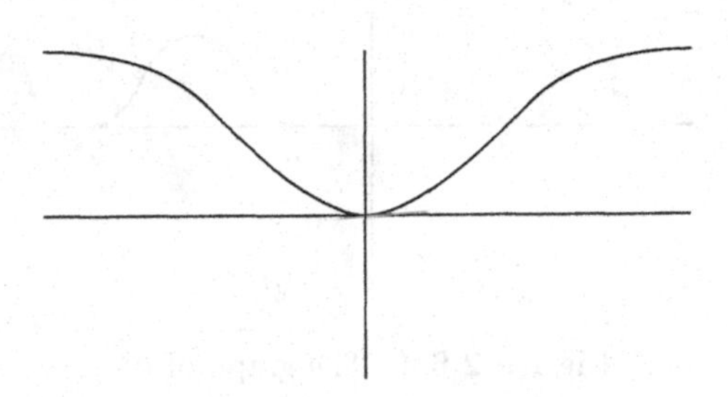

Figure 2.6.1. The graph of h

The functions $h_\pm : \mathbb{R} \to [0,1)$ are defined by

$$\begin{aligned} h_+(t) &= \begin{cases} e^{-1/t^2}, & t > 0, \\ 0, & t \leq 0, \end{cases} \\ h_-(t) &= \begin{cases} e^{-1/t^2}, & t < 0, \\ 0, & t \geq 0. \end{cases} \end{aligned}$$

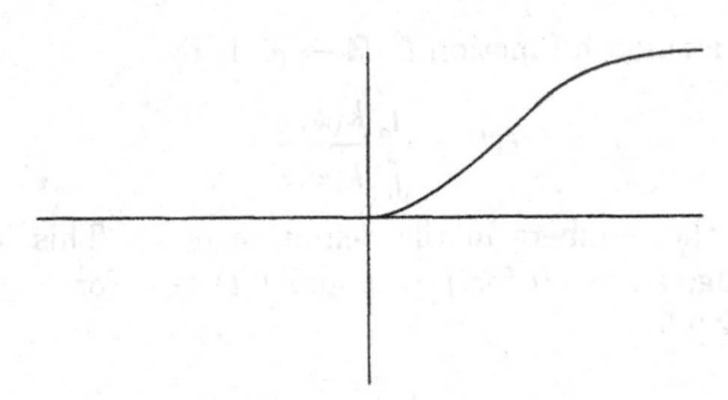

Figure 2.6.2. The graph of h_+

These are smooth and C^∞-flat at $t = 0$ by exactly the same reasons that h is. The graphs are depicted in Figures 2.6.2 and 2.6.3 respectively.

Combining these functions produces a C^∞ function $k : \mathbb{R} \to [0,1)$,

$$k(t) = h_-(t-b)h_+(t-a),$$

where $a < b$. This function is C^∞. It is positive exactly for $a < t < b$ (see Figure 2.6.4).

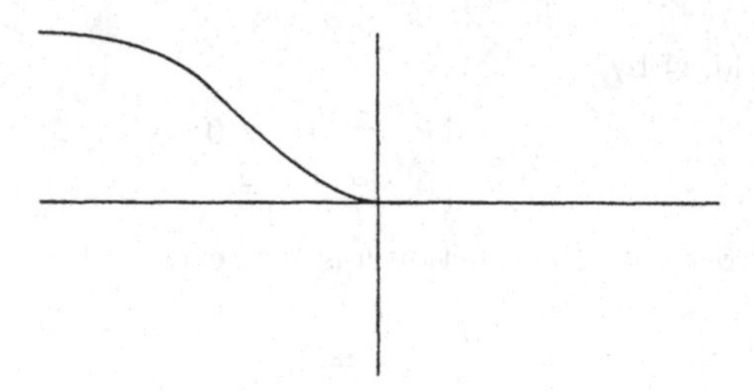

Figure 2.6.3. The graph of h_-

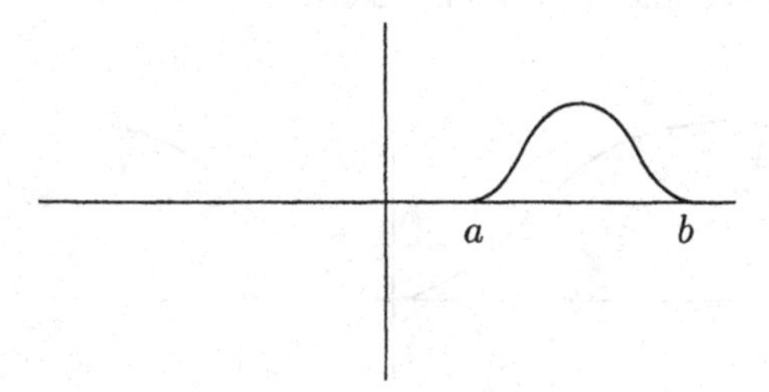

Figure 2.6.4. The graph of k

Lemma 2.6.3. *Let $A = (a_1, b_1) \times \cdots \times (a_n, b_n) \subset \mathbb{R}^n$ be an open, bounded, n-dimensional interval. Then there is a smooth function $g : \mathbb{R}^n \to [0, 1)$ such that $g > 0$ on A and $g|(\mathbb{R}^n \setminus A) \equiv 0$.*

Proof. The definition of k gives functions k_i, by taking $a = a_i$ and $b = b_i$ in that definition, $1 \le i \le n$. Then

$$g(x^1, x^2, \ldots, x^n) = k_1(x^1)k_2(x^2) \cdots k_n(x^n)$$

is as desired. □

Next we define a smooth function $\ell : \mathbb{R} \to [0, 1]$ by

$$\ell(t) = \frac{\int_a^t k(x)\, dx}{\int_a^b k(x)\, dx},$$

where a and b are the numbers in the definition of k. This function is weakly monotonic increasing, $\ell(t) \equiv 0$ for $t \le a$ and $\ell(t) \equiv 1$ for $t \ge b$. The graph is depicted in Figure 2.6.5.

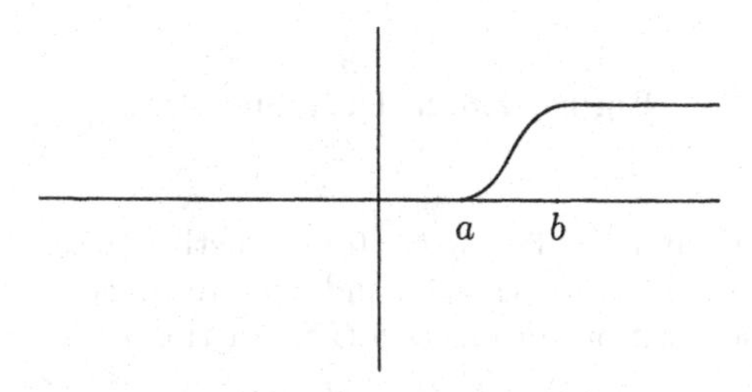

Figure 2.6.5. The graph of ℓ

Proof of Theorem 2.6.1. Let K and U be as in the statement of the theorem. For each $x \in K$, let A_x be an open, bounded, n-dimensional interval, centered at x and having $\overline{A}_x \subset U$. Apply Lemma 2.6.3 to obtain a smooth function $g_x : \mathbb{R}^n \to [0,1)$, strictly positive on A_x and vanishing identically outside of A_x. Since K is compact, it is covered by finitely many $A_{x_1}, \dots, A_{x_q}$. The function $G = g_{x_1} + \cdots + g_{x_q}$ is C^∞ on $\mathbb{R}^n$, strictly positive on K, and has

$$\operatorname{supp}(G) = \overline{A}_{x_1} \cup \cdots \cup \overline{A}_{x_q} \subset U.$$

Since K is compact, we can also find $\min(G|K) = \delta > 0$. In the definition of $\ell : \mathbb{R} \to [0,1]$, take $a = 0$ and $b = \delta$. Then $f = \ell \circ G : \mathbb{R}^n \to [0,1]$ is smooth, $\operatorname{supp}(f) \subset U$, and $f|K \equiv 1$. □

Exercise 2.6.4. Let $U \subseteq \mathbb{R}^n$ be open, $f : U \to \mathbb{R}$ smooth, and $p \in U$. Prove that there is a C^∞ function $\tilde{f} : \mathbb{R}^n \to \mathbb{R}$ such that $[f]_p = [\tilde{f}]_p$ in $\mathfrak{G}_p$. Conclude that, for arbitrary $p \in \mathbb{R}^n$, $\mathfrak{G}_p$ can be identified canonically with the set of germs at p of globally defined, smooth, real valued functions on $\mathbb{R}^n$.

Exercise 2.6.5. Let $C \subset \mathbb{R}^n$ be a closed subset, $U \subseteq \mathbb{R}^n$ an open neighborhood of C. Show that there is a smooth, nonnegative function $f : \mathbb{R}^n \to \mathbb{R}$ such that $f|C > 0$ and $\operatorname{supp}(f) \subset U$.

2.7. Smooth Vector Fields

Let $U \subseteq \mathbb{R}^n$ be an open subset. In particular, this is a smooth submanifold and we consider the set of smooth vector fields on U (Definition 2.5.13). This set is commonly denoted by $\mathfrak{X}(U)$. It is a vector space over $\mathbb{R}$ under the pointwise operations. If $X \in \mathfrak{X}(U)$, its value at a point $x \in U$ is commonly denoted by X_x. Throughout this section, the preferred way to think of $X_x \in T_x(U)$ will be as a derivative of the algebra $\mathfrak{G}_x$ of germs (Definition 2.2.15). For intuition, however, it is helpful to identify $T_x(U) = \mathbb{R}^n$, viewing X_x as a column vector. Thus,

$$X_x = \begin{bmatrix} f^1(x) \\ f^2(x) \\ \vdots \\ f^n(x) \end{bmatrix} = \sum_{i=1}^n f^i(x) D_{i,x},$$

where $D_{i,x} = \partial/\partial x^i\big|_x$ is the ith partial derivative at x. As x varies, each $f^i(x)$ varies smoothly, so

$$X = \begin{bmatrix} f^1 \\ f^2 \\ \vdots \\ f^n \end{bmatrix} = \sum_{i=1}^n f^i D_i,$$

where $D_i = \partial/\partial x^i$ and $f^i \in C^\infty(U)$, $1 \le i \le n$. Using the column vector interpretation, we can picture X as a smoothly varying field of directed line segments (arrows), issuing from points of U and possibly degenerating to zero length segments somewhere (Figure 2.7.1). While this makes sense only in Euclidean space, the second interpretation makes X a *first order differential operator on the space* $C^\infty(U)$ and will continue to make sense on manifolds.

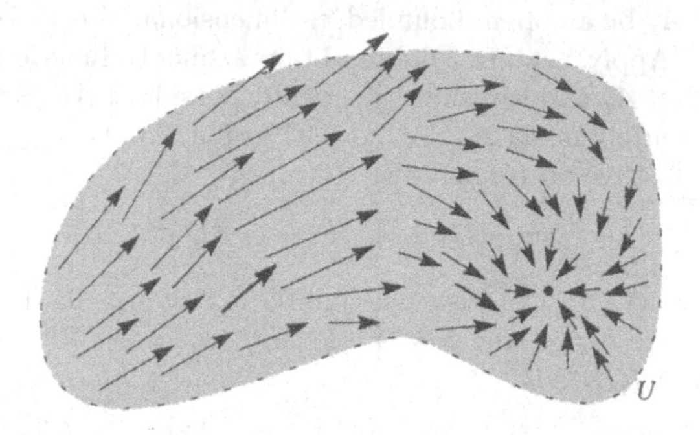

Figure 2.7.1. The vector field X can be pictured as a smooth field of arrows on U.

Interpreting $X \in \mathfrak{X}(U)$ as an operator, we write

$$X(g) = \sum_{i=1}^{n} f^i \frac{\partial g}{\partial x^i}, \quad \forall g \in C^\infty(U).$$

We will give an abstract, purely algebraic definition of the term "first order differential operator" and then show that such operators are exactly the elements of $\mathfrak{X}(U)$.

We view $C^\infty(U)$ as an algebra over $\mathbb{R}$ (commutative and associative, with unity the constant function 1). While $\mathfrak{X}(U)$ is a vector space over $\mathbb{R}$, it has more algebraic structure than that. For example, it is a *module* over the algebra $C^\infty(U)$ under pointwise scalar multiplication:

$$C^\infty(U) \times \mathfrak{X}(U) \to \mathfrak{X}(U),$$
$$(f, X) \mapsto fX,$$

where $(fX)_x = f(x)X_x$, for each $x \in U$. The formal definition of a module over an algebra is as follows.

Definition 2.7.1. Let F be an algebra over $\mathbb{R}$, $\mathcal{M}$ a vector space over $\mathbb{R}$. Suppose that there is an $\mathbb{R}$-bilinear map

$$F \times \mathcal{M} \to \mathcal{M},$$
$$(\rho, \mu) \mapsto \rho \cdot \mu$$

such that

$$(\rho\sigma) \cdot \mu = \rho \cdot (\sigma \cdot \mu), \quad \forall \rho, \sigma \in F, \quad \forall \mu \in \mathcal{M}.$$

Then $\mathcal{M}$ is said to be a *module* over F. If the algebra F has a unity $\iota \in F$, it is further required that

$$\iota \cdot \mu = \mu, \quad \forall \mu \in \mathcal{M}.$$

The module is *free* if there is a subset $\{\mu_\alpha\}_{\alpha \in \mathfrak{A}} \subset \mathcal{M}$, called a *basis* of $\mathcal{M}$ over F, such that each $\mu \in \mathcal{M}$ has a unique representation

$$\mu = \sum_{i=1}^{r} \rho_i \cdot \mu_{\alpha_i},$$

with coefficients $\rho_i \in F$, $1 \leq i \leq r$.

Example 2.7.2. It should be clear that $\mathfrak{X}(U)$ is a module over $C^\infty(U)$. Because U is an open subset of $\mathbb{R}^n$, this module is free with canonical basis $\{D_1, D_2, \dots, D_n\}$. While $\mathfrak{X}(U)$ will continue to be a module over $C^\infty(U)$ when U is an open subset of a manifold, it will not be true, generally, that this module is free. Thus, while a module over an algebra is analogous to a vector space over a field, one must not press this analogy too far.

We are going to give a deeper algebraic interpretation of $\mathfrak{X}(U)$. For this, we need some definitions.

Definition 2.7.3. Let F be an (associative) $\mathbb{R}$-algebra with unity. A linear map $\Delta : F \to F$ such that $\Delta(fg) = \Delta(f)g + f\Delta(g)$, $\forall\, f, g \in F$, is called a derivation of F. Derivations of $C^\infty(U)$ are also called first order differential operators. The set of derivations of F will be denoted by $\mathfrak{D}(F)$.

Lemma 2.7.4. *The set of derivations $\mathfrak{D}(F)$ is a vector space over $\mathbb{R}$ under the linear operations*

$$(a\Delta_1 + b\Delta_2)(f) = a\Delta_1(f) + b\Delta_2(f),$$

$\forall\, a, b \in \mathbb{R}$, $\forall\, \Delta_1, \Delta_2 \in \mathfrak{D}(F)$, $\forall\, f \in F$.

The proof of this is completely elementary and is left to the reader.

If the algebra F is commutative as well as associative, define an operation of "scalar" multiplication

$$F \times \mathfrak{D}(F) \to \mathfrak{D}(F),$$

by

$$(f\Delta)(g) = f(\Delta(g)), \quad \forall\, f, g \in F, \quad \forall\, \Delta \in \mathfrak{D}(F).$$

Exercise 2.7.5. If F is commutative, prove that $f\Delta$ is an element of $\mathfrak{D}(F)$, $\forall\, f \in F$, $\forall\, \Delta \in \mathfrak{D}(F)$. This makes $\mathfrak{D}(F)$ a module over the algebra F.

Lemma 2.7.6. *The space $\mathfrak{X}(U)$ is a $C^\infty(U)$-submodule of $\mathfrak{D}(C^\infty(U))$.*

Proof. Indeed, given $X \in \mathfrak{X}(U)$, write it as

$$X = \sum_{i=1}^{n} f^i D_i,$$

where $f^i \in C^\infty(U)$, $1 \leq i \leq n$. The Leibnitz rule for each partial derivative D_i implies that

$$X(hg) = X(h)g + hX(g),$$

$\forall\, h, g \in C^\infty(U)$. □

The main goal in this section is to prove that all derivations of $C^\infty(U)$ are vector fields.

Theorem 2.7.7. *The inclusion map $\mathfrak{X}(U) \hookrightarrow \mathfrak{D}(C^\infty(U))$ is surjective.*

Before commencing the proof, we consider another algebraic structure on the module of derivations of an algebra.

Definition 2.7.8. If $\Delta_1, \Delta_2 \in \mathcal{D}(F)$, then the Lie bracket

$$[\Delta_1, \Delta_2] : F \to F$$

is the operator defined by

$$[\Delta_1, \Delta_2](f) = \Delta_1(\Delta_2(f)) - \Delta_2(\Delta_1(f)),$$

$\forall f \in F$. This is also called the commutator of Δ_1 and Δ_2.

Exercise 2.7.9. Prove that the Lie bracket satisfies the following properties:

1. $[\Delta_1, \Delta_2] \in \mathcal{D}(F)$, $\forall \Delta_1, \Delta_2 \in \mathcal{D}(F)$;
2. the operation $[\cdot, \cdot] : \mathcal{D}(F) \times \mathcal{D}(F) \to \mathcal{D}(F)$ is $\mathbb{R}$- bilinear;
3. $[\Delta_1, \Delta_2] = -[\Delta_2, \Delta_1]$, $\forall \Delta_1, \Delta_2 \in \mathcal{D}(F)$ (anticommutativity);
4. $[\Delta_1, [\Delta_2, \Delta_3]] = [[\Delta_1, \Delta_2], \Delta_3] + [\Delta_2, [\Delta_1, \Delta_3]]$, $\forall \Delta_1, \Delta_2, \Delta_3 \in \mathcal{D}(F)$ (the Jacobi identity).

Thus, we can think of the operation

$$[\cdot, \cdot] : \mathcal{D}(F) \times \mathcal{D}(F) \to \mathcal{D}(F)$$

as a bilinear multiplication making $\mathcal{D}(F)$ into a kind of $\mathbb{R}$-algebra. This algebra is nonassociative, however, with the Jacobi identity replacing the associative law. The algebra is also anticommutative and does not have a unity.

Remark. One way to remember the Jacobi identity is to notice that, by this identity, the operator $[\Delta, \cdot] : \mathcal{D}(F) \to \mathcal{D}(F)$ is a derivation of the (nonassociative) algebra $\mathcal{D}(F)$, $\forall \Delta \in \mathcal{D}(F)$.

Definition 2.7.10. A nonassociative algebra having the properties in Exercise 2.7.9 is called a Lie algebra.

Exercise 2.7.11. It will follow from Theorem 2.7.7 that $\mathfrak{X}(U)$ is a Lie algebra. Here, you are to prove directly that

$$[X, Y] \in \mathfrak{X}(U), \quad \forall X, Y \in \mathfrak{X}(U).$$

For this, note that the composed operators $X \circ Y$ and $Y \circ X$ are *second order* differential operators, but show by direct calculation that the commutator $X \circ Y - Y \circ X$ is first order.

Lemma 2.7.12. *If $\iota \in F$ is the unity, then $\Delta(c\iota) = 0$, $\forall \Delta \in \mathcal{D}(F)$ and $\forall c \in \mathbb{R}$.*

The proof is exactly like that of Lemma 2.2.19.

In order to prove Theorem 2.7.7, we must establish the reverse inclusion

$$\mathcal{D}(C^\infty(U)) \subseteq \mathfrak{X}(U).$$

For this, we need to show that a derivation of $C^\infty(U)$ can be *localized* to a derivative of the germ algebra $\mathfrak{G}_x$, at each $x \in U$. This is by no means evident.

Lemma 2.7.13 (Key Lemma). *Let $\Delta \in \mathcal{D}(C^\infty(U))$, $f \in C^\infty(U)$, and suppose that $V \subseteq U$ is an open set such that $f|V \equiv 0$. Then $\Delta(f)|V \equiv 0$.*

Proof. Let $x \in V$. By Theorem 2.6.1, we find $\varphi \in C^\infty(U)$ such that

$$\varphi(x) = 0,$$
$$\varphi|(U \smallsetminus V) \equiv 1.$$

Indeed, since $\{x\}$ is compact, we find $\psi \in C^\infty(U)$ such that $\psi(x) = 1$ and $\text{supp}(\psi) \subset V$. Then $\varphi = 1 - \psi$ is as desired. Since $f|V \equiv 0$, we see that $\varphi f = f$. Thus,

$$\Delta(f) = \Delta(\varphi f) = \Delta(\varphi)f + \varphi\Delta(f).$$

Hence

$$\Delta(f)(x) = \Delta(\varphi)(x)f(x) + \varphi(x)\Delta(f)(x).$$

But $f(x) = 0 = \varphi(x)$, and so $\Delta(f)(x) = 0$. Since $x \in V$ is arbitrary, $\Delta(f)|V \equiv 0$. □

Corollary 2.7.14. *Let $\Delta \in \mathcal{D}(C^\infty(U))$, $f \in C^\infty(U)$, and $x \in U$. Then $\Delta(f)(x)$ depends only on Δ and the germ $[f]_x \in \mathfrak{G}_x$.*

Proof. Let $f, g \in C^\infty(U)$ have the same germ $[f]_x = [g]_x$. Choose an open neighborhood $W \subseteq U$ of x such that $f|W = g|W$. By the Key Lemma 2.7.13,

$$(\Delta(f) - \Delta(g))|W = \Delta(f - g)|W \equiv 0,$$

and so $\Delta(f)(x) = \Delta(g)(x)$. □

Given $\omega \in \mathfrak{G}_x$, $x \in U$, there exists $f \in C^\infty(U)$ such that $\omega = [f]_x$. This is by Exercise 2.6.4. This allows us to make the following definition.

Definition 2.7.15. Given $\Delta \in \mathcal{D}(C^\infty(U))$ and $x \in U$, $\Delta_x : \mathfrak{G}_x \to \mathbb{R}$ is given by

$$\Delta_x(\omega) = \Delta(f)(x), \quad \forall \omega = [f]_x \in \mathfrak{G}_x,$$

where $f \in C^\infty(U)$.

By the above discussion, it is clear that Δ_x is well-defined, $\forall x \in U$, $\forall \Delta \in \mathcal{D}(C^\infty(U))$.

Proposition 2.7.16. *If $\Delta \in \mathcal{D}(C^\infty(U))$ and $x \in U$, then $\Delta_x \in T_x(U)$.*

Proof. Let $[f]_x, [g]_x \in \mathfrak{G}_x$, where $f, g \in C^\infty(U)$. Then,

$$\begin{aligned}
\Delta_x([f]_x[g]_x) &= \Delta_x[fg]_x \\
&= \Delta(fg)(x) \\
&= (\Delta(f)g + f\Delta(g))(x) \\
&= \Delta(f)(x)g(x) + f(x)\Delta(g)(x) \\
&= \Delta_x[f]_x e_x[g]_x + e_x[f]_x \Delta_x[g]_x.
\end{aligned}$$

It is clear that $\Delta_x : \mathfrak{G}_x \to \mathbb{R}$ is linear; so $\Delta_x \in T_x(U)$. □

Given $\Delta \in \mathcal{D}(C^\infty(U))$, define $\widetilde{\Delta} : U \to T(U)$ by $\widetilde{\Delta}(x) = \Delta_x \in T_x(U)$. This function satisfies $p \circ \widetilde{\Delta} = \text{id}_U$. (Maps with this property are called *sections* of the tangent bundle.) If this section is smooth, then $\widetilde{\Delta} \in \mathfrak{X}(U)$. Write

$$\widetilde{\Delta} = \sum_{i=1}^n \tilde{f}^i D_i,$$

remarking that the smoothness of $\widetilde{\Delta}$ is equivalent to

$$\tilde{f}^i \in C^\infty(U), \quad 1 \le i \le n.$$

The following proposition completes the proof of Theorem 2.7.7.

Proposition 2.7.17. *If $\Delta \in \mathcal{D}(C^\infty(U))$, then $\widetilde{\Delta} \in \mathfrak{X}(U)$ and, as a derivation of $C^\infty(U)$, the vector field $\widetilde{\Delta}$ is identical with Δ.*

Proof. Consider the coordinate functions $x^i \in U$, $1 \le i \le n$. Let

$$f^i = \Delta(x^i) \in C^\infty(U).$$

Then

$$f^i = \left(\sum_{j=1}^{n} \tilde{f}^j D_j\right)(x^i) = \tilde{f}^i,$$

$1 \le i \le n$. □

From now on, we think of vector fields either as sections of $T(U)$ or as derivations of $C^\infty(U)$, but we denote the Lie algebra and $C^\infty(U)$-module of all such fields only by $\mathfrak{X}(U)$.

Example 2.7.18. One is often interested in certain Lie subalgebras of $\mathfrak{X}(U)$. We give here an example on the group manifold $\mathrm{Gl}(n)$ to which we will be returning later.

Write $T(\mathrm{Gl}(n)) = \mathrm{Gl}(n) \times \mathfrak{M}(n)$. A vector field $X \in \mathfrak{X}(\mathrm{Gl}(n))$ can be written $X_Q = (Q, \widetilde{X}_Q)$, where $\widetilde{X} : \mathrm{Gl}(n) \to \mathfrak{M}(n)$ is smooth. Such a field is *left-invariant* if, for each $P \in \mathrm{Gl}(n)$,

$$L_{P*}(X) = X.$$

This means that $L_{P*Q}(X_Q) = X_{PQ}$, for all $Q \in \mathrm{Gl}(n)$, or equivalently, that $P\widetilde{X}_Q = \widetilde{X}_{PQ}$. In particular, such a field is completely determined by its value $A = \widetilde{X}_I$ at I. We have

$$\widetilde{X}_P = P\widetilde{X}_I = PA = R_A(P), \quad \forall\, P \in \mathrm{Gl}(n).$$

That is, a left-invariant vector field is identified with

$$R_A : \mathrm{Gl}(n) \to \mathfrak{M}(n),$$

where $A \in \mathfrak{M}(n)$. As a vector space, the set $\mathfrak{gl}(n) \subset \mathfrak{X}(\mathrm{Gl}(n))$ of left-invariant vector fields is identified with $\mathfrak{M}(n)$ via $R_A \leftrightarrow A$.

In $\mathfrak{M}(n)$, define the bracket to be the usual commutator of matrices

$$[A, B] = AB - BA, \quad \forall\, A, B \in \mathfrak{M}(n).$$

Exercise 2.7.19. Prove that the commutator operation makes $\mathfrak{M}(n)$ into a Lie algebra and that the canonical identification $\mathfrak{M}(n) = \mathfrak{gl}(n)$ turns the commutator into the Lie bracket of vector fields. More precisely,

$$[R_A, R_B] = R_{[A,B]}, \quad \forall\, A, B \in \mathfrak{M}(n);$$

so $\mathfrak{gl}(n)$ is a Lie subalgebra of $\mathfrak{X}(\mathrm{Gl}(n))$ isomorphic to the Lie algebra $\mathfrak{M}(n)$ of $n \times n$ matrices under the commutator bracket.

As noted earlier, a diffeomorphism $z : U \to V$ onto an open subset $V \subseteq \mathbb{R}^n$ can be thought of as a change of coordinates. Recall that such a change of coordinates translates formulas in differential calculus to new formulas relative to the new coordinates. In the present context, vector fields $X \in \mathfrak{X}(U)$, viewed as first order differential operators, are carried to vector fields $z_*(X) \in \mathfrak{X}(V)$. One can think of this as a change of formula for the operator X in terms of the new coordinate system $\{V, z^1, \dots, z^n\}$.

We analyze this "push forward" of vector fields more carefully. Recall the pullback operation

$$z^* : C^\infty(V) \to C^\infty(U),$$
$$z^*(f) = f \circ z.$$

The following is completely straightforward to check.

Lemma 2.7.20. *If $z : U \to V$ is a diffeomorphism between open subsets of $\mathbb{R}^n$, then $z^* : C^\infty(V) \to C^\infty(U)$ is an isomorphism of algebras.*

Definition 2.7.21. If $z : U \to V$ is a diffeomorphism between open subsets of $\mathbb{R}^n$, then $z_* : \mathfrak{X}(U) \to \mathfrak{X}(V)$ is defined by setting

$$z_*(X)(f) = (z^{-1})^*(X(z^*(f))) = X(f \circ z) \circ z^{-1},$$

for all $X \in \mathfrak{X}(U)$ and all $f \in C^\infty(V)$.

The fact that $z_*(X) \in \mathfrak{X}(V)$ is elementary, as is the following. These are left as exercises for the reader.

Lemma 2.7.22. *If $z : U \to V$ is a diffeomorphism between open subsets of $\mathbb{R}^n$, then $z_* : \mathfrak{X}(U) \to \mathfrak{X}(V)$ is an isomorphism of Lie algebras.*

Exercise 2.7.23. Show that, if $X \in \mathfrak{X}(U)$ is viewed as a smooth section of the tangent bundle $p : T(U) \to U$, then the section $Z = z_*(X)$ of the tangent bundle $\pi : T(V) \to V$ is given by $Z_\zeta = dz_{z^{-1}(\zeta)}(X_{z^{-1}(\zeta)})$, $\forall \zeta \in V$.

2.8. Local Flows

At this point we have two ways of viewing a vector field $X \in \mathfrak{X}(U)$. It is a smooth section of the tangent bundle and it is a derivation of the algebra $C^\infty(U)$. In this section, we view X as a field of infinitesimal curves and show that, in this guise, X is equivalent to a system of ordinary differential equations (O.D.E.) on U.

Definition 2.8.1. Let $s : (a, b) \to U$ be smooth. The velocity vector of s at $t_0 \in (a, b)$ is

$$\dot{s}(t_0) = s_{*t_0}\left(\left.\frac{d}{dt}\right|_{t_0}\right) \in T_{s(t_0)}(U).$$

Remark. If $\tau = t+t_0$, $a-t_0 < t < b-t_0$, then $\sigma(t) = s(\tau)$ has velocity $\dot{\sigma}(t) = \dot{s}(\tau) = \dot{s}(t+t_0)$. In particular, $\dot{s}(t_0) = \dot{\sigma}(0) = \langle\sigma\rangle_{s(t_0)}$. We will write this infinitesimal curve as $\langle s\rangle_{s(t_0)}$.

Definition 2.8.2. The map $\dot{s} : (a, b) \to T(U)$ is called the velocity field of the smooth curve $s : (a, b) \to U$.

Remark that $p \circ \dot{s} = s$, where $p : T(U) \to U$ is the bundle projection. Also remark that $\dot{s}$ is smooth.

Definition 2.8.3. Let $X \in \mathfrak{X}(U)$ and $x_0 \in U$. An integral curve to X through x_0 is a smooth curve $s : (-\delta, \epsilon) \to U$, defined for suitable $\delta, \epsilon > 0$, such that $s(0) = x_0$ and $\dot{s}(t) = X_{s(t)}$, $-\delta < t < \epsilon$.

Suppose that s is an integral curve to $X \in \mathfrak{X}(U)$ through $x_0 \in U$. Write

$$X = \sum_{i=1}^{n} f^i D_i$$

and

$$s(t) = (x^1(t), \dots, x^n(t)).$$

At each $t \in (-\delta, \epsilon)$, the Jacobian matrix of s is

$$Js(t) = \begin{bmatrix} \frac{dx^1}{dt}(t) \\ \vdots \\ \frac{dx^n}{dt}(t) \end{bmatrix}.$$

The vector $\frac{d}{dt}\big|_t \in T_t(-\delta, \epsilon) = \mathbb{R}^1$ is the canonical basis element $1 \in \mathbb{R}^1$. So

$$\begin{aligned} ds_t\left(\frac{d}{dt}\bigg|_t\right) &= \begin{bmatrix} \frac{dx^1}{dt}(t) \\ \vdots \\ \frac{dx^n}{dt}(t) \end{bmatrix} \in \mathbb{R}^n \\ &= \sum_{i=1}^{n} \frac{dx^i}{dt}(t) D_{i,s(t)} \in T_{s(t)}(U). \end{aligned}$$

But

$$X_{s(t)} = \sum_{i=1}^{n} f^i(x^1(t), \dots, x^n(t)) D_{i,s(t)}.$$

Thus, s is an integral curve to X if and only if

$$\frac{dx^i}{dt}(t) = f^i(x^1(t), \dots, x^n(t)),$$

$-\delta < t < \epsilon$, $1 \le i \le n$. This is an (autonomous) system of O.D.E. with solution s subject to the initial condition $s(0) = x_0$. The existence and uniqueness of integral curves is guaranteed by the following theorem.

Theorem 2.8.4. *Let $V \subseteq \mathbb{R}^r$ and $U \subseteq \mathbb{R}^n$ be open, let $c > 0$, let*

$$f^i \in C^\infty((-c, c) \times V \times U), \quad 1 \le i \le n,$$

and consider the system of O.D.E. with parameters $b = (b^1, \dots, b^r) \in V$

$$(*) \qquad \frac{dx^i}{dt} = f^i(t, b, x^1(t, b), \dots, x^n(t, b)), \quad 1 \le i \le n.$$

If $a = (a^1, \dots, a^n) \in U$, there are smooth functions

$$x^i(t, b), \quad 1 \le i \le n,$$

defined on some nondegenerate interval $[-\delta, \epsilon]$ about 0, that satisfy the system $(*)$ and the initial condition

$$x^i(0, b) = a^i, \quad 1 \le i \le n.$$

Furthermore, if the functions $\tilde{x}_i(t, b)$, $1 \le i \le n$, give another solution, defined on $[-\tilde{\delta}, \tilde{\epsilon}]$ and satisfying the same initial condition, then these solutions agree on $[-\delta, \epsilon] \cap [-\tilde{\delta}, \tilde{\epsilon}]$. Finally, if we write these solutions as $x^i = x^i(t, b, a)$ (in order to emphasize the dependence on the initial condition), there is a neighborhood W of *a in U, a neighborhood B of b in V, and a choice of $\epsilon > 0$ such that the solutions $x^i(t, z, x)$ are defined and smooth on the open set $(-\epsilon, \epsilon) \times B \times W \subseteq \mathbb{R}^{n+r+1}$.*

This is the well-known theorem giving the existence, uniqueness, and smooth dependence on initial conditions and parameters of solutions of systems of ordinary differential equations. The proof is given in Appendix C (where the smooth dependence on initial conditions will require some elementary facts from calculus in Banach spaces).

Definition 2.8.5. The system $(*)$ is autonomous if the functions f^i do not depend on t, $1 \leq i \leq n$.

The problem of finding integral curves to a vector field is the problem of solving an *autonomous* system of O.D.E. *without parameters.* The proof of the existence, uniqueness, and smooth dependence of these curves on the initial conditions, however, involves an inductive step that requires the general formulation of Theorem 2.8.4. From now on, we consider only autonomous systems without parameters.

Corollary 2.8.6. *If $X \in \mathfrak{X}(U)$ and $x \in U$, then there is an integral curve to X through x. Any two such curves agree on their common domain.*

Definition 2.8.7. A local flow Φ around $x_0 \in U$ is a smooth map

$$\Phi : (-\epsilon, \epsilon) \times W \to U$$

(written $\Phi(t, x) = \Phi_t(x)$), where W is a suitable open neighborhood of x_0 in U, such that

1. $\Phi_0 : W \to U$ is the inclusion $W \hookrightarrow U$;
2. $\Phi_{t_1+t_2}(x) = \Phi_{t_1}(\Phi_{t_2}(x))$ whenever both sides of this equation are defined.

If $z \in U$, the flow line through z is the curve $\sigma(t) = \Phi_t(z)$, $-\epsilon < t < \epsilon$.

Theorem 2.8.8. *Let $X \in \mathfrak{X}(U)$ and $x_0 \in U$. Then there is a local flow around x_0 such that the flow lines are integral curves to X. Two such local flows agree on their common domain.*

Proof. By Theorem 2.8.4, we find an open neighborhood W of x_0 in U and a number $\epsilon > 0$ such that the integral curve $s_z(t)$ through z is defined for $-\epsilon < t < \epsilon$ and for all $z \in W$. By the smooth dependence on initial conditions, we define a smooth map

$$\Phi : (-\epsilon, \epsilon) \times W \to U$$

by

$$\Phi(t, z) = s_z(t).$$

Theorem 2.8.4 also assures us that, if $\widetilde{\Phi}$ is another local flow around x_0 with flow lines integral to X, then Φ and $\widetilde{\Phi}$ agree on their common domain. We must show that Φ is a local flow.

Since $\Phi_0(z) = s_z(0) = z$, it is clear that Φ_0 is the inclusion map $W \hookrightarrow U$.

Fix $t_0 \in (-\epsilon, \epsilon)$ and $z \in W$ and define the curves

$$s(t) = s_z(t + t_0) = \Phi_{t+t_0}(z)$$

and

$$\sigma(t) = s_{s_z(t_0)}(t) = \Phi_t(s_z(t_0)) = \Phi_t(\Phi_{t_0}(z)).$$

These are defined for small values of t and are both integral to X. They satisfy $s(0) = s_z(t_0)$ and $\sigma(0) = s_{s_z(t_0)}(0) = s_z(t_0)$; so $s(t) = \sigma(t)$ whenever both sides are defined. That is, $\Phi_{t+t_0}(z) = \Phi_t(\Phi_{t_0}(z))$ whenever both sides are defined. □

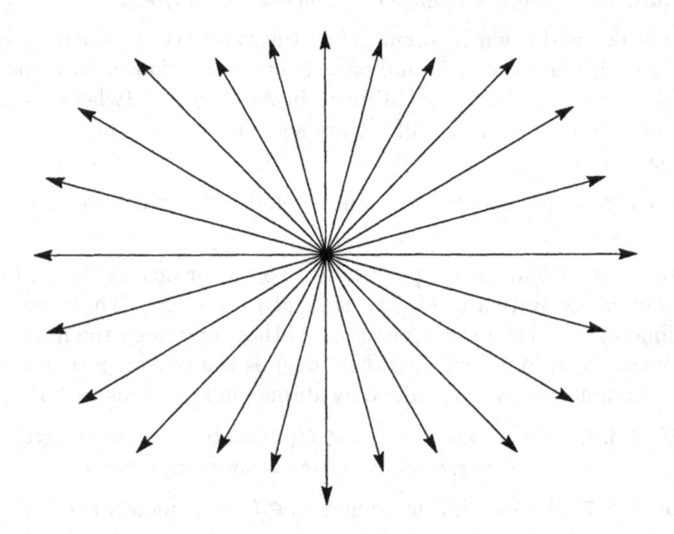

Figure 2.8.1. Flow lines for $X = x^1 D_1 + x^2 D_2$

Definition 2.8.9. The local flow Φ associated to $X \in \mathfrak{X}(U)$ as in Theorem 2.8.8 is said to be generated by X. Also, the vector field X is called the infinitesimal generator of the local flow Φ.

Example 2.8.10. Let $U = \mathbb{R}^2$ and $X = x^1 D_1 + x^2 D_2$. The integral curves $s(t) = (x^1(t), x^2(t))$ will satisfy

$$\frac{dx^1}{dt} = x^1,$$
$$\frac{dx^2}{dt} = x^2.$$

The solution curve with initial condition $s(0) = (a^1, a^2)$ is

$$s(t) = (a^1 e^t, a^2 e^t)$$

(see Figure 2.8.1). All of these solutions are defined for $-\infty < t < \infty$. Remark that

$$\Phi_{t_1+t_2}(a^1, a^2) = (a^1 e^{t_1+t_2}, a^2 e^{t_1+t_2}) = \Phi_{t_1}(\Phi_{t_2}(a^1, a^2)).$$

This curve is stationary for $(a^1, a^2) = (0, 0)$ and in all other cases follows a radial trajectory out of the origin, but is not parametrized linearly. The "speed" of the trajectory increases proportionally with the distance from the origin.

Example 2.8.11. Let $U = \mathbb{R}^2$ and $X = x^1 D_1 - x^2 D_2$. The integral curves $s(t) = (x^1(t), x^2(t))$ will satisfy

$$\frac{dx^1}{dt} = x^1,$$
$$\frac{dx^2}{dt} = -x^2.$$

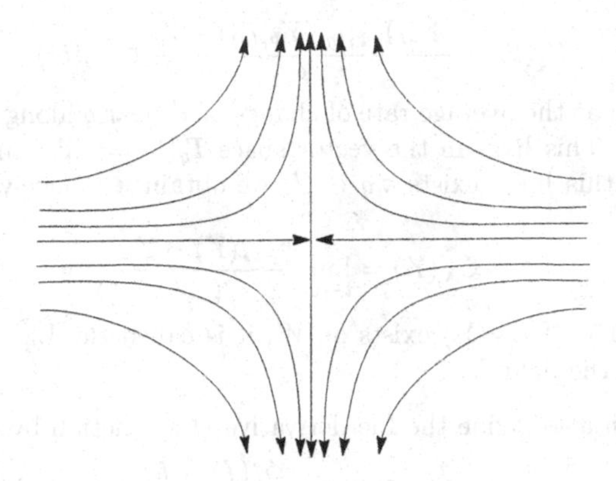

Figure 2.8.2. Flow lines for $X = x^1 D_1 - x^2 D_2$

The solution curve with initial condition $s(0) = (a^1, a^2)$ is

$$s(t) = (a^1 e^t, a^2 e^{-t})$$

(see Figure 2.8.2). All of these solutions are defined for $-\infty < t < \infty$. Again

$$\Phi_{t_1+t_2}(a^1, a^2) = (a^1 e^{t_1+t_2}, a^2 e^{-t_1-t_2}) = \Phi_{t_1}(\Phi_{t_2}(a^1, a^2)).$$

This curve is stationary for $(a^1, a^2) = (0,0)$. Trajectories of points on the x^2-axis (other than the origin) stay on the x^2-axis and head toward the origin. Trajectories of points on the x^1-axis (other than the origin) stay on that axis and head away from the origin. The remaining trajectories follow hyperbolic paths asymptotic to the coordinate axes.

Exercise 2.8.12. As in Examples 2.8.10 and 2.8.11, discuss the vector field $X = x^2 D_1 - x^1 D_2$ on $\mathbb{R}^2$. In particular, sketch the flow lines and show that they are defined for $-\infty < t < \infty$.

Exercise 2.8.13. On $U = \mathbb{R}$, consider the vector field $X = e^x \frac{d}{dx}$. Let $a \in \mathbb{R}$ and compute the integral curve s_a to X through a. Be sure to find the *largest* open interval (η, ω) on which $s_a(t)$ is defined.

Exercise 2.8.14. Let $X \in \mathfrak{X}(U)$ generate a local flow

$$\Phi : (-\epsilon, \epsilon) \times W \to U.$$

Prove that, for each $q \in W$,

$$(\Phi_t)_{*q}(X_q) = X_{\Phi_t(q)}, \quad -\epsilon < t < \epsilon.$$

So far, we have used vector fields to differentiate functions. Now we will show how a vector field can be used to differentiate another vector field. Let $X \in \mathfrak{X}(U)$,

$q \in U$, and let $\Phi : (-\epsilon, \epsilon) \times W \to U$ be a local flow about q generated by X. Let $Y \in \mathfrak{X}(U)$. Generally, the vector

$$(\Phi_{-t})_{*\Phi_t(q)}(Y_{\Phi_t(q)}) \in T_q(U),$$

defined for $-\epsilon < t < \epsilon$, differs from Y_q, although they are equal if $Y = X$ (Exercise 2.8.14). The difference quotient

$$Z_q(t) = \frac{(\Phi_{-t})_{*\Phi_t(q)}(Y_{\Phi_t(q)}) - Y_q}{t} \in T_q(U)$$

can be thought of as the average rate of change of Y near q along the integral curve to X through q. This lives in the vector space $T_q(U) = \mathbb{R}^n$, and so $\lim_{t\to 0} Z_q(t)$ makes sense. If this limit exists $\forall q \in U$, we obtain a (conceivably not smooth) vector field

$$\mathcal{L}_X(Y) = \lim_{t\to 0} \frac{\Phi_{-t*}(Y) - Y}{t}.$$

Definition 2.8.15. If $\mathcal{L}_X(Y)$ exists on W, it is called the Lie derivative (on W) of the field Y by the field X.

Remark. One can also define the Lie derivative of a function by the formula

$$\mathcal{L}_X(f) = \lim_{t\to 0} \frac{\Phi_t^*(f) - f}{t}.$$

The use of Φ_t instead of Φ_{-t} is due to the fact that Φ_t^* *pulls* functions *back*, while Φ_{t*} *pushes* vector fields *forward*. The reader should have no trouble seeing that $\mathcal{L}_X(f) = X(f)$.

Theorem 2.8.16. *If $X, Y \in \mathfrak{X}(U)$, the Lie derivative of Y by X is defined and smooth throughout U and*

$$\mathcal{L}_X(Y) = [X, Y].$$

The proof requires a couple of lemmas. We will fix $q \in U$ and a neighborhood W of q so that $\Phi : (-\epsilon, \epsilon) \times W \to U$ is defined. Write

$$Z_q(t) = \sum_{i=1}^n \zeta^i(t) D_{i,q}$$

and remark that $\lim_{t\to 0} Z_q(t)$ exists if and only if $\lim_{t\to 0} \zeta^i(t)$ exists, for $1 \le i \le n$.

Lemma 2.8.17. *The limit $A_q = \lim_{t\to 0} Z_q(t)$ exists if and only if, for each $f \in C^\infty(U)$, $\lim_{t\to 0} Z_q(t)(f)$ exists, in which case this limit is $A_q(f)$.*

Proof. Assume that the limit $A_q = \sum_{i=1}^n a^i D_{i,q}$ exists. That is,

$$a^i = \lim_{t\to 0} \zeta^i(t),$$

$1 \le i \le n$. Then, for arbitrary $f \in C^\infty(U)$, it is clear that

$$\lim_{t\to 0} Z_q(t)(f) = \lim_{t\to 0} \sum_{i=1}^n \zeta^i(t) \frac{\partial f}{\partial x^i}(q) = \sum_{i+1}^n a^i \frac{\partial f}{\partial x^i}(q) = A_q(f)$$

exists.

For the converse, suppose that, for each $f \in C^\infty(U)$,

$$\lim_{t\to 0} Z_q(t)(f) = \lim_{t\to 0} \sum_{i=1}^n \zeta^i(t) \frac{\partial f}{\partial x^i}(q)$$

exists. Choose $f = x^j$ and define

$$a^j = \lim_{t \to 0} Z_q(t)(x^j) = \lim_{t \to 0} \zeta^j(t),$$

$1 \le j \le n$. Thus, $\lim_{t \to 0} Z_q(t) = A_q$ exists and, as noted above, it follows that

$$\lim_{t \to 0} Z_q(t)(f) = A_q(f),$$

for arbitrary $f \in C^\infty(U)$. □

Lemma 2.8.18. *Given $f \in C^\infty(U)$, there is a function*

$$g \in C^\infty((-\epsilon, \epsilon) \times W)$$

such that

1. $f(\Phi_{-t}(x)) = f(x) - tg(t, x)$, $\forall x \in W$, $-\epsilon < t < \epsilon$;
2. $X_x(f) = g(0, x)$, $\forall x \in W$.

Proof. Define

$$h(t, x) = f(\Phi_{-t}(x)) - f(x);$$

so $h \in C^\infty((-\epsilon, \epsilon) \times W)$ and $h(0, x) = 0$. To simplify notation, we denote the partial of h with respect to t by $\dot{h}(t, x)$. Define $g \in C^\infty((-\epsilon, \epsilon) \times W)$ by

$$g(t, x) = -\int_0^1 \dot{h}(tu, x)\, du.$$

Then

$$\begin{aligned} -tg(t, x) &= \int_0^1 \dot{h}(tu, x) t\, du \\ &= \int_0^t \dot{h}(v, x)\, dv \\ &= h(t, x) - h(0, x) \\ &= f(\Phi_{-t}(x)) - f(x), \end{aligned}$$

giving the first assertion. For the second, consider

$$\begin{aligned} g(0, x) &= \lim_{t \to 0} g(t, x) \\ &= \lim_{t \to 0} \frac{f(\Phi_{-t}(x)) - f(x)}{-t} \\ &= \lim_{t \to 0} \frac{f(\Phi_t(x)) - f(x)}{t} \\ &= X_x(f). \end{aligned}$$

□

Proof of Theorem 2.8.16. Let $f \in C^\infty(U)$ and let g be as in the preceding lemma. Let $g_t : W \to U$ be given by $g_t(x) = g(t, x)$. Thus, $X(f) = g_0 = \lim_{t \to 0} g_t$, and so

$$\lim_{t \to 0} Y_{\Phi_t(q)}(g_t) = Y_q(X(f)).$$

We now compute

$$\begin{aligned}
\lim_{t\to 0} Z_q(t)(f) &= \lim_{t\to 0}\left(\frac{(\Phi_{-t})_{*\Phi_t(q)}(Y_{\Phi_t(q)}) - Y_q}{t}\right)(f) \\
&= \lim_{t\to 0}\frac{Y_{\Phi_t(q)}(f\circ\Phi_{-t}) - Y_q(f)}{t} \\
&= \lim_{t\to 0}\frac{Y_{\Phi_t(q)}(f - tg_t) - Y_q(f)}{t} \\
&= \lim_{t\to 0}\frac{Y_{\Phi_t(q)}(f) - Y_q(f)}{t} - \lim_{t\to 0} Y_{\Phi_t(q)}(g_t) \\
&= \lim_{t\to 0}\frac{Y(f)(\Phi_t(q)) - Y(f)(q)}{t} - \lim_{t\to 0} Y_{\Phi_t(q)}(g_t) \\
&= X_q(Y(f)) - Y_q(X(f)) \\
&= [X,Y]_q(f).
\end{aligned}$$

Since f is arbitrary, Lemma 2.8.17 gives the desired conclusion. □

Let $X, Y \in \mathfrak{X}(U)$ and let Φ, Ψ denote the respective local flows generated by these fields about some point $q \in U$. We can choose $\delta_q > 0$ so that $\Phi_t\Psi_s(q)$ and $\Psi_s\Phi_t(q)$ are defined, $-\delta_q < s, t < \delta_q$. As $q \in U$ varies, these bounds δ_q will also vary. The local flows vary too, but they agree on overlaps by Theorem 2.8.4.

Definition 2.8.19. The local flows of X and Y commute on U if

$$\Phi_t\Psi_s(q) = \Psi_s\Phi_t(q), \quad -\delta_q < s, t < \delta_q, \quad \forall\, q \in U.$$

The vector fields themselves commute on U if $[X, Y] \equiv 0$ on U.

Theorem 2.8.20. *The vector fields X and Y commute on U if and only if their local flows commute on U.*

Proof. If the local flows commute on U, then, for $-\delta_x < t < \delta_x$, Φ_t carries any flow line $\{\Psi_s(x) \mid -\delta_x < s < \delta_x\}$ of Ψ onto another flow line of Ψ. By taking the infinitesimal curve point of view, we see immediately that $(\Phi_t)_{*x}(Y_x) = Y_{\Phi_t(x)}$, $-\delta_x < t < \delta_x$, $\forall\, x \in U$. That is,

$$[X, Y] = \lim_{t\to 0}\frac{\Phi_{-t*}(Y) - Y}{t} = \lim_{t\to 0}\frac{Y - Y}{t} = 0$$

throughout U.

For the converse, we assume that $[X, Y] \equiv 0$ on U and deduce that the local flows commute. Let $q \in U$, fix $s \in (-\delta_q, \delta_q)$, and let $q' = \Psi_s(q)$. Define $v : (-\delta_q, \delta_q) \to T_{q'}(U)$ by the formula $v(t) = \Phi_{-t*}(Y_{\Phi_t(q')})$. (Here and elsewhere, in an attempt to streamline notation, we drop the subscript ξ on differentials $f_{*\xi}$.) Then

$v(t)$ is a differentiable curve in the vector space $T_{q'}(U) = \mathbb{R}^n$, and

$$\begin{aligned}\frac{dv}{dt} &= \lim_{h\to 0} \frac{(\Phi_{-t-h})_*(Y_{\Phi_{t+h}(q')}) - \Phi_{-t*}(Y_{\Phi_t(q')})}{h} \\ &= \lim_{h\to 0} \Phi_{-t*} \frac{\Phi_{-h*}(Y_{\Phi_{t+h}(q')}) - Y_{\Phi_t(q')}}{h} \\ &= \Phi_{-t*} \underbrace{\lim_{h\to 0} \frac{\Phi_{-h*}(Y_{\Phi_h(\Phi_t(q'))}) - Y_{\Phi_t(q')}}{h}}_{\in T_{\Phi_t(q')}(U)} \\ &= \Phi_{-t*}[X,Y]_{\Phi_t(q')} \\ &= 0,\end{aligned}$$

$-\delta_q < t < \delta_q$. It follows that $v(t)$ is constant on $(-\delta_q, \delta_q)$; so

$$\Phi_{-t*}(Y_{\Phi_t(q')}) = Y_{q'}, \quad -\delta_q < t < \delta_q.$$

But q' ranges over $\sigma(s) = \Psi_s(q)$, $-\delta_q < s < \delta_q$, an integral curve to Y. Thus, $\dot{\sigma}(s) = Y_{\sigma(s)}$ and $\Phi_{t*}(\dot{\sigma}(s)) = Y_{\Phi_t(\sigma(s))}$ as s and t range independently over $(-\delta_q, \delta_q)$. Therefore, $\Phi_t \circ \sigma$ is also an integral curve to Y with initial condition $\Phi_t(\sigma(0)) = \Phi_t(q)$. But $\Phi_t \Psi_s(q) = \Psi_s \Phi_t(q)$, $-\delta_q < s, t < \delta_q$, by the uniqueness part of Theorem 2.8.4. □

Exercise 2.8.21. Given $A \in \mathfrak{M}(n)$, view the right translation operation

$$R_A : \mathrm{Gl}(n) \to \mathfrak{M}(n)$$

as a vector field $R_A \in \mathfrak{X}(\mathrm{Gl}(n))$.

(1) For

$$A = \begin{bmatrix} 0 & 1 \\ 0 & 0 \end{bmatrix},$$

show that the local flow generated by R_A has the formula

$$\Phi_t(Q) = Q \cdot \begin{bmatrix} 1 & t \\ 0 & 1 \end{bmatrix},$$

$\forall t \in \mathbb{R}$, $\forall Q \in \mathrm{Gl}(2)$. In particular, we obtain a *global* flow

$$\Phi : \mathbb{R} \times \mathrm{Gl}(2) \to \mathrm{Gl}(2).$$

(2) Note that the formal definition

$$e^{tA} = I + tA + \frac{t^2}{2!}A^2 + \cdots + \frac{t^n}{n!}A^n + \cdots$$

yields

$$e^{tA} = \begin{bmatrix} 1 & t \\ 0 & 1 \end{bmatrix}.$$

Compute e^{tB} for the matrix

$$B = \begin{bmatrix} 0 & 1 \\ -1 & 0 \end{bmatrix}$$

and make an educated guess of a flow on $\mathrm{Gl}(2)$ generated by R_B. Prove that your guess is correct.

2.9. Critical Points and Critical Values

Let $U \subseteq \mathbb{R}^n$ and $V \subseteq \mathbb{R}^m$ be open and let $\Phi : U \to V$ be smooth.

Definition 2.9.1. A point $x \in U$ is a regular point of Φ if

$$\Phi_{*x} : T_x(U) \to T_{\Phi(x)}(V)$$

is surjective. Otherwise, x is a critical point.

Thus, smooth maps from lower dimensions to higher dimensions have only critical points. At the other extreme, if Φ has only regular points, it is a submersion.

Definition 2.9.2. A point $y \in V$ is a critical value of Φ if $\Phi^{-1}(y)$ contains at least one critical point of Φ. Otherwise, y is a regular value of Φ.

You must take care with these terms. If $\Phi^{-1}(y) = \emptyset$, then this set contains no critical points. That is, a point $y \in V$ that is not a value of Φ at all is a *regular value* of Φ!

2.9.A. Sard's theorem. In this subsection, we prove the following fundamental result of Sard. It has important applications in topology, some of which will be treated in the next chapter.

Theorem 2.9.3. *If $\Phi : U \to V$ is a smooth map, then the set of critical values has Lebesgue measure zero.*

We will understand the term "almost every" to mean "Lebesgue almost every". Thus, almost every point of V is a regular value. Our proof will follow closely that given by J. Milnor [**30**]. First, however, we consider some examples and corollaries.

Example 2.9.4. It is well known that one can construct a *continuous* surjection $s : \mathbb{R} \to \mathbb{R}^2$ (a "space filling curve"). However, if s is smooth, every true value of s is a critical value; so $s(\mathbb{R}) \subset \mathbb{R}^2$ has measure zero. Smooth curves cannot be space filling. More generally, smooth maps from lower to higher dimensions always have images of measure zero.

Corollary 2.9.5. *Let $\{\Phi_i : U_i \to V\}_{i=1}^N$, $1 \leq N \leq \infty$, be an at most countable family of smooth maps, each $U_i \subseteq \mathbb{R}^{n_i}$ being open. Then almost every $y \in V$ is simultaneously a regular value of Φ_i, $1 \leq i < N+1$.*

Proof. Let C_i denote the set of critical values of Φ_i, $1 \leq i < N+1$. Then $C = \bigcup_{i=1}^N C_i$ is a set of Lebesgue measure zero and the complement of C is the set of simultaneous regular values. □

The fact that almost every point is a regular value says that the situation in the following theorem is somehow "generic".

Theorem 2.9.6. *If $\Phi : U \to V$ is a smooth map and if $y \in V$ is a regular value, then $\Phi^{-1}(y)$ is a smooth submanifold of U of dimension $n - m$.*

Proof. If $\Phi^{-1}(y) = \emptyset$, this is a submanifold of U of any desired dimension and we are done. We consider the interesting case, therefore, in which the regular value y is actually a value of Φ. Let $x_0 \in \Phi^{-1}(y)$. Since $J\Phi(x_0)$ has maximum rank m, there is a neighborhood W_{x_0} of x_0 in U such that $J\Phi(z)$ has rank m, $\forall\, z \in W_{x_0}$. Let

$$W = \bigcup_{x \in \Phi^{-1}(y)} W_x,$$

an open neighborhood of $\Phi^{-1}(y)$ in U on which Φ has rank m. The assertion follows from Theorem 2.5.3. □

We turn to the proof of Theorem 2.9.3. Accordingly, let $\Phi : U \to V$ be a smooth map, where $U \subseteq \mathbb{R}^n$ and $V \subseteq \mathbb{R}^m$ are open. The critical set $C \subseteq U$ consists of those points x for which $\operatorname{rank} J\Phi_x < m$ and we must prove that $\Phi(C) \subseteq V$ has Lebesgue measure zero. As usual, write $\Phi = (\Phi^1, \dots, \Phi^m)$.

Definition 2.9.7. For each integer $k \geq 1$, $C_k \subseteq C$ is the set of points $x \in U$ such that all mixed partials of Φ^i of order $\leq k$ vanish at x, $1 \leq i \leq m$.

It is clear that we obtain a nest

$$C \supseteq C_1 \supseteq C_2 \supseteq \cdots \supseteq C_k \supseteq \cdots$$

of closed subsets of U. The basic estimates behind Sard's theorem are contained in the following proof.

Proposition 2.9.8. *For $k \geq 1$ sufficiently large, the set $\Phi(C_k)$ has Lebesgue measure zero.*

Proof. Let $Q \subset U$ be a compact cube. Since $C_k \cap U$ is covered by countably many such cubes, it will be enough to find a value of k, depending only on m and n (not on Q) such that $\Phi(C_k \cap Q)$ has Lebesgue measure zero. Let δ denote the edge length of Q.

Let p range over $C_k \cap Q$. The kth order Taylor series, expanded about p, takes the form

$$\Phi(p+v) - \Phi(p) = R(p,v),$$

where the remainder term satisfies a uniform estimate

$$\|R(p,v)\| \leq c\|v\|^{k+1},$$

for all $p \in C_k \cap Q$ and all $v \in \mathbb{R}^n$ such that $p+v \in Q$. Subdivide Q into r^n subcubes of edge length δ/r and let Q' be one of these subcubes containing a point $p \in C_k$. Every point of Q' has the form $p+v$, where

$$\|v\| \leq \frac{\delta\sqrt{n}}{r}.$$

By the above estimate on the remainder term, we see that $\Phi(Q')$ lies in a cube, centered at $\Phi(p)$ and having edge length ϵ/r^{k+1}, with $\epsilon = 2c(\delta\sqrt{n})^{k+1}$. Thus, $\Phi(C_k \cap Q)$ is covered by a union of at most r^n cubes with total measure

$$V(r) \leq \epsilon^m r^{n-(k+1)m}.$$

For large enough k, the exponent of r is negative and $\lim_{r\to\infty} V(r) = 0$. □

Sard's theorem is trivial when $n = 0$, so we make the inductive assumption that the theorem has been proven for the case $U \subseteq \mathbb{R}^{n-1}$, some $n \geq 1$, and deduce the case $U \subseteq \mathbb{R}^n$.

Lemma 2.9.9. *Let $p \in C \setminus C_1$. Then there are coordinate neighborhoods of p in U and of $\Phi(p)$ in V relative to which the formula for Φ becomes*

$$\Phi(y^1, \dots, y^n) = (y^1, \Phi^2(y^1, \dots, y^n), \dots, \Phi^m(y^1, \dots, y^n)).$$

Proof. By the definition of C_1, $p \notin C_1$ implies that some first order partial of some coordinate function of Φ fails to vanish at y. Suitably permuting the coordinates in $\mathbb{R}^n$ and $\mathbb{R}^m$, we lose no generality in assuming that

$$\frac{\partial \Phi^1}{\partial x^1}(p) \neq 0.$$

Thus, by an application of the inverse function theorem, the map

$$(x^1, x^2, \dots, x^n) \mapsto (\Phi^1(x^1, x^2, \dots, x^n), x^2, \dots, x^n) = (y^1, y^2, \dots, y^n)$$

is a diffeomorphism of some open neighborhood W of p onto an open subset $\Phi(W) \subseteq \mathbb{R}^n$. This defines a coordinate system with the required property. □

Proposition 2.9.10. *The set $\Phi(C \setminus C_1)$ has Lebesgue measure zero.*

Proof. It will be enough to show that, for each $p \in C \setminus C_1$, there is a neighborhood W_p of p in U such that $\Phi(C \cap W_p)$ has measure zero. We assume that $m \geq 2$ since, when $m = 1$, $C = C_1$ and the assertion is vacuously true.

By Lemma 2.9.9, we can assume that there is a neighborhood W_p in which Φ carries each point $(t, x^2, \dots, x^n)$ into the hyperplane $\{t\} \times \mathbb{R}^{m-1}$. For fixed t, the restriction of Φ to $W_p \cap (\{t\} \times \mathbb{R}^{n-1})$ can be viewed as a map Φ_t of that set into the hyperplane $\{t\} \times \mathbb{R}^{m-1}$. Since the coordinate t is preserved, the critical set of Φ_t is $C \cap W_p \cap (\{t\} \times \mathbb{R}^{n-1})$. Hence, by the inductive hypothesis, this is carried by Φ_t onto a set of $(m-1)$-dimensional measure zero. Integrate with respect to the t coordinate, concluding by Fubini's theorem that $\Phi(C \cap W_p)$ has m-dimensional measure zero. □

Proposition 2.9.11. *For each integer $k \geq 1$, the set $\Phi(C_k \setminus C_{k+1})$ has Lebesgue measure zero.*

Proof. Again, it will be enough to show that, for each $p \in C_k \setminus C_{k+1}$, there is a neighborhood W_p of p in U such that $\Phi(C_k \cap W_p)$ has measure zero.

Let $\varphi : U \to \mathbb{R}$ be a kth order partial of a coordinate function Φ^r such that some first order partial of φ fails to vanish at p. Without loss of generality, we assume that

$$\frac{\partial \varphi}{\partial x^1}(p) \neq 0.$$

Of course, φ vanishes identically on C_k. The inverse function theorem again gives a change of coordinates

$$(x^1, x^2, \dots, x^n) \mapsto (\varphi(x^1, \dots, x^n), x^2, \dots, x^n) = (y^1, y^2, \dots, y^n),$$

defined on some neighborhood W_p of p, thereby coordinatizing $C_k \cap W_p$ as a subset of the hyperplane $\{0\} \times \mathbb{R}^{n-1}$. Every point in this set is a critical point of the restriction Φ_0 of Φ to $U \cap (\{0\} \times \mathbb{R}^{n-1})$, so the inductive hypothesis implies that $\Phi(C_k \cap W_p) = \Phi_0(C_k \cap W_p)$ has Lebesgue measure zero. □

Corollary 2.9.12. *For each integer $k \geq 1$, the set $\Phi(C \setminus C_k)$ has Lebesgue measure zero.*

Proof. Indeed, $C \setminus C_k = (C \setminus C_1) \cup (C_1 \setminus C_2) \cup \cdots \cup (C_{k-1} \setminus C_k)$. □

By this corollary and Proposition 2.9.8, the proof of the inductive step is complete.

2.9.B. Nondegenerate critical points*. Of special note are the critical points of a smooth, real-valued function. Suppose that $f : U \to \mathbb{R}$ is such a map, where $U \subseteq \mathbb{R}^n$ is open. We present some facts that are the beginnings of "Morse theory", a remarkable application of critical point theory to topology due to M. Morse. This introduction to Morse theory will be continued in Sections 3.10 and 4.2.

Exercise 2.9.13. Let $p \in U$ be a critical point of f. If $X, Y \in \mathfrak{X}(U)$, then $X_p(Y(f)) = Y_p(X(f))$, and this number depends only on the tangent vectors X_p, Y_p, not on their extensions to fields on U.

Each element $Y_p \in T_p(U)$ can be extended to a vector field $Y \in \mathfrak{X}(U)$. Indeed, if $Y_p = \sum_{i=1}^n a^i D_{i,p}$, we can define $Y = \sum_{i=1}^n a^i D_i$. This remark, together with the exercise, insures that the following definition makes sense.

Definition 2.9.14. If $p \in U$ is a critical point of f, the Hessian of f at p is the symmetric, bilinear form $H_p(f) : T_p(U) \times T_p(U) \to \mathbb{R}$ defined by

$$H_p(f)(X_p, Y_p) = X_p(Y(f)).$$

Definition 2.9.15. The critical point $p \in U$ of f is nondegenerate if the symmetric matrix representing the Hessian $H_p(f)$, relative to some choice of basis, is nonsingular. The (Morse) index λ of the critical point is the number of negative eigenvalues of this matrix.

Exercise 2.9.16. Show that, relative to the standard basis of $T_p(U)$, the matrix representing the Hessian is the matrix

$$\left[\frac{\partial^2 f}{\partial x^i \partial x^j}(p)\right]$$

of 2nd partials of f at p.

It is straightforward to check that a matrix representing the Hessian with respect to some coordinate system is nondegenerate of index λ if and only if this is true relative to *every* coordinate system.

Example 2.9.17. Let $U = \mathbb{R}^n$ and let $f : \mathbb{R}^n \to \mathbb{R}$ have the formula

$$(*) \qquad f(z^1, \dots, z^n) = f(0) - \sum_{i=1}^{\lambda} (z^i)^2 + \sum_{i=\lambda+1}^{n} (z^i)^2,$$

relative to suitable coordinates. Then 0 is the only critical point and the Hessian at 0 is represented by the matrix

$$H = 2\begin{bmatrix} -I_\lambda & 0 \\ 0 & I_{n-\lambda} \end{bmatrix}.$$

It is evident that this matrix is nondegenerate of index λ.

The following, due to M. Morse, asserts that this is essentially the only example.

Theorem 2.9.18 (The Morse Lemma). *Let $p \in U$ be a nondegenerate critical point of index λ of the smooth function $f : U \to \mathbb{R}$. Then there is an open neighborhood U_p of p in U and a smooth change of coordinates (i.e., diffeomorphism) $z : U_p \to W$ onto an open neighborhood W of 0 in $\mathbb{R}^n$ such that, relative to the new coordinates $z = (z^1, \dots, z^n)$,*

(1) $p = 0$;
(2) *f has the formula* $(*)$.

We need a preliminary lemma. By a translation, we assume that the critical point $p = 0$.

Lemma 2.9.19. *In a suitable neighborhood V of 0 in U, there are defined smooth, real-valued functions $h_{ij} = h_{ji}$, $1 \le i, j \le n$, such that*

$$f(x^1, \dots, x^n) = f(0) + \sum_{i,j=1}^n x^i x^j h_{ij}(x^1, \dots, x^n)$$

and such that the matrix $2[h_{ij}(0)]$ represents the Hessian $H_0(f)$.

Proof. Choose V to be the open ϵ-ball centered at 0, where $\epsilon > 0$ is small enough that $V \subseteq U$. Write

$$\begin{aligned} f(x^1, \dots, x^n) - f(0) &= \int_0^1 \frac{df}{dt}(tx^1, \dots, tx^n)\, dt \\ &= \sum_{i=1}^n \int_0^1 x^i \frac{\partial f}{\partial x^i}(tx^1, \dots, tx^n)\, dt. \end{aligned}$$

Thus, setting

$$g_i(x^1, \dots, x^n) = \int_0^1 \frac{\partial f}{\partial x^i}(tx^1, \dots, tx^n)\, dt,$$

we write

$$f(x^1, \dots, x^n) - f(0) = \sum_{i=1}^n x^i g_i(x^1, \dots, x^n).$$

By differentiating the formula for g_i under the integral sign, we compute

$$\frac{\partial g_i}{\partial x^j}(0) = \frac{1}{2} \frac{\partial^2 f}{\partial x^j \partial x^i}(0).$$

Note that

$$g_i(0) = \frac{\partial f}{\partial x^i}(0) = 0, \quad 1 \le i \le n,$$

since 0 is a critical point of f. Thus, we can apply the same construction to each g_i, obtaining smooth functions g_{ij} such that

$$f(x^1, \dots, x^n) - f(0) = \sum_{i,j=1}^n x^i x^j g_{ij}(x^1, \dots, x^n).$$

Here,

$$g_{ij}(x^1, \dots, x^n) = \int_0^1 \frac{\partial g_i}{\partial x^j}(tx^1, \dots, tx^n)\, dt,$$

so

$$g_{ij}(0) = \frac{1}{2} \frac{\partial^2 f}{\partial x^j \partial x^i}(0).$$

If we set $h_{ij} = (g_{ij} + g_{ji})/2$, all assertions follow. □

Proof of Theorem 2.9.18. Suppose inductively that there are coordinates $u = (u^1, \dots, u^n)$ in a neighborhood $V \subseteq U$ of 0 such that

$$(**) \qquad f(u) - f(0) = \pm(u^1)^2 \pm \cdots \pm (u^{r-1})^2 + \sum_{i,j \geq r} u^i u^j H_{ij}(u)$$

on V, where the matrix $[H_{ij}]$ is symmetric and nonsingular. Lemma 2.9.19 gives the case $r = 1$, while the case $r = n$ implies Theorem 2.9.18 by a permutation of coordinates. It remains that we prove the inductive step.

As in the standard proof that symmetric matrices can be diagonalized, a suitable linear change in the last $n - r + 1$ coordinates allows us to assume that $H_{rr}(0) \neq 0$, hence that $|H_{rr}(u)| > 0$ and is smooth throughout a neighborhood $V' \subseteq V$ of 0. We set $h(u) = |H_{rr}(u)|^{1/2}$ on V' and introduce new coordinates $v = v(u)$ by setting

$$v^i = \begin{cases} u^i, & i \neq r, \\ h(u)\left(u^r + \sum_{i>r} u^i H_{ir}(u)/H_{rr}(u)\right), & i = r. \end{cases}$$

Then $v(0) = 0$ and $\det J(v)(0) = h(0) > 0$, so the inverse function theorem guarantees that $v = (v^1, \dots, v^n)$ is a coordinate system on some neighborhood $V'' \subseteq V'$ of 0. We leave to the reader the rather tedious exercise of substituting $u^i = v^i$, $i \neq r$, and

$$u^r = \frac{v^r}{h(u(v))} - \sum_{i>r} v^i \frac{H_{ir}(u(v))}{H_{rr}(u(v))}$$

into the equation (**) and collecting terms. This exercise yields

$$f(v) - f(0) = \pm(v^1)^2 \pm \cdots \pm (v^r)^2 + \sum_{i,j>r} v^i v^j \overline{H}_{ij}(v),$$

for suitable smooth functions $\overline{H}_{ij}$. Symmetrizing these coefficients by

$$H_{ij}(v) = \frac{\overline{H}_{ij}(v) + \overline{H}_{ji}(v)}{2}$$

completes the inductive step. □

Corollary 2.9.20. *Nondegenerate critical points are isolated.*

By contrast, degenerate critical points may easily fail to be isolated. For example, the function $f(x, y) = x^2$ has the entire y-axis as its set of critical points, while $f(x, y) = x^n y^m$, $n, m > 1$, has the union of both axes as its critical set. Examples of isolated, but degenerate, critical points include $f(x) = x^3$, having 0 as its sole critical point, and the "monkey saddle" $f(x, y) = x^3 - 3xy^2$. This latter has only the origin as critical point and the Hessian there is the zero matrix.

CHAPTER 3

The Global Theory of Smooth Functions

Our present goal is to extend the theory of smooth functions, developed on open subsets of $\mathbb{R}^n$ in Chapter 2, to arbitrary differentiable manifolds. Geometric topology becomes an essential feature.

3.1. Smooth Manifolds and Mappings

Let M be a topological manifold of dimension n. The locally Euclidean property allows us to choose local coordinates in any small region of M.

Definition 3.1.1. A coordinate chart on M is a pair (U, φ), where U is an open subset of M and $\varphi : U \to \mathbb{R}^n$ is a homeomorphism onto an open subset of $\mathbb{R}^n$.

One often writes $\varphi(p) = (x^1(p), \dots, x^n(p))$, viewing this as the coordinate n-tuple of the point $p \in U$. Relative to such a coordinatization, one can do calculus in the region U of M. The problem is that the point p will generally belong to infinitely many different coordinate charts and calculus in one of these coordinatizations about p might not agree with calculus in another. One needs the coordinate systems to be smoothly compatible in the following sense.

Definition 3.1.2. Two coordinate charts, (U, φ) and (V, ψ) on M are said to be C^∞-related if either $U \cap V = \emptyset$ or

$$\varphi \circ \psi^{-1} : \psi(U \cap V) \to \varphi(U \cap V)$$

is a diffeomorphism (between open subsets of $\mathbb{R}^n$).

This is illustrated in Figure 3.1.1. We think of $\varphi \circ \psi^{-1}$ as a smooth change of coordinates on $U \cap V$. Thus, on $U \cap V$, functions are smooth relative to one coordinate system if and only if they are smooth relative to the other. Indeed, differential calculus carried out in $U \cap V$ via the coordinates of $\varphi(U \cap V)$ is equivalent to the calculus carried out via the coordinates of $\psi(U \cap V)$. The explicit formulas will, of course, change from the one coordinate system to the other. Furthermore, piecing together these local calculi produces a global calculus on M. The concept that allows us to make these remarks precise is that of a *smooth atlas*.

Definition 3.1.3. A C^∞ atlas on M is a collection $\mathcal{A} = \{(U_\alpha, \varphi_\alpha)\}_{\alpha \in \mathfrak{A}}$ of coordinate charts such that

1. $(U_\alpha, \varphi_\alpha)$ is C^∞-related to (U_β, φ_β), $\forall\, \alpha, \beta \in \mathfrak{A}$;
2. $M = \bigcup_{\alpha \in \mathfrak{A}} U_\alpha$.

Definition 3.1.4. Two C^∞ atlases $\mathcal{A}$ and $\mathcal{A}'$ on M are equivalent if $\mathcal{A} \cup \mathcal{A}'$ is also a C^∞ atlas on M.

It will be seen that global calculus carried out relative to $\mathcal{A}$ will be identical to global calculus carried out relative to the equivalent atlas $\mathcal{A}'$.

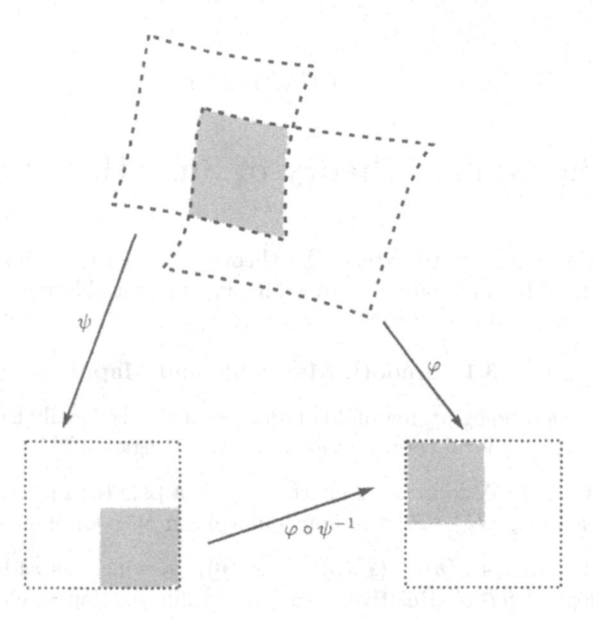

Figure 3.1.1. The smooth coordinate change $\varphi \circ \psi^{-1}$

Exercise 3.1.5. Equivalence of C^∞ atlases is an equivalence relation. Each C^∞ atlas on M is equivalent to a unique maximal C^∞ atlas on M.

Definition 3.1.6. A maximal C^∞ atlas $\mathcal{A}$ on M is called a smooth structure on M (also called a differentiable structure or a C^∞ structure). The pair $(M, \mathcal{A})$ is called a smooth (or differentiable or C^∞) n-manifold.

By a typical abuse of notation, we usually write M for the smooth manifold, the presence of the differentiable structure $\mathcal{A}$ being understood. By Exercise 3.1.5, any C^∞ atlas (not necessarily maximal) on M completely determines the differentiable structure. Note that the dimension n of a smooth n-manifold is well-defined by Proposition 2.3.12.

Example 3.1.7. The manifold $\mathbb{R}^n$ has a canonical smooth structure, namely the set $\mathcal{A}_n$ of all pairs (U, φ) where $U \subseteq \mathbb{R}^n$ is open and $\varphi : U \to \mathbb{R}^n$ is a diffeomorphism onto an open set $\varphi(U) \subseteq \mathbb{R}^n$.

Example 3.1.8. If M and N are smooth manifolds, $\dim M = m$ and $\dim N = n$, with respective smooth structures $\mathcal{A} = \{(U_\alpha, \varphi_\alpha)\}_{\alpha \in \mathfrak{A}}$ and $\mathcal{B} = \{(V_\beta, \psi_\beta)\}_{\beta \in \mathfrak{B}}$, then $M \times N$ is canonically a smooth $(m+n)$-manifold. Indeed,

$$\mathcal{A} \times \mathcal{B} = \{(U_\alpha \times V_\beta, \varphi_\alpha \times \psi_\beta)\}_{(\alpha,\beta) \in \mathfrak{A} \times \mathfrak{B}}$$

is a C^∞ atlas, determining uniquely a maximal one, called the Cartesian product of the two smooth structures.

Example 3.1.9. If $W \subseteq M$ is an open subset of a smooth n-manifold, then W is a smooth n-manifold in a natural way. Details are left as an easy exercise.

Remark. By substituting C^k for C^∞ in the above discussion, one obtains the notion of a C^k manifold, $1 \leq k < \infty$. Similarly, one defines the notion of a *real analytic* (C^ω) manifold. The reader should have no trouble in adapting the following discussion to these cases.

Let M be a smooth n-manifold with a smooth atlas $\mathcal{A} = \{(U_\alpha, \varphi_\alpha)\}_{\alpha \in \mathfrak{A}}$. We do not require this atlas to be maximal. Set

$$g_{\alpha\beta} = \varphi_\alpha \circ \varphi_\beta^{-1} : \varphi_\beta(U_\alpha \cap U_\beta) \to \varphi_\alpha(U_\alpha \cap U_\beta).$$

These local diffeomorphisms in $\mathbb{R}^n$ satisfy the *cocycle* conditions

(1) $g_{\alpha\beta} \circ g_{\beta\gamma} = g_{\alpha\gamma}$ on $\varphi_\gamma(U_\alpha \cap U_\beta \cap U_\gamma)$,
(2) $g_{\alpha\alpha} = \mathrm{id}_{\varphi_\alpha(U_\alpha)}$,
(3) $g_{\beta\alpha} = g_{\alpha\beta}^{-1}$.

It should be noted that properties (2) and (3) follow from property (1).

Definition 3.1.10. The system $\{g_{\alpha\beta}\}_{\alpha,\beta \in \mathfrak{A}}$ is called a structure cocycle for the smooth manifold M.

The term "cocycle" is borrowed from algebraic topology due to certain formal similarities to cocycles in Čech cohomology.

Remark. It will be useful to see how to "reassemble" M out of the data $\{\widetilde{U}_\alpha = \varphi_\alpha(U_\alpha), g_{\alpha\beta}\}_{\alpha,\beta \in \mathfrak{A}}$. On the disjoint union

$$\widetilde{M} = \bigsqcup_{\alpha \in \mathfrak{A}} \widetilde{U}_\alpha$$

define the relation

$$x \sim y \Leftrightarrow \exists\, \alpha, \beta \in \mathfrak{A} \text{ such that } x \in \widetilde{U}_\alpha, \quad y \in \widetilde{U}_\beta \text{ and } y = g_{\beta\alpha}(x).$$

By properties (1), (2), and (3), this is an equivalence relation, so we form the topological quotient space $\widetilde{M}/\sim$. We will show that this space is homeomorphic to M and exhibit a natural smooth structure on it.

Let $[z] \in \widetilde{M}/\sim$ denote the equivalence class of $z \in \widetilde{M}$. Define

$$\varphi : M \to \widetilde{M}/\sim$$

by setting $\varphi(x) = [\varphi_\alpha(x)]$ if $x \in U_\alpha$. If $x \in U_\beta$ also, then

$$g_{\alpha\beta}(\varphi_\beta(x)) = \varphi_\alpha(x),$$

so φ is well defined. It is also continuous. The map from $\widetilde{M}$ to M that takes $z \in \widetilde{U}_\alpha$ to $\varphi_\alpha^{-1}(z)$ respects the equivalence relation, hence passes to a continuous map

$$\psi : \widetilde{M}/\sim \to M.$$

It is easy to see that φ and ψ are mutually inverse, so M and $\widetilde{M}/\sim$ are canonically homeomorphic. Each $\widetilde{U}_\alpha$ imbeds canonically in $\widetilde{M}/\sim$ as an open subset and $\mathrm{id}_\alpha : \widetilde{U}_\alpha \to \widetilde{U}_\alpha \subseteq \mathbb{R}^n$ defines a coordinate chart $(\widetilde{U}_\alpha, \mathrm{id}_\alpha)$ on $\widetilde{M}/\sim$. These charts are C^∞-related via the cocycle $\{g_{\alpha\beta}\}_{\alpha,\beta \in \mathfrak{A}}$, so $\widetilde{M}/\sim$ is canonically identified with M as a smooth manifold via the mutually inverse *diffeomorphisms* φ and ψ (see Definition 3.1.18).

Exercise 3.1.11. Prove that the topological n-manifold P^n (see Exercise 1.3.26) is a smooth n-manifold.

Exercise 3.1.12. Show that the manifold S^n can be assembled from a C^∞ atlas with just two charts.

We turn to the smooth maps defined on a manifold.

Definition 3.1.13. A function $f : M \to \mathbb{R}$ is said to be smooth if, for each $x \in M$, there is a chart $(U, \varphi) \in \mathcal{A}$ such that $x \in U$ and

$$f \circ \varphi^{-1} : \varphi(U) \to \mathbb{R}$$

is smooth. The set of all smooth, real valued functions on M will be denoted by $C^\infty(M)$.

The definition only requires us to be able to find *some* such chart about each point $x \in M$, but the following assures us that *all* charts will then work.

Lemma 3.1.14. *The function $f : M \to \mathbb{R}$ is smooth if and only if*

$$f \circ \varphi_\alpha^{-1} : \varphi_\alpha(U_\alpha) \to \mathbb{R}$$

is smooth, $\forall\, (U_\alpha, \varphi_\alpha) \in \mathcal{A}$.

Proof. Clearly this condition implies that f is smooth. For the converse, suppose that f is smooth and let $x \in U_\alpha$ where $(U_\alpha, \varphi_\alpha) \in \mathcal{A}$. By Definition 3.1.13, choose $(U_\beta, \varphi_\beta) \in \mathcal{A}$ such that $x \in U_\beta$ and

$$f \circ \varphi_\beta^{-1} : \varphi_\beta(U_\beta) \to \mathbb{R}$$

is smooth. Then,

$$f \circ \varphi_\alpha^{-1} : \varphi_\alpha(U_\alpha \cap U_\beta) \to \mathbb{R}$$

is given by the composition

$$\varphi_\alpha(U_\alpha \cap U_\beta) \xrightarrow{g_{\beta\alpha}} \varphi_\beta(U_\alpha \cap U_\beta) \xrightarrow{f\circ\varphi_\beta^{-1}} \mathbb{R}.$$

As a composition of smooth maps, this is smooth. That is,

$$f \circ \varphi_\alpha^{-1} : \varphi_\alpha(U_\alpha) \to \mathbb{R}$$

is smooth on some neighborhood of the point $\varphi_\alpha(x)$. But $x \in U_\alpha$ is arbitrary, so $f \circ \varphi_\alpha^{-1}$ is smooth on all of $\varphi_\alpha(U_\alpha)$. □

We think of $f \circ \varphi_\alpha^{-1}$ as a *formula* for $f|U_\alpha$ relative to the coordinate system $\varphi_\alpha = (x_\alpha^1, \dots, x_\alpha^n)$. We generally write $(U_\alpha, x_\alpha^1, \dots, x_\alpha^n)$ or (U_α, x_α) for $(U_\alpha, \varphi_\alpha)$.

Definition 3.1.15. Let M and N be C^∞ manifolds with respective smooth structures $\mathcal{A}$ and $\mathcal{B}$. A map $f : M \to N$ is said to be smooth if, for each $x \in M$, there are $(U_\alpha, \varphi_\alpha) \in \mathcal{A}$ and $(V_\beta, \psi_\beta) \in \mathcal{B}$ such that $x \in U_\alpha$, $f(U_\alpha) \subseteq V_\beta$, and

$$\psi_\beta \circ f \circ \varphi_\alpha^{-1} : \varphi_\alpha(U_\alpha) \to \psi_\beta(V_\beta)$$

is smooth.

Lemma 3.1.16. *The map $f : M \to N$ is smooth if and only if, for all choices of $(U_\alpha, \varphi_\alpha) \in \mathcal{A}$ and $(V_\beta, \psi_\beta) \in \mathcal{B}$ such that $f(U_\alpha) \subseteq V_\beta$, the map*

$$\psi_\beta \circ f \circ \varphi_\alpha^{-1} : \varphi_\alpha(U_\alpha) \to \psi_\beta(V_\beta)$$

is smooth.

The proof is similar to the previous one and is left to the reader. Again, we think of $\psi_\beta \circ f \circ \varphi_\alpha^{-1}$ as a local coordinate formula for f.

These two lemmas give an important part of the content of our remark that differential calculus in one coordinate chart is equivalent, in overlaps, to differential calculus in C^∞-related neighboring charts.

Lemma 3.1.17. *If $f : M \to N$ and $g : N \to P$ are smooth maps between manifolds, then $g \circ f : M \to P$ is smooth.*

This is also elementary and is left to the reader.

Definition 3.1.18. A map $f : M \to N$ between smooth manifolds is a diffeomorphism if it is smooth and there is a smooth map $g : N \to M$ such that $f \circ g = \mathrm{id}_N$ and $g \circ f = \mathrm{id}_M$.

Example 3.1.19. The maps

$$\psi : \widetilde{M}/\!\sim \to M,$$
$$\varphi : M \to \widetilde{M}/\!\sim$$

are mutually inverse diffeomorphisms (see the *remark* following Definition 3.1.0).

Lemma 3.1.20. *If $(M, \mathcal{A})$ is a smooth n-manifold, $U \subseteq M$ an open subset, and $\varphi : U \to \mathbb{R}^n$ a diffeomorphism of U onto an open subset $\varphi(U)$ of $\mathbb{R}^n$, then $(U, \varphi) \in \mathcal{A}$.*

Proof. By the definition of diffeomorphism, (U, φ) is C^∞-related to every $(U_\alpha, \varphi_\alpha) \in \mathcal{A}$, so $(U, \varphi) \in \mathcal{A}$ by the maximality of this atlas. □

If we write (U, φ) as $(U, x^1, x^2, \dots, x^n)$, the above discussion allows us to write $f(x^1, \dots, x^n)$ for $f|U$, whenever $f : M \to N$ is smooth. This is logically a bit sloppy, but it is *psychologically* helpful.

Exercise 3.1.21. Suppose that M is a smooth n-manifold and that

$$\pi : M' \to M$$

is a covering space. Prove that M' has a unique smooth structure relative to which the projection π is locally a diffeomorphism.

If $p \in M$, it makes good sense to talk about germs at p of real valued C^∞ functions defined on open neighborhoods of p. As before, these form an associative algebra $\mathfrak{G}_p$ over $\mathbb{R}$. The evaluation map $e_p : \mathfrak{G}_p \to \mathbb{R}$ is defined exactly as before.

Definition 3.1.22. A derivative of $\mathfrak{G}_p$ is an $\mathbb{R}$-linear map

$$D : \mathfrak{G}_p \to \mathbb{R}$$

such that

$$D(\xi\zeta) = D(\xi)e_p(\zeta) + e_p(\xi)D(\zeta),$$

$\forall \xi, \zeta \in \mathfrak{G}_p$. This operator D is also called a tangent vector to M at p and the vector space $T_p(M)$ of all derivatives of $\mathfrak{G}_p$ is called the tangent space to M at p.

Definition 3.1.23. If $f : M \to N$ is a smooth map between manifolds and if $p \in M$, the differential

$$f_{*p} = df_p : T_p(M) \to T_{f(p)}(N)$$

is the linear map defined by

$$(f_{*p}(D))[g]_{f(p)} = D[g \circ f]_p,$$

for all $D \in T_p(M)$ and all $[g]_{f(p)} \in \mathfrak{G}_{f(p)}$.

Lemma 3.1.24 (Global chain rule). *If $f : M \to N$ and $g : N \to P$ are smooth maps between manifolds and $x \in M$, then $d(g \circ f)_x = dg_{f(x)} \circ df_x$.*

Proof. Consider

$$\begin{aligned} ((g \circ f)_{*p}(D))[h]_{g(f(p))} &= D[h \circ g \circ f]_p \\ &= (f_{*p}(D)[h \circ g]_{f(p)}) \\ &= (g_{*f(p)}(f_{*p}(D)))[h]_{g(f(p))}. \end{aligned}$$

Since $[h]_{g(f(p))} \in \mathfrak{G}_{g(f(p))}$ and $D \in T_p(M)$ are arbitrary, the assertion follows. □

It is clear that $\mathrm{id}_{*p} = \mathrm{id} : T_p(M) \to T_p(M)$, so the chain rule has the following consequence.

Corollary 3.1.25. *If $f : M \to N$ is a diffeomorphism, then*

$$f_{*p} : T_p(M) \to T_{f(p)}(N)$$

is an isomorphism of real vector spaces, $\forall\, p \in M$.

Corollary 3.1.26. *If M is a smooth manifold of dimension n, then $T_x(M)$ is a real vector space of dimension n, $\forall\, x \in M$.*

Proof. Let (U, φ) be a coordinate patch on M with $x \in U$. Then, $T_x(U) = T_x(M)$ and, by the previous corollary,

$$\varphi_{*x} : T_x(U) \to T_{\varphi(x)}(\varphi(U))$$

is an $\mathbb{R}$-linear isomorphism. Since $\varphi(U) \subseteq \mathbb{R}^n$ is open, we know that

$$T_{\varphi(x)}(\varphi(U)) = \mathbb{R}^n.$$

□

The same kind of argument gives the following.

Corollary 3.1.27. *If $f : M \to N$ is a diffeomorphism, then* $\dim M = \dim N$.

Remark. We do not have a *canonical* basis for $T_x(M)$ since there is no preferred choice of local coordinates about x. Thus, we cannot write $T_x(M) = \mathbb{R}^n$.

The notion of infinitesimal curve $\langle s \rangle_p$ makes sense in our context, as does the derivative $D_{\langle s \rangle_p} \in T_p(M)$. Viewing tangent vectors as infinitesimal curves is preferable from an intuitive point of view, making the differential of a map "visible" and making the chain rule evident. If one is developing a theory of C^k manifolds, defining tangent vectors to be infinitesimal curves rather than first order differential operators is essential [**33**]. The following is evident via local coordinates and the corresponding facts in $\mathbb{R}^n$.

Lemma 3.1.28. *The correspondence $\langle s \rangle_p \leftrightarrow D_{\langle s \rangle_p}$ is a one-to-one correspondence between the set of infinitesimal curves at $p \in M$ and $T_p(M)$. Furthermore, if $f : M \to N$ is a smooth mapping between manifolds, the differential $f_{*p} : T_p(M) \to T_{f(p)}(N)$ is given, in terms of infinitesimal curves, by*

$$f_{*p}(\langle s \rangle_p) = \langle f \circ s \rangle_{f(p)}.$$

Finally, for smooth maps $f : M \to N$, we define the notions of regular point, critical point, critical value, and regular value exactly as before.

3.2. Diffeomorphic Structures

This section is really an extended remark on some very deep theorems, the point of which can now be easily appreciated.

Let M be a differentiable manifold with smooth structure

$$\mathcal{A} = \{(U_\alpha, \varphi_\alpha)\}_{\alpha \in \mathfrak{A}}.$$

Let $\Phi : M \to M$ be any *homeomorphism*. Set

$$\mathcal{A}_\Phi = \{(\Phi^{-1}(U_\alpha), \varphi_\alpha \circ \Phi)\}_{\alpha \in \mathfrak{A}}.$$

Proposition 3.2.1. *The set $\mathcal{A}_\Phi$ is a C^∞ structure on M having the same structure cocycle as $\mathcal{A}$.*

Proof. Indeed,

$$g_{\alpha\beta} = \varphi_\alpha \circ \varphi_\beta^{-1} = (\varphi_\alpha \circ \Phi) \circ (\varphi_\beta \circ \Phi)^{-1},$$

and this map carries the set

$$(\varphi_\beta \circ \Phi)(\Phi^{-1}(U_\alpha) \cap \Phi^{-1}(U_\beta)) = \varphi_\beta(U_\alpha \cap U_\beta)$$

onto the set

$$(\varphi_\alpha \circ \Phi)(\Phi^{-1}(U_\alpha) \cap \Phi^{-1}(U_\beta)) = \varphi_\alpha(U_\alpha \cap U_\beta).$$

Finally, the maximality of the C^∞ atlas $\mathcal{A}_\Phi$ follows from that of $\mathcal{A}$ and the fact that $\mathcal{A}_{\Phi\Phi^{-1}} = \mathcal{A}$. □

Corollary 3.2.2. *The smooth manifold $\widetilde{M/\sim}$, whcther defined from the C^∞ structure $\mathcal{A}$ or $\mathcal{A}_\Phi$, is the same. Consequently, $(M, \mathcal{A})$ and $(M, \mathcal{A}_\Phi)$ are canonically diffeomorphic.*

We will let M_Φ denote the smooth manifold $(M, \mathcal{A}_\Phi)$.

Definition 3.2.3. Two C^∞ structures $\mathcal{A}$ and $\mathcal{B}$, defined on the same topological manifold M, are said to be diffeomorphic structures if $\mathcal{B} = \mathcal{A}_\Phi$, for some homeomorphism $\Phi : M \to M$.

It is clear that diffeomorphism is an equivalence relation on the set of smooth structures on a topological manifold M. It is natural to ask how many diffeomorphism classes of smooth structures a given topological manifold can support. The following examples are the deep facts, referred to at the beginning of this section, that we cannot prove here. The methods of proof are quite advanced.

Example 3.2.4. If $n \neq 4$, any two smooth structures on $\mathbb{R}^n$ are diffeomorphic [**42**].

Example 3.2.5. The case of $\mathbb{R}^4$ was cracked by the combined work of a topologist, M. Freedman, and a global analyst, S. Donaldson, showing that $\mathbb{R}^4$ admits a differentiable structure not diffeomorphic to the usual one. In this structure, it is possible to find a compact set that cannot be surrounded by any smoothly imbedded S^3! (For a discussion of this, see [**11**, Section 1].) Subsequently, various researchers found more "exotic" differentiable structures on $\mathbb{R}^4$ (the first of these was R. Gompf, who found two new structures [**12**]). It is now known that the number of distinct diffeomorphism classes of differentiable structures on $\mathbb{R}^4$ is uncountably infinite. There is even a "universal" smooth $(\mathbb{R}^4, \mathcal{A}_u)$ such that every other smooth $(\mathbb{R}^4, \mathcal{A})$ smoothly imbeds as an open subset of $(\mathbb{R}^4, \mathcal{A}_u)$.

Example 3.2.6. Let $\sigma(n)$ denote the number of diffeomorphism classes of differentiable structures on S^n, up to *oriented diffeomorphism* (see Example 3.4.13). It was long known that $\sigma(n) = 1$ for $n = 1, 2, 3$. The value of $\sigma(4)$ remains a mystery. The following table was computed by M. Kervaire and J. Milnor [**21**].

n	5	6	7	8	9	10	11	12	13	14	15	16	17	18
$\sigma(n)$	1	1	28	2	8	6	992	1	3	2	16,256	2	16	16

Table 1. Oriented differentiable structures on spheres

Example 3.2.7. A topological manifold is said to be smoothable if it can be given a differentiable structure. For $n = 1, 2, 3$, it is known that all topological n-manifolds are smoothable. The first dimension in which there exist nonsmoothable manifolds is $n = 4$ [**11**, p. 23].

Exercise 3.2.8. Let $\Phi : \mathbb{R} \to \mathbb{R}$ be the homeomorphism $\Phi(x) = x^3$. Show that the identity map, viewed as $\mathrm{id} : \mathbb{R}_\Phi \to \mathbb{R}$, is not a diffeomorphism (although it is clearly a homeomorphism). On the other hand, show that, for any homeomorphism $\Phi : M \to M$ of a differentiable manifold M, $\Phi : M_\Phi \to M$ is a diffeomorphism.

3.3. The Tangent Bundle

Let M be a C^∞ n-manifold with smooth structure $\{(U_\alpha, \varphi_\alpha)\}_{\alpha\in\mathfrak{A}}$. Consider the *set*

$$T = \bigsqcup_{x\in M} T_x(M),$$

a disjoint union with, as yet, no topological structure. For each U_α, $\alpha \in \mathfrak{A}$, define

$$T(U_\alpha) = \bigsqcup_{x\in U_\alpha} T_x(M) \subseteq T.$$

Then the individual linear maps $d\varphi_{\alpha x}$, $x \in U_\alpha$, unite to define a set map

$$d\varphi_\alpha : T(U_\alpha) \to T(\varphi_\alpha(U_\alpha)) = \varphi_\alpha(U_\alpha) \times \mathbb{R}^n \subseteq \mathbb{R}^{2n}.$$

More precisely, if v_x denotes a tangent vector to M at $x \in U_\alpha$,

$$d\varphi_\alpha(v_x) = (\varphi_\alpha(x), d\varphi_{\alpha x}(v_x)),$$

and this defines a bijection of $T(U_\alpha)$ onto an open subset of $\mathbb{R}^{2n}$. Whenever $U_\alpha \cap U_\beta \neq \emptyset$, consider

$$d\varphi_\alpha \circ d\varphi_\beta^{-1} : T(\varphi_\beta(U_\alpha \cap U_\beta)) \to T(\varphi_\alpha(U_\alpha \cap U_\beta)).$$

By the chain rule, this is

$$dg_{\alpha\beta} : d\varphi_\beta(T(U_\alpha) \cap T(U_\beta)) \to d\varphi_\alpha(T(U_\alpha) \cap T(U_\beta)),$$

a C^∞ diffeomorphism between open subsets of $\mathbb{R}^{2n}$.

We topologize the set T. If

$$W \subseteq d\varphi_\alpha(T(U_\alpha)) = T(\varphi_\alpha(U_\alpha)) \subseteq \mathbb{R}^{2n}$$

is an open set, then $d\varphi_\alpha^{-1}(W)$ is to be an open subset of T.

Exercise 3.3.1. Prove that the above sets form the base of a topology on T and that, in this topology, T is a topological manifold of dimension $2n$. Furthermore, show that the system $\{(T(U_\alpha), d\varphi_\alpha)\}_{\alpha\in\mathfrak{A}}$ is a (not maximal) C^∞ atlas on T determining a maximal such atlas $\mathcal{A}$. Finally, if $T(M)$ denotes the differentiable manifold $(T, \mathcal{A})$, show that the map

$$\pi : T(M) \to M,$$
$$\pi(v) = x \Leftrightarrow v \in T_x(M),$$

is smooth.

Definition 3.3.2. The system $\pi : T(M) \to M$ is called the tangent bundle of M. The total space is $T(M)$, the base space is M, and π is called the bundle projection.

Remark. It is often convenient to replace $\varphi_\alpha(U_\alpha) \times \mathbb{R}^n$ with $U_\alpha \times \mathbb{R}^n$, identifying $d\varphi_\alpha$ with the map $v_x \mapsto (x, d\varphi_{\alpha x}(v_x))$. This minor abuse of notation will be a major convenience in what follows.

For each $\alpha \in \mathfrak{A}$, we get a commutative diagram

$$\begin{array}{ccc} T(U_\alpha) & \xrightarrow{d\varphi_\alpha} & U_\alpha \times \mathbb{R}^n \\ {\scriptstyle\pi}\downarrow & & \downarrow{\scriptstyle p_1} \\ U_\alpha & \xrightarrow[\text{id}]{} & U_\alpha \end{array}$$

where p_1 denotes projection onto the first factor and $d\varphi_\alpha$ is a diffeomorphism that restricts to be a linear isomorphism $T_x(M) \to \{x\} \times \mathbb{R}^n$, $\forall x \in U_\alpha$. Thus, $T(M)$ is "locally" a Cartesian product of M and $\mathbb{R}^n$, the projection π being "locally" the projection of the Cartesian product onto the first factor, and the fiber $\pi^{-1}(x) = T_x(M)$ has a canonical vector space structure, $\forall x \in M$.

Definition 3.3.3. A vector field on M is a smooth map $X : M \to T(M)$ $(p \mapsto X_p)$ such that $\pi \circ X = \mathrm{id}_M$. The set of all vector fields on M is denoted by $\mathfrak{X}(M)$.

Remark. Let X be a vector field on M, $(U, x^1, \dots, x^n)$ a coordinate chart on M. By this point in the book, the reader should be able to justify writing

$$X|U = \sum_{i=1}^{n} f^i \frac{\partial}{\partial x^i}$$

where $f^i : U \to \mathbb{R}$ is smooth, $1 \le i \le n$.

Tangent bundles are examples of vector bundles. Vector bundles play a very important role in manifold theory, so we close this section with a brief discussion of the general theory.

Definition 3.3.4. Let M be a smooth m-manifold, E a smooth manifold of dimension $(m+n)$, and $\pi : E \to M$ a smooth map. This will be called an n-plane bundle over M (or a vector bundle over M of fiber dimension n) if the following properties hold.

(1) For each $x \in M$, $E_x = \pi^{-1}(x)$ has the structure of a real, n-dimensional vector space.

(2) There is an open cover $\{W_j\}_{j\in J}$ of M, together with commutative diagrams

$$\begin{array}{ccc} \pi^{-1}(W_j) & \xrightarrow{\psi_j} & W_j \times \mathbb{R}^n \\ \pi \big\downarrow & & \big\downarrow p_1 \\ W_j & \xrightarrow[\text{id}]{} & W_j \end{array}$$

such that ψ_j is a diffeomorphism, $\forall j \in J$.

(3) For each $j \in J$ and $x \in W_j$, $\psi_{jx} = \psi_j|E_x$ maps the vector space E_x isomorphically onto the vector space $\{x\} \times \mathbb{R}^n$.

As with tangent bundles, we call E the total space, M the base space, and π the bundle projection. We also call each W_j a trivializing neighborhood for the bundle and $\{W_j\}_{j\in J}$ a locally trivializing cover (of M) for E.

An obvious example of an n-plane bundle is given by $p_1 : M \times \mathbb{R}^n \to M$. Here, M itself is a trivializing neighborhood and the bundle is said to be trivial.

Definition 3.3.5. Let $\pi_1 : E_1 \to M$ and $\pi_2 : E_2 \to M$ be n-plane bundles over M. A bundle isomorphism is a commutative diagram

$$\begin{array}{ccc} E_1 & \xrightarrow{\varphi} & E_2 \\ \pi_1 \big\downarrow & & \big\downarrow \pi_2 \\ M & \xrightarrow[\text{id}]{} & M \end{array}$$

such that φ is bijective, smooth, and carries E_{1x} isomorphically (as a vector space) onto E_{2x}, $\forall x \in M$. If $E_2 = M \times \mathbb{R}^n$ with π_2 the canonical projection, the isomorphism φ is called a trivialization of E_1.

Note that we did not explicitly require that φ^{-1} be smooth. One needs this to be true, however, in order that bundle isomorphism be an equivalence relation. The following lemma comes to the rescue. Remark that it is useful not to be required to check in each instance that φ is a diffeomorphism.

Lemma 3.3.6. *The map φ in* Definition 3.3.5 *is necessarily a diffeomorphism.*

Proof. Find a locally trivializing cover $\mathfrak{U}$ of M for both E_1 and E_2 simultaneously. If $U \in \mathfrak{U}$, we use the local trivializations

$$\pi_i^{-1}(U) \cong U \times \mathbb{R}^n, \quad i = 1, 2,$$

and the bundle isomorphism

$$\varphi|\pi_1^{-1}(U) : \pi_1^{-1}(U) \to \pi_2^{-1}(U)$$

to induce a commutative diagram

$$\begin{array}{ccc} U \times \mathbb{R}^n & \xrightarrow{\widetilde{\varphi}} & U \times \mathbb{R}^n \\ p_1 \big\downarrow & & \big\downarrow p_1 \\ U & \xrightarrow[\text{id}]{} & U \end{array}$$

which is also a bundle isomorphism. It will be enough to prove that $\widetilde{\varphi}^{-1}$ is smooth. But

$$\widetilde{\varphi}(x, v) = (x, \gamma(x) \cdot v),$$

where $\gamma : U \to \mathrm{Gl}(n)$.

We claim that γ is smooth. Indeed, let $e_k \in \mathbb{R}^n$ denote the column vector with 1 in the kth position, 0s elsewhere. Then the map $\varphi_k : U \to U \times \mathbb{R}^n$, defined by

$$\varphi_k(x) = \widetilde{\varphi}(x, e_k) = (x, \gamma(x) \cdot e_k),$$

is smooth. In particular, the kth column of $\gamma(x)$ is $\gamma(x) \cdot e_k$, hence depends smoothly on x. Since k is arbitrary, γ is smooth.

The operation of taking the inverse of a matrix defines a map

$$\iota : \mathrm{Gl}(n) \to \mathrm{Gl}(n)$$

the coordinate functions of which are rational functions of the coordinates. This map is smooth, so

$$\widetilde{\varphi}^{-1}(x, w) = (x, \iota \circ \gamma(x) \cdot w)$$

is also smooth. □

Definition 3.3.7. An n-plane bundle is trivial if it is isomorphic to the product bundle $p_1 : M \times \mathbb{R}^n \to M$.

Definition 3.3.8. A section of the n-plane bundle $\pi : E \to M$ is a smooth map $s : M \to E$ such that $\pi \circ s = \mathrm{id}_M$. The set of all such sections is denoted by $\Gamma(E)$.

Thus, $\mathfrak{X}(M) = \Gamma(T(M))$. The following is elementary and is left to the reader.

Lemma 3.3.9. *The set $\Gamma(E)$ is a $C^\infty(M)$-module under the pointwise operations*

$$(f^1 s_1 + f^2 s_2)(x) = f^1(x) s_1(x) + f^2(x) s_2(x) \in E_x, \quad \forall x \in M,$$

where $f^i \in C^\infty(M)$ and $s_i \in \Gamma(E)$, $i = 1, 2$.

Definition 3.3.10. The manifold M is parallelizable if there are fields

$$X_1, X_2, \dots, X_n \in \mathfrak{X}(M)$$

such that $\{X_{1x}, X_{2x}, \dots, X_{nx}\}$ is a basis of $T_x(M)$, $\forall x \in M$.

Proposition 3.3.11. *The manifold M is parallelizable if and only if $T(M)$ is a trivial bundle.*

Exercise 3.3.12. Prove that the n-plane bundle $\pi : E \to M$ is trivial if and only if there exist $s_1, \dots, s_n \in \Gamma(E)$ such that $\{s_1(x), \dots, s_n(x)\}$ is a basis of E_x, $\forall x \in M$. In particular, this proves Proposition 3.3.11.

So far, the only real examples of vector bundles that we have seen are tangent bundles and trivial bundles. The following is the least complicated example of a nontrivial vector bundle.

Example 3.3.13. We give an example of a 1-plane bundle (a "line" bundle) over the circle, known as the Möbius bundle. On $\mathbb{R} \times \mathbb{R}$, define the equivalence relation $(s, t) \sim (s + n, (-1)^n t)$, $n \in \mathbb{Z}$. Remark that $t \mapsto (-1)^n t$ is a linear automorphism of $\mathbb{R}$. The projection $(s, t) \mapsto s$ passes to a well-defined map $\pi : (\mathbb{R} \times \mathbb{R})/(\sim) \to \mathbb{R}/\mathbb{Z} = S^1$. It should be clear, intuitively, that this is a vector bundle over S^1 of fiber dimension 1, but a rigorous proof of this involves checking many details.

Exercise 3.3.14. Give a careful proof that the Möbius bundle (Example 3.3.13) is truly a line bundle over S^1. Show that this bundle is not trivial.

3.4. Cocycles and Geometric Structures

This section is somewhat philosophical. The main point is to give the reader some insight into geometric structures on a manifold determined by subgroups $G \subset \mathrm{Gl}(n)$. A more elegant formulation of these ideas, using the notion of a *principal bundle*, will be taken up in Chapter 11.

Let $\pi : E \to M$ be an n-plane bundle and let $\{W_j\}_{j\in J}$ be a locally trivializing open cover for E, the trivializations being

$$\psi_j : \pi^{-1}(W_j) \to W_j \times \mathbb{R}^n.$$

If $W_i \cap W_j \neq \emptyset$, consider

$$(W_i \cap W_j) \times \mathbb{R}^n \xrightarrow{\psi_i^{-1}} \pi^{-1}(W_i \cap W_j) \xrightarrow{\psi_j} (W_i \cap W_j) \times \mathbb{R}^n.$$

This composition must have the form

$$\psi_j\psi_i^{-1}\left(x, \begin{bmatrix} a^1 \\ \vdots \\ a^n \end{bmatrix}\right) = \left(x, \gamma_{ji}(x) \cdot \begin{bmatrix} a^1 \\ \vdots \\ a^n \end{bmatrix}\right),$$

where $\gamma_{ji}(x) \in \mathrm{Gl}(n)$, $\forall x \in W_i \cap W_j$. The following is proven by exactly the argument employed in the proof of Lemma 3.3.6.

Lemma 3.4.1. *The map* $\gamma_{ji} : W_i \cap W_j \to \mathrm{Gl}(n)$ *is smooth.*

These smooth maps have the "cocycle" property

$$\gamma_{kj}(x) \cdot \gamma_{ji}(x) = \gamma_{ki}(x), \tag{3.1}$$

$\forall x \in W_i \cap W_j \cap W_k$, $\forall i, j, k \in J$. As usual, this property implies also, for all appropriate choices of x and indices $i, j \in J$,

$$\gamma_{ii}(x) = I_n, \tag{3.2}$$

$$\gamma_{ij}(x) = (\gamma_{ji}(x))^{-1}. \tag{3.3}$$

Again, this "cocycle" terminology comes from cohomology theory (see the *remark* following Exercise 3.3.14).

Definition 3.4.2. A $\mathrm{Gl}(n)$-cocycle on M is a family $\gamma = \{W_j, \gamma_{ji}\}_{i,j\in J}$ such that $\{W_j\}_{j\in J}$ is an open cover of M and $\gamma_{ji} : W_i \cap W_j \to \mathrm{Gl}(n)$ is a smooth map, $\forall i, j \in J$, all subject to the cocycle condition (3.1). If the cocycle γ arises as above from an n-plane bundle E, it is said to be a structure cocycle for E.

Just as a structure cocycle $g = \{U_\alpha, g_{\alpha\beta}\}_{\alpha,\beta\in\mathfrak{A}}$ for M gave all the data necessary for reassembling the smooth manifold, up to diffeomorphism, so a structure cocycle γ for E gives all the data necessary for reassembling the bundle, up to bundle isomorphism. Indeed, given any $\mathrm{Gl}(n)$-cocycle $\gamma = \{W_j, \gamma_{ji}\}_{i,j\in J}$, one assembles an n-plane bundle for which it is a structure cocycle. Here is a quick sketch of the procedure.

Set

$$\widetilde{E}_\gamma = \bigsqcup_{j\in J} W_j \times \mathbb{R}^n$$

and define on $\widetilde{E}_\gamma$ an equivalence relation by setting $(x, v) \sim (y, w)$ whenever $(x, v) \in W_j \times \mathbb{R}^n$, $(y, w) \in W_i \times \mathbb{R}^n$, $x = y \in W_j \cap W_i$ and $w = \gamma_{ij}(x) \cdot v$. The fact that this is an equivalence relation is an obvious consequence of equations (3.1), (3.2), and

(3.3). The standard projections $W_j \times \mathbb{R}^n \to W_j$ then fit together as maps into M to define a map $\tilde{\pi} : \widetilde{E}_\gamma \to M$ that respects the equivalence relation. We obtain

$$\pi : E_\gamma = \widetilde{E}_\gamma/\!\sim \,\to M,$$

and the reader can check that this is a smooth n-plane bundle. In case the cocycle came from local trivializations $\psi_j : \pi^{-1}(W_j) \to W_j \times \mathbb{R}^n$ of a bundle $\pi : E \to M$, the diffeomorphisms ψ_j^{-1} fit together to define $\widetilde{\psi} : \widetilde{E}_\gamma \to E$, again respecting the equivalence relation, and this defines a bundle isomorphism

$$\psi : E_\gamma \to E,$$

as the reader again can check.

Definition 3.4.3. Two $\mathrm{Gl}(n)$-cocycles

$$\gamma = \{W_j, \gamma_{ij}\}_{i,j \in J} \text{ and } \theta = \{V_a, \theta_{ab}\}_{a,b \in A}$$

on the same manifold M are equivalent if they are contained in a common $\mathrm{Gl}(n)$-cocycle on M. The equivalence class of γ will be denoted by $[\gamma]$.

In order for $[\gamma]$ to make sense, one must prove that equivalence of cocycles is an equivalence relation. The following exercise will be useful for this.

Exercise 3.4.4. If two $\mathrm{Gl}(n)$-cocycles on the same manifold contain a common $\mathrm{Gl}(n)$-cocycle, show that they are contained in some common $\mathrm{Gl}(n)$-cocycle.

Lemma 3.4.5. *Equivalence of* $\mathrm{Gl}(n)$*-cocycles is an equivalence relation.*

Proof. The only problem is transitivity. If $\gamma \sim \theta$ and $\theta \sim \delta$, let ψ be a cocycle containing both γ and θ, φ a cocycle containing both θ and δ. Then ψ and φ both contain θ, so Exercise 3.4.4 guarantees that they are contained in a common cocycle ρ. Then $\gamma \subseteq \rho$ and $\delta \subseteq \rho$, so $\gamma \sim \delta$. □

Let $\mathrm{Vect}_n(M)$ denote the set of isomorphism classes $[E]$ of n-plane bundles E on M and let $H^1(M; \mathrm{Gl}(n))$ denote the set of equivalence classes of $\mathrm{Gl}(n)$-cocycles (notation borrowed from algebraic topology, again because of formal analogies with Čech cohomology).

Exercise 3.4.6. (Bundle classification) If γ is a $\mathrm{Gl}(n)$-cocycle on M, prove that the isomorphism class $[E_\gamma] \in \mathrm{Vect}_n(M)$ depends only on the equivalence class $[\gamma] \in H^1(M; \mathrm{Gl}(n))$. This defines a canonical bijective correspondence

$$\mathrm{Vect}_n(M) \leftrightarrow H^1(M; \mathrm{Gl}(n)).$$

By this exercise we identify $\mathrm{Vect}_n(M)$ with $H^1(M; \mathrm{Gl}(n))$.

In the case of the tangent bundle $T(M)$, any smooth atlas $\{(U_\alpha, \varphi_\alpha)\}_{\alpha \in \mathfrak{A}}$, with associated structure cocycle $\{g_{\alpha\beta}\}_{\alpha,\beta \in \mathfrak{A}}$ for M, provides a structure cocycle $\{U_\alpha, Jg_{\alpha\beta}\}_{\alpha,\beta \in \mathfrak{A}}$. There are, of course, structure cocycles for $T(M)$ that are not obtained in this way, but these special cocycles tie together the bundle structure of $T(M)$ and the smooth structure of M in an important way.

Definition 3.4.7. A structure cocycle $\{U_\alpha, Jg_{\alpha\beta}\}_{\alpha,\beta \in \mathfrak{A}}$ for $T(M)$, associated to a smooth atlas on M, will be called a Jacobian cocycle.

Example 3.4.8. If $T(M)$ admits a cocycle $\{U_\alpha, \gamma_{\alpha\beta}\}_{\alpha,\beta\in\mathfrak{A}}$ such that $\gamma_{\alpha\beta} \equiv I$, for all $\alpha, \beta \in \mathfrak{A}$, it turns out that M is parallelizable (Exercise 3.4.9). A much stronger condition would be that $T(M)$ admits a *Jacobian* cocycle $\{U_\alpha, Jg_{\alpha\beta}\}_{\alpha,\beta\in\mathfrak{A}}$ such that $Jg_{\alpha\beta} \equiv I$, for all $\alpha, \beta \in \mathfrak{A}$. In this case, M is said to be *integrably parallelizable.* We will see in the next chapter that this forces M to be diffeomorphic to $T^k \times \mathbb{R}^{n-k}$, for some nonnegative integer $k \leq n$.

Exercise 3.4.9. Prove that the following are equivalent for an n-plane bundle

$$\pi : E \to M.$$

(1) The bundle $\pi : E \to M$ is trivial.
(2) There is a $\mathrm{Gl}(n)$-cocycle $\{W_j, \gamma_{ji}\}_{j,i\in J}$ for the bundle such that $\gamma_{ji}(x) = I_n$, $\forall\, x \in W_j \cap W_i$, $\forall\, i, j \in J$.
(3) There is a smooth function $f : E \to \mathbb{R}^n$ such that

$$f_x = f|E_x : E_x \to \mathbb{R}^n$$

is a linear isomorphism, $\forall\, x \in M$.

In particular, setting $E = T(M)$ and appealing to Proposition 3.3.11, we see that these conditions are equivalent to parallelizability of M.

Definition 3.4.10. Let $G \subseteq \mathrm{Gl}(n)$ be a subgroup and let $\pi : E \to M$ be an n-plane bundle. We say that the structure group of E can be reduced to G if there is a $\mathrm{Gl}(n)$-cocycle $\{W_j, \gamma_{ji}\}_{i,j\in J}$ representing the isomorphism class of E such that $\mathrm{im}(\gamma_{ji}) \subseteq G$, $\forall\, j, i \in J$. Such a cocycle will be called a G-cocycle for E.

Equivalence of G-cocycles can be defined exactly as for the case that $G = \mathrm{Gl}(n)$, giving a "cohomology set" $H^1(M; G)$. Distinct elements of this set may correspond to distinct bundles or to the same bundle with inequivalent G-reductions.

Definition 3.4.11. If E admits a G-cocycle γ, then $[\gamma] \in H^1(M; G)$ is called a G-reduction of E.

Definition 3.4.12. For a subgroup $G \subseteq \mathrm{Gl}(n)$, an *infinitesimal G-structure* on the n-manifold M is a G-reduction $[\gamma]$ of $T(M)$. If $[\gamma]$ contains a Jacobian cocycle, then $[\gamma]$ is said to be *integrable* and will be called simply a G-structure or a geometric structure on M.

In the case of the trivial subgroup $I = \{I_n\} \subset \mathrm{Gl}(n)$, M is parallelizable if there is an infinitesimal I-structure $[\gamma]$ on M. The manifold is integrably parallelizable if there is an integrable $[\gamma] \in H^1(M; I)$.

Example 3.4.13. Let $\mathrm{Gl}_+(n) \subset \mathrm{Gl}(n)$ be the subgroup of matrices with positive determinant. Two ordered bases $(v_1, \dots, v_n)$ and $(w_1, \dots, w_n)$ of an n-dimensional vector space V are said to have the same orientation if the unique matrix $A \in \mathrm{Gl}(n)$ such that

$$(v_1, \dots, v_n)A = (w_1, \dots, w_n)$$

belongs to $\mathrm{Gl}_+(n)$. This is an equivalence relation having exactly two equivalence classes, called *orientations* of V.

A linear isomorphism $L : V \to W$ between n-dimensional vector spaces carries each orientation μ of V to an orientation $L(\mu)$ of W. Indeed, if $(v_1, \dots, v_n)$ represents μ, we take for $L(\mu)$ the orientation represented by $(L(v_1), \dots, L(v_n))$. By linearity, the basis $(v_1, \dots, v_n)A$ will be taken to the basis $(L(v_1), \dots, L(v_n))A$, $\det(A) > 0$, so $L(\mu)$ is well defined.

The *standard orientation* μ_n of $\mathbb{R}^n$ is the orientation class of the standard ordered basis $(e_1, \dots, e_n)$. The other orientation of $\mathbb{R}^n$ will be denoted by the symbol $-\mu_n$. A linear automorphism L of $\mathbb{R}^n$ is orientation-preserving if $L(\mu_n) = \mu_n$ and is orientation-reversing if $L(\mu_n) = -\mu_n$. Clearly, L is orientation-preserving if and only if $\det(L) > 0$.

Let $\pi : E \to M$ be an n-plane bundle and let μ_x be an orientation of E_x, $\forall x \in M$. We will say that μ_x depends continuously on x if, for each $x \in M$, there is a trivialization $\psi : E|U \to U \times \mathbb{R}^n$, $x \in U$, such that $\psi_{*y}(\mu_y) = \mu_n$, $\forall y \in U$. If $\mu = \{\mu_x\}_{x \in M}$ depends continuously on x, we say that μ is an orientation of E and that E is orientable. Given μ, there is always the opposite orientation $-\mu$.

An orientation μ of $T(M)$ is also called an orientation of M. If such exists, M is *orientable* and the pair (M, μ) is an *oriented manifold.* If (M, μ) and (N, ν) are oriented n-manifolds, a diffeomorphism $f : M \to N$ is *orientation-preserving* (or, simply, *oriented*) if $f_{*x}(\mu_x) = \nu_{f(x)}$, $\forall x \in M$.

An orientation of E determines a $\mathrm{Gl}_+(n)$-reduction of E. Indeed, cover M with local trivializations (U_i, ψ_i) as above and remark that the associated cocycle $\gamma_{ij}(x)$ carries the orientation μ_n of $\{x\} \times \mathbb{R}^n$ to itself, $\forall x \in U_i \cap U_j$. That is, $\gamma_{ij}(x) \in \mathrm{Gl}_+(n)$. Conversely, given a $\mathrm{Gl}_+(n)$-reduction, the reader should be able to produce an associated orientation of E. Thus, an infinitesimal $\mathrm{Gl}_+(n)$-structure on M is an orientation of M.

Every infinitesimal $\mathrm{Gl}_+(n)$-structure is integrable. Indeed, fix an orientation μ of $T(M)$ and let $\{U_\alpha, Jg_{\alpha\beta}\}_{\alpha,\beta \in \mathfrak{A}}$ be any Jacobian cocycle such that each coordinate chart (U_α, x_α) is connected. The n-tuple of fields

$$\left(\frac{\partial}{\partial x^1_\alpha}, \dots, \frac{\partial}{\partial x^n_\alpha}\right)$$

defines a continuous orientation μ_α of $T(U_\alpha)$. Since U_α is connected, either $\mu_\alpha = \mu|U_\alpha$ or $\mu_\alpha = -\mu|U_\alpha$. In the latter case, replace the coordinate x^1_α with $-x^1_\alpha$. This is an orientation reversing change of coordinates and the new coordinates give the correct orientation μ_x to $T_x(M)$, $\forall x \in U_\alpha$. Carrying this out for each $\alpha \in \mathfrak{A}$, we produce an atlas on M with associated Jacobian cocycle $\mathrm{Gl}_+(n)$-valued.

Exercise 3.4.14. Let M be connected and set $\mathcal{O}(M) = \{\pm\mu_x \mid x \in M\}$.

(1) Put a topology and differentiable structure on the set $\mathcal{O}(M)$ such that the projection $\pi : \mathcal{O}(M) \to M$ sending $\pm\mu_x \mapsto x$, for each $x \in M$, is a smooth covering map.
(2) Prove that $\mathcal{O}(M)$ is connected if and only if M is nonorientable. In the orientable case, $\mathcal{O}(M)$ falls into two components, each carried diffeomorphically onto M.
(3) Prove that the manifold $\mathcal{O}(M)$ is orientable.
(4) If M is simply connected, prove that M is orientable.

Example 3.4.15. Let $G = \mathrm{O}(n)$. An infinitesimal $\mathrm{O}(n)$-structure is called a *Riemannian structure* and an n-manifold M with a Riemannian structure is called a *Riemannian manifold.* This is the starting point of Riemannian geometry, a subject that we will treat in Chapter 10. A Riemannian structure enables one to define lengths of tangent vectors and of curves, angles between vectors tangent at the same point and between curves through a point, curves that locally minimize length (geodesics), curvature of the manifold, etc.

Exercise 3.4.16. Prove that the manifold M admits an infinitesimal $\mathrm{O}(n)$-structure if and only if there is a positive definite inner product $\langle \cdot, \cdot \rangle_x$ on $T_x(M)$, $\forall\, x \in M$, that varies smoothly with x in the following sense: given any local coordinate chart $(U, x^1, \dots, x^n)$ in the C^∞ structure of M,

$$h_{ij}(x) = \left\langle \left.\frac{\partial}{\partial x^i}\right|_x, \left.\frac{\partial}{\partial x^j}\right|_x \right\rangle_x$$

is smooth on U, $1 \le i, j \le n$.

This smoothly varying inner product on the fibers of $T(M)$ is called a *Riemannian metric* on M. Using this metric, one defines the norm

$$\|\cdot\| : T(M) \to \mathbb{R}^+$$

by $\|v\| = \|v\|_x = \sqrt{\langle v, v \rangle_x}$, whenever $v \in T_x(M)$. Similarly, angles between elements of $T_x(M)$ are defined in the usual way, using the fiberwise inner product. Lengths of smooth curves $s : [a, b] \to M$ are defined by

$$L(s) = \int_a^b \|\dot{s}(t)\|\, dt.$$

After it has been proven that open covers admit smooth, subordinate partitions of unity (Theorem 3.5.4), you will be invited to show that all manifolds admit Riemannian metrics (Exercise 3.5.9). For this, the fact that manifolds are 2nd countable is essential. On the contrary, very few manifolds admit integrable Riemannian structures ("flat" Riemannian manifolds). The obstruction to integrability is the Riemann curvature tensor. This will be treated in Chapter 10, where we will show (Theorem 10.6.7) that curvature 0 is equivalent to the geometry being locally Euclidean.

Definition 3.4.17. Let M and M' be Riemannian manifolds with associated Riemannian metrics $\langle \cdot, \cdot \rangle$ and $\langle \cdot, \cdot \rangle'$. A diffeomorphism $f : M' \to M$ is said to be an *isometry* if, for each $x \in M'$ and all $v, w \in T_x(M')$,

$$\langle f_{*x}(v), f_{*x}(w) \rangle_{f(x)} = \langle v, w \rangle'_x \,.$$

A local diffeomorphism with this property is called a local isometry.

Exercise 3.4.18. If M is a Riemannian manifold and $\pi : M' \to M$ is a covering space, prove that there is a unique Riemannian metric on M' relative to which π is a local isometry.

Example 3.4.19. One can take $G = \mathrm{O}(k, n-k) \subset \mathrm{Gl}(n)$ the group of matrices that leave invariant the quadratic form

$$Q_k(x^1, \dots, x^n) = (x^1)^2 + \cdots + (x^k)^2 - (x^{k+1})^2 - \cdots - (x^n)^2.$$

A discussion similar to the one above shows that such an infinitesimal structure corresponds to a nondegenerate, indefinite inner product or metric (also called a pseudo-Riemannian metric or structure) in each fiber $T_x(M)$, varying smoothly with x. Manifolds equipped with such metrics are called pseudo-Riemannian manifolds. If the infinitesimal $\mathrm{O}(k, n-k)$-structure is integrable, the pseudo-Riemannian manifold and metric are said to be flat.

One can again define the notions of "isometry" and "local isometry" between pseudo-Riemannian manifolds and prove that covering spaces of pseudo-Riemannian manifolds are uniquely pseudo-Riemannian in such a way that the covering projections are local isometries.

The Lorentzian manifolds in relativity theory are 4-manifolds (space-time) with an infinitesimal $O(3,1)$-structure. It is no longer true that all manifolds admit such structures. Integrability places an even more severe restriction on the manifold. The obstruction to integrability is again a (Lorentz) curvature tensor, the physical interpretation of curvature being gravity. Flat Lorentzian manifolds are those in which the curvature is everywhere 0 (no gravity), these space-times being locally equivalent to special relativity.

Example 3.4.20. Consider the subgroup $\mathrm{Gl}(k, n-k) \subset \mathrm{Gl}(n)$ consisting of all matrices of the form

$$\begin{bmatrix} A & B \\ 0 & C \end{bmatrix},$$

where $A \in \mathrm{Gl}(k)$, $C \in \mathrm{Gl}(n-k)$, and B is an arbitrary $k \times (n-k)$ matrix. Since the linear action $\mathrm{Gl}(k, n-k) \times \mathbb{R}^n \to \mathbb{R}^n$ carries the subspace $\mathbb{R}^k \subset \mathbb{R}^n$ isomorphically onto itself, a $\mathrm{Gl}(k, n-k)$-reduction γ of the n-plane bundle $\pi : E \to M$ selects a k-dimensional subspace $F_x \subset E_x$, $\forall x \in M$. Indeed, the equivalence relation $\sim$ on $\widetilde{E}_\gamma$ matches the fiber $\{x\} \times \mathbb{R}^k \subset W_j \times \mathbb{R}^n$ isomorphically to $\{x\} \times \mathbb{R}^k \subset W_i \times \mathbb{R}^n$, whenever $x \in W_j \cap W_i$, so the subspace $\{x\} \times \mathbb{R}^k$ passes to a well-defined subspace $F_x \subset (E_\gamma)_x$ in the quotient.

Definition 3.4.21. A k-plane subbundle F of an n-plane bundle E over M is a k-plane bundle, together with a commutative diagram

$$\begin{array}{ccc} F & \xrightarrow{\ i\ } & E \\ {\scriptstyle\pi}\downarrow & & \downarrow{\scriptstyle\pi} \\ M & \xrightarrow[\mathrm{id}]{} & M \end{array}$$

such that $i_x : F_x \to E_x$ is a linear monomorphism, $\forall x \in M$. A k-plane subbundle of $T(M)$ is also called a k-plane distribution on M.

We can view i as an inclusion map $F \hookrightarrow E$. The choices $\{F_x\}_{x \in M}$, defined by a $\mathrm{Gl}(k, n-k)$-reduction, are the fibers of a k-plane subbundle.

Exercise 3.4.22. Prove that a $\mathrm{Gl}(k, n-k)$-reduction of the bundle E defines a k-plane subbundle F of E with fibers F_x as above. Show that, given a k-plane subbundle $F \hookrightarrow E$, there is a $\mathrm{Gl}(k, n-k)$-reduction of E that gives back this subbundle.

Thus, an infinitesimal $\mathrm{Gl}(k, n-k)$-structure on M is a k-plane distribution on M. If this structure is integrable, it will be called a *foliation* of M of dimension k. These geometric structures will be studied in the next chapter, together with the integrability condition, the Frobenius theorem.

Example 3.4.23. The complex general linear group $\mathrm{Gl}(n, \mathbb{C})$ is the group of nonsingular, $n \times n$ matrices with complex entries. If we write the elements of this group as $A + \sqrt{-1}B$, where A and B are real matrices, then one can check that

$$\begin{bmatrix} A & B \\ -B & A \end{bmatrix} \in \mathrm{Gl}(2n).$$

In fact, this realizes the complex general linear group as a subgroup

$$\mathrm{Gl}(n, \mathbb{C}) \subset \mathrm{Gl}(2n).$$

A $2n$-manifold M, together with an infinitesimal $\mathrm{Gl}(n, \mathbb{C})$-structure, is known as an *almost complex manifold.* In this case, one can define a bundle isomorphism

$$J : T(M) \to T(M)$$

such that $J^2 = -\,\mathrm{id}_{T(M)}$ (the reader who has been successful with Exercises 3.4.16 and 3.4.22 will be able to check this). One extends the fiberwise scalar multiplication

$$\mathbb{R} \times T(M) \to T(M)$$

to a complex scalar multiplication

$$\mathbb{C} \times T(M) \to T(M)$$

by the formula

$$(a + \sqrt{-1}b)v = av + bJ(v),$$

$\forall\, a, b \in \mathbb{R}, \forall\, v \in T_x(M), \forall\, x \in M$. Effectively, the tangent bundle becomes a vector bundle over $\mathbb{C}$ rather than $\mathbb{R}$, of complex fiber dimension n, called the complex tangent bundle.

If the almost complex structure is integrable, it can be shown that the manifold admits an atlas $(U_\alpha, \varphi_\alpha)$ where $\varphi_\alpha : U_\alpha \to \mathbb{C}^n$ and the local coordinate changes $g_{\alpha\beta}$ are complex analytic. That is, M has the structure of a complex analytic manifold and the integrable almost complex structure will be called a complex analytic structure. One defines a "holomorphic tangent bundle" and shows that it is isomorphic, as a complex vector bundle, to the complex tangent bundle associated as above to the almost complex structure. The details of all this, as well as the integrability condition, are advanced topics that will not be treated in this book.

Exercise 3.4.24. If $\pi : M' \to M$ is a covering space and if M admits an infinitesimal G-structure, show that this structure "lifts" canonically to an infinitesimal G-structure on M'. Prove that the lifted structure is integrable if the one on M is integrable.

3.5. Global Constructions of Smooth Functions

The proof of Theorem 2.6.1 adapts easily to the global case.

Proposition 3.5.1. *Let M be an n-manifold, let $U \subseteq M$ be an open subset and $K \subset U$ a compact subset. Then there is a smooth function $f : M \to \mathbb{R}$ such that $f|K \equiv 1$ and* $\mathrm{supp}(f) \subset U$.

Proof. For each $p \in K$, choose a coordinate neighborhood (U_p, z_p) about p, $U_p \subseteq U$, and an open, n-dimensional interval A_p with $\overline{A}_p \subset U_p$, centered at p. Since K is compact, finitely many of the A_p cover K. The proof of Theorem 2.6.1 now goes through unchanged. □

This is not quite the global C^∞ Urysohn Lemma (Corollary 3.5.5) in which the set K is assumed only to be closed, not compact. Actually, Proposition 3.5.1 suffices for most purposes and we will use one of its standard consequences, the existence of smooth partitions of unity, in order to prove the general version.

One useful application of Proposition 3.5.1 is the following.

Lemma 3.5.2. *Let M be a smooth manifold and let $x \in M$. Then the natural map*

$$C^\infty(M) \to \mathfrak{G}_x$$

that carries $f \mapsto [f]_x$ is surjective.

Proof. Indeed, given $[g]_x \in \mathfrak{G}_x$, find $\varphi \in C^\infty(M)$ with

$$\operatorname{supp}(\varphi) \subset \operatorname{dom}(g)$$

and $\varphi \equiv 1$ on some compact neighborhood of x in $\operatorname{dom}(g)$. Then φg extends by 0 to a smooth function f on M and $[f]_x = [g]_x$. □

Another application is to identify $\mathfrak{X}(M)$ with the Lie algebra $\mathfrak{D}(C^\infty(M))$ of derivations of the function algebra $C^\infty(M)$. Indeed, each $X \in \mathfrak{X}(M)$ is a derivation $X : C^\infty(M) \to C^\infty(M)$ in the obvious way. In the local case, $M = U \subseteq \mathbb{R}^n$, the proof of the reverse inclusion used Theorem 2.6.1 to show how to localize an arbitrary operator $D \in \mathfrak{D}(C^\infty F(U))$ to be a derivative $D_x \in T_x(U)$, and one uses Proposition 3.5.1 in the same way for $D \in \mathfrak{D}(C^\infty(M))$.

Proposition 3.5.3. *The set $\mathfrak{X}(M)$ is the Lie algebra of derivations of the algebra $C^\infty(M)$.*

One advantage to this point of view is that the Lie bracket is defined intrinsically on $\mathfrak{X}(M)$. The alternative is to use the local definitions of bracket in coordinate charts and show that the formulas for the bracket transform correctly under coordinate changes so that the local definitions fit together to give $[X, Y] \in \mathfrak{X}(M)$ globally.

Theorem 3.5.4. *If $\mathcal{U} = \{U_\alpha\}_{\alpha\in\mathfrak{A}}$ is an open cover of M, there is a C^∞ partition of unity $\{\lambda_\alpha\}_{\alpha\in\mathfrak{A}}$ subordinate to $\mathcal{U}$.*

Proof. First remark that, if $\mathcal{W} = \{W_\beta\}_{\beta\in\mathfrak{B}}$ is a locally finite refinement of $\mathcal{U}$, a smooth partition of unity subordinate to $\mathcal{W}$ induces a smooth partition of unity subordinate to $\mathcal{U}$. Indeed, let $i : \mathfrak{B} \to \mathfrak{A}$ be a map such that $W_\beta \subseteq U_{i(\beta)}$, $\forall\, \beta \in \mathfrak{B}$. If $\{\mu_\beta\}_{\beta\in\mathfrak{B}}$ is a partition of unity subordinate to $\mathcal{W}$, define $\lambda_\alpha = \sum_{\beta\in i^{-1}(\alpha)} \mu_\beta$, $\forall\, \alpha \in \mathfrak{A}$. If $i^{-1}(\alpha) = \emptyset$, we understand that $\lambda_\alpha \equiv 0$. Since $\mathcal{W}$ is locally finite, $\bigcup_{\beta\in i^{-1}(\alpha)} \operatorname{supp} \mu_\beta$ is closed (Exercise 1.4.2), hence is the support of λ_α. It should be clear, then, that $\{\lambda_\alpha\}_{\alpha\in\mathfrak{A}}$ is a partition of unity subordinate to $\mathcal{U}$.

By the above remark and the fact that manifolds are locally compact, we lose no generality in assuming that each U_α has compact closure in M. Thus, the precise refinement $\mathcal{V} = \{V_\alpha\}_{\alpha\in\mathfrak{A}}$ found in Lemma 1.4.8 has the property that $\overline{V}_\alpha \subset U_\alpha$ is a compact subset, $\forall\, \alpha \in \mathfrak{A}$. In the proof of the existence of continuous partitions of unity (Theorem 1.4.11), appeal was made to the general form of Urysohn's lemma. Since each $\overline{V}_\alpha$ is compact, Proposition 3.5.1 can be used in the same way to complete the proof of existence of smooth partitions of unity. □

At this point, we can remove the compactness hypothesis in Proposition 3.5.1.

Corollary 3.5.5 (The global smooth Urysohn lemma). *Let $U \subseteq M$ be an open subset of a smooth manifold and $K \subseteq U$ a subset that is closed in M. Then there is a smooth function $f : M \to \mathbb{R}$ such that $f|K \equiv 1$ and $\operatorname{supp}(f) \subset U$.*

Proof. Let $W = M \setminus K$. Then $\{U, W\}$ is an open cover of M, and we take a subordinate smooth partition of unity $\{\lambda_U, \lambda_W\}$. Since

$$\lambda_W | K \equiv 0,$$
$$\lambda_U + \lambda_W \equiv 1,$$

the function $f = \lambda_U$ is as required. □

Recall that, if $X \subset \mathbb{R}^n$ is an arbitrary subset, a function $f : X \to \mathbb{R}^k$ is said to be smooth if it extends to a smooth function $\tilde{f} : U \to \mathbb{R}^k$, where U is some open neighborhood of X in $\mathbb{R}^n$. This definition continues to work well when $X \subset M$, but it is not very convenient when the target space is a smooth manifold that is not an open subset of Euclidean space. The following result, another application of smooth partitions of unity, suggests a slightly different, backwardly compatible definition of smoothness that is more useful.

Proposition 3.5.6. *If M is a smooth manifold and $X \subseteq M$, then*

$$f : X \to \mathbb{R}^k$$

is smooth if and only if, $\forall x \in X$, $\exists$ an open neighborhood $U_x \subseteq M$ of x, and a smooth map $f_x : U_x \to \mathbb{R}^k$ such that $f_x|(U_x \cap X) = f|(U_x \cap X)$.

Proof. This property clearly follows from our definition of smoothness. We must recover our definition from this property. Let $U = \bigcup_{x \in X} U_x$. Then there is a smooth partition of unity $\{\lambda_x\}_{x \in X}$ on U, subordinate to the open cover $\{U_x\}_{x \in X}$ of the manifold U. Since each f_x is $\mathbb{R}^k$-valued, $\lambda_x f_x$ makes sense and can be interpreted as a smooth map of U into $\mathbb{R}^k$. Then define

$$\tilde{f} = \sum_{x \in X} \lambda_x f_x,$$

a smooth map of U into $\mathbb{R}^k$. Evidently,

$$\tilde{f}(y) = \left(\sum_{x \in X} \lambda_x(y) \right) f(y) = 1 \cdot f(y) = f(y),$$

$\forall y \in X$, so $\tilde{f}$ is the required smooth extension of f to the neighborhood U of X. □

Definition 3.5.7. A function $f : X \to Y$ from a subset $X \subseteq M$ of a smooth manifold M into a subset $Y \subseteq N$ of a smooth manifold N is said to be smooth if, for each $x \in X$, there is an open neighborhood $U_x \subseteq M$ of x and a smooth map $f_x : U_x \to N$ such that $f_x|(U_x \cap X) = f|(U_x \cap X)$. Such a map is a diffeomorphism of X onto Y if it is bijective and both f and f^{-1} are smooth.

As an application, we prove and globalize a smooth version of Theorem 1.1.9.

Theorem 3.5.8 (Smooth invariance of domain). *Let M and N be C^∞ manifolds of the same dimension n. If $U \subseteq M$ is open, if $X \subseteq N$, and if $\varphi : U \to X$ is a diffeomorphism, then X is open in N.*

Proof. Let $x_0 \in U$ and $\varphi(x_0) \in X$. Since $\varphi^{-1} : X \to U$ is smooth, we can produce an open neighborhood V of $\varphi(x_0)$ in N and a smooth extension $\psi : V \to M$ of $\varphi^{-1}|(V \cap X)$. Since $\varphi : U \to X$ is continuous, $\widetilde{V} = \varphi^{-1}(V \cap X)$ is an open neighborhood of x_0 in U and

$$\psi \circ \varphi | \widetilde{V} = \varphi^{-1} \circ \varphi | \widetilde{V} = \mathrm{id}_{\widetilde{V}}.$$

Since $\varphi : U \to N$ is smooth in the usual sense, the chain rule gives

$$d\psi_{\varphi(x_0)} \circ d\varphi_{x_0} = \mathrm{id}_{T_{x_0}(M)},$$

so $d\varphi_{x_0} : T_{x_0}(M) \to T_{\varphi(x_0)}(N)$ is a linear isomorphism. By the inverse function theorem, there is an open neighborhood $W \subseteq \widetilde{V} \subseteq U$ of x_0 that is carried by φ diffeomorphically onto an open subset $\varphi(W) \subseteq N$. But $\varphi(x_0)$ is an arbitrary point of X and $\varphi(x_0) \in \varphi(W) \subseteq X$, so X is an open subset of N. □

We close this section with some exercises that can be solved using the tools we have developed.

Exercise 3.5.9. Use the existence of partitions of unity to prove that every smooth manifold M admits a Riemannian structure. Explain why it is impossible to generalize this argument to prove the existence of an infinitesimal $\mathrm{O}(k, n-k)$-structure.

Exercise 3.5.10. Prove that a compact n-manifold M cannot be diffeomorphic to any subset of $\mathbb{R}^n$.

Exercise 3.5.11. Let M be a smooth manifold, $K \subset M$ a closed subset, $U \supset K$ an open neighborhood of K, $v \in \mathfrak{X}(U)$. Prove that $v|K$ extends to a smooth vector field on all of M.

Exercise 3.5.12. Let $\mathfrak{U} = \{U_\alpha\}_{\alpha\in\mathfrak{A}}$ be an open cover of the smooth manifold M. For each $\alpha \in \mathfrak{A}$, let $\varphi_\alpha : M \to \mathbb{R}$ have a constant value $c_\alpha > 0$ on U_α and be identically 0 on the complement $M \smallsetminus U_\alpha$. Set

$$\varphi = \sum_{\alpha\in\mathfrak{A}} \varphi_\alpha : M \to [0, \infty]$$

and construct a smooth function $f : M \to \mathbb{R}$ such that $0 < f < \varphi$ everywhere on M.

3.6. Smooth Manifolds with Boundary

Manifolds with boundary were introduced in Section 1.6 from the purely topological point of view. Here we introduce the smooth version.

Since Euclidean half-space $\mathbb{H}^n$ is a subset of the smooth manifold $\mathbb{R}^n$, Definition 3.5.7 allows us to talk about smooth maps and diffeomorphisms between open subsets of $\mathbb{H}^n$. Thus, if M is a topological manifold with boundary, we can define the notion of C^∞-relatedness of $\mathbb{H}^n$-charts on M in the obvious way. So we can define a differentiable structure on M to be a maximal $\mathbb{H}^n$-atlas $\mathcal{A}$ of C^∞-related charts. As in the topological case, we define

$$\partial M = \{x \in M \mid \exists (U_\alpha, \varphi_\alpha) \in \mathcal{A},\ x \in U_\alpha,\ \varphi_\alpha(x) \in \partial\mathbb{H}^n\},$$
$$\mathrm{int}(M) = \{x \in M \mid \exists (U_\alpha, \varphi_\alpha) \in \mathcal{A},\ x \in U_\alpha,\ \varphi_\alpha(U_\alpha) \subseteq \mathrm{int}(\mathbb{H}^n)\}.$$

The pair $(M, \mathcal{A})$ is a (smooth) n-manifold with boundary ∂M. Of course, all smooth n-manifolds without boundary are special cases, as is $\mathbb{H}^n$ itself. The notion of smooth maps between manifolds with boundary is defined exactly as in the boundaryless case.

The following is an immediate corollary of Theorem 3.5.8.

Proposition 3.6.1. *If $U \subseteq \mathbb{H}^n \smallsetminus \partial\mathbb{H}^n$ is open, then U is not diffeomorphic to an open subset $V \subseteq \mathbb{H}^n$ such that $V \cap \partial\mathbb{H}^n \neq \emptyset$.*

Corollary 3.6.2. *If M is a manifold with boundary, then*

$$M \smallsetminus \partial M = \operatorname{int}(M).$$

Proof. Indeed, the proposition shows that

$$\partial M = \{x \in M \mid \nexists\, (U_\alpha, \varphi_\alpha) \in \mathcal{A},\, x \in U_\alpha,\, \varphi_\alpha(U_\alpha) \subseteq \operatorname{int}(\mathbb{H}^n)\}.$$

□

Corollary 3.6.3. *If $\varphi : M \to N$ is a diffeomorphism, then*

$$\varphi(\operatorname{int}(M)) = \operatorname{int}(N),$$
$$\varphi(\partial M) = \partial N.$$

Corollary 3.6.4. *A smooth n-manifold M with boundary is also a smooth n-manifold $\Leftrightarrow M = \operatorname{int}(M) \Leftrightarrow \partial M = \emptyset$.*

Corollary 3.6.5. *If M is a smooth n-manifold with boundary, then ∂M is a smooth $(n-1)$-manifold and $\operatorname{int}(M)$ is a smooth n-manifold.*

Exercise 3.6.6. Let M and N be smooth manifolds of dimension m and n respectively. If $\partial M \neq \emptyset = \partial N$, show how to use the differentiable structures on these two manifolds to give $M \times N$ the structure of a smooth $(m+n)$-manifold with nonempty boundary. If both M and N have nonempty boundary, show that $M \times N$ is a *topological* manifold with boundary. Discuss the problem you encounter in trying to give $M \times N$ a natural *smooth* structure of manifold with boundary. This is, in fact, an example of what is called a smooth manifold with corners.

For $x \in \operatorname{int}(M)$, the n-dimensional tangent space $T_x(M)$ is defined as before. If $x \in \partial M$, $T_x(\partial M)$ is only $(n-1)$-dimensional, but we are going to define an n-dimensional tangent space $T_x(M)$ having $T_x(\partial M)$ as a subspace.

The more intuitively appealing approach is that of infinitesimal curves. Indeed, the notion of smooth curve

$$s : [0, \epsilon) \to M, \;\; s(0) = x,$$

makes good sense and we can define the corresponding infinitesimal curve $\langle s \rangle_x$. This can be thought of as a tangent vector to M at x. In this way, one obtains all the vectors tangent to the boundary and all the vectors that point "into" M. Unfortunately, the negatives of the inward vectors are not so obtained and the resulting set of tangent vectors is not a vector space. Alternatively, the definition of tangent vectors as derivatives of the algebra of germs of C^∞ functions at x, while less intuitive, yields a vector space exactly as before. However, in order to show that this vector space is n-dimensional, we will find it convenient to turn to infinitesimal curves.

If $x \in M$, whether or not $x \in \partial M$, the set $C^\infty(M, x)$ of smooth, real valued functions defined in neighborhoods of x is defined and there is no problem defining germinal equivalence on this set. Thus, we obtain the $\mathbb{R}$-algebra $\mathfrak{G}_x$ of germs. We define $T_x(M)$ to be the vector space of derivatives $D : \mathfrak{G}_x \to \mathbb{R}$. If $x \in \operatorname{int}(M)$, this agrees with our usual definition and is n-dimensional.

If $f : M \to N$ is a smooth map between manifolds with boundary, the differentials

$$df_x = f_{*x} : T_x(M) \to T_{f(x)}(N)$$

are defined by the usual formula

$$df_x(D)[g]_{f(x)} = D[g \circ f]_x$$

and are linear. The proof of the global chain rule (Lemma 3.1.24) goes through equally well in our present context.

Lemma 3.6.7. *If $f : M \to N$ and $g : N \to P$ are smooth maps between manifolds with boundary, then $g \circ f$ is smooth and, for each $x \in M$,*

$$d(g \circ f)_x = dg_{f(x)} \circ df_x.$$

Corollary 3.6.8. *If $f : M \to N$ is a diffeomorphism between manifolds with boundary, then $df_x : T_x(M) \to T_{f(x)}(N)$ is a linear isomorphism, $\forall\, x \in M$.*

We have reached the point where a little work has to be done. We must show that $T_x(\mathbb{H}^n)$ is n-dimensional, even when $x \in \partial\mathbb{H}^n$. The above considerations will then extend this property to arbitrary manifolds with boundary. When $x \in \partial\mathbb{H}^n \subset \mathbb{R}^n$, we will use the notation $\mathfrak{G}_x(\mathbb{R}^n)$ for the algebra of germs of $C^\infty(\mathbb{R}^n, x)$ and $\mathfrak{G}_x(\mathbb{H}^n)$ for the germ algebra of $C^\infty(\mathbb{H}^n, x)$.

Lemma 3.6.9. *Let $x \in \partial\mathbb{H}^n$ and let $\rho : \mathfrak{G}_x(\mathbb{R}^n) \to \mathfrak{G}_x(\mathbb{H}^n)$ be defined by $\rho[f]_x = [f|(\mathbb{H}^n \cap \operatorname{dom}(f))]_x$. Then ρ is a surjection.*

Proof. Let $U \subseteq \mathbb{H}^n$ be an open neighborhood of x. If $g : U \to \mathbb{R}$ is smooth, there is a neighborhood V of x in $\mathbb{R}^n$ and a smooth extension $\tilde{g} : V \to \mathbb{R}$ of $g|(V \cap U \cap \mathbb{H}^n)$. Then $[\tilde{g}]_x \in \mathfrak{G}_x(\mathbb{R}^n)$ and $\rho[\tilde{g}]_x = [g]_x$. □

For $x \in \partial\mathbb{H}^n$, define

$$\rho^* : T_x(\mathbb{H}^n) \to T_x(\mathbb{R}^n)$$

by setting

$$\rho^*(D)[f]_x = D(\rho[f]_x).$$

It is elementary that this is linear.

Lemma 3.6.10. *ρ^* is bijective.*

Proof. We prove that ρ^* is one to one. If $\rho^*(D_1) = \rho^*(D_2)$, then $D_1(\rho[f]_x) = D_2(\rho[f]_x)$, $\forall\, [f]_x \in \mathfrak{G}_x(\mathbb{R}^n)$. Since ρ is surjective, it follows that $D_1[g]_x = D_2[g]_x$, $\forall\, [g]_x \in \mathfrak{G}_x(\mathbb{H}^n)$, so $D_1 = D_2$.

We prove that ρ^* is onto. Let $v \in T_x(\mathbb{R}^n) = \mathbb{R}^n$. As an infinitesimal curve, this vector is represented by $s(t) = x + tv$. As an operator on germs, $v = D_{\langle s \rangle_x}$. Either v points into $\mathbb{H}^n$ (we intend this to include the case that v is tangent to $\partial\mathbb{H}^n$) or v points out of $\mathbb{H}^n$, in which case $-v$ points into $\mathbb{H}^n$.

If v points into $\mathbb{H}^n$, then $s(t) \in \mathbb{H}^n$, $\forall\, t \geq 0$. Define $D : \mathfrak{G}_x(\mathbb{H}^n) \to \mathbb{R}$ by

$$D[g]_x = \lim_{t \to 0+} \frac{g(s(t)) - g(x)}{t}.$$

It is elementary that $D \in T_x(\mathbb{H}^n)$ and that $\rho^*(D) = D_{\langle s \rangle_x} = v$.

If v points out of $\mathbb{H}^n$, then $s(t) \in \mathbb{H}^n$, $\forall\, t \leq 0$. Define $D : \mathfrak{G}_x(\mathbb{H}^n) \to \mathbb{R}$ by

$$D[g]_x = \lim_{t \to 0^-} \frac{g(s(t)) - g(x)}{t}.$$

Again, $D \in T_x(\mathbb{H}^n)$ and $\rho^*(D) = v$. □

Corollary 3.6.11. *The vector space $T_x(\mathbb{H}^n)$ is n-dimensional, for all $x \in \partial\mathbb{H}^n$.*

Corollary 3.6.12. *Let M be a smooth n-manifold with boundary and let $x \in \partial M$. Then the vector space $T_x(M)$ is n-dimensional.*

Proof. Let (U, φ) be an $\mathbb{H}^n$-coordinate chart about x. Then

$$\varphi_{*x} : T_x(U) \to T_{\varphi(x)}(\varphi(U))$$

is an isomorphism. But $T_x(U) = T_x(M)$ and $T_{\varphi(x)}(\varphi(U)) = T_{\varphi(x)}(\mathbb{H}^n)$. This latter is n-dimensional. □

At this point, one can define the tangent bundle $\pi : T(M) \to M$

Exercise 3.6.13. For an n-manifold M with boundary, mimic the construction of the tangent bundle

$$\pi : T(M) \to M,$$

showing that one obtains a smooth $2n$-manifold with boundary, the projection π being identified locally with the canonical projection

$$p_1 : \mathbb{H}^n \times \mathbb{R}^n \to \mathbb{H}^n.$$

Vector fields on a manifold M with boundary are smooth sections of $T(M)$. The smooth Urysohn lemma and its consequences extend to this context. In particular, vector fields are derivations of the function algebra $C^\infty(M)$ and open covers always admit smooth, subordinate partitions of unity.

Exercise 3.6.14. Let M be a manifold with nonempty boundary. Show that there is a smooth function

$$f : M \to [0, \infty)$$

such that $\partial M = f^{-1}(0)$.

3.7. Smooth Submanifolds

We give a definition of "submanifold" that applies to manifolds with boundary.

Definition 3.7.1. Let M be an m-manifold, possibly with boundary. A subset $X \subset M$ is a properly imbedded submanifold of dimension n if X is closed in M and, for each $p \in X$, there is an $\mathbb{H}^m$-coordinate chart (U, φ) about p in M in which $\varphi(U \cap X) = \varphi(U) \cap \mathbb{H}^n$, where $\mathbb{H}^n \subset \mathbb{H}^m$ is the (image of the) standard inclusion.

Remark that, in the above definition, $(U \cap X, \varphi|(U \cap X))$ can be viewed as an $\mathbb{H}^n$-coordinate chart on X and that the collection of all such charts makes X a smooth n-manifold with boundary $\partial X = X \cap \partial M$. Thus, if $\partial M = \emptyset$, then $\partial X = \emptyset$ also. Note also that X cannot be tangent to ∂M at any point of ∂X. If $\partial M = \emptyset = \partial X$ and we drop the requirement that X be a closed subset of M, but keep the requirement on local charts, X will be called simply a submanifold of M.

Example 3.7.2. The image of the standard inclusion $\mathbb{H}^n \hookrightarrow \mathbb{H}^m$ is a properly imbedded submanifold.

3.7.A. Regular values and submanifolds. The following is a globalization of Theorem 2.9.6.

Theorem 3.7.3. *Let $f : M \to N$ be a smooth map between manifolds of respective dimensions m and n, and assume that $\partial N = \emptyset$. If $y \in N$ is a regular value simultaneously for f and for $\partial f = f|\partial M$, then $f^{-1}(y)$ is a properly imbedded submanifold of dimension $m - n$.*

Proof. Since f is continuous, $f^{-1}(y)$ is a closed subset of M. Let $p \in f^{-1}(y)$, and find a suitable coordinate chart about p in M. There are two cases.

Case 1. Suppose $p \in \operatorname{int}(M)$. Choose a coordinate neighborhood (U, x) about p such that $U \subseteq \operatorname{int}(M)$. Then y is also a regular value of $f|U$, so Theorem 2.9.6 implies that $f^{-1}(y) \cap U$ is a smooth submanifold of U of dimension $m - n$.

Case 2. Suppose $p \in \partial M$ and let $(U, x^1, x^2, \dots, x^m)$ be an $\mathbb{H}^m$-chart about p in M. Assume that $f(U) \subset W$, where $(W, y^1, \dots, y^n)$ is a coordinate chart about y in which $y = 0$. Let $\partial f|(U \cap \partial M)$ be denoted by $\varphi(x^2, \dots, x^m)$ with component functions $\varphi^1, \dots, \varphi^n$ relative to the coordinates of W. Since p is a regular point for φ, U can be chosen so small that the matrix

$$\begin{bmatrix} \frac{\partial \varphi^1}{\partial x^2} & \cdots & \frac{\partial \varphi^1}{\partial x^m} \\ \vdots & & \vdots \\ \frac{\partial \varphi^n}{\partial x^2} & \cdots & \frac{\partial \varphi^n}{\partial x^m} \end{bmatrix}$$

has constant rank n on $U \cap \partial M$. By a permutation of the coordinates $x^2, \dots, x^m$, it can be assumed that the last $n \times n$ block

$$\begin{bmatrix} \frac{\partial \varphi^1}{\partial x^{m-n+1}} & \cdots & \frac{\partial \varphi^1}{\partial x^m} \\ \vdots & & \vdots \\ \frac{\partial \varphi^n}{\partial x^{m-n+1}} & \cdots & \frac{\partial \varphi^n}{\partial x^m} \end{bmatrix}$$

is nonsingular on $U \cap \partial M$. Choosing U even smaller, if necessary, the corresponding $n \times n$ block in the matrix

$$\begin{bmatrix} \frac{\partial f^1}{\partial x^1} & \frac{\partial f^1}{\partial x^2} & \cdots & \frac{\partial f^1}{\partial x^m} \\ \vdots & \vdots & & \vdots \\ \frac{\partial f^n}{\partial x^1} & \frac{\partial f^n}{\partial x^2} & \cdots & \frac{\partial f^n}{\partial x^m} \end{bmatrix}$$

is also nonsingular. We then resort to the trick of recoordinatizing U near p by setting $z^i = x^i$, $1 \le i \le m - n$, and $z^{m-n+j} = f^j$, $1 \le j \le n$. The inverse function theorem shows, by the above remarks, that this will define an $\mathbb{H}^n$-chart on a small enough neighborhood (again called U) of p. But, relative to these coordinates,

$$f(z^1, z^2, \dots, z^m) = (z^{m-n+1}, \dots, z^m).$$

Then

$$f^{-1}(y) \cap U = \{(z^1, \dots, z^{m-n}, 0, \dots, 0)\}.$$

That is,

$$f^{-1}(y) \cap U = \mathbb{H}^{m-n} \cap U.$$

□

Example 3.7.4. Let $f : \mathbb{H}^{n+1} \to \mathbb{R}$ be given by

$$f(x^1, \dots, x^{n+1}) = \sum_{i=1}^{n+1} (x^i)^2.$$

Then $1 \in \mathbb{R}$ is a regular value both for f and for ∂f. The hemisphere $f^{-1}(1)$ is the intersection $S^n \cap \mathbb{H}^{n+1}$ and is an n-manifold with boundary $f^{-1}(1) \cap \partial \mathbb{H}^{n+1} = S^{n-1}$.

The following lemma shows that there are plenty of regular values as in Theorem 3.7.3.

Lemma 3.7.5. *If $\partial M = \emptyset$ and $f : N \to M$ is smooth, then the set of points in M that are simultaneous regular values for f and ∂f is dense in M.*

Proof. Clearly, if $p \in \partial N$ is a regular point for ∂f, it is also a regular point for f. Thus, $y \in M$ is a regular value both of f and ∂f precisely when it is a regular value both of $f|\operatorname{int}(N)$ and of ∂f. Use countable coordinate coverings $\{U_i\}_{i \in I}$ of $\operatorname{int}(N)$, $\{V_j\}_{j \in J}$ of ∂N, and $\{W_k\}_{k \in K}$ of M. For each $k \in K$, consider the countable family of smooth maps

$$f_{ik} : U_i \cap f^{-1}(W_k) \longrightarrow W_k,$$
$$\partial f_{jk} : V_j \cap \partial f^{-1}(W_k) \longrightarrow W_k$$

obtained by restrictions. By Corollary 2.9.5, almost every $y \in W_k$ is a common regular value of all these maps. Doing this for each $k \in K$, we complete the proof. □

Exercise 3.7.6. Suppose that U and V are open subsets of $\mathbb{H}^n$ and that $f : U \to V$ is a smooth map such that $f(\partial U) \subset \partial V$. If $x \in \partial U$ and $f_{*x} : T_x(U) \to T_{f(x)}(V)$ is an isomorphism, prove that f restricts to a diffeomorphism of some neighborhood U' of x onto some neighborhood V' of $f(x)$. This extends the inverse function theorem to open subsets of $\mathbb{H}^n$, hence to manifolds with boundary.

3.7.B. Maps of constant rank and submanifolds.

For simplicity, the discussion in this subsection will be carried out only for the case of manifolds with empty boundary.

Definition 3.7.7. A smooth map $f : N \to M$ of an n-manifold into an m-manifold has constant rank r if, for each $p \in N$, the rank of the linear map f_{*p} is r. The map is an immersion if f has constant rank n and it is a submersion if it has constant rank m.

Exactly as Theorem 2.9.6 globalizes to Theorem 3.7.3, so does the constant rank theorem (Theorem 2.4.6). The statement follows and details are left to the reader.

Theorem 3.7.8 (Global constant rank theorem). *If $f : N \to M$ has constant rank r and if $p \in f(N)$, then $f^{-1}(p) \subseteq N$ is a smooth, properly imbedded submanifold of dimension $n - r$.*

This result is not very striking for immersions and, for submersions, it is just Theorem 3.7.3 for maps with no critical values. Since we are assuming empty boundaries, the term "proper imbedding" refers to a smooth imbedding with closed image. Note that this agrees with the usual topological notion of a "proper map", this being a map that pulls back compact sets to compact sets.

Definition 3.7.9. If $i : N \to M$ is a one-to-one immersion, $i(N)$ is called an immersed submanifold of M.

The reader should be warned that many authors call an immersed submanifold $i(N) \subset M$ simply a submanifold. This is misleading because the relative topology inherited from M may not agree with the manifold topology of N.

Exercise 3.7.10. Let $i : N \to M$ be a one-to-one immersion, let X be a manifold, and let $f : X \to M$ be a smooth map with $f(X) \subseteq i(N)$.

(1) Show by an example that $i^{-1} \circ f : X \to N$ may fail to be continuous.
(2) If $i^{-1} \circ f$ is continuous, prove that it is smooth.

3.7.C. Imbeddings in Euclidean space*. First, we note that the existence of smooth partitions of unity allows us to adapt the proof of Theorem 1.5.7 without serious change to prove the following.

Theorem 3.7.11. *If M is a compact, differentiable n-manifold without boundary, then there is a smooth imbedding $i : M \to \mathbb{R}^k$, for some integer $k > n$.*

In fact, the following much more general theorem, due to H. Whitney [**50**], is known. We will prove it for compact manifolds. For a proof of the general case, see [**2**, Chapter 6].

Theorem 3.7.12 (Whitney imbedding theorem). *If M is an arbitrary differentiable n-manifold, then there is a smooth, proper imbedding of M into $\mathbb{H}^{2n+1}$.*

Proof for M compact and $\partial M = \emptyset$. Since $\partial M = \emptyset$, we imbed in $\mathbb{R}^{2n+1}$. By Theorem 3.7.11, we choose a smooth imbedding $M \subset \mathbb{R}^k$ for a suitably large value of $k \geq 2n+1$. If $k = 2n+1$, we are done. We assume that $k > 2n+1$ and show that M imbeds smoothly in $\mathbb{R}^{k-1}$. Finite repetition of this argument then yields the theorem for the compact case.

View $\mathbb{R}^{k-1} \subset \mathbb{R}^k$ as the subspace $x^k = 0$ and let $p : \mathbb{R}^k \to \mathbb{R}$ be projection onto the kth component. For each unit vector $v \in S^{k-1} \smallsetminus \mathbb{R}^{k-1}$, define

$$p_v : \mathbb{R}^k \to \mathbb{R}^{k-1}$$

by

$$p_v(w) = w - \frac{p(w)}{p(v)} v.$$

That is, p_v is the linear projection of $\mathbb{R}^k$ onto $\mathbb{R}^{k-1}$ along v. The idea will be to choose v so that $p_v|M$ actually imbeds M in $\mathbb{R}^{k-1}$.

To begin with, we choose v so that $p_v|M$ is injective. Consider the *diagonal*

$$\Delta = \{(x, x) \mid x \in M\} \subset M \times M$$

and the map

$$f : M \times M \smallsetminus \Delta \to S^{k-1}$$

defined by

$$f(x, y) = \frac{x - y}{\|x - y\|},$$

where $\|w\|$ denotes the usual Euclidean norm of $w \in \mathbb{R}^k$. Since $k - 1 > 2n$, Sard's theorem guarantees that f is not surjective (*cf.*Example 2.9.4), so we choose $v \in S^{k-1}$ not in the image of f. That is, v is not a scalar multiple of $x - y$, for any two distinct points $x, y \in M$. Thus, $p_v(x - y) \neq 0$, whenever x and y are distinct points of M, proving that $p_v|M$ is injective.

If we can prove that $p_v|M$ is an immersion, then compactness of M and injectivity of $p_v|M$ implies that this map is an imbedding. Equivalently, we must find v as above such that, for every nonzero tangent vector $w \in T(M)$, $v \neq w/\|w\|$. Let

$$S(M) = \{w \in T(M) \mid \|w\| = 1\},$$

the so-called *unit tangent bundle* of M. It is convenient to view this as a subset of $M \times S^{k-1}$ and, in Exercise 3.7.13, you will show that it is a smooth, compact submanifold of dimension $2n - 1$. The canonical projection of $M \times S^{k-1}$ onto S^{k-1} restricts to a smooth map

$$g : S(M) \to S^{k-1}$$

that can be viewed as parallel translation in $\mathbb{R}^k \times \mathbb{R}^k$ of unit tangent vectors to M to vectors issuing from the origin. Again, by the dimension hypothesis and Sard's theorem, there is $v \in \operatorname{im} g \cup \operatorname{im} f$. Since p_v is linear, $p_{v*} = p_v$ at each point of M, so p_v is both one-to-one and an immersion. □

Exercise 3.7.13. Prove that $S(M)$ is a smooth submanifold of

$$T(M) \subset M \times \mathbb{R}^k.$$

(Hint. Find a suitable map $\nu : T(M) \to \mathbb{R}$ having 1 as a regular value.)

Exercise 3.7.14. If M is a compact n-manifold without boundary, show that it admits a smooth immersion into $\mathbb{R}^{2n}$.

Remark. The existence of proper imbeddings $M \hookrightarrow \mathbb{H}^{2n+1}$ of compact n-manifolds with boundary is proven by modifying carefully the above proof (*cf.* [**16**, Theorem 4.3 on page 31], where proper imbeddings are called "neat imbeddings"). The noncompact case is proven by suitable modifications of the treatment in [**2**, Chapter 6].

Exercise 3.7.15. Let $M \subset \mathbb{H}^k$ be a properly imbedded n-manifold with boundary and prove that there is a Riemannian metric on $\mathbb{R}^k$, agreeing with the standard Euclidean metric outside of a neighborhood of $\partial\mathbb{H}^k$ and such that, at each point x of ∂M, the orthogonal complement of $T_x(M)$ in $T_x(\mathbb{R}^k)$ lies in $T_x(\partial\mathbb{H}^k)$. We say that, relative to this metric, M meets $\partial\mathbb{H}^k$ orthogonally along its boundary.

Example 3.7.16. If $M \subset \mathbb{H}^n$ is the hemisphere $S^{n-1} \cap \mathbb{H}^n$, then M is properly imbedded in this half-space and meets $\partial\mathbb{H}^n$ orthogonally along its boundary. Here it is not necessary to change the standard Euclidean metric on $\mathbb{R}^n$ near $\partial\mathbb{H}^n$.

If $i : M \to \mathbb{H}^k$ is a smooth, proper imbedding of an n-manifold, we routinely identify M with $i(M)$, as above, realizing $T(M) \subset M \times \mathbb{R}^k$ as a subbundle in the usual way. Via the inclusion $\mathbb{H}^k \subset \mathbb{R}^k$, we view $M \subset \mathbb{R}^k$ wherever convenient. Modifying the Euclidean metric in $\mathbb{R}^k$ as in Exercise 3.7.15, we define

$$\nu(M) = \{(x, v) \in M \times \mathbb{R}^k \mid v \perp T_x(M)\}.$$

If $\partial M = \emptyset$, we view $M \subset \mathbb{R}^k$ and define $\nu(M)$ via the standard Euclidean metric.

Exercise 3.7.17. Prove that the product projection $M \times \mathbb{R}^k \to M$ restricts to a map $\pi : \nu(M) \to M$ that is the projection map of a vector bundle of fiber dimension $k - n$. This is called the *normal bundle* of M in $\mathbb{H}^k$.

Remark that the normal bundle $\nu(M)$ is a manifold of dimension k, generally with boundary $\nu(M)|\partial M$, and that this boundary is exactly the normal bundle of ∂M in $\partial \mathbb{H}^k = \mathbb{R}^{k-1}$. Define a smooth map

$$\varphi : \nu(M) \to \mathbb{H}^k,$$
$$\varphi(x, v) = x + v,$$

using the additive structure of $\mathbb{R}^k$. Note that we can identify the (image of) the zero section $\{(x,0)|x \in M\} \subset \nu(M)$ with M and that, under this identification, $\varphi|M = \mathrm{id}_M$.

Proposition 3.7.18. *If $M \subset \mathbb{H}^k$ is a smooth, properly imbedded submanifold of dimension n, there is an open neighborhood U of M in $\nu(M)$ that is carried diffeomorphically by φ onto an open neighborhood $\varphi(U)$ of M in $\mathbb{H}^k$. If M is compact, this neighborhood can be taken to be of the form $U(\epsilon) = \{(x,v) \in \nu(M) \mid \|v\| < \epsilon\}$, for suitably small $\epsilon > 0$.*

Proof. We give the proof for the case that M is compact, leaving the general case as Exercise 3.7.19. Let $(x,0) \in M \subset \nu(M)$ and remark that there is a natural identification

$$T_{(x,0)}(\nu(M)) = T_x(M) \oplus \nu_x(M).$$

Relative to this identification, we can write

$$\varphi_{*(x,0)} = \mathrm{id}_{T_x(M)} \oplus \mathrm{id}_{\nu_x(M)},$$

so the inverse function theorem (if $x \in \partial M$, use Exercise 3.7.6) guarantees that there is a neighborhood U_x of $(x,0)$ in $\nu(M)$ carried diffeomorphically by φ onto a neighborhood of x in $\mathbb{H}^k$. This neighborhood can be taken to be of the form

$$U_x(\epsilon_x) = \{(y,v) \in \nu(M) \mid y \in W_x, \|v\| < \epsilon_x\},$$

where W_x is an open neighborhood of x in M and $\epsilon_x > 0$ is small enough.

Cover M with finitely many sets of the form $U_x(\epsilon_x)$, let ϵ be the smallest ϵ_x that occurs, and consider the union of the corresponding neighborhoods $U_x(\epsilon)$. This is an open neighborhood of M in $\nu(M)$ of the form

$$U(\epsilon) = \{(x,v) \in \nu(M) \mid \|v\| < \epsilon\}.$$

Although φ is locally a diffeomorphism on $U(\epsilon)$, it might fail to be globally one-to-one. We claim that, by choosing $\epsilon > 0$ smaller, if necessary, we can make sure that φ is one-to-one on $U(\epsilon)$. If not, we could choose sequences $(x_n, v_n) \neq (y_n, w_n)$ in $U(\epsilon)$ such that

$$\varphi(x_n, v_n) = \varphi(y_n, w_n)$$

while $\|v_n\| \le 1/n$ and $\|w_n\| \le 1/n$. Passing to a subsequence, if necessary, we assume that $x_n \to x$ and $y_n \to y$ in M, hence that $(x_n, v_n) \to (x,0)$ and $(y_n, w_n) \to (y,0)$ as $n \to \infty$. By continuity, $\varphi(x,0) = \varphi(y,0)$ and, since φ is one-to-one on M, we conclude that $x = y$. Thus, all (x_n, v_n) and (y_n, w_n) ultimately belong to a neighborhood $U_x(\epsilon_x)$ as in the first paragraph, contradicting the fact that φ is one-to-one on this neighborhood. □

Exercise 3.7.19. Extend the above proof to properly imbedded noncompact manifolds. For this, use local compactness and 2nd countability to express M as a countable increasing union of open, relatively compact submanifolds W_i. The neighborhood U will be a union of neighborhoods $U_i(\epsilon_i)$ of W_i in $\nu(M)$ with $\epsilon_i \downarrow 0$ as $i \to \infty$.

Remark that the bundle projection of $\nu(M)$ onto M induces a map

$$\pi : U \to M$$

such that $\pi|M = \mathrm{id}_M$. Such a map is called a *retraction* and M is said to be a *retract* of U. The triple (U, π, M) is called a *normal neighborhood* of M in $\mathbb{R}^k$, although one usually calls U itself the normal neighborhood.

3.8. Smooth Homotopy and Smooth Approximations

We will study maps $f : M \to N$ between manifolds. The set of all such smooth maps will be denoted by $C^\infty(M, N)$, the set of continuous ones by $C^0(M, N)$. We will show that continuous maps admit arbitrarily small perturbations (homotopies) to smooth ones and that, in this way, the continuous homotopy classes of continuous maps correspond one-to-one to the smooth homotopy classes of smooth maps.

The proof of the smoothing theorem and the equivalence of smooth and continuous homotopy can be omitted in a first reading without seriously disrupting later topics. The following subsection, however, contains basic definitions and results that should not be omitted.

3.8.A. Smooth homotopies. To begin with, we assume that $\partial M = \emptyset$. This avoids manifolds with corners in the following definition. An alternative and equivalent definition of smooth homotopy will then be given that accomodates manifolds with boundary.

Definition 3.8.1. Elements $f_0, f_1 \in C^\infty(M, N)$ are said to be smoothly homotopic if there is a smooth map $H : M \times [0, 1] \to N$ such that

(1) $f_0(x) = H(x, 0)$, $\forall x \in M$;
(2) $f_1(x) = H(x, 1)$, $\forall x \in M$.

We write $f_0 \sim f_1$. The map H is called a (smooth) homotopy between f_0 and f_1.

We frequently drop the words "smoothly" and "smooth", when this qualification is clear from the context. One should think of a homotopy as a deformation of one smooth map to another through smooth maps. It can be thought of as a "smooth curve" in $C^\infty(M, N)$ connecting f_0 to f_1. We write $f_t(x) = H(x, t)$, $0 \le t \le 1$. Similarly, if $\mathrm{Diff}(M)$ denotes the set of all diffeomorphisms of M to itself, "smooth curves" in $\mathrm{Diff}(M)$ will be smooth deformations, called isotopies, of one diffeomorphism to another through diffeomorphisms.

Definition 3.8.2. If $f_0, f_1 \in \mathrm{Diff}(M)$, a (smooth) homotopy f_t between f_0 and f_1 will be called a (smooth) isotopy of f_0 to f_1 if $f_t \in \mathrm{Diff}(M)$, $0 \le t \le 1$. If such an isotopy exists, we say that f_0 is isotopic to f_1 and we write $f_0 \approx f_1$.

Exercise 3.8.3. If $f_0 \sim f_1$ (respectively, $f_0 \approx f_1$), prove that there exists a homotopy (respectively, an isotopy) $H : M \times [0, 1] \to N$ such that $f_t = f_0$, $0 \le t \le \epsilon$, and $f_t = f_1$, $1 - \delta \le t \le 1$, for suitably small $\epsilon > 0$ and $\delta > 0$. Use this to prove that homotopy (respectively, isotopy) is an equivalence relation on $C^\infty(M, N)$ (respectively, on $\mathrm{Diff}(M)$). The equivalence classes for these relations will be called, respectively, homotopy classes and isotopy classes.

Definition 3.8.4. A diffeomorphism $f \in \mathrm{Diff}(M)$ is compactly supported if there is a compact subset $K \subseteq M$ such that $f|(M \smallsetminus K) = \mathrm{id}_{M \smallsetminus K}$. The set of all compactly supported diffeomorphisms is denoted $\mathrm{Diff}_c(M)$. A compactly supported isotopy between $f_0, f_1 \in \mathrm{Diff}_c(M)$ is an isotopy such that there is a compact subset $C \subseteq M$ with $f_t|(M \smallsetminus C) = \mathrm{id}_{M \smallsetminus C}$, $0 \le t \le 1$.

Exercise 3.8.5. Prove that the set $\mathrm{Diff}_c(M)$ is a group under composition.

Exercise 3.8.6. Prove that compactly supported isotopy is an equivalence relation on the group $\mathrm{Diff}_c(M)$.

Remark. For a compactly supported isotopy, f_t belongs to $\mathrm{Diff}_c(M)$, $0 \le t \le 1$.

Theorem 3.8.7 (Homogeneity lemma). *If N is connected, boundaryless, and $x, y \in N$, then there is $f \in \mathrm{Diff}_c(N)$ and a compactly supported isotopy f_t such that $f(x) = y$ and $f_0 = \mathrm{id}_N$, $f_1 = f$.*

The proof of this theorem uses flows and will be deferred until the next chapter.

Corollary 3.8.8. *If $g \in C^\infty(M, N)$, $y \in N$, and N is connected and boundaryless, then $g \sim \widetilde{g}$ such that y is a regular value of $\widetilde{g}$.*

Proof. By Lemma 3.7.5, we choose a regular value $x \in N$ of g. Let f and H be as in Theorem 3.8.7. Since $f(x) = y$ and f is a diffeomorphism, it follows that y is a regular value of $\widetilde{g} = f \circ g$. But $f_t \circ g$ is a homotopy of $g = f_0 \circ g$ with $\widetilde{g} = f_1 \circ g$. □

The definition of smooth homotopy and isotopy that we have given does not adapt nicely to manifolds with boundary. The problem is that $[0, 1]$ is itself a manifold with boundary, hence $M \times [0, 1]$ will be a manifold with corners when $\partial M \neq \emptyset$. This minor difficulty can be overcome by slightly modifying our definitions.

Definition 3.8.9. If $f_0, f_1 \in C^\infty(M, N)$, these maps are (smoothly) homotopic if there is a smooth map $H : M \times \mathbb{R} \to N$ such that

(1) $H(x, 0) = f_0(x)$, $\forall\, x \in M$;
(2) $H(x, 1) = f_1(x)$, $\forall\, x \in M$.

As usual, we set

$$f_t(x) = H(x, t)$$

and say that $f_0, f_1 \in \mathrm{Diff}(M)$ are isotopic if there is a homotopy between them such that $f_t \in \mathrm{Diff}(M)$, $\forall\, t \in \mathbb{R}$.

Exercise 3.8.10. Prove that, under the second definition of homotopy and isotopy, these continue to be equivalence relations. If $\partial M = \emptyset$, prove that the second definition of homotopy and isotopy is equivalent to the first.

Definition 3.8.11. A map $f \in C^\infty(M, N)$ is a (smooth) homotopy equivalence if there is $g \in C^\infty(N, M)$ such that $f \circ g \sim \mathrm{id}_N$ and $g \circ f \sim \mathrm{id}_M$. In this case, we say that M and N are homotopy equivalent manifolds and that f and g are homotopy inverses of one another.

Example 3.8.12. Let $M \subset \mathbb{H}^k$ be a compact, proper submanifold with normal bundle $\pi : \nu(M) \to M$, and let $U(\epsilon)$ be the open neighborhood of the zero section $M \subset \nu(M)$ as in Proposition 3.7.18. Then

$$\pi|U(\epsilon) \in C^\infty(U(\epsilon), M)$$

is a homotopy equivalence with the inclusion $i : M \hookrightarrow U(\epsilon)$ as a homotopy inverse. Indeed, $\pi|U(\epsilon) \circ i = \mathrm{id}_M$, so we investigate $i \circ \pi|U(\epsilon) : U(\epsilon) \to U(\epsilon)$. But $H : U(\epsilon) \times I \to U(\epsilon)$, defined by

$$H((x, v), t) = (x, tv) = \pi_t(x, v),$$

is clearly a homotopy $i \circ \pi | U(\epsilon) = \pi_0 \sim \pi_1 = \mathrm{id}_{U(\epsilon)}$. Notice that each stage π_t of this homotopy fixes M pointwise. In this situation, M is called a *deformation retract* of $U(\epsilon)$. This is a very special type of homotopy equivalence. Since the normal neighborhood U of M in $\mathbb{H}^k$ is just the image $\varphi(U(\epsilon))$ under the diffeomorphism φ of Proposition 3.7.18, we have shown that M is a deformation retract of its normal neighborhood. Intuitively, we can "shrink" the normal neighborhood U down to M while fixing M itself pointwise.

3.8.B. Smooth approximations*. Topologists usually formulate homotopy theory purely in the topological category. Differential topologists, on the other hand, like to take advantage of the differentiable structure of manifolds in using homotopy theory. The fact that smooth homotopy theory is equivalent (on manifolds) to the purely topological version is due to the approximation theory that we now develop. The key to this is the following classical result.

Theorem 3.8.13 (Stone–Weierstrass Theorem). *Let X be a locally compact topological space, $C(X)$ the algebra of real-valued, continuous functions on X. Let $A \subseteq C(X)$ be a subalgebra containing the constant functions and separating points. That is, for arbitrary $x, y \in X$ such that $x \neq y$, there is $f \in A$ such that $f(x) \neq f(y)$. Then, for each $f \in C(X)$, each compact subset $K \subseteq X$, and each $\epsilon > 0$, there exists $g \in A$ such that $|f(x) - g(x)| < \epsilon$, for all $x \in K$.*

For a proof, see [**8**].

Corollary 3.8.14. *If W and V are open, relatively compact subsets of a manifold N such that $\overline{W} \subset V$, if $\epsilon > 0$ and if $f \in C^0(N, \mathbb{H}^k)$, there is $\widetilde{f} \in C^0(N, \mathbb{H}^k)$, uniformly ϵ-close to f, smooth on W and equal to f on the complement of V.*

Proof. Consider first the case $k = 1$. By the Stone–Weierstrass theorem, we can find $\widehat{f} \in C^\infty(M)$ that is ϵ-close to f on $\overline{V}$. Let $O = N \smallsetminus \overline{W}$ and let $\{\lambda_V, \lambda_O\}$ be a smooth partition of unity subordinate to the open cover $\{V, O\}$ of N. Then

$$\widetilde{f} = \lambda_V \widehat{f} + \lambda_O f$$

has the desired properties. For the general case, apply this argument to each of the coordinate functions of $f = (f^1, \dots, f^k)$, replacing ϵ by $\epsilon/\sqrt{k}$. □

We consider maps $f \in C^0(N, M)$, where N and M are both manifolds. By an appeal to Theorem 3.7.12, we imbed M as a proper submanifold of $\mathbb{H}^k$. In fact, this does not use the full force of the Whitney imbedding theorem since any large enough dimension k will do. Define a topological metric ρ on M by restricting the Euclidean metric of $\mathbb{H}^k$. That is,

$$\rho(x, y) = \|x - y\|.$$

By an ϵ-small perturbation of f, we will mean a C^0 homotopy f_t such that $f_0 = f$ and $\rho(f(x), f_t(x)) < \epsilon$, uniformly for $x \in N$ and $0 \leq t \leq 1$. When we say "there is an arbitrarily small perturbation such that ...", this should be read: "for each $\epsilon > 0$, there is an ϵ-small perturbation such that ...".

We also fix a choice of normal neighborhood $\pi : U \to M$ for the imbedded submanifold M.

Proposition 3.8.15. *If W and V are open, relatively compact subsets of N such that $\overline{W} \subset V$ and if $f \in C^0(N, M)$, there is an arbitrarily small perturbation f_t*

of f such that f_1 is smooth on W and f_t agrees with f on the complement of V, $0 \le t \le 1$.

Proof. By Corollary 3.8.14, produce a map $\widetilde{f} \in C^0(M, \mathbb{R}^k)$ that is smooth on W, agrees with f on the complement of V, and is uniformly so close to f on $\overline{V}$ that the image of $\widetilde{f}$ lies entirely in the normal neighborhood U. Set $\widetilde{f}_t = t\widetilde{f} + (1-t)f$ and note that

$$\|f - \widetilde{f}_t\| \le \|f - \widetilde{f}\|, \quad 0 \le t \le 1,$$

uniformly on N. In the case that $\partial N \neq \emptyset$, t should be replaced with $\tau(t)$, a smooth, $[0,1]$-valued function on $\mathbb{R}$ that is identically 0 on $(-\infty, \delta)$ and identically 1 on $(1-\delta, \infty)$, for some small $\delta > 0$. In any case, the homotopy $\widetilde{f}_t$ takes its values entirely in U, so we can define $f_t = \pi \circ \widetilde{f}_t$. Since π is smooth, f_1 will be smooth on W and the perturbation can be made as small as desired by choosing $\widetilde{f}$ uniformly sufficiently close to f. Since $\widetilde{f}$ agrees with f outside of V, so does f_t, $0 \le t \le 1$. □

Theorem 3.8.16. *Given $f \in C^0(N, M)$, there is an arbitrarily small perturbation of f to a map $\widetilde{f} \in C^\infty(N, M)$ and, if $f, g \in C^0(M, N)$ are homotopic, then $\widetilde{f}$ and $\widetilde{g}$ will be smoothly homotopic. Finally, if f is already smooth on some neighborhood of a closed subset $X \subset N$, then $\widetilde{f}$ can be chosen to agree with f on a smaller neighborhood of X.*

Indeed, if M is compact, the first assertion is an immediate corollary of Proposition 3.8.15. The second assertion follows from the first by smoothly approximating the homotopy $H \in C^0(N \times \mathbb{R}, M)$. Details of the full proof are left to the following exercise.

Exercise 3.8.17. Prove the general case of Theorem 3.8.16 by an infinite sequence of applications of Proposition 3.8.15.

Corollary 3.8.18. *The set of homotopy classes in $C^0(N, M)$ and the set of smooth homotopy classes in $C^\infty(N, M)$ are canonically the same.*

Proof. If $f \in C^\infty(N, M)$, its smooth homotopy class $[f]_\infty$ is a subset of its continuous homotopy class $[f]_0$. This defines a map $\iota : [f]_\infty \mapsto [f]_0$. By Theorem 3.8.16, if $f \in C^0(N, M)$, there is a smooth map $\widetilde{f} \in [f]_0$, so ι is surjective. If $f, g \in C^\infty(N, M)$ are continuously homotopic, we can choose the homotopy to be constant in t near the values 0 and 1, hence smooth on an open neighborhood of the closed subset $X = N \times \{0, 1\}$ of $N \times \mathbb{R}$. By Theorem 3.8.16, we approximate this homotopy by a smooth one that is unchanged on a neighborhood of X. That is, f and g are smoothly homotopic, so the map ι is injective. □

Definition 3.8.19. The set of homotopy classes of maps $f : N \to M$, in either the topological or smooth category, will be denoted by $\pi[N, M]$.

By the above corollary, this definition introduces no ambiguity.

3.9. Degree Theory Modulo 2*

Throughout this section, $\dim M = \dim N > 0$ and $\partial M = \emptyset = \partial N$. The manifold M will be compact and N will be connected.

If $f \in C^\infty(M, N)$, choose a regular value $y \in N$ of f. By Theorem 3.7.3, $f^{-1}(y)$ is a 0-dimensional submanifold, hence a set of isolated points. Being a closed subset of a compact space, it must therefore be a finite set. Let $k = |f^{-1}(y)|$ denote the cardinality of this set, an integer ≥ 0.

Theorem 3.9.1 (Stack of records theorem). *If, under the above hypotheses, $k > 0$, then there exists an open connected neighborhood U of y in N that is evenly covered by f. Indeed, $f^{-1}(U)$ falls into k connected components, each carried by f diffeomorphically onto U.*

Proof. By the inverse function theorem, choose an open neighborhood W_i of p_i in M that is carried by f diffeomorphically onto an open neighborhood $f(W_i)$ of y in N, $1 \leq i \leq k$. Since M is Hausdorff, the neighborhoods W_i can be assumed to be pairwise disjoint. Since M is compact, so is $X = f(M \setminus \bigcup_{i=1}^{k} W_i)$, and this set does not contain y. Then $f(W_1) \cap \cdots \cap f(W_k) \setminus X$ is an open neighborhood of y in N and we let U be the component containing y. Let $U_i = f^{-1}(U) \cap W_i$, $1 \leq i \leq k$. It is obvious that f carries each U_i diffeomorphically onto U and that $\bigcup_{i=1}^{k} U_i \subseteq f^{-1}(U)$. But, if $x \in f^{-1}(U) \setminus \bigcup_{i=1}^{k} U_i$, then $f(x) \in X$, contradicting the fact that $X \cap U = \emptyset$. □

Corollary 3.9.2. *The set R of regular values of f is an open, dense subset of N. The function $\lambda_f : R \to \mathbb{Z}^+$, defined by $\lambda_f(y) = |f^{-1}(y)|$, is constant on each connected component of R.*

Indeed, we already know that R is dense and Theorem 3.9.1 shows that it is open. The theorem also shows that λ_f is locally constant, hence constant on each component.

Definition 3.9.3. If $y \in N$ is a regular value of f, then $\deg_2(f, y) \in \mathbb{Z}_2$ is the mod 2 residue class of $\lambda_f(y)$.

Lemma 3.9.4 (Homotopy lemma). *If $f, g \in C^\infty(M, N)$ are smoothly homotopic and if $y \in N$ is a regular value for both f and g, then $\deg_2(f, y) = \deg_2(g, y)$.*

Proof. Let H be a smooth homotopy of f to g. We consider two cases.

Case 1. The point y is also a regular value of $H : M \times [0, 1] \to N$. By Theorem 3.7.3, $H^{-1}(y)$ is a properly imbedded 1-manifold in $M \times [0, 1]$. This submanifold is compact (as a closed subset of $M \times [0, 1]$) and

$$\partial H^{-1}(y) = H^{-1}(y) \cap (M \times \{0\} \cup M \times \{1\}) = f^{-1}(y) \times \{0\} \cup g^{-1}(y) \times \{1\}.$$

It follows, by Corollary 1.6.15, that $\lambda_f(y) + \lambda_g(y)$ is an even integer, hence that $\deg_2(f, y) = \deg_2(g, y)$.

Case 2. The point y is not a regular value of H. It is, however, a regular value for both f and g, so Corollary 3.9.2 implies that there is an open neighborhood W of y in N on which both λ_f and λ_g are defined and constant. The set of regular values of H is dense, so we choose such a regular value $z \in W$. By Case 1, we get $\deg_2(f, z) = \deg_2(g, z)$, but $\lambda_f(y) = \lambda_f(z)$ and $\lambda_g(y) = \lambda_g(z)$, so $\deg_2(f, y) = \deg_2(g, y)$. □

Let $z \in N$. By Corollary 3.8.8, choose $\tilde{f} \sim f$ for which z is a regular value. Then $\deg_2(\tilde{f}, z)$ is independent of this choice, so we set

$$\deg_2(f, z) = \deg_2(\tilde{f}, z)$$

unambiguously, obtaining a function

$$\deg_2(f) : N \to \mathbb{Z}_2.$$

By Theorem 3.9.1, this function is locally constant. By the connectivity of N, $\deg_2(f)$ is constant.

Definition 3.9.5. The element $\deg_2(f) \in \mathbb{Z}_2$ is called the degree (mod 2) of $f \in C^\infty(M, N)$.

Corollary 3.9.6. *If $f, g \in C^\infty(M, N)$ are homotopic, then*

$$\deg_2(f) = \deg_2(g).$$

Lemma 3.9.7. *If $f : M \to N$ is not surjective, then* $\deg_2(f) = 0$.

Proof. Any $z \in M$ that is not a value of f is a regular value, so $\deg_2(f)$ is the residue class mod 2 of $\lambda_f(z) = |\emptyset| = 0$. □

Corollary 3.9.8. *If N is not compact,* $\deg_2(f) = 0$.

Definition 3.9.9. A map $f \in C^\infty(M, N)$ is essential if it is not homotopic to a constant map. A manifold M is contractible if id_M is not essential.

Corollary 3.9.10. *If $\deg_2(f) \neq 0$, then f is essential.*

Corollary 3.9.11. *If M is compact and connected with empty boundary, then M is not contractible.*

Proof. Indeed, $\deg_2(\mathrm{id}_M) = 1$, so id_M is essential. □

Theorem 3.9.12 (Boundary theorem). *Suppose that $M = \partial W$ for a compact manifold W and let $g \in C^\infty(M, N)$. If g extends to a smooth map*

$$G : W \to N,$$

then $\deg_2(g) = 0$.

Proof. Let $y \in N$ be regular, both for G and $g = \partial G$. Then $G^{-1}(y)$ is a compact, one-dimensional manifold with $\partial G^{-1}(y) = G^{-1}(y) \cap \partial W = g^{-1}(y)$. As usual, this set has an even number of elements, so $\deg_2(g) = 0$. □

Example 3.9.13. Let $f : \mathbb{C} \to \mathbb{C}$ be smooth and let $W \subset \mathbb{C}$ be a compact region bounded by smooth, closed curves. A basic question is whether or not f has a zero in W, assuming that f has no zeros on ∂W. Define

$$g : \partial W \to S^1$$

by

$$g(z) = \frac{f(z)}{|f(z)|}, \quad \forall\, z \in \partial W,$$

a smooth function between compact 1-manifolds, S^1 being connected. If $f(z) \neq 0$, $\forall\, z \in W$, then g extends smoothly to $G : W \to S^1$ by

$$G(z) = \frac{f(z)}{|f(z)|}, \quad \forall\, z \in W,$$

hence $\deg_2(g) = 0$. This will be enough to prove "half" of the fundamental theorem of algebra.

Theorem 3.9.14 (Fundamental theorem of algebra). *If $f : \mathbb{C} \to \mathbb{C}$ is a polynomial of odd degree m, then f has a zero in $\mathbb{C}$.*

Proof. No generality is lost in assuming that f has leading coefficient 1. Write

$$f(z) = z^m + a_1 z^{m-1} + \cdots + a_m,$$

and define a homotopy by

$$H(z,t) = f_t(z) = tf(z) + (1-t)z^m = z^m + t(a_1 z^{m-1} + \cdots + a_m).$$

Then, $f_0(z) = z^m$ and $f_1(z) = f(z)$.

Suppose that $W_r \subset \mathbb{C}$ is a closed disk, centered at 0 and of radius $r > 0$. We claim that, for r sufficiently large, f_t has no zeros on ∂W_r, $0 \le t \le 1$. Indeed,

$$\frac{f_t(z)}{z^m} = 1 + t\left(\frac{a_1}{z} + \frac{a_2}{z^2} + \cdots + \frac{a_m}{z^m}\right),$$

and the term in the parentheses converges to 0 as $z \to \infty$.

Thus, for $r > 0$ sufficiently large, we define

$$G : \partial W_r \times [0,1] \to S^1$$

by

$$G(z,t) = \frac{H(z,t)}{|H(z,t)|}.$$

This is a homotopy between

$$G_1(z) = \frac{f(z)}{|f(z)|}$$

and

$$G_0(re^{i\theta}) = e^{im\theta},$$

so $\deg_2(G_0) = \deg_2(G_1)$. But $G_0^{-1}(y)$ contains exactly m points, $\forall y \in S^1$, so $\deg_2(G_1) = 1$ since m is odd. It follows, by Example 3.9.13, that f has a zero in W_r. □

There is an integer valued degree for maps $f \in C^\infty(M,N)$ when M and N are both orientable. This can be used to give a proof of the full fundamental theorem of algebra. We will take this up in Chapter 8 when we study differential forms and de Rham cohomology.

Definition 3.9.15. Let W be a compact manifold, possibly with boundary, and let $X \subsetneq W$. A retraction of W to X is a smooth map $f : W \to X$ such that $f|X = \mathrm{id}_X$.

Theorem 3.9.16. *If W is a compact manifold with $\partial W \neq \emptyset$ connected, then there is no retraction $f : W \to \partial W$.*

Proof. Indeed, $\deg_2(f|\partial W) = 1$, since $f|\partial W = \mathrm{id}_{\partial W}$, and this contradicts the existence of the extension $f : W \to \partial W$ by Theorem 3.9.12. □

Corollary 3.9.17 (Brouwer fixed point theorem). *If*

$$f : D^n \to D^n$$

is smooth, there is a point $x \in D^n$ such that $f(x) = x$.

Proof. Suppose f has no fixed point. Define $g : D^n \to \partial D^n = S^{n-1}$ as follows. For each $x \in D^n$, construct the ray R_x starting at $f(x)$ and passing through $x \neq f(x)$. Let $g(x)$ be the unique point $R_x \cap S^{n-1}$. If $x \in \partial D^n$, it is clear that $g(x) = x$, so if g is smooth, we have contradicted Theorem 3.9.16. The smoothness of g is left for the reader to check. □

Exercise 3.9.18. Prove that Corollary 3.9.17 holds when f is assumed only to be continuous.

Another famous theorem that can be proven using mod 2 degree is the Jordan–Brouwer separation theorem (smooth version). We introduce the key idea, that of "winding number", and then, in a series of exercises, lead you through a proof of the separation theorem in the plane (the smooth version of the Jordan curve theorem).

Definition 3.9.19. Let $f : S^1 \to \mathbb{R}^2$ be a smooth map and let $p \in \mathbb{R}^2 \smallsetminus f(S^1)$. Define

$$f_p : S^1 \to S^1$$

by the formula

$$f_p(z) = \frac{f(z) - p}{\|f(z) - p\|},$$

where $\|\cdot\|$ denotes the usual Euclidean norm. Then the (mod 2) winding number of the closed curve f around p is

$$w_2(f, p) = \deg_2(f_p).$$

Remark that the winding number is defined for an arbitrary smooth closed curve f. It is not required that f be an imbedding or even an immersion. In the case that f is a diffeomorphic imbedding (a smooth Jordan curve), you will show in the exercises that the open set $\mathbb{R}^2 \smallsetminus f(S^1)$ has exactly two components, distinguished from one another by the fact that $w_2(f, p) = 0$, for every point p in one component, and $w_2(f, p) = 1$, for every point p in the other. The component in which $w_2(f, p) = 0$ is unbounded (and called the "outside" of $f(S^1)$), while the other component is bounded (the "inside").

Exercise 3.9.20. Let f be a smooth Jordan curve, let U be a connected component of $\mathbb{R}^2 \smallsetminus f(S^1)$, and let $p, q \in U$. Prove that $w_2(f, p) = w_2(f, q)$. (Hint: The mod 2 degree is a homotopy invariant.)

Definition 3.9.21. If $p \in \mathbb{R}^2$, the ray in $\mathbb{R}^2$ out of p and having direction given by the unit vector $v \in S^1$ will be denoted by $R_p(v)$.

Exercise 3.9.22. If $f : S^1 \to \mathbb{R}^2$ is a smooth Jordan curve and $p \in \mathbb{R}^2 \smallsetminus f(S^1)$, prove that $v \in S^1$ is a critical value of $f_p : S^1 \to S^1$ if and only if the ray $R_p(v)$ is somewhere tangent to the Jordan curve f.

Exercise 3.9.23. Prove the smooth Jordan curve theorem: *If f is a smooth Jordan curve, then $\mathbb{R}^2 \smallsetminus f(S^1)$ has exactly two components, one of which (called the inside of f) is bounded (i.e., has compact closure) and the other of which (the outside of f) is unbounded. For every point p in the outside of f, $w_2(f, p) = 0$, and for every p on the inside, $w_2(f, p) = 1$. Finally, $f(S^1)$ is the set-theoretic boundary of each of these components.* Proceed as follows.

(1) If $p \in \mathbb{R}^2 \smallsetminus f(S^1)$, prove that $w_2(f, p)$ is the number of points mod 2 in $R_p(v) \cap f(S^1)$, for $v \in S^1$ any regular value of f_p. (Hint: Use Exercise 3.9.22.)
(2) Use (1) to prove that there are points $p, q \in \mathbb{R}^2 \smallsetminus f(S^1)$ such that $w_2(f, p) \neq w_2(f, q)$. By Exercise 3.9.20, conclude that $\mathbb{R}^2 \smallsetminus f(S^1)$ has at least two components. Also remark that the winding number is 0 about points in at least one of these components and is 1 about points in at least one of the other components.

(3) Using the fact that f is a smooth imbedding, choose a coordinate chart (U, u, w) about any point $f(z)$ in which

$$U = \{(u, w) | -2 \le u \le 2, -2 \le w \le 2\}$$

and $f(S^1) \cap U$ is just a horizontal line segment $w \equiv 0$. Show that every point of $\mathbb{R}^2 \smallsetminus f(S^1)$ can be connected by a continuous path in $\mathbb{R}^2 \smallsetminus f(S^1)$ either to the point $(0, 1) \in U$ or $(0, -1) \in U$. This proves that $\mathbb{R}^2 \smallsetminus f(S^1)$ has at most two connected components.

(4) Prove that one of these components is bounded, that the other is not, and that the winding number of f is 0 about the points in the unbounded component.

(5) Show that each point of $f(S^1)$ lies on an arbitrarily short arc that meets both components. Conclude that $f(S^1)$ is the common set-theoretic boundary of these components.

3.10. Morse Functions*

In Subsection 2.9.B, we defined the notion of a nondegenerate critical point of a function $f \in C^\infty(U)$, where U is an open subset of Euclidean space. Via local coordinates, this notion carries over to functions $f \in C^\infty(M)$ on arbitrary manifolds. Since the definition of the Hessian (Definition 2.9.14) is coordinate-free, the actual choice of coordinates about the critical point is immaterial.

Definition 3.10.1. If $f \in C^\infty(M)$ and $a \in \mathbb{R}$,

$$M_f^a = \{x \in M \mid f(x) \le a\}.$$

Definition 3.10.2. A function $f \in C^\infty(M)$ is a Morse function if all of its critical points are nondegenerate and, for each $a \in \mathbb{R}$, M_f^a is compact.

Example 3.10.3. We give a simple intuitive example of a Morse function. View the 4-holed torus M as imbedded in $\mathbb{R}^3$ as in Figure 3.10.1. The function $f : M \to \mathbb{R}$ defined by orthogonal projection to the vertical axis has critical values as indicated. The ten corresponding critical points are all Morse singularities. The eight intermediate critical points have index 1 (saddle points), the minimum has index 0 and the maximum has index 2. Since M is compact, the requirement on M_f^a in the definition is automatic. It should be noted that, as a increases from its minimum value of 0 to its maximum value of 1, M_f^a undergoes a change in topology exactly when a critical level is passed. This sort of behavior, typical for Morse functions, will be examined further in Section 4.2 and illustrates the fundamental importance of such functions in differential topology.

By the Morse lemma (Theorem 2.9.18), the critical points of a Morse function are isolated. In particular, Morse functions on compact manifolds have only finitely many critical points. In this section, we will demonstrate the following result.

Theorem 3.10.4. *If M is a manifold without boundary, it admits a Morse function.*

In fact, this will be improved considerably in Exercise 3.10.18, where you will show that every smooth function on M can be approximated arbitrarily well on compact sets, together with its derivatives, by a Morse function.

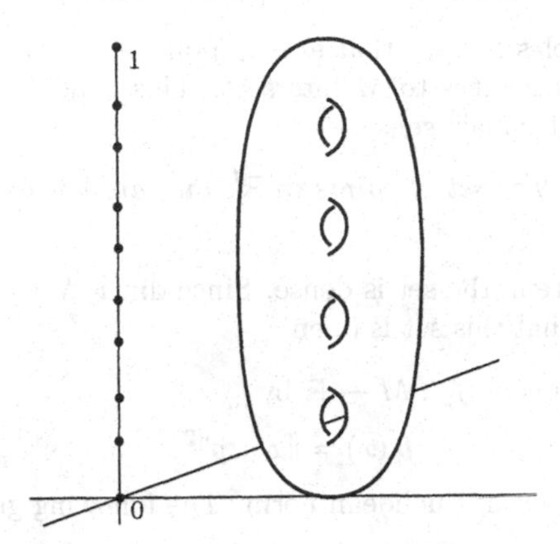

Figure 3.10.1. The height function for the 4-holed torus M

The importance of Theorem 3.10.4 will become apparent in Section 4.2, where we will sketch the proof that a Morse function defines a decomposition of a compact manifold into a finite union of "handles", joined together in a very regular way. This so-called "handle-body decomposition" has deep topological applications, including S. Smale's proof of the Poincaré conjecture in dimensions ≥ 5 [**38**].

We consider $M \subset \mathbb{R}^k$ via a proper imbedding. This does not use the full force of Theorem 3.7.12 since the exact value of k is not important. The normal bundle $\nu(M)$is a properly imbedded submanifold of dimension k. Recall that

$$\nu(M) = \{(x, v) \mid x \in M, \text{ and } v \perp T_x(M)\}.$$

We have earlier used the map

$$\varphi : \nu(M) \to \mathbb{R}^k,$$
$$\varphi(x, v) = x + v$$

to imbed a neighborhood of the zero section into $\mathbb{R}^k$, but now we will be interested in this map on the entire manifold $\nu(M)$. Intuitively, we see v as a line segment issuing from x and perpendicular to M and take $\varphi(x, v)$ to be the terminal point of this segment.

Definition 3.10.5. A point $p \in \mathbb{R}^k$ is a focal point of M if it is a critical value of φ. If $p = \varphi(q, v)$ is such a focal point, where the nullity of the linear map $\varphi_{*(q,v)}$ is $\mu > 0$, we say that p is a focal point of (M, q) having multiplicity μ.

Example 3.10.6. Let $M = S^1 \times \mathbb{R} \subset \mathbb{R}^3$ be the right circular cylinder with axis the z-axis and radius 1. Then this axis is exactly the set of focal points of M. If $p = (0, 0, z)$ is one of these focal points, then $\varphi^{-1}(p)$ is the set of inwardly pointing unit normal vectors (q, v) at the points $q \in S^1 \times \{z\}$. Since $T_{(q,v)}(\nu(M))$ decomposes naturally into direct summands tangent to M and perpendicular to M, respectively, the reader should have little trouble seeing that the nullity of $\varphi_{*(q,v)}$ is 1.

Example 3.10.7. The unit sphere $S^{n-1} \subset \mathbb{R}^n$ has just one focal point, the origin. The preimage $\varphi^{-1}(0)$ is the set of inwardly pointing unit normal vectors (q, v), $q \in S^{n-1}$, and the multiplicity of the focal point at each $q \in S^{n-1}$ is $n-1$.

Remark. These examples suggest that a focal point of M should be a point where arbitrarily nearby normal lines to M intersect. This is not exactly right, but it is correct in some "infinitesimal" sense.

Proposition 3.10.8. *The set of points in $\mathbb{R}^k$ that are not focal points of M is an open, dense subset.*

Proof. By Sard's theorem, the set is dense. Since $\dim \nu(M) = k$, the inverse function theorem implies that this set is open. □

Given $p \in \mathbb{R}^k$, we define $f_p : M \to \mathbb{R}$ by

$$f_p(x) = \|x - p\|^2,$$

where $\|\cdot\|$ denotes the usual Euclidean norm. The following gives Theorem 3.10.4.

Theorem 3.10.9. *If $p \in \mathbb{R}^k$ is not a focal point of M, then f_p is a Morse function.*

In light of Proposition 3.10.8, it will follow that there are uncountably many Morse functions on M. This will be made sharper in Exercise 3.10.18. The proof of Theorem 3.10.9 will also show how to compute the index of a critical point of f_p (Exercise 3.10.17).

The following is evident.

Lemma 3.10.10. *For fixed $p \in \mathbb{R}^k$ and arbitrary $a \in \mathbb{R}$, $M^a_{f_p}$ is compact.*

It will be helpful to write the proper imbedding explicitly as $\xi : M \hookrightarrow \mathbb{R}^k$. By the usual abuse, we write M for $\xi(M)$. Relative to coordinate neighborhoods $(U, u^1, \dots, u^n)$ on M, the vector-valued first derivatives $\partial\xi/\partial u^i$, $1 \le i \le n$, form a basis of $T_{\xi(u)}(M)$. The formula for f_p in the coordinates $u = (u^1, \dots, u^n)$ is

$$f(u) = f_p(\xi(u)) = \xi(u) \cdot \xi(u) - 2\xi(u) \cdot p - p \cdot p.$$

Correspondingly, we have derivative formulas

$$(*) \qquad \frac{\partial f}{\partial u^i} = 2\frac{\partial \xi}{\partial u^i} \cdot (\xi - p).$$

Lemma 3.10.11. *For fixed $p \in \mathbb{R}^k$, a point $q \in M$ is a critical point of f_p if and only if there is a vector $v_0 \perp T_q(M)$ in $T_q(\mathbb{R}^k)$, $\|v_0\| = 1$, and a real number $\lambda \ge 0$ such that $p = q + \lambda v_0$ $(= \varphi(q, \lambda v_0))$.*

Proof. Let $(U, u^1, \dots, u^n)$ be a coordinate neighborhood centered at q. That is, $u^i(q) = 0$, $1 \le i \le n$. By the local formula $(*)$, q is a critical point of f_p if and only if

$$\frac{\partial \xi}{\partial u^i}(0) \perp \xi(0) - p = q - p, \quad 1 \le i \le n.$$

Equivalently, $q - p \perp T_q(M)$. □

Exercise 3.10.12. Fix a choice of unit vector $v_0 \perp T_q(M)$. If the coordinate neighborhood U of q in M is small enough, we can extend v_0 to a smooth unit normal field v on U. That is, $\sigma(u) = (u, v(u))$ is a section of the normal bundle such that $\|v(u)\| \equiv 1$. Here, in terms of the local coordinates, $q = 0$ and $v(0) = v_0$.

If $w \in T_q(M)$, the derivative $D_w v \in T_q(\mathbb{R}^k)$ is taken by applying the directional derivative D_w to the coordinate functions of $v(u)$. For $w_1, w_2 \in T_q(M)$, prove that

$$\mathfrak{L}_{v_0}(w_1, w_2) = (D_{w_1} v) \cdot w_2$$

defines a symmetric bilinear form on $T_q(M)$ that depends only on v_0, not on the choice of extension v. (Hint. Extend w_1 and w_2 locally to tangent fields and use the Leibnitz rule for differentiating the dot product.)

Definition 3.10.13. The symmetric bilinear form $\mathfrak{L}_{v_0}$ is called the second fundamental form of M at q in the normal direction v_0.

As the language *second* fundamental form suggests, there is also a *first* one. In the terminology of classical differential geometry, the first fundamental form is just the Euclidean dot product restricted to $T_q(M)$. This defines the intrinsic geometry of M. The second fundamental form detects the way in which M relates to the surrounding Euclidean space, the so-called "extrinsic geometry" (*cf.* Definition 10.2.14 ff.). The symmetric $n \times n$ matrix of "metric coefficients"

$$g_{ij}(u) = \frac{\partial \xi}{\partial u^i} \cdot \frac{\partial \xi}{\partial u^j},$$

evaluated at $u = 0$, gives the matrix for the first fundamental form at q relative to the local coordinates. The following exercise gives the matrix of the second fundamental form relative to these coordinates.

Exercise 3.10.14. Let v extend $v_0 = v(0)$ to a unit normal field on the coordinate neighborhood U as in Exercise 3.10.12. Set

$$\ell_{ij}(u) = \frac{\partial^2 \xi}{\partial u^i \partial u^j} \cdot v.$$

Prove that the numbers $\ell_{ij}(0)$ are the entries of the matrix for the second fundamental form $\mathfrak{L}_{v_0}$ of M at q.

We fix the critical point $q \in M$ and the chart $(U, u^1, \dots, u^n)$ as above. By a linear change of coordinates, we assume that

$$\frac{\partial \xi}{\partial u^1}(0), \dots, \frac{\partial \xi}{\partial u^n}(0)$$

is an orthonormal basis of $T_q(M)$. That is, the metric coefficients satisfy $g_{ij}(0) = \delta_{ij}$.

Lemma 3.10.15. *If $p = q + \lambda v_0$ is as above, then p is a focal point of (M, q) of multiplicity μ if and only if $1/\lambda$ is an eigenvalue of the matrix $[\ell_{ij}(0)]$ of multiplicity μ. (In classical terminology, $\lambda \geq 0$ is a radius of curvature at q relative to the normal direction v_0.)*

Proof. The coordinate neighborhood U can be chosen to trivialize the normal bundle, so we choose linearly independent sections

$$\sigma_\alpha : U \to \nu(M), \quad 1 \leq \alpha \leq k - n.$$

Written in terms of the coordinates, these have the form

$$\sigma_\alpha(u) = (u, v_\alpha(u)).$$

By the Gram–Schmidt process, we arrange that $\{v_1, \ldots, v_{k-n}\}$ be everywhere orthonormal. We can also arrange that $v_1(0) = v_0$. Since

$$\frac{\partial \xi}{\partial u^1}(0), \ldots, \frac{\partial \xi}{\partial u^n}(0), v_1(0), \ldots, v_{k-n}(0)$$

form an orthonormal basis of $\mathbb{R}^k$, we can rotate so that they are the standard coordinate basis. We then coordinatize $\pi^{-1}(U) \subset \nu(M)$ by

$$(\underbrace{u^1, \ldots, u^n}_{u}, \underbrace{t^1, \ldots, t^{k-n}}_{t}) \leftrightarrow \left(\xi(u), \sum_{\alpha=1}^{k-n} t^\alpha v_\alpha(u)\right).$$

Relative to these coordinates, we write

$$\varphi(u,t) = \xi(u) + \sum_{\alpha=1}^{k-} t^\alpha v_\alpha(u)$$

and

$$\frac{\partial \varphi}{\partial u^i} = \frac{\partial \xi}{\partial u^i} + \sum_{\alpha=1}^{k-} t^\alpha \frac{\partial v_\alpha}{\partial u^i}, \qquad 1 \le i \le n,$$

$$\frac{\partial \varphi}{\partial t^\beta} = v_\beta, \qquad 1 \le \beta \le k-n.$$

Consider the matrix function

$$A(u,t) = \begin{bmatrix} \begin{bmatrix} \frac{\partial \varphi}{\partial u^i} \cdot \frac{\partial \xi}{\partial u^j} \\ \frac{\partial \varphi}{\partial t^\beta} \cdot \frac{\partial \xi}{\partial u^j} \end{bmatrix} & \begin{bmatrix} \frac{\partial \varphi}{\partial u^i} \cdot v_\gamma \\ \frac{\partial \varphi}{\partial t^\beta} \cdot v_\gamma \end{bmatrix} \end{bmatrix}.$$

At the point

$$(u,t) = (\underbrace{0, \ldots, 0}_{q}, \underbrace{\lambda, 0, \ldots, 0}_{\lambda v_0})$$

the matrix $A(u,t)$ is simply the Jacobian matrix of φ. Explicitly,

$$J\varphi_{(q,\lambda v_0)} = \begin{bmatrix} \left[\delta_{ij} + \lambda \frac{\partial v_1}{\partial u^i}(0) \cdot \frac{\partial \xi}{\partial u^j}(0)\right] & * \\ 0 & I_{k-n} \end{bmatrix}.$$

This matrix is singular if and only if $q + \lambda v_0$ is a focal point of (M, q), the nullity μ of the matrix being the multiplicity of the focal point. This is the same as the nullity of the upper left-hand corner, which we claim is just the multiplicity of $1/\lambda$ as an eigenvalue of the matrix $[\ell_{ij}(0)]$. Indeed,

$$0 \equiv \underbrace{\frac{\partial}{\partial u^i}(v_1 \cdot \frac{\partial \xi}{\partial u^j})}_{\equiv 0} = \frac{\partial v_1}{\partial u^i} \cdot \frac{\partial \xi}{u^j} + \ell_{ij},$$

and so we want the nullity of the matrix

$$[\delta_{ij} - \lambda \ell_{ij}(0)].$$

But this is exactly the multiplicity of $1/\lambda$ as an eigenvalue of the matrix $[\ell_{ij}(0)]$. □

Let $q \in M$, $v_0 \perp T_q(M)$, $\|v_0\| = 1$ and $p = q + \lambda v_0$. By Lemma 3.10.11, q is a critical point of f_p.

Lemma 3.10.16. *The critical point q of f_p is nondegenerate if and only if p is not a focal point of (M, q).*

Proof. We write $f(u) = f_p(\xi(u))$ and differentiate equation $(*)$ to obtain the matrix equation

$$(**) \qquad \left[\frac{\partial^2 f}{\partial u^i \partial u^j}\right] = 2\left[\frac{\partial \xi}{\partial u^i} \cdot \frac{\partial \xi}{\partial u^j} + \frac{\partial^2 \xi}{\partial u^i \partial u^j} \cdot (\xi - p)\right].$$

At the critical point $q = \xi(0)$, the left-hand side of this equation becomes the Hessian matrix for f_p (Exercise 2.9.16) and the right-hand side becomes

$$2[\delta_{ij} - \lambda \ell_{ij}(0)].$$

The nullity of this matrix is exactly the multiplicity of $1/\lambda$ as an eigenvalue of $[\ell_{ij}(0)]$. By Lemma 3.10.15, this is zero if and only if p is not a focal point of (M, q). □

Proof of Theorem 3.10.9. If $p \in \mathbb{R}^k$ is not a focal point of M, then Lemma 3.10.11 and Lemma 3.10.16 imply that every critical point of f_p is nondegenerate. Together with Lemma 3.10.10, this proves that f_p is a Morse function. □

Exercise 3.10.17. Using the formula $(**)$ for the Hessian, prove that the number of negative eigenvalues of this matrix (the Morse index of the critical point q) is equal to the number of focal points of (M, q), counted with multiplicity, on the line joining q to p.

Exercise 3.10.18. Let $g \in C^\infty(M)$, let $K \subseteq M$ be compact, let $\epsilon > 0$ and let $m \geq 0$ be an integer. Prove that there is a Morse function h such that h and its derivatives of order $\leq m$ are uniformly close, respectively, to g and its derivatives of order $\leq m$ on the compact set K. Proceed as follows.

(1) Given an imbedding $\zeta : M \hookrightarrow \mathbb{R}^{k-1}$, define
$$\xi = (g, \zeta) : M \to \mathbb{R} \times \mathbb{R}^{k-1} = \mathbb{R}^k.$$
Prove that this is also a smooth imbedding.

(2) Choose $p = (\epsilon_1 - c, \epsilon_2, \ldots, \epsilon_k) \in \mathbb{R}^k$ so that p is not a focal point of M. Note that c can be chosen as large as desired and the ϵ_is can be chosen as small as desired.

(3) Show that $h = (f_p - c^2)/2c$ is the desired approximating Morse function, for suitable choices of c and ϵ_i, $1 \leq i \leq k$.

In particular, if M is compact, any smooth function can be approximated uniformly well by a Morse function and the approximation can be made uniformly close in as many derivatives as desired.

[illegible] is an [illegible]

[illegible]

CHAPTER 4

Flows and Foliations

In this chapter, we investigate the global theory of ordinary differential equations (flows), referred to as O.D.E., and the Frobenius integrability condition for k-plane distributions (foliations). Although this latter topic concerns global partial differential equations, our approach will be largely qualitative, with very few explicit partial differential equations in evidence. Unless otherwise indicated, *all manifolds will have empty boundary.*

4.1. Complete Vector Fields

The space $\mathfrak{X}(M)$ of smooth sections of the tangent bundle is a module over the algebra $C^\infty(M)$ and a vector space over $\mathbb{R}$. Viewed as the space of derivatives of $C^\infty(M)$, $\mathfrak{X}(M)$ is a Lie algebra over $\mathbb{R}$. By the local theory of O.D.E., for each $q \in M$, a vector field $X \in \mathfrak{X}(M)$ generates a local flow $\Phi : (-\epsilon, \epsilon) \times V \to U$, where $(U, x^1, \dots, x^n)$ is a local coordinate chart about q, V is an open neighborhood of q with compact closure in U, and $\epsilon > 0$ is sufficiently small. Any two local flows generated by X agree wherever both are defined. The notion of a local flow about a point makes sense even when V and U are not coordinate neighborhoods. A system of suitably coherent local flows covering a manifold M will be called a local flow on M. Here is the precise definition.

Definition 4.1.1. A local flow Φ on M is a family of smooth maps

$$\{\Phi^\alpha : (-\epsilon_\alpha, \epsilon_\alpha) \times V_\alpha \to U_\alpha \mid \epsilon_\alpha > 0\}_{\alpha \in \mathfrak{A}},$$

written $\Phi^\alpha(t, x) = \Phi^\alpha_t(x)$, such that

(1) $V_\alpha \subseteq U_\alpha \subseteq M$ are open sets and $\{V_\alpha\}_{\alpha \in \mathfrak{A}}$ covers M;
(2) $\Phi^\alpha_0 : V_\alpha \to U_\alpha$ is the inclusion map, $\forall\, \alpha \in \mathfrak{A}$;
(3) $\Phi^\alpha_{t_1+t_2} = \Phi^\beta_{t_1} \circ \Phi^\alpha_{t_2}$, wherever both sides are defined, $\forall\, \alpha, \beta \in \mathfrak{A}$.

Definition 4.1.2. If Φ is a local flow on M and $q \in M$, a curve of the form $s^\alpha_q(t) = \Phi^\alpha_t(q)$, $-\epsilon_\alpha < t < \epsilon_\alpha$, where $q \in V_\alpha$, is called a flow line of Φ through q or the *orbit* of the flow through q.

Exercise 4.1.3. If s^α_q and s^β_q are flow lines through q, show that they agree on their common domain $(-\epsilon_\alpha, \epsilon_\alpha) \cap (-\epsilon_\beta, \epsilon_\beta)$. Thus, the velocity vector

$$X_q = \dot{s}^\alpha_q(0) \in T_q(M)$$

is well defined, $\forall\, q \in M$. Prove that this defines a smooth field $X \in \mathfrak{X}(M)$ and that

$$X_{s^\alpha_q(t)} = \dot{s}^\alpha_q(t), \forall\, t \in (-\epsilon_\alpha, \epsilon_\alpha).$$

Definition 4.1.4. The vector field X obtained from the local flow Φ as above is called the infinitesimal generator of Φ.

Exercise 4.1.5. Show that every vector field $X \in \mathfrak{X}(M)$ is the infinitesimal generator of a local flow Φ on M. If two local flows Φ and Ψ have the same infinitesimal generator X, prove that $\Phi \cup \Psi$ is a local flow with the same infinitesimal generator X.

Consequently, by partially ordering local flows by inclusion we see that, given a local flow Φ on M, there is one and only one maximal local flow on M containing Φ.

Corollary 4.1.6. *There is a one-to-one correspondence between maximal local flows Φ on M and vector fields $X \in \mathfrak{X}(M)$ given by letting X be the infinitesimal generator of Φ.*

Working with local flows can be somewhat uncomfortable. Happily, there are natural situations in which the maximal local flow of a vector field contains an honest (*i.e.*, global) flow.

Definition 4.1.7. A (global) flow on M is a smooth map

$$\Phi : \mathbb{R} \times M \to M,$$

written $\Phi_t(x) = \Phi(t, x)$, such that

(1) $\Phi_0 = \mathrm{id}_M$;
(2) $\Phi_{t_1+t_2} = \Phi_{t_1} \circ \Phi_{t_2}$, $\forall t_1, t_2 \in \mathbb{R}$.

If the maximal local flow of a vector field $X \in \mathfrak{X}(M)$ contains a global flow, we say that X is a complete vector field. In this case, X is also called the infinitesimal generator of the global flow.

Example 4.1.8. We define a global flow on $T^2 = S^1 \times S^1$. Fix $\rho \in \mathbb{R}$ and, for each $z = (e^{2\pi i a}, e^{2\pi i b}) \in T^2$ and $t \in \mathbb{R}$, define

$$\Phi_t^\rho(z) = (e^{2\pi i(a+t)}, e^{2\pi i(b+\rho t)}).$$

It is clear that this defines a global flow Φ^ρ on T^2. Of some interest is the way in which the qualitative behavior of this flow depends on the value of the constant ρ. If ρ is rational, there is a least positive integer k such that ρk is also an integer. Thus,

$$\Phi_{t+k}^\rho(e^{2\pi i a}, e^{2\pi i b}) = \Phi_t^\rho(e^{2\pi i a}, e^{2\pi i b}).$$

One concludes rather easily that each flow line $\Phi^\rho(\mathbb{R} \times \{z\})$ is an imbedded circle, the flow being periodic of period k (*cf.* Exercise 4.1.9). By contrast, if ρ is irrational, each flow line $\Phi^\rho(\mathbb{R} \times \{z\})$ is a one-to-one immersed copy of $\mathbb{R}$ that is everywhere dense in T^2. This is the two-dimensional version of a theorem of Kronecker that we will prove in Chapter 5 (Example 5.3.9). The two-dimensional version will also follow from a result to be proven in Section 4.4 (*cf.* Corollary 4.4.10 and Exercise 4.4.12).

Remark that the infinitesimal generator X of the flow Φ^ρ "lifts" to a well-defined vector field on $\mathbb{R}^2$ relative to the canonical projection

$$p : \mathbb{R}^2 \to \mathbb{R}^2/\mathbb{Z}^2 = T^2,$$
$$p(x, y) = (e^{2\pi i x}, e^{2\pi i y}).$$

More precisely, the constant vector field $\widetilde{X} \in \mathfrak{X}(\mathbb{R}^2)$, defined by

$$\widetilde{X}_{(x,y)} \equiv v_\rho = (1, \rho)$$

satisfies

$$p_{*(x,y)}\widetilde{X}_{(x,y)} = X_{p(x,y)},$$

for all $(x,y) \in \mathbb{R}^2$. The reader can check this easily. It is noteworthy that the vector field $\widetilde{X}$ on $\mathbb{R}^2$ is also complete, generating the translation flow

$$\widetilde{\Phi}^\rho_t(w) = w + tv_\rho.$$

This lifted flow is quite tame, regardless of whether ρ is rational or irrational.

Exercise 4.1.9. Let Φ be a flow on M and let $x \in M$ be a point not fixed by the flow, but such that $\Phi_c(x) = x$ for some $c > 0$. Prove that there is a number $c_0 > 0$ and a smooth imbedding

$$\varphi : S^1 \to M$$

such that

$$\Phi_t(x) = \varphi(e^{2\pi i t/c_0}), \quad -\infty < t < \infty.$$

In this case, the flow line R_x is the imbedded circle $\varphi(S^1)$ and we say that x is a *periodic point* of the flow of period c_0.

Not every vector field is complete. For example, Exercise 2.8.13 showed that $e^t \frac{d}{dt} \in \mathfrak{X}(\mathbb{R})$ is not complete.

Lemma 4.1.10. *If the maximal local flow of $X \in \mathfrak{X}(M)$ contains an element of the form*

$$\Phi : (-\epsilon, \epsilon) \times M \to M$$

with $\epsilon > 0$, then X is complete.

Proof. Let $t \in \mathbb{R}$. Then one can find $k \in \mathbb{Z}$ and $r \in (-\epsilon/2, \epsilon/2)$ such that $t = r + k \cdot \epsilon/2$. Given $x \in M$, define

$$\Phi_t(x) = \begin{cases} \underbrace{\Phi_{\epsilon/2} \circ \cdots \circ \Phi_{\epsilon/2}}_{k} \circ \Phi_r(x), & k > 0, \\ \underbrace{\Phi_{-\epsilon/2} \circ \cdots \circ \Phi_{-\epsilon/2}}_{-k} \circ \Phi_r(x), & k < 0, \\ \Phi_r(x), & k = 0. \end{cases}$$

If $\Phi_t(x)$ is *well defined* by this formula, $-\infty < t < \infty$, then it will be an integral curve to X. To see this, remark that $\Phi_r(x)$, $-\frac{\epsilon}{2} < r < \epsilon/2$, is integral to X and use the fact that $(\Phi_{\pm\epsilon/2})_*(X) = X$ (Exercise 2.8.14).

We show that $\Phi_t(x)$ is well defined. For simplicity, let $t > 0$. Obvious modifications of the argument give the general case. Suppose that

$$r + k \cdot \epsilon/2 = t = s + q \cdot \epsilon/2,$$

where $s, r \in (-\epsilon/2, \epsilon/2)$ and $k, q \in \mathbb{Z}$. It follows that $r - s \in (-\epsilon, \epsilon)$, hence that $q - k = 0, 1$, or -1. If $q - k = 0$, then $r - s = 0$ and we are done. Assume, therefore, that $q - k = \pm 1$. Without loss of generality, take $q - k = 1$. Then, $r - s = \epsilon/2$, so

$$\begin{aligned} \underbrace{\Phi_{\epsilon/2} \circ \cdots \circ \Phi_{\epsilon/2}}_{k} \circ \Phi_r(x) &= \underbrace{\Phi_{\epsilon/2} \circ \cdots \circ \Phi_{\epsilon/2}}_{k} \circ \Phi_{s+\epsilon/2}(x) \\ &= \underbrace{\Phi_{\epsilon/2} \circ \cdots \circ \Phi_{\epsilon/2}}_{k+1=q} \circ \Phi_s(x). \end{aligned}$$

Thus, $\Phi_t(x)$ is well defined, $-\infty < t < \infty$, for each $x \in M$, and is an integral curve to X.

We must show that $\Phi : \mathbb{R} \times M \to M$ is smooth. Let $(t_0, x_0) \in \mathbb{R} \times M$. For small enough $\eta > 0$, we fix $k_0 \in \mathbb{Z}$ such that $t = r + k_0 \cdot \epsilon/2$, for each $t \in (t_0 - \eta, t_0 + \eta)$ and suitable $r \in (-\epsilon/2, \epsilon/2)$. Then, $(t_0 - \eta, t_0 + \eta) \times M$ is an open neighborhood of (t_0, x_0) in $\mathbb{R} \times M$ on which

$$\Phi_t(x) = \begin{cases} \underbrace{\Phi_{\epsilon/2} \circ \cdots \circ \Phi_{\epsilon/2}}_{k_0} \circ \Phi_r(x), & k_0 > 0, \\ \underbrace{\Phi_{-\epsilon/2} \circ \cdots \circ \Phi_{-\epsilon/2}}_{-k_0} \circ \Phi_r(x), & k_0 < 0, \\ \Phi_r(x), & k_0 = 0. \end{cases}$$

This is a smooth function of $(r, x) = (t - k_0\epsilon/2, x)$, hence a smooth function of (t, x). □

Theorem 4.1.11. *If $X \in \mathfrak{X}(M)$ has compact support, then X is complete.*

Proof. Since $\operatorname{supp}(X)$ is compact, cover it with finitely many open subsets $U_1, \dots, U_r$ of M such that the local flow of X contains elements

$$\Phi^i : (-\epsilon_i, \epsilon_i) \times U_i \to M,$$

$1 \le i \le r$. Let $U_0 = M \setminus \operatorname{supp}(X)$, an open set with $X|U_0 \equiv 0$. Define

$$\Phi^0 : \mathbb{R} \times U_0 \to M$$

by $\Phi^0_t(x) = x$, $\forall x \in U_0$, $\forall t \in \mathbb{R}$. Since $\{U_i\}_{i=0}^r$ covers M and the Φ^i agree on overlaps, we have a local flow on M generated by X. Let $\epsilon = \min_{1 \le i \le r} \epsilon_i$ and fit the elements of the local flow together to get

$$\Phi : (-\epsilon, \epsilon) \times M \to M$$

generated by X. By Lemma 4.1.10, X is complete. □

Corollary 4.1.12. *If M is compact, every vector field $X \in \mathfrak{X}(M)$ is complete.*

Exercise 4.1.13. Let $\varphi : [0, 1] \to [0, \frac{\pi}{2}]$ be a smooth map such that $\varphi(x) \equiv 0$ on $[0, \frac{1}{5}] \cup [\frac{4}{5}, 1]$ and $\varphi(x) \equiv \frac{\pi}{2}$ on $[\frac{2}{5}, \frac{3}{5}]$ (smooth Urysohn lemma). Extend this to a smooth map $\varphi : \mathbb{R} \to [0, \frac{\pi}{2}]$ by requiring periodicity: $\varphi(x + 1) = \varphi(x)$. This extension is clearly smooth. In $\mathfrak{X}(\mathbb{R})$ define

$$\begin{aligned} X &= x^2 \cos^2 \varphi(x) \frac{d}{dx}, \\ Y &= x^2 \sin^2 \varphi(x) \frac{d}{dx}. \end{aligned}$$

Prove that X and Y are complete, but that $X + Y$ is not.

As another application of Theorem 4.1.11, we return to the homogeneity lemma (Theorem 3.8.7). Let

$$\operatorname{Diff}^0_c(M) \subseteq \operatorname{Diff}_c(M)$$

denote the compactly supported isotopy class of id_M. We prove a local version of Theorem 3.8.7.

Lemma 4.1.14 (Local homogeneity lemma). *Let $U = \operatorname{int}(D^n)$ and let $x_0, y_0 \in U$. Then there is $\varphi \in \operatorname{Diff}^0_c(U)$ such that $\varphi(x_0) = y_0$.*

Proof. Use the ordinary Euclidean norm from $\mathbb{R}^n$ and choose points $\epsilon, \eta \in (0,1)$ such that

$$0 \le \max\{\|x_0\|, \|y_0\|\} < \eta < \epsilon.$$

Let $f : U \to \mathbb{R}$ be smooth such that

$$f(x) \equiv \begin{cases} 1, & 0 \le \|x\| \le \eta, \\ 0, & \epsilon \le \|x\| < 1. \end{cases}$$

Let

$$y_0 - x_0 = v = (c^1, c^2, \dots, c^n) \in \mathbb{R}^n$$

and define

$$X = \sum_{i=1}^n c^i f \frac{\partial}{\partial x^i} \in \mathfrak{X}(U).$$

Since $\operatorname{supp}(X) \subseteq \operatorname{supp}(f)$ is closed in U and bounded away from ∂D^n, it is compact. By Theorem 4.1.11, X is a complete vector field, and we let

$$\Phi : \mathbb{R} \times U \to U$$

be the flow it generates. This flow is stationary outside the compact set $\operatorname{supp}(X)$.

Let $s(t) = x_0 + tv = x_0 + (tc^1, tc^2, \dots, tc^n)$, $0 \le t \le 1$. Then

$$\|s(t)\| = \|ty_0 + (1-t)x_0\| \le \max\{\|x_0\|, \|y_0\|\} < \eta,$$

so $f(s(t)) \equiv 1$, $0 \le t \le 1$. Thus,

$$X_{s(t)} = \sum_{i=1}^n c^i \left.\frac{\partial}{\partial x^i}\right|_{s(t)} = \dot{s}(t),$$

$0 \le t \le 1$. Therefore, $s(t)$ is integral to X, $0 \le t \le 1$, and it follows that $\Phi_1(x_0) = y_0$. Then Φ_t, $0 \le t \le 1$, defines a compactly supported isotopy of $\Phi_0 = \mathrm{id}_M$ to $\Phi_1 = \varphi$, where $\varphi \in \operatorname{Diff}^0_c(U)$ and $\varphi(x_0) = y_0$. □

The proof of Theorem 3.8.7 is a fairly easy consequence of the local homogeneity lemma (Exercise 4.1.16). The idea is to show that any two points in the same connected component of M are isotopic in the following sense.

Definition 4.1.15. Let $x_0, y_0 \in M$. We say that x_0 is isotopic to y_0, and write $x_0 \sim_I y_0$, if there is $\varphi \in \operatorname{Diff}^0_c(M)$ such that $\varphi(x_0) = y_0$.

Exercise 4.1.16. Prove the homogeneity lemma, Theorem 3.8.7, proceeding as follows.

(1) Show that $\operatorname{Diff}^0_c(M)$ is a (normal) subgroup of $\operatorname{Diff}_c(M)$.
(2) Show that $\sim_I$ is an equivalence relation on M.
(3) Prove that the $\sim_I$ equivalence classes are exactly the connected components of M.

We close this section with a few more exercises.

Exercise 4.1.17. Let $\Phi : \mathbb{R} \times M \to M$ be a flow. A subset $C \subseteq M$ is said to be *Φ-invariant* if $\Phi_t(x) \in C$, $\forall x \in C$, $\forall t \in \mathbb{R}$. If $x \in M$ and $R_x = \{\Phi_t(x)\}_{t \in \mathbb{R}}$ is the flow line through x, prove that the closure $\overline{R_x}$ is a Φ-invariant set.

Exercise 4.1.18. Let Φ be a flow on M. A subset $C \subseteq M$ is said to be a *minimal set* of Φ if C is a nonempty, closed, Φ-invariant set containing no proper subset with these same properties. For example, if x is a periodic point (Exercise 4.1.9), the flow line R_x is a minimal set. Prove that, if M is compact, every closed, nonempty, Φ-invariant subset of M contains at least one minimal set. (In particular, by Exercise 4.1.17, every flow line approaches at least one minimal set.) Show by an example that M itself may be a minimal set.

Exercise 4.1.19. Let Φ be a flow on M. One defines the α-limit set and the ω-limit set of a flow line R_x as follows:

$$\alpha(x) = \{y \in M \mid \exists t_k \downarrow -\infty \text{ such that } \lim_{k\to\infty} \Phi_{t_k}(x) = y\},$$
$$\omega(x) = \{y \in M \mid \exists t_k \uparrow \infty \text{ such that } \lim_{k\to\infty} \Phi_{t_k}(x) = y\}.$$

If M is compact, prove that each of these limit sets is a compact, nonempty, Φ-invariant set. Show by examples that $\alpha(x)$ and $\omega(x)$ may or may not be equal and may or may not be minimal. (Remark: The α- and ω-limit set terminology is standard and seems to have its origin in a biblical quotation (Revelations 1:8).)

4.2. The Gradient Flow and Morse Functions*

This section is really an extended example, showing how a certain flow associated to a Morse function on M leads to a detailed topological analysis of M. Together with Subsection 2.9.B and Section 3.10, this will complete a brief introduction to Morse theory.

In multivariable calculus, the gradient ∇f of a function f on $\mathbb{R}^n$ is defined as the vector field

$$\nabla f = \left(\frac{\partial f}{\partial x^1}, \dots, \frac{\partial f}{\partial x^n}\right).$$

It is characterized as the field perpendicular to the level hypersurfaces $f^{-1}(a)$ and such that its dot product with an arbitrary vector $v \in T_p(\mathbb{R}^n)$ is the derivative $D_v f$. In order to generalize this notion to manifolds M, we must fix some choice of Riemannian metric $\langle\cdot,\cdot\rangle$ on M. This can always be done by gluing together local choices of metric with a smooth partition of unity (Exercise 3.5.9).

Lemma 4.2.1. *Let $f \in C^\infty(M)$. Then, relative to a choice of Riemannian metric, there is a unique field $\nabla f \in \mathfrak{X}(M)$ such that $\langle \nabla f, X\rangle = X(f)$, for every field $X \in \mathfrak{X}(M)$. The field ∇f is called the gradient of f relative to the metric.*

Indeed, $X_p \mapsto X_p(f)$ defines a linear functional L on $T_p(M)$. By nonsingularity of the Riemannian metric, there is a unique vector $\nabla f_p \in T_p(M)$ such that $L = \langle\cdot, \nabla f_p\rangle$. Smoothness of the assignment $p \mapsto \nabla f_p$ is elementary and left to the reader. When the metric is fixed throughout a discussion, we will refer to ∇f simply as the gradient of f.

The following is another elementary observation.

Lemma 4.2.2. *A point $p \in M$ is a critical point of f if and only if $\nabla f_p = 0$. Furthermore, if $a \in \mathbb{R}$ is a regular value of f and $p \in f^{-1}(a)$, then $\nabla f_p \perp T_p(f^{-1}(a))$.*

The gradient will be used to prove the following theorem. Recall the definition of the set M^a_f (Definition 3.10.1). If a is a regular value of f, this is a manifold with $\partial M^a_f = f^{-1}(a)$. In any event, we fix the assumption that all of the sets M^a_f

are compact. Recall (Definition 3.10.2) that this is part of the requirement that f be a Morse function.

Theorem 4.2.3. *Suppose that $a < b$ and that the level sets $f^{-1}(a)$ and $f^{-1}(b)$ are nonempty. If $f^{-1}[a,b]$ contains no critical points of f, then there is a diffeomorphism $\varphi : M_f^b \to M_f^a$ that is the identity on $M_f^{a-\epsilon}$, where $\epsilon > 0$ is as small as desired.*

Proof. If we choose $\epsilon > 0$ small enough, f will have no critical points in $f^{-1}[a-\epsilon, b+\epsilon]$ and, by Lemma 4.2.2, ∇f is nowhere zero on this compact set. Thus, we can choose a smooth, nonnegative function $\rho \in C^\infty(M)$, identically 0 on $M_f^{a-\epsilon}$ and on $M \setminus M_f^{b+\epsilon}$ and such that

$$\rho | f^{-1}[a,b] = \frac{1}{\langle \nabla f, \nabla f \rangle}.$$

The vector field $X = -\rho \nabla f$ has compact support, hence generates a flow Φ_t on M. Since X vanishes identically on $M_f^{a-\epsilon}$, this flow fixes every point of that set. Also, for each $q \in M$,

$$\begin{aligned} \frac{d}{dt} f(\Phi_t(q)) &= X_{\Phi_t(q)}(f) \\ &= \langle X, \nabla f \rangle_{\Phi_t(q)} \\ &= -\rho(\Phi_t(q)) \, \langle \nabla f, \nabla f \rangle_{\Phi_t(q)} . \end{aligned}$$

In particular,

$$\begin{aligned} \frac{d}{dt} f(\Phi_t(q)) &\le 0, \qquad \forall t \in \mathbb{R}, \\ \frac{d}{dt} f(\Phi_t(q)) &= -1, \qquad \text{if } \Phi_t(q) \in f^{-1}[a,b]. \end{aligned}$$

Thus, $f(\Phi_t(q))$ is nonincreasing as a function of t and, if $q \in f^{-1}(b)$, then

$$f(\Phi_t(q)) = b - t, \quad 0 \le t \le b - a.$$

It follows that the diffeomorphism

$$\Phi_{b-a} : M \to M$$

carries M_f^b onto M_f^a and fixes every point in $M_f^{a-\epsilon}$. □

Now suppose that f is a Morse function (Definition 3.10.2). In particular, M_f^a is always compact and the critical points of f are isolated. Note that, by 2nd countability, there can be at most a countable infinity of critical points and only finitely many of these can lie in any particular level set $f^{-1}(a)$.

Exercise 4.2.4. Using arbitrarily small "bump functions" in neighborhoods of critical points, show that the Morse function f can be slightly perturbed to a Morse function $\widetilde{f}$ for which distinct critical points p_i lie in distinct levels $\widetilde{f}^{-1}(a_i)$.

In light of this exercise, we assume that distinct critical points of f lie at distinct levels. Fix a critical point p and, replacing f with $f - f(p)$, assume that $p \in f^{-1}(0)$. For $\epsilon > 0$ sufficiently small, p will be the only critical point of f in $f^{-1}[-\epsilon, \epsilon]$. We intend to analyze the change in the topology of M_f^a as a varies from $-\epsilon$ to ϵ. For this, we need the notion of a *handle*.

Definition 4.2.5. Let $0 \le \lambda \le n$. If B^λ and $B^{n-\lambda}$ are spaces homeomorphic to the closed unit balls in $\mathbb{R}^\lambda$ and $\mathbb{R}^{n-\lambda}$, respectively, then the Cartesian product $B^\lambda \times B^{n-\lambda}$ will be called a λ-handle of dimension n.

If N is a topological n-manifold and

$$\varphi : (\partial B^\lambda) \times B^{n-\lambda} \to \partial N$$

is a homeomorphism onto a closed subset of ∂N, one forms the quotient space of $N \sqcup B^\lambda \times B^{n-\lambda}$ that identifies points $\varphi(y)$ and y. The resulting space, denoted by $N \cup_\varphi B^\lambda \times B^{n-\lambda}$, is said to be the result of attaching a λ-handle to N. It is again an n-manifold. Note that, for a 0-handle, $(\partial B^0) \times B^n$ is empty and we agree that such handles are "attached" via disjoint union. In the case of a 1-handle, $(\partial B^1) \times B^{n-1}$ has two components, the "ends" of the handle, but in all other cases, handles are attached along connected subsets of their boundary.

Return now to the consideration of the sole critical point $p \in f^{-1}[-\epsilon, \epsilon]$, $f(p) = 0$ and let λ be the index of this critical point (Definition 2.9.15).

Theorem 4.2.6. *The manifold M_f^ϵ is homeomorphic to the space obtained by attaching a λ-handle of dimension n to $M_f^{-\epsilon}$.*

Remark. If the attaching map is a smooth imbedding, one can put a differentiable structure on the resulting manifold. The differentiable structure generally depends on the attaching map. Thus, for instance, the result of attaching an n-handle to a 0-handle (both of dimension n) by a diffeomorphism of boundaries is a space homeomorphic to S^n. Milnor's constructions of exotic differentiable structures on spheres (Example 3.2.6) proceeded by suitably choosing these attaching maps.

We will give a detailed sketch of the proof of Theorem 4.2.6. The first step is to use the Morse lemma (Theorem 2.9.18) to define a coordinate chart $(U, z^1, \dots, z^n)$ about p such that $f|U$ has the formula

$$f(z^1, \dots, z^n) = -x + y$$

where

$$x = \sum_{i=1}^{\lambda} (z^i)^2,$$

$$y = \sum_{j=\lambda+1}^{n} (z^j)^2.$$

In these coordinates, p is the origin.

We consider three cases:

1. $\lambda = 0$;
2. $\lambda = n$;
3. $0 < \lambda < n$.

Case 1. In this case, $x = 0$ and it is clear that p is a local minimum. Then, if $\epsilon > 0$ is sufficiently small, we obtain an n-ball $B^n = U \cap M_f^\epsilon$ which is a connected component of M_f^ϵ. Also, $U \cap M_f^{-\epsilon} = \emptyset$. By Theorem 4.2.3, the manifold $M_f^\epsilon \smallsetminus B^n$ is diffeomorphic to $M_f^{-\epsilon}$, and so

$$M_f^\epsilon \cong M_f^{-\epsilon} \sqcup B^n,$$

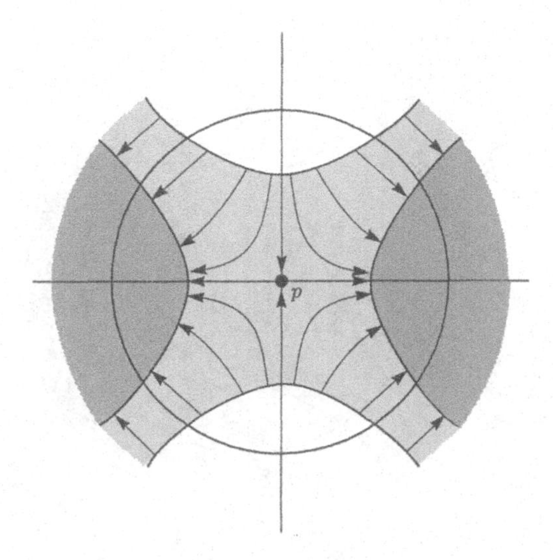

Figure 4.2.1. The neighborhood U of p

a disjoint union. That is, M_f^ϵ is obtained by attaching a 0-handle of dimension n to $M_f^{-\epsilon}$.

Case 2. Here, $y = 0$ and p is a local maximum. Evidently,

$$\begin{aligned} M_f^\epsilon \cap U &= U, \\ M_f^{-\epsilon} \cap U &= \text{ the complement in } U \text{ of an open ball,} \end{aligned}$$

where the second equality requires that $\epsilon > 0$ be sufficiently small. Clearly, M_f^ϵ is obtained from $M_f^{-\epsilon}$ by attaching an n-handle of dimension n.

Case 3. This is the interesting case. To begin with, choose $\epsilon > 0$ small enough that the coordinate chart $(U, z^1, \dots, z^n)$ contains the closed ball B defined by

$$\sum_{i=1}^{n} (z^i)^2 \leq 2\epsilon.$$

In Figure 4.2.1, we give a 2-dimensional schematic drawing of the neighborhood U of p, representing B as the disk bounded by the circle and representing the hypersurfaces $f^{-1}(\pm\epsilon)$ as hyperbolas. The horizontal axis represents the subspace on which the last $n - \lambda$ coordinates vanish, the vertical axis representing the space on which the first λ coordinates vanish. The darker shading represents $M^{-\epsilon}$ and the lighter shading represents $f^{-1}[-\epsilon, \epsilon]$. The flow of the gradient field is also indicated and it should be intuitively clear that, by adjusting the time parameter suitably, this flow can be used to deform M_f^ϵ to a subspace consisting of $M_f^{-\epsilon}$ with a λ-handle attached as in Figure 4.2.2. As indicated, the critical point p will be in the interior of this handle.

The idea (following [**28**, §3]) for a more rigorous approach, is to replace f with a function F that agrees with f outside a small ellipsoidal neighborhood of p in U

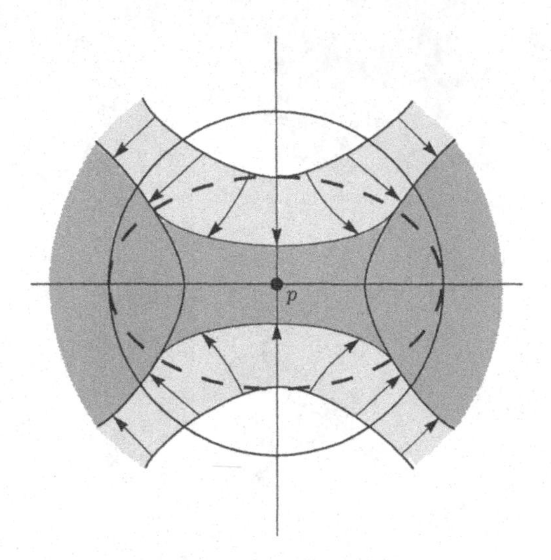

Figure 4.2.2. Attaching a handle to $M_f^{-\epsilon}$

(see Figure 4.2.2), has p as sole critical point in U, and satisfies

$$M_F^\epsilon = M_f^\epsilon,$$
$$M_F^{-\epsilon} = M_f^{-\epsilon} \cup_\varphi B^\lambda \times B^{n-\lambda},$$

where p lies in the interior of the handle. Since there are no critical points in $F^{-1}[-\epsilon, \epsilon]$, an application of Theorem 4.2.3 proves that

$$M_F^\epsilon \cong M_F^{-\epsilon},$$

giving Theorem 4.2.6. The flow in the proof of that theorem is indicated in Figure 4.2.2. More details follow.

We are free to choose the Riemannian metric in any convenient way, patching together local choices by a partition of unity. Accordingly, we choose the metric in U so that, with respect to the coordinates z^i, it is just the Euclidean metric. Thus, the gradient of any smooth function on U can be computed exactly as in standard multivariable calculus.

Construct a smooth function $\mu : \mathbb{R} \to [0, \infty)$ with the following properties:

- $\mu(0) > \epsilon$,
- $\mu(t) = 0$, $2\epsilon \le t < \infty$,
- $-1 < \mu'(t) \le 0$, $\forall t \in \mathbb{R}$.

Define F to coincide with f on $M \smallsetminus U$ and, in U, let

$$F = -x + y - \mu(x + 2y),$$

where x and y are the functions defined earlier. Since $\mu(x + 2y)$ vanishes for $x + 2y \ge 2\epsilon$, it vanishes outside the ellipsoidal ball $E \subset B$ defined by $x + 2y \le 2\epsilon$ and the local definitions of F fit together smoothly. The solid E is indicated in Figure 4.2.2 by a dashed ellipse, in which the handle is shown to be inscribed.

Claim 1. $M_f^\epsilon = M_F^\epsilon$.

Proof. Everywhere, $F \leq f$, and so $M_f^a \subseteq M_F^a$, for every value of a. Since F and f are equal outside the ellipsoidal ball E, we have

$$(*) \qquad M_f^a \setminus E = M_F^a \setminus E.$$

Inside E, $F \leq f \leq (x+2y)/2 \leq \epsilon$, and so $E \subseteq M_f^\epsilon \cap M_F^\epsilon$. Thus,

$$M_f^\epsilon \subseteq (M_f^\epsilon \setminus E) \cup E \subseteq (M_f^\epsilon \setminus E) \cup M_f^\epsilon \subseteq M_f^\epsilon,$$
$$M_F^\epsilon \subseteq (M_F^\epsilon \setminus E) \cup E \subseteq (M_F^\epsilon \setminus E) \cup M_F^\epsilon \subseteq M_F^\epsilon.$$

These inclusions must all be equalities and, by $(*)$, $M_f^\epsilon = M_F^\epsilon$. □

Claim 2. The function F has exactly the same critical points as f.

Proof. Since F and f coincide outside of U, it is enough to show that $F|U$ has p as its sole critical point. Write $F = -x + y - \mu(x+2y)$ in U and compute

$$\nabla F = \frac{\partial F}{\partial x}\nabla x + \frac{\partial F}{\partial y}\nabla y,$$

this being obvious since, in U, the formula for ∇F is the classical one. Furthermore, the partials

$$\frac{\partial F}{\partial x} = -1 - \mu'(x+2y) < -1 + 1 = 0,$$
$$\frac{\partial F}{\partial y} = 1 - 2\mu'(x+2y) \geq 1 + 0 = 1,$$

are never 0, so $\nabla f|U$ vanishes only where both ∇x and ∇y vanish, namely, at the origin p. □

One notes that $F(0) = -\mu(0) < -\epsilon$, hence that there are no critical points of F in $F^{-1}[-\epsilon, \epsilon]$. By Theorem 4.2.3, applied to the function F, we obtain

Claim 3. $M_F^\epsilon \cong M_F^{-\epsilon}$.

Evidently, we can write

$$M_F^{-\epsilon} = M_f^{-\epsilon} \cup H,$$

where H is a compact set inscribed in E as in Figure 4.2.2. Thus, the final step in the demonstration of Theorem 4.2.6 would be to prove the following intuitively plausible fact (*cf.* [**38**]). We omit this step.

Claim 4. The space H is a λ-handle of dimension n.

Remark. For compact manifolds, Morse theory gives a description of M as a space obtained by gluing together finitely many handles of dimension n. Each handle is attached only along the part $(\partial B^\lambda) \times B^{n-\lambda}$ of its boundary, hence it is possible to "flatten" it appropriately, proving that M_f^ϵ has the homotopy type of the space $M_f^{-\epsilon} \cup_\psi B^\lambda$, obtained by attaching the λ-cell B^λ by an imbedding $\psi : \partial B^\lambda \to \partial M_f^{-\epsilon}$. This leads to a description of M, up to homotopy, as a finite "cell complex". For many applications of algebraic topology, this is more useful than the precise decomposition of M into handles. For more details, see [**28**].

4.3. The Lie Bracket

As we have already remarked, $\mathfrak{X}(M)$ is a Lie algebra over $\mathbb{R}$ under the Lie bracket. Vector fields D, when viewed as derivations of the function algebra $C^\infty(M)$, localize to open subsets $U \subset M$. This localization is equivalent to the restriction $D|U$ of D as a smooth section of the tangent bundle, so it is elementary to check the following.

Lemma 4.3.1. *If $X, Y \in \mathfrak{X}(M)$ and if $U \subseteq M$ is open, then*

$$[X|U, Y|U] = [X, Y]|U.$$

In particular, properties of the bracket that were proven with coordinates can be extended to global properties on M. We apply this remark to Lie derivatives.

Since every vector field $X \in \mathfrak{X}(M)$ generates a local flow Φ on M, the definition of the Lie derivative

$$\mathfrak{L}_X(Y) = \lim_{t\to 0} \frac{\Phi_{-t*}(Y) - Y}{t}$$

makes sense pointwise on M and defines a new field $\mathfrak{L}_X(Y) \in \mathfrak{X}(M)$. The proof that $\mathfrak{L}_X(Y) = [X, Y]$ (Theorem 2.8.16) that was given in $\mathbb{R}^n$ can be carried out in local coordinate charts hence, by Lemma 4.3.1, globalizes.

Theorem 4.3.2. *If $X, Y \in \mathfrak{X}(M)$, the Lie derivative of Y by X is defined and smooth throughout M and*

$$\mathfrak{L}_X(Y) = [X, Y].$$

Similarly, Theorem 2.8.20 globalizes. Here, commutativity of local flows $\Phi = \{\Phi^\alpha\}_{\alpha\in\mathfrak{A}}$ and $\Psi = \{\Psi^\beta\}_{\beta\in\mathfrak{B}}$ means that $\Psi_t^\beta \circ \Phi_s^\alpha = \Phi_s^\alpha \circ \Psi_t^\beta$ wherever both sides are defined. Commutativity of vector fields $X, Y \in \mathfrak{X}(M)$ means that $[X, Y] \equiv 0$ on M.

Theorem 4.3.3. *Vector fields X, Y on M commute if and only if the local flows they generate on M commute.*

Corollary 4.3.4. *Complete vector fields $X, Y \in \mathfrak{X}(M)$ commute if and only if the flows that they generate commute.*

We are going to be interested in Lie subalgebras of $\mathfrak{X}(M)$. Of course, $\mathfrak{F} \subseteq \mathfrak{X}(M)$ is a Lie subalgebra if it is closed under the vector space operations and the bracket. If $F \subseteq T(M)$ is a k-plane subbundle (which we also refer to as a k-plane distribution on M), $\Gamma(F) \subseteq \mathfrak{X}(M)$ is a $C^\infty(M)$-submodule and a real vector subspace. It is not generally a Lie subalgebra but, when it is, there are important geometric consequences.

Definition 4.3.5. The k-plane distribution $F \subseteq T(M)$ is a Frobenius distribution (or an *involutive* distribution) if $\Gamma(F)$ is a Lie subalgebra of $\mathfrak{X}(M)$.

Remark. If $f, g \in C^\infty(M)$ and $X, Y \in \mathfrak{X}(M)$, it is easy to verify the identity

$$[fX, gY] = fX(g)Y - gY(f)X + fg[X, Y].$$

Consequently, if $F \subseteq T(M)$ is a k-plane distribution and if the fields $X_1, X_2, \dots, X_r \in \Gamma(F)$ span $\Gamma(F)$ over $C^\infty(M)$, then F will be a Frobenius distribution if and only if $[X_i, X_j] \in \Gamma(F)$, $1 \le i, j \le r$.

Example 4.3.6. On the group manifold $\mathrm{Gl}(n)$, we will define a couple of interesting Frobenius distributions. First recall (Example 2.7.18) that the space of left-invariant vector fields $\mathfrak{gl}(n) \subset \mathfrak{X}(\mathrm{Gl}(n))$ is a finite dimensional Lie subalgebra, canonically identified with the Lie algebra $\mathfrak{M}(n)$ of $n \times n$ real matrices under the commutator bracket. Let $\mathfrak{sl}(n) \subset \mathfrak{gl}(n)$ be the set of matrices of trace 0 and let $\mathfrak{o}(n) \subset \mathfrak{gl}(n)$ be the subset of skew symmetric matrices. These are clearly vector subspaces and the reader should have little difficulty in computing

$$\dim \mathfrak{sl}(n) = n^2 - 1,$$
$$\dim \mathfrak{o}(n) = \frac{n(n-1)}{2}.$$

Since $\mathrm{tr}(AB) = \mathrm{tr}(BA)$, we see that $\mathrm{tr}[A, B] = 0$, hence $\mathfrak{sl}(n)$ is a Lie subalgebra of $\mathfrak{gl}(n)$. If $A, B \in \mathfrak{o}(n)$, then

$$[A, B]^{\mathrm{T}} = B^{\mathrm{T}} A^{\mathrm{T}} - A^{\mathrm{T}} B^{\mathrm{T}} = [B, A] = -[A, B],$$

so $\mathfrak{o}(n) \subset \mathfrak{gl}(n)$ is also a Lie subalgebra.

If $\{S_1, \dots, S_{n^2-1}\}$ is a basis of $\mathfrak{sl}(n)$, extend it to a basis of $\mathfrak{gl}(n)$ by adjoining a suitable vector S_{n^2} and view $\{S_i\}_{i=1}^{n^2}$ as a set of left-invariant vector fields on $\mathrm{Gl}(n)$. It is clear that, at each point $P \in \mathrm{Gl}(n)$, these fields give a basis $\{PS_1, \dots, PS_{n^2}\}$ of $T_P(\mathrm{Gl}(n))$. Let $S_P \subset T_P(\mathrm{Gl}(n))$ be the $(n^2 - 1)$-dimensional subspace spanned by $\{PS_i\}_{i=1}^{n^2-1}$ and remark that this subspace does not depend on the choice of basis of $\mathfrak{sl}(n)$. Let

$$S = \bigcup_{P \in \mathrm{Gl}(n)} S_P.$$

This is an $(n^2 - 1)$-plane distribution on $\mathrm{Gl}(n)$. Indeed, the vector fields $\{S_i\}_{i=1}^{n^2}$ define an explicit trivialization of $T(\mathrm{Gl}(n)) \cong \mathrm{Gl}(n) \times \mathfrak{gl}(n)$ relative to which S becomes the trivial subbundle $\mathrm{Gl}(n) \times \mathfrak{sl}(n)$. (Warning: We are *not* using the standard trivialization of $T(\mathrm{Gl}(n)) = \mathrm{Gl}(n) \times \mathfrak{M}(n)$. This would give an imbedding $\mathrm{Gl}(n) \times \mathfrak{sl}(n) \subset T(\mathrm{Gl}(n))$ different from the one we have defined and not very interesting.) Similarly, extending a basis of $\mathfrak{o}(n)$ to one of $\mathfrak{gl}(n)$ and viewing these as left-invariant vector fields on $\mathrm{Gl}(n)$, we obtain a distribution $O \subset T(\mathrm{Gl}(n))$ of fiber dimension $n(n-1)/2$ and independent of the choices. By the remark preceding this example, the fact that $\mathfrak{sl}(n)$ and $\mathfrak{o}(n)$ are closed under the bracket implies that S and O are Frobenius distributions on $\mathrm{Gl}(n)$. Also, by our construction, if $P \in \mathrm{Gl}(n)$, then

$$L_{P*}(S_I) = S_P,$$
$$L_{P*}(O_I) = O_P.$$

Recall that the *special linear group* is the subgroup $\mathrm{Sl}(n) \subset \mathrm{Gl}(n)$ consisting of the matrices of determinant 1 (Example 2.5.5). In Example 2.5.8, we showed that $T_I(\mathrm{Sl}(n))$ is the subspace of $T_I(\mathrm{Gl}(n)) = \mathfrak{M}(n)$ consisting of the matrices of trace 0. For each $Q \in \mathrm{Sl}(n)$, L_Q carries $\mathrm{Sl}(n)$ onto itself, so it follows from our construction that $T_Q(\mathrm{Sl}(n)) = S_Q$. Also, for arbitrary $P \in \mathrm{Gl}(n)$,

$$T_{PQ}(P \cdot \mathrm{Sl}(n)) = L_{P*}T_Q(\mathrm{Sl}(n)) = L_{P*}(S_Q) = S_{PQ}.$$

We say that the left cosets $P \cdot \mathrm{Sl}(n)$ are *integral submanifolds* to the distribution S. In exactly the same way, using Exercise 2.5.9, we see that the left cosets $P \cdot \mathrm{O}(n)$ of the orthogonal group are integral submanifolds to the distribution $O \subset T(\mathrm{Gl}(n))$.

These are the first examples of *foliations* in this book. The connected components of the left cosets of these groups are the *leaves* of the foliation integral to the respective distributions. A distribution F on M with (one-to-one immersed) integral submanifolds through each point of M is said to be *integrable*. We will see that the Frobenius property is precisely the integrability condition (*cf.* Example 3.4.20). This is the theorem of Clebsch, Deahna, and Frobenius, commonly called the Frobenius theorem.

Generally speaking, a smooth map $f : M \to N$ does not push a vector field $X \in \mathfrak{X}(M)$ forward to a vector field $f_*(X) \in \mathfrak{X}(N)$. There are two problems. If f is not surjective, there would be points of N where $f_*(X)$ would not even be defined. If f is not injective, there could be points of N where $f_*(X)$ would be multiply defined. Nevertheless, there are situations in which f fails to be bijective, but the following concept makes sense.

Definition 4.3.7. If $f : M \to N$ is a smooth map between manifolds (possibly with boundary), vector fields $X \in \mathfrak{X}(M)$ and $Y \in \mathfrak{X}(N)$ are said to be f-related if, for each $q \in M$, $f_{*q}(X_q) = Y_{f(q)}$.

Example 4.3.8. It is possible that $X \in \mathfrak{X}(M)$ is not f-related to any $Y \in \mathfrak{X}(N)$. For example, let $f : \mathbb{R} \to S^1$ be the map $f(t) = e^{2\pi i t}$. Then

$$f_{*t}\left(t\frac{d}{dt}\right) = t f_{*t}\left(\frac{d}{dt}\right) = 2\pi i t e^{2\pi i t} \in T_{f(t)}(S^1) \subset T_{f(t)}(\mathbb{C}) = \mathbb{C}.$$

There is clearly no vector field on S^1 satisfying this.

Example 4.3.9. It is possible that $X \in \mathfrak{X}(M)$ may be f-related to many vector fields in $\mathfrak{X}(N)$. For example, let $f : S^1 \to \mathbb{C}$ be the inclusion map. Let $U \subset \mathbb{C}$ be an open neighborhood of S^1 such that $U \neq \mathbb{C}$. Let $\lambda : \mathbb{C} \to [0,1]$ be smooth such that $\operatorname{supp}(\lambda) \subset U$ and $\lambda|S^1 \equiv 1$. Define the vector field $Y_z = iz$ on $\mathbb{C}$. Since $z \perp iz$, we obtain $X \in \mathfrak{X}(S^1)$ by setting $X_z = iz$, $\forall\, z \in S^1$. Clearly, X is f-related to both Y and λY and these are distinct fields on $\mathbb{C}$.

Proposition 4.3.10. *Let $f : M \to N$ be smooth and let $X, Y \in \mathfrak{X}(M)$ be f-related to $\widetilde{X}, \widetilde{Y} \in \mathfrak{X}(N)$, respectively. Then $[X,Y]$ is f-related to $[\widetilde{X}, \widetilde{Y}]$. Equivalently, $\mathcal{L}_X(Y)$ is f-related to $\mathcal{L}_{\widetilde{X}}(\widetilde{Y})$.*

Proof. For each $h \in C^\infty(N)$ and for each $q \in M$,

$$Y(h \circ f)(q) = Y_q(h \circ f) = f_{*q}(Y_q)(h) = \widetilde{Y}_{f(q)}(h) = (\widetilde{Y}(h) \circ f)(q).$$

That is,

$$Y(h \circ f) = \widetilde{Y}(h) \circ f, \quad \forall\, h \in C^\infty(N).$$

There is a similar relation between X and $\widetilde{X}$. Thus,

$$\begin{aligned}
[\widetilde{X}, \widetilde{Y}]_{f(q)}(h) &= \widetilde{X}_{f(q)}(\widetilde{Y}(h)) - \widetilde{Y}_{f(q)}(\widetilde{X}(h)) \\
&= f_{*q}(X_q)(\widetilde{Y}(h)) - f_{*q}(Y_q)(\widetilde{X}(h)) \\
&= X_q(\widetilde{Y}(h) \circ f) - Y_q(\widetilde{X}(h) \circ f) \\
&= X_q(Y(h \circ f)) - Y_q(X(h \circ f)) \\
&= [X,Y]_q(h \circ f) \\
&= f_{*q}([X,Y]_q)(h),
\end{aligned}$$

$\forall h \in C^\infty(N)$. That is,

$$[\widetilde{X}, \widetilde{Y}]_{f(q)} = f_{*q}([X,Y]_q), \quad \forall q \in M.$$

This is the assertion of the proposition. □

Exercise 4.3.11. Let $f : M \to N$ be a one-to-one immersion and let $Y \in \mathfrak{X}(N)$. Prove that there exists a field $X \in \mathfrak{X}(M)$ that is f-related to Y if and only if $Y_{f(q)} \in f_{*q}(T_q(M))$, $\forall q \in M$. In this case, prove that X is unique (we call X the *restriction* of Y to the immersed submanifold).

Exercise 4.3.12. Let $f : S^{2n-1} \hookrightarrow \mathbb{R}^{2n}$ be the inclusion (an imbedding, hence a one-to-one immersion). Find $Y \in \mathfrak{X}(\mathbb{R}^{2n})$ that restricts, as above, to a nowhere zero vector field $X \in \mathfrak{X}(S^{2n-1})$. (By Theorem 1.2.16, this is false for the inclusion $f : S^{2n} \hookrightarrow \mathbb{R}^{2n+1}$. This fact is more elementary than Theorem 1.2.16, however, and will be proven later (Theorem 8.7.5).)

4.4. Commuting Flows

An $\mathbb{R}^n$-chart $(U, x^1, \dots, x^n)$ about $q \in M$ determines n commuting vector fields

$$\left\{\frac{\partial}{\partial x^1}, \dots, \frac{\partial}{\partial x^n}\right\} \subset \mathfrak{X}(U).$$

The corresponding local flows about q are of the form

$$\Phi^i_t(x^1, \dots, x^i, \dots, x^n) = (x^1, \dots, x^i + t, \dots, x^n).$$

Conversely, the following implies that commuting, linearly independent vector fields correspond to a coordinate chart.

Theorem 4.4.1. *Let M be an n-manifold without boundary, let $q \in M$, let U be an open neighborhood of q, and let $X^1, \dots, X^k \in \mathfrak{X}(U)$. If these vector fields commute and if $\{X^1_q, \dots, X^k_q\}$ is linearly independent, then there is a local coordinate chart (W, φ) about q such that*

$$\varphi_*(X^i|W) = \frac{\partial}{\partial x^i}, \quad 1 \le i \le k.$$

Proof. Making U smaller, if necessary, we assume that it is a coordinate chart, hence view it as an open subset of $\mathbb{R}^n$. We can do this so that q becomes the origin 0 and so that the vectors

$$X^1_0, \dots, X^k_0, \left.\frac{\partial}{\partial x^{k+1}}\right|_0, \dots, \left.\frac{\partial}{\partial x^n}\right|_0$$

form a basis of $T_0(U)$. Let

$$\Phi^i : (-\epsilon, \epsilon) \times \widetilde{W} \to U$$

be a local flow about $q = 0$ generated by X^i, $1 \le i \le k$. We can choose $\widetilde{W}$ to be of the form $(-\epsilon, \epsilon)^n$ with $\epsilon > 0$ so small that the formula

$$\theta(x^1, \dots, x^n) = \Phi^1_{x^1} \circ \cdots \circ \Phi^k_{x^k}(0, \dots, 0, x^{k+1}, \dots, x^n)$$

is defined, $\forall (x^1, \dots, x^n) \in \widetilde{W}$. This defines a smooth map

$$\theta : \widetilde{W} \to U.$$

Since $[X^i, X^j] \equiv 0$ on U, $1 \le i, j \le k$, we have

$$\theta(x^1, \dots, x^n) = \Phi^{\sigma(1)}_{x^{\sigma(1)}} \circ \cdots \circ \Phi^{\sigma(k)}_{x^{\sigma(k)}}(0, \dots, 0, x^{k+1}, \dots, x^n)$$

for each permutation σ of $\{1, 2, \dots, k\}$. Also,

$$\theta(0, \dots, 0) = (0, \dots, 0),$$

$$\theta_{*0}\left(\left.\frac{\partial}{\partial x^i}\right|_0\right) = \left.\frac{\partial}{\partial x^i}\right|_0, \quad k+1 \le i \le n.$$

For $r = (r_1, \dots, r_n) \in \widetilde{W}$ and $1 \le i \le k$,

$$\begin{aligned}
\theta_{*r}\left(\left.\frac{\partial}{\partial x^i}\right|_r\right) &= \left.\frac{\partial \theta}{\partial x^i}\right|_r \\
&= \left.\frac{d}{dx^i}\right|_{r_i} (\Phi^1_{r_1} \circ \cdots \circ \Phi^i_{x^i} \circ \cdots \circ \Phi^k_{r_k}(0, \dots, 0, r_{k+1}, \dots, r_n)) \\
&= \left.\frac{d}{dx^i}\right|_{r_i} (\Phi^i_{x^i}(\Phi^1_{r_1} \circ \cdots \widehat{\Phi^i_{r_i}} \cdots \circ \Phi^k_{r_k}(0, \dots, 0, r_{k+1}, \dots, r_n))) \\
&= X^i_{\theta(r)}.
\end{aligned}$$

Here, the notation $\widehat{\Phi^i_{r_i}}$ is a common device for indicating omission of the term $\Phi^i_{r_i}$. In particular, $\theta_{*0} : \mathbb{R}^n \to \mathbb{R}^n$ is nonsingular, so we can assume (choosing $\epsilon > 0$ smaller, if necessary) that $\theta : \widetilde{W} \to U$ carries $\widetilde{W}$ diffeomorphically onto an open neighborhood W of 0. By the above,

$$\theta_*\left(\frac{\partial}{\partial x^i}\right) = X^i|W, \quad 1 \le i \le k.$$

The desired coordinate chart (W, φ) is obtained by setting $\varphi = \theta^{-1}$. □

In particular, if X is a nonzero vector field defined in a neighborhood of $q \in M$, then there is a coordinate chart $(U, x^1, \dots, x^n)$ about q such that

$$X|U = \frac{\partial}{\partial x^1}.$$

Definition 4.4.2. A k-flow on M is a smooth map $\Psi : \mathbb{R}^k \times M \to M$, written $\Psi(v, x) = \Psi_v(x)$, such that

1. $\Psi_0 = \mathrm{id}_M$,
2. $\Psi_{v+w} = \Psi_v \circ \Psi_w$, $\forall\, v, w \in \mathbb{R}^k$.

It follows that $\Psi_{-v} = \Psi_v^{-1}$, so $\Psi_v \in \mathrm{Diff}(M)$, $\forall\, v \in \mathbb{R}^k$. The map

$$\mathbb{R}^k \to \mathrm{Diff}(M),$$

defined by $v \mapsto \Psi_v$, is a homomorphism of the additive group $\mathbb{R}^k$ into the group $\mathrm{Diff}(M)$. We think of Ψ as a smooth action of the group $\mathbb{R}^k$ on M.

Given $q \in M$ define

$$\Psi^q : \mathbb{R}^k \to M$$

by setting

$$\Psi^q(v) = \Psi_v(q).$$

In particular, $\Psi^q(0) = q$.

Definition 4.4.3. Given a k-flow Ψ, the Ψ-orbit of $q \in M$ is $\Psi^q(\mathbb{R}^k)$. When a choice of Ψ is fixed, the Ψ-orbit is also called the $\mathbb{R}^k$-orbit of q.

Remark. Points $p, q \in M$ are said to be equivalent under the k-flow if and only if there exists $v \in \mathbb{R}^k$ such that $\Psi_v(p) = q$. The fact that this is an equivalence relation is a trivial consequence of Definition 4.4.2. The $\mathbb{R}^k$-orbits are the equivalence classes.

Definition 4.4.4. The k-flow Ψ is nonsingular if

$$\Psi^q_{*v} : T_v(\mathbb{R}^k) \to T_{\Psi^q(v)}(M)$$

is one-to-one, $\forall\, v \in \mathbb{R}^k$, $\forall\, q \in M$.

Lemma 4.4.5. *The k-flow Ψ is nonsingular if and only if*

$$\Psi^q_{*0} : T_0(\mathbb{R}^k) \to T_q(M)$$

is one-to-one, $\forall\, q \in M$.

Proof. For fixed $v \in \mathbb{R}^k$, let $\tau_v : \mathbb{R}^k \to \mathbb{R}^k$ denote translation by v. Then,

$$\Psi^q = \Psi_v \circ \Psi^q \circ \tau_{-v},$$

$\forall v \in \mathbb{R}^k, \forall\, q \in M$. By the chain rule,

$$\Psi^q_{*v} = (\Psi_v)_{*q} \circ \Psi^q_{*0} \circ (\tau_{-v})_{*v}.$$

But $(\tau_{-v})_{*v} = \mathrm{id}_{\mathbb{R}^k}$ and $(\Psi_v)_{*q}$ is bijective, so Ψ^q_{*v} is one-to-one if and only if Ψ^q_{*0} is one-to-one. □

Proposition 4.4.6. *The set of k-tuples $(X^1, \dots, X^k)$ of complete, commuting vector fields on M is in natural, one-to-one correspondence with the set of k-flows Ψ on M. The fields $X^1, \dots, X^k$ are pointwise linearly independent if and only if Ψ is nonsingular.*

Proof. Given the k-tuple $(X^1, \dots, X^k)$ of complete, commuting vector fields, let $\Phi^1, \dots, \Phi^k$ be the corresponding flows. Since these flows commute, the formula

$$\Psi_{(x^1,\dots,x^k)} = \Phi^1_{x^1} \circ \cdots \circ \Phi^k_{x^k}$$

defines a k-flow on M. Conversely, given the k-flow Ψ, let $\{e_1, \dots, e_k\}$ be the standard basis of $\mathbb{R}^k$ and set

$$\Phi^i_t = \Psi_{te_i},$$

$1 \le i \le k$. This defines k commuting flows and their corresponding infinitesimal generators, $X^1, \dots, X^k$ are complete, commuting vector fields on M. Finally, these fields are linearly independent at $q \in M$ if and only if Ψ^q_{*0} is one-to-one. Thus, the previous lemma gives the final assertion. □

Theorem 4.4.7. *Let M be a connected n-manifold. If there exists a nonsingular n-flow on M, then M is diffeomorphic to $T^k \times \mathbb{R}^{n-k}$ for some integer $k = 0, 1, \dots, n$.*

Modulo one technical point, the proof of Theorem 4.4.7 is quite straightforward. Nonsingularity of the n-flow implies that, for each $q \in M$, the smooth map

$$\Psi^q : \mathbb{R}^n \to M$$

has constant rank n. Consequently, each $\mathbb{R}^n$-orbit is an open subset of M. Being equivalence classes, distinct orbits are disjoint, so the connectivity of M implies that there is only one $\mathbb{R}^n$-orbit and Ψ^q is a surjection.

Fix $q \in M$. If $v, w \in \mathbb{R}^n$, then

$$\begin{aligned}\Psi^q(v) = \Psi^q(w) &\Leftrightarrow \Psi_v(q) = \Psi_w(q)\\ &\Leftrightarrow \Psi_{v-w}(q) = q\\ &\Leftrightarrow \Psi^q(v-w) = q\\ &\Leftrightarrow v - w \in (\Psi^q)^{-1}(q).\end{aligned}$$

It is clear that $G = (\Psi^q)^{-1}(q)$ is an additive subgroup of $\mathbb{R}^n$, so Ψ^q passes to a well-defined homeomorphism

$$\psi : \mathbb{R}^n/G \to M.$$

If G were the subgroup $\mathbb{Z}^k \times \{0\} \subset \mathbb{R}^k \times \mathbb{R}^{n-k}$, we would have a natural structure of smooth n-manifold on $\mathbb{R}^n/G = T^k \times \mathbb{R}^{n-k}$ and, Ψ^q being locally a diffeomorphism, the homeomorphism

$$\psi : T^k \times \mathbb{R}^{n-k} \to M$$

would be a diffeomorphism. Therefore, it will be enough to find a linear automorphism $A : \mathbb{R}^n \to \mathbb{R}^n$ such that $A(G) = \mathbb{Z}^k \times \{0\}$. This is the technical point mentioned above.

Definition 4.4.8. An additive subgroup $G \subset \mathbb{R}^n$ is a k-dimensional lattice if it is generated by a linearly independent subset $\{v_1, \dots, v_k\} \subset \mathbb{R}^n$. If G is a k-dimensional lattice for some $k = 0, 1, \dots, n$, then G is called a lattice subgroup.

For example, $\mathbb{Z}^n \subset \mathbb{R}^n$ is an n-dimensional lattice. More generally, if $0 \leq k \leq n$, then $\mathbb{Z}^k \times \{0\} \subset \mathbb{R}^k \times \mathbb{R}^{n-k}$ is a k-dimensional lattice. In fact, if G is as in Definition 4.4.8, a linear automorphism of $\mathbb{R}^n$ taking v_i to the standard basis vector e_i, $1 \leq i \leq k$, carries G to $\mathbb{Z}^k \times \{0\}$.

Theorem 4.4.9. *A nontrivial, additive subgroup $G \subset \mathbb{R}^n$ is a lattice subgroup if and only if there is a neighborhood $U \subset \mathbb{R}^n$ of the origin such that $G \cap U = \{0\}$.*

By the above remarks, this theorem will complete the proof of Theorem 4.4.7. Indeed, the fact that Ψ^q is locally a diffeomorphism implies immediately that there is a neighborhood of $0 \in \mathbb{R}^n$ meeting $G = (\Psi^q)^{-1}(q)$ only in the point 0.

Before proving Theorem 4.4.9, we discuss some other consequences.

Corollary 4.4.10. *If $G \subset \mathbb{R}$ is an additive subgroup that is isomorphic to $\mathbb{Z}^k$, some $k \geq 2$, then G is dense in $\mathbb{R}$.*

Proof. For dimension reasons, G is not a lattice subgroup. Given $\epsilon > 0$, one can find $a \in G$ such that $|a| < \epsilon$ (Theorem 4.4.9). Thus, $\{ra\}_{r \in \mathbb{Z}} \subset G$ partitions $\mathbb{R}$ into intervals of length $< \epsilon$, so every $t \in \mathbb{R}$ lies within ϵ of a point of G and G is dense in $\mathbb{R}$. □

Corollary 4.4.11. *If $G \subset S^1$ is a subgroup that is isomorphic to $\mathbb{Z}^k$, some $k \geq 1$, then G is dense in S^1.*

Proof. Indeed, the standard projection $p : \mathbb{R} \to S^1$ is a group surjection and $p^{-1}(G)$ is an additive subgroup of $\mathbb{R}$ that is isomorphic to $\mathbb{Z}^{k+1}$. By Corollary 4.4.10, $p^{-1}(G)$ is dense in $\mathbb{R}$ and the assertion follows. □

For the proof of Theorem 4.4.9, we need two lemmas.

Lemma 4.4.13. *If there exists a neighborhood U as in* Theorem 4.4.9, *then every bounded subset of G is finite.*

Proof. If $B \subseteq G$ is bounded, let $\{g_i\}_{i=1}^{\infty} \subseteq B$. Since B is bounded, we can assume, without loss of generality, that this sequence is Cauchy. Let $\epsilon > 0$ be so small that the ϵ-neighborhood of 0 in $\mathbb{R}^n$ is contained in U and choose $r > 0$ such that $\|g_i - g_j\| < \epsilon$, $\forall i, j \geq r$. Then $i, j \geq r \Rightarrow g_i - g_j \in G \cap U = \{0\}$, so the sequence must have only finitely many distinct terms. □

Lemma 4.4.14. *If U is as in* Theorem 4.4.9 *and $n = 1$, then G is infinite cyclic.*

Proof. Let $g \in G \cap (0, \infty)$ be the element closest to 0. If there were no such element, we could produce an infinite, strictly decreasing sequence in $G \cap (0, \infty)$, contradicting Lemma 4.4.13. Let $\Lambda = \{mg \mid m \in \mathbb{Z}\}$, an infinite cyclic subgroup of G. We claim that $\Lambda = G$. Otherwise, find $f \in G \smallsetminus \Lambda$ and $m \in \mathbb{Z}$ such that $mg < f < (m+1)g$. Then $0 < f - mg < g$ and $f - mg \in G$, contradicting the choice of g. □

Proof of theorem 4.4.9. If $G \subset \mathbb{R}^n$ is a lattice subgroup generated by the linearly independent vectors $v_1, \dots, v_k$, there is a nonsingular linear automorphism $L : \mathbb{R}^n \to \mathbb{R}^n$ such that $L(v_i) = e_i$, the ith standard basis vector, $1 \leq i \leq k$. Since L is also a homeomorphism, we lose no generality in assuming $v_i = e_i$, $1 \leq i \leq k$. In this case, elementary geometry shows that $\|g\| \geq 1$, $\forall g \in G \smallsetminus \{0\}$. Thus, $U = \{v \in \mathbb{R}^n \mid \|v\| < 1\}$ has the property that $U \cap G = \{0\}$.

For the converse, suppose U exists as desired and proceed by induction on n. The case $n = 1$ is true by Lemma 4.4.14. For the inductive step, assume the truth of the theorem for some $n \geq 1$ and suppose that $G \subset \mathbb{R}^{n+1}$ and $U \subset \mathbb{R}^{n+1}$ satisfy the hypotheses of the theorem. Let $0 \neq g \in G$ and let $V \subset \mathbb{R}^{n+1}$ be the one-dimensional vector subspace spanned by g. The major step in our proof will be to show, via the inductive hypothesis, that $G/(G \cap V)$ is a lattice subgroup of $\mathbb{R}^{n+1}/V \cong \mathbb{R}^n$.

By Lemma 4.4.14, $G \cap V$ is infinite cyclic, generated by some $g_0 \in G \cap V$. Let $\{\overline{f}_i\}_{i=1}^{\infty}$ be a sequence in $G/(G \cap V) \subset \mathbb{R}^{n+1}/V$ converging to 0 in that vector space. Write $\overline{f}_i = f_i + (G \cap V)$, $f_i \in G \subset \mathbb{R}^{n+1}$. Then the distance of f_i from the line V approaches 0 as $i \to \infty$. Thus, for some constant $c > 0$, one can find $m_i \in \mathbb{Z}$ such that $\|f_i - m_i g_0\| < c$, $\forall i \geq 1$. By Lemma 4.4.13, $\{f_i - m_i g_0\}_{i=1}^{\infty} \subset G$ contains only finitely many distinct elements. That is, $\overline{f}_i = f_i + (G \cap V) = (f_i - m_i g_0) + (G \cap V)$ assumes only finitely many distinct values as $i \to \infty$. Since $\lim_{i \to \infty} \overline{f}_i = 0$, we conclude that $\overline{f}_i = 0$ for large enough values of i. Therefore, there is a neighborhood $U' \subset \mathbb{R}^{n+1}/V$ of 0 such that $U' \cap (G/(G \cap V)) = \{0\}$. By the inductive hypothesis, $G/(G \cap V)$ is a lattice generated by a linearly independent set $\{g_i + (G \cap V)\}_{i=1}^{\ell} \subset \mathbb{R}^{n+1}/V$.

Given $g \in G$, write

$$g + (G \cap V) = \sum_{i=1}^{\ell} r_i g_i + (G \cap V), \quad r_i \in \mathbb{Z}.$$

That is, there is $r_0 \in \mathbb{Z}$ such that

$$g - \sum_{i=1}^{\ell} r_i g_i = r_0 g_0.$$

Thus, $\{g_0, g_1, \dots, g_\ell\}$ generates G and this set is clearly linearly independent in $\mathbb{R}^{n+1}$. □

Exercise 4.4.15. Let M be an n-manifold, $\partial M = \emptyset$. Prove that M is integrably parallelizable (Example 3.4.8) if and only if there exist pointwise linearly independent, commuting vector fields $X^1, \dots, X^n$ (not necessarily complete). Use this to prove that a compact n-manifold is integrably parallelizable if and only if each component is diffeomorphic to T^n.

Remark. Following J. Milnor [**29**], one defines the *rank* of an n-manifold M to be the maximum number of everywhere linearly independent, commuting vector fields that the manifold admits. By Proposition 4.4.6, the rank of a compact n-manifold M is the largest integer $r \le n$ for which there exists a nonsingular r-flow on M. It is a celebrated theorem of E. Lima [**25**] that the rank of S^3 is 1. Since S^3 is parallelizable, it admits a nowhere 0 vector field, hence a nonsingular 1-flow. On the other hand, Theorem 4.4.7 implies that the only compact 3-manifold of rank 3 is T^3. The hard part of Lima's theorem is to show that S^3 does not have rank 2.

4.5. Foliations

Let $F \subseteq T(M)$ be a k-plane distribution on M. For simplicity, we consider only the case in which $\partial M = \emptyset$.

Definition 4.5.1. An integral manifold of F through $q \in M$ is a one-to-one immersion $i : N \to M$ of a k-manifold such that $q \in i(N)$ and

$$i_{*x}(T_x(N)) = F_{i(x)}, \quad \forall x \in N.$$

We generally identify N and $i(N)$. This is similar to the customary identification of a curve $s : [a,b] \to M$ with its image. The correct topology on the subset $i(N)$ of M is the manifold topology of N, not the relative topology.

Definition 4.5.2. A k-plane distribution F on M is said to be *integrable* if, through each $q \in M$, there passes an integral manifold of F.

Exercise 4.5.3. Let $M = \mathbb{R}^3 \setminus \{0\}$and define

$$F_v = \{w \in T_v(M) \mid w \perp v\},$$

$\forall v \in M$. Prove that $F = \bigcup_{v \in M} F_v$ is an integrable distribution and describe the integral manifolds.

Example 4.5.4. Take $M = \mathbb{R}^3$ and let F be the 2-plane distribution spanned by the pointwise linearly independent vector fields

$$X = \frac{\partial}{\partial x} + g(x,y)\frac{\partial}{\partial z},$$
$$Y = \frac{\partial}{\partial y} + h(x,y)\frac{\partial}{\partial z}.$$

Let $\pi : \mathbb{R}^3 \to \mathbb{R}^2$ be the projection $\pi(x, y, z) = (x, y)$. Since

$$\pi_{*q}(X_q) = \left.\frac{\partial}{\partial x}\right|_{\pi(q)},$$

$$\pi_{*q}(Y_q) = \left.\frac{\partial}{\partial y}\right|_{\pi(q)},$$

an integral manifold of F through q will be carried by π locally diffeomorphically onto an open subset of $\mathbb{R}^2$. That is, an integral manifold of F is locally a graph $z = f(x, y)$ and

$$X_{(x,y,f(x,y))} = \begin{bmatrix} 1 \\ 0 \\ g(x,y) \end{bmatrix} = \begin{bmatrix} 1 \\ 0 \\ \frac{\partial f}{\partial x}(x,y) \end{bmatrix},$$

$$Y_{(x,y,f(x,y))} = \begin{bmatrix} 0 \\ 1 \\ h(x,y) \end{bmatrix} = \begin{bmatrix} 0 \\ 1 \\ \frac{\partial f}{\partial y}(x,y) \end{bmatrix}.$$

That is, $f(x, y)$ solves the system

$$\frac{\partial f}{\partial x} = g(x, y),$$

$$\frac{\partial f}{\partial y} = h(x, y).$$

This overdetermined system of P.D.E. implies that

$$\frac{\partial g}{\partial y} = \frac{\partial^2 f}{\partial y\,\partial x} = \frac{\partial^2 f}{\partial x\,\partial y} = \frac{\partial h}{\partial x}.$$

That is, a necessary condition for F to be integrable is that

$$\frac{\partial g}{\partial y} = \frac{\partial h}{\partial x}.$$

It turns out, as we will see, that this is also a sufficient condition for integrability. This integrability condition can also be written in terms of brackets. Indeed,

$$[X, Y] = \left(\frac{\partial h}{\partial x} - \frac{\partial g}{\partial y}\right)\frac{\partial}{\partial z},$$

so the integrability condition becomes

$$[X, Y] = 0.$$

By our theory of commuting vector fields, this condition implies that there is a local coordinate chart (U, u, v, w) about q in which

$$X = \frac{\partial}{\partial u},$$

$$Y = \frac{\partial}{\partial v},$$

and, in this coordinate system, the integral manifolds are given by the equations $w =$ const. Finally, in this coordinate neighborhood, arbitrary fields $Z_1, Z_2 \in \Gamma(F|U)$ are linear function combinations of X, Y, hence integrability implies that $[Z_1, Z_2] \in \Gamma(F|U)$. This is true in suitable coordinate charts about each point of $\mathbb{R}^3$, hence $\Gamma(F) \subset \mathfrak{X}(\mathbb{R}^3)$ is a Lie subalgebra and F is a Frobenius 2-plane distribution on

$\mathbb{R}^3$. This exemplifies the following theorem, due to Clebsch, Deahna, and Frobenius, but generally credited only to the last of this trio.

Theorem 4.5.5 (The Frobenius theorem). *If $F \subseteq T(M)$ is a k-plane distribution on M, the following are equivalent.*

(1) *F is integrable.*
(2) *F is a Frobenius distribution* (Definition 4.3.5).
(3) *About each $q \in M$ there is a coordinate chart $(W, x^1, \dots, x^n)$ such that*

$$\frac{\partial}{\partial x^i} \in \Gamma(F|W), \quad 1 \le i \le k.$$

Exercise 4.5.6. Define

$$E_v = \{(a,b,c) \in T_v(M) \mid (b,c,a) \perp v\},$$

$\forall\, v \in M$. Prove that $E = \bigcup_{v\in M} E_v$ is a 2-plane distribution and decide whether or not it is integrable.

The proof of the Frobenius theorem is the primary goal of this section. Before giving a proof, however, we discuss some of the consequences.

A coordinate chart as in (3) of Theorem 4.5.5 will be called a *Frobenius chart.* Theorem 4.5.5 allows us to find a C^∞ atlas $\{(U_\alpha, \varphi_\alpha)\}_{\alpha\in\mathfrak{A}}$ on M such that the associated family of local trivializations

$$\{d\varphi_\alpha : T(U_\alpha) \to U_\alpha \times \mathbb{R}^n\}_{\alpha\in\mathfrak{A}}$$

is contained in the $\mathrm{Gl}(k, n-k)$-reduction of $T(M)$ corresponding to F. This means that the associated Jacobian cocycle satisfies

$$Jg_{\alpha\beta} : U_\alpha \cap U_\beta \to \mathrm{Gl}(k, n-k),$$

$\forall\, \alpha, \beta \in \mathfrak{A}$. That is, the infinitesimal $\mathrm{Gl}(k, n-k)$-structure determined by F is integrable in the sense of Definition 3.4.12.

Fix a coordinate cover $\{V_\lambda, x^1_\lambda, \dots, x^n_\lambda\}_{\lambda\in\mathfrak{L}}$ satisfying property (3) in Theorem 4.5.5 and such that

1. the index set $\mathfrak{L}$ is at most countably infinite and $\{V_\lambda\}_{\lambda\in\mathfrak{L}}$ is a locally finite cover of M;
2. x^i_λ ranges over the open interval $(-2, 2)$, $\forall\, \lambda \in \mathfrak{L}$, $1 \le i \le n$;
3. if $W_\lambda \subset V_\lambda$ is defined by the inequalities $-1 < x^i_\lambda < 1$, $1 \le i \le n$, then $\{W_\lambda\}_{\lambda\in\mathfrak{L}}$ is an open cover of M.

This is possible since M is 2nd countable and paracompact. We are primarily interested in the coordinate neighborhoods W_λ, with the V_λ's playing an auxiliary role.

Definition 4.5.7. A coordinate cover $\{W_\lambda, x^1_\lambda, \dots, x^n_\lambda\}_{\lambda\in\mathfrak{L}}$ with all of the above properties is called a regular cover for the integrable distribution F.

If $a = (a^{k+1}, \dots, a^n)$ where $a^i \in (-1, 1)$, $k+1 \le i \le n$, the equations

$$x^i_\lambda = a^i, \quad k+1 \le i \le n,$$

define an integral manifold $P_{\lambda,a}$ to $F|W_\lambda$, called a *plaque* of F in W_λ. Similarly, there are plaques $\widetilde{P}_{\lambda,a}$ of F in V_λ with $P_{\lambda,a} \subset \widetilde{P}_{\lambda,a}$. Remark that, by the definition of regular cover, the closure $\overline{W}_\lambda$ is a compact subset of V_λ and $\overline{P}_{\lambda,a}$ is a compact subset of $\widetilde{P}_{\lambda,a}$. In fact, $\overline{W}_\lambda = [-1, 1]^n$ in the coordinates of V_λ.

Lemma 4.5.8. *Suppose that F is an integrable distribution on M and that $(W, x^1, \dots, x^n)$ is a Frobenius chart W_λ or V_λ as above. Then every connected integral manifold to $F|W$ lies in some plaque of W. Consequently, if $N_1, N_2 \subset M$ are arbitrary integral manifolds of F, then $N_1 \cap N_2$ is an integral manifold of F.*

Proof. Let $p \in W$ and let P_a be the plaque through p. If $N \subset W$ is also a connected integral manifold through p and if $q \in N$, there is a smooth curve $s : [0,1] \to N$ with $s(0) = p$ and $s(1) = q$. Then $\dot{s}(t) \in T_{s(t)}(N) = F_{s(t)}$, $0 \leq t \leq 1$, so

$$\dot{s}(t) = \sum_{i=1}^{k} f_i(t) \left.\frac{\partial}{\partial x^i}\right|_{s(t)}.$$

Thus, if $k+1 \leq j \leq n$ and $0 \leq t \leq 1$,

$$\frac{d}{dt} x^j(s(t)) = \dot{s}(t)(x^j) = \sum_{i=1}^{k} f_i(t) \frac{\partial x^j}{\partial x^i} = 0.$$

That is,

$$x^j(s(t)) \equiv x^j(p) = a^j,$$

a constant, $k+1 \leq j \leq n$, $0 \leq t \leq 1$. This proves that $q \in P_a$, $\forall q \in N$.

Let N_1 and N_2 be arbitrary integral manifolds to F. If $N_1 \cap N_2 = \emptyset$, there is nothing to prove. Let $p \in N_1 \cap N_2$ and choose W about p as above. Then, the component of p in $N_1 \cap W$ lies in a plaque P_a. Since $\dim N_1 = k$, this component is an open subset of the plaque P_a. Similarly, the component of p in $N_2 \cap W$ is an open subset of P_a. Therefore, the component of p in $N_1 \cap N_2 \cap W$ is an open subset of P_a, hence is an integral manifold to F. Since $p \in N_1 \cap N_2$ is arbitrary, $N_1 \cap N_2$ is a (possibly disconnected) integral manifold to F. □

Lemma 4.5.9. *A plaque $P_{\lambda,a}$ of F in W_λ meets at most countably many plaques of the regular cover.*

Proof. Since the cover $\{W_\lambda\}_{\lambda \in \mathfrak{L}}$ is locally finite, each point $x \in \overline{P}_{\lambda,a}$ has a connected neighborhood U_x in $\widetilde{P}_{\lambda,a}$ that meets only finitely many W_{λ_i}, $1 \leq i \leq r$. The intersections of U_x with distinct plaques of W_{λ_i} are disjoint open (by Lemma 4.5.8) subsets of U_x, hence second countability of U_x implies that it meets at most countably many of the plaques of W_{λ_i}, $1 \leq i \leq r$. Since $\overline{P}_{\lambda,a}$ is compact, it can be covered by finitely many of these neighborhoods U_x and the assertion follows. □

Remark. With a little more care, one can guarantee that each plaque meets only finitely many other plaques, but we do not need this.

Definition 4.5.10. If F is an integrable distribution on M, then points $x, y \in M$ are said to be F-equivalent, and we write $x \sim_F y$, if there exist connected integral manifolds $N_1, N_2, \dots, N_r$ of F such that $x \in N_1$, $y \in N_r$, and $N_i \cap N_{i+1} \neq \emptyset$, $1 \leq i < r$.

Proposition 4.5.11. *Let F be an integrable k-plane distribution on M. Then the relation $\sim_F$ is an equivalence relation on M and the equivalence classes are the maximal connected integral manifolds to F.*

Proof. The fact that $\sim_F$ is an equivalence relation is practically immediate. We show that each $\sim_F$ equivalence class $L \subseteq M$ is (the image of) a one-to-one immersed submanifold integral to F.

Fix $p \in L$. Given $q \in L$, choose connected integral manifolds

$$N_1, N_2, \dots, N_r$$

such that $p \in N_1$, $q \in N_r$, and $N_i \cap N_{i+1} \neq \emptyset$, $1 \leq i \leq r-1$. By the definition of $\sim_F$, $N_i \subset L$, $1 \leq i \leq r$. In particular, L is a union of integral manifolds to F and we can topologize L by letting the open subsets $N \subseteq L$ be exactly the unions of integral manifolds to F that lie entirely in L. Then L itself is open, the empty subset is open by default, arbitrary unions of open sets are open and Lemma 4.5.8 guarantees that finite intersections of open sets are open.

This is a locally Euclidean topology of dimension k. Indeed, the topology in each of the integral manifolds in L is the usual manifold topology (open subsets of integral manifolds are integral manifolds). Since connected integral manifolds are path-connected, the definition of $\sim_F$ implies that L is path-connected in this topology.

The fact that the topology is Hausdorff is easy. We prove that L is 2nd countable. Let $\{(W_\alpha, x_\alpha^1, \dots, x_\alpha^n)\}_{\alpha \in \mathfrak{A}}$ be a regular cover of M. Each plaque $P_a \subset W_\alpha$ is connected, hence each plaque that meets L lies entirely in L. Thus, L is a union of plaques and the chains $N_1, \dots, N_r$ in the definition of F-equivalence can be taken so that each N_i is a plaque. It will be enough to prove that the set of plaques in L, defined by the given regular cover, is at most countably infinite.

Fix a plaque P_0 with $p \in P_0 \subseteq L$. Recall (Lemma 4.5.9) that each plaque meets at most countably many other plaques. Thus, for a fixed integer $r > 0$, there can be only countably many plaque chains $P_0, P_1, \dots, P_r$ with $P_{i-1} \cap P_i \neq \emptyset$, $0 < i \leq r$. Thus, as r ranges over the positive integers, the number of such plaque chains, starting at the fixed P_0, is at most countable. Since every plaque in L is reached by a finite plaque chain from P_0, there are at most countably many distinct such plaques.

We have shown that L is a connected topological k-manifold. But property (3) in Theorem 4.5.5 provides a smooth atlas on L and shows that this manifold is smoothly immersed in M. Being locally integral to F, L is itself a connected integral manifold, obviously maximal with this property by the definition of $\sim_F$. □

Definition 4.5.12. The decomposition of M into F-equivalence classes is called a foliation $\mathcal{F}$ of M. Each F-equivalence class L is called a leaf of the foliation $\mathcal{F}$. If $\dim M = n$ and the leaves of $\mathcal{F}$ are k-dimensional, the dimension of the foliation is $\dim \mathcal{F} = k$ and $\operatorname{codim} \mathcal{F} = n - k$ is called the codimension of the foliation.

Example 4.5.13. If

$$\Psi : \mathbb{R}^k \times M \to M$$

is a nonsingular k-flow, the $\mathbb{R}^k$-orbits are the leaves of a k-dimensional foliation $\mathcal{F}$ of M. Indeed, by Proposition 4.4.6, the k-flow is generated by a family $\{X^1, \dots, X^k\}$ of everywhere linearly independent, commuting vector fields and these fields span a k-plane distribution $F \subseteq T(M)$. Since the fields commute, F is a Frobenius distribution, hence integrable, and there is a corresponding k-dimensional foliation $\mathcal{F}$ of M. For each $q \in M$,

$$\Psi^q : \mathbb{R}^k \to M$$

is an immersion and $G = (\Psi^q)^{-1}(q)$ is a lattice subgroup of $\mathbb{R}^k$. Thus, $\mathbb{R}^k/G$ is a smooth k-manifold and Ψ^q passes to a one-to-one immersion

$$\psi : \mathbb{R}^k/G \to M.$$

This immersed submanifold is evidently a connected integral manifold to F, hence is an open neighborhood of q in the leaf L_q of $\mathfrak{F}$ through q. The image of ψ is, in fact, the $\mathbb{R}^k$-orbit of q. If it were not all of L_q, then this leaf would be the disjoint union of two or more such orbits, each open in L_q, contradicting the fact that L_q is connected.

As we have seen, the foliation $\mathfrak{F}$ can be quite complicated. As in Example 4.1.8, there is such a one-dimensional foliation of T^2 with each leaf everywhere dense in T^2. In fact, we will see that this is true for T^n, $\forall n \geq 2$. (Example 5.3.9). In the same way, higher dimensional nonsingular k-flows on T^n are readily produced having everywhere dense leaves (*cf.* Exercise 4.5.14 below).

Exercise 4.5.14. Construct a nonsingular 2-flow on T^3 having each $\mathbb{R}^2$-orbit diffeomorphic to the cylinder $S^1 \times \mathbb{R}$ and everywhere dense in T^3.

Exercise 4.5.15. Let M be an n-manifold without boundary. Let the map $f : M \to N$ have constant rank k and prove that, as y ranges over $f(M)$, the connected components of the level sets $f^{-1}(y)$ range over the leaves of a foliation $\mathfrak{F}$ of M of dimension $n - k$.

Exercise 4.5.16. Let $f : \mathbb{R}^3 \to \mathbb{R}$ be the submersion

$$f(x, y, z) = (1 - x^2 - y^2)e^z.$$

By Exercise 4.5.15, there is an associated two-dimensional foliation $\mathfrak{F}$ of $\mathbb{R}^3$ (see Figure 4.5.1). Show the following.

(1) The cylinder $x^2 + y^2 = 1$ is a leaf L_0 of $\mathfrak{F}$.
(2) The leaves interior to this cylinder L_0 are diffeomorphic to $\mathbb{R}^2$.
(3) The leaves exterior to L_0 are diffeomorphic to cylinders. (But they are *not* geometric cylinders.)
(4) The foliation $\mathfrak{F}$ is invariant under translations in the z-coordinate.

Exercise 4.5.17. Let $\mathfrak{F}$ be a foliation of M and let $i : L \hookrightarrow M$ be the one-to-one immersion of a leaf. Let X be a manifold and $f : X \to M$ a smooth map such that $f(X) \subseteq i(L)$. Then $i^{-1} \circ f : X \to L$ is smooth. (Hint: *cf.* Exercise 3.7.10.)

We turn to the proof of Theorem 4.5.5.

Proposition 4.5.18. *If F is an integrable distribution, then F is Frobenius.*

Proof. Let $X, Y \in \Gamma(F)$, $q \in M$, and let $i : N \to M$ be an integral manifold of F through q. Let $\widetilde{X}, \widetilde{Y} \in \mathfrak{X}(N)$ be the unique restrictions of X, Y to N (Exercise 4.3.11). Then $[\widetilde{X}, \widetilde{Y}]$ is i-related to $[X, Y]$. In particular, if $q = i(p)$,

$$[X, Y]_q = i_{*p}[\widetilde{X}, \widetilde{Y}]_p \in i_{*p}(T_p(N)) = F_q.$$

□

This proves that (1) ⇒ (2) in Theorem 4.5.5. The following gives the implication (3) ⇒ (1).

Proposition 4.5.19. *Let F be a k-plane distribution on M. If each point $q \in M$ has a coordinate neighborhood $(U, x^1, \dots, x^n)$ such that*

$$\frac{\partial}{\partial x^i} \in \Gamma(F|U), \quad 1 \leq i \leq k,$$

then F is integrable.

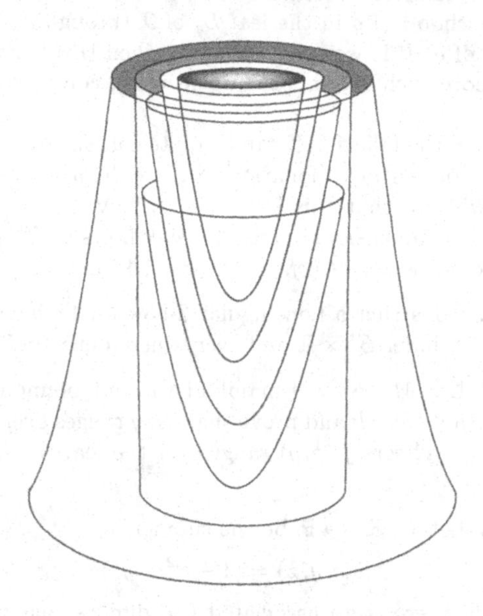

Figure 4.5.1. Foliation by the level sets $(1 - x^2 - y^2)e^z = c$

Proof. Indeed, if $a^i = x^i(q)$, $k+1 \leq i \leq n$, the level set

$$\{(x^1, \dots, x^n) \mid x^i = a^i, k+1 \leq i \leq n\}$$

is a k-dimensional integral manifold of F through q. □

It remains that we prove (2) ⇒ (3). This is the hard part.

Lemma 4.5.20. *Let $U \subseteq \mathbb{R}^n$ be an open subset and let F be a k-plane distribution on U. Let*

$$\pi : \mathbb{R}^n \to \mathbb{R}^k$$

be the projection $(x^1, \dots, x^n) \mapsto (x^1, \dots, x^k)$. Let $p \in U$ and suppose that $\pi_{*p} : F_p \to T_{\pi(p)}(\mathbb{R}^k)$ is bijective. Then there is an open neighborhood W of p in U such that $\pi_{*x} : F_x \to T_{\pi(x)}(\mathbb{R}^k)$ is bijective, $\forall x \in W$.

Proof. Let V be an open neighborhood of p in U such that there are fields $X^1, \dots,$ $X^k \in \Gamma(F|U)$ which give a basis of F_x, $\forall x \in V$. Write

$$X^i = \sum_{j=1}^{n} f_j^i \frac{\partial}{\partial x^j}, \quad 1 \leq i \leq k.$$

Then

$$\pi_{*x}(X_x^i) = \sum_{j=1}^{k} f_j^i(x) \left.\frac{\partial}{\partial x^j}\right|_{\pi(x)}, \quad \forall x \in V, \quad 1 \leq i \leq k.$$

Consider the $k \times k$ matrix

$$A(x) = \begin{bmatrix} f_1^1(x) & \cdots & f_k^1(x) \\ \vdots & & \vdots \\ f_1^k(x) & \cdots & f_k^k(x) \end{bmatrix}.$$

By assumption, $\det A(p) \neq 0$, so, for a small enough open neighborhood W of p in V, $\det A(x) \neq 0$, $\forall\, x \in W$. That is, $\pi_{*x} : F_x \to T_{\pi(x)}(\mathbb{R}^k)$ is bijective, $\forall\, x \in W$. □

Lemma 4.5.21. *Let F be a k-plane distribution on M and let $q \in M$. Then there is a coordinate chart $(W, x^1, \dots, x^n)$ about q such that the map*

$$\pi : W \to \mathbb{R}^k,$$

given by $\pi(x^1, \dots, x^n) = (x^1, \dots, x^k)$, has

$$\pi_{*x} : F_x \to T_{\pi(x)}(\mathbb{R}^k)$$

bijective, $\forall\, x \in W$.

Proof. By Lemma 4.5.20, we must choose the coordinates so that

$$\pi_{*q} : F_q \to T_{\pi(q)}(\mathbb{R}^k)$$

is bijective. Then we restrict to a smaller neighborhood, if necessary. Let $(U, x^1, \dots, x^n)$ be a coordinate chart about q and let $X^1, \dots, X^k \in \Gamma(F|U)$ give a basis of F_q. Permuting the coordinates suitably, we can assume that

$$\left\{ X_q^1, \dots, X_q^k, \left.\frac{\partial}{\partial x^{k+1}}\right|_q, \dots, \left.\frac{\partial}{\partial x^n}\right|_q \right\}$$

is a basis of $T_q(M)$. Then the surjection $\pi_{*q} : T_q(M) \to T_{\pi(q)}(\mathbb{R}^k)$ annihilates the last $n-k$ of these vectors, hence it carries $\{X_q^1, \dots, X_q^k\}$ to a basis of $T_{\pi(q)}(\mathbb{R}^k)$. □

Lemma 4.5.22. *Let F, $q \in M$, and*

$$\pi : W \to \mathbb{R}^k$$

all be as in Lemma 4.5.21. *Given $x \in W$ and $1 \leq i \leq k$, let $Z_x^i \in F_x$ be the unique vector such that*

$$\pi_{*x}(Z_x^i) = \left.\frac{\partial}{\partial x^i}\right|_{\pi(x)}.$$

Then $Z^1, \dots, Z^k \in \Gamma(F|W)$.

Proof. The only problem is to prove that Z^i is smooth at each $x \in W$, $1 \leq i \leq k$. We can assume that there is a trivialization $T(W) \cong W \times \mathbb{R}^n$ relative to which $F|W \cong W \times \mathbb{R}^k$. The standard trivialization of $T(\mathbb{R}^k)$ is given by the coordinate fields

$$\frac{\partial}{\partial x^1}, \dots, \frac{\partial}{\partial x^k}.$$

We express the linear map $\pi_{*x} : T_x(W) \to T_{\pi(x)}(\mathbb{R}^k)$ relative to these trivializations by a matrix $[A(x), B(x)]$ where $A(x)$ is $k \times k$ and $B(x)$ is $k \times (n-k)$. Since π_{*x} carries $F_x = \{x\} \times \mathbb{R}^k$ bijectively onto $\{\pi(x)\} \times \mathbb{R}^k$, we see that $A(x) \in \mathrm{Gl}(k)$ and depends smoothly on x. Thus, $A(x)^{-1}$ is also smooth in x and

$$\begin{bmatrix} A(x)^{-1} \\ 0 \end{bmatrix} : \{\pi(x)\} \times \mathbb{R}^k \to \{x\} \times \mathbb{R}^n$$

has image F_x. The ith column of this matrix is Z_x^i, $1 \le i \le k$, so this vector depends smoothly on x. □

We can now complete the proof of Theorem 4.5.5.

Proposition 4.5.23. *If F is a Frobenius distribution on M, then about each $q \in M$ there is a coordinate chart $(U, y^1, \dots, y^n)$ such that*

$$\frac{\partial}{\partial y^i} \in \Gamma(F|U), \quad 1 \le i \le k.$$

Proof. Let $(W, x^1, \dots, x^n)$ be a coordinate neighborhood of q and $Z^1, \dots, Z^k \in \Gamma(F|W)$, all as in Lemma 4.5.22. Since Z^i is π-related to the ith coordinate field, $1 \le i \le k$, it follows that $\pi_*[Z^i, Z^j] \equiv 0$, $1 \le i, j \le k$. But the Frobenius condition implies that $[Z^i, Z^j] \in \Gamma(F|W)$ and, on $F|W$, π_* is one-to-one, so $[Z^i, Z^j] \equiv 0$, $1 \le i, j \le k$. Since these fields are pointwise linearly independent, Theorem 4.4.1 furnishes a coordinate chart $(U, y^1, \dots, y^n)$ around $q \in W$ such that

$$Z^i|U = \frac{\partial}{\partial y^i}, \quad 1 \le i \le k.$$

□

Exercise 4.5.24. Although we have considered foliations only on manifolds M without boundary, one often relaxes this assumption by requiring special behavior for foliations near ∂M. In the case of foliations of codimension one, it is natural to require that each component of ∂M be a leaf. Use Exercise 4.5.16 to produce a foliation of $D^2 \times S^1$ having the boundary torus $S^1 \times S^1$ as a leaf and having all other leaves diffeomorphic to $\mathbb{R}^2$. (This is the famous "Reeb foliation" of the solid torus, pictured in Figure 4.5.2 and discovered by G. Reeb [**37**].)

Nonsingular flows become foliations of dimension one if we forget the parametrization of the leaves. That is, the nonsingular vector field X that generates the flow is replaced by the one-dimensional, integrable distribution spanned by this field. The study of flows belongs to the branch of mathematics called "dynamical systems". Notions such as "minimal set" (see Exercise 4.1.18) and "limit set" (Exercise 4.1.19) are fundamental to the theory of dynamical systems and they carry over nicely to foliation theory, where they are likewise fundamental. The following exercises introduce these important ideas.

Exercise 4.5.25. Let $\mathcal{F}$ be a foliation of M. A subset $C \subseteq M$ is said to be $\mathcal{F}$-*saturated* if, for each $x \in C$, the entire leaf L_x through x lies in C. Prove that the closure $\overline{C}$ of an $\mathcal{F}$-saturated set is an $\mathcal{F}$-saturated set.

Exercise 4.5.26. Let $\mathcal{F}$ be a foliation of M. A subset $C \subseteq M$ is said to be a *minimal set* of $\mathcal{F}$ if

(a) $C \ne \emptyset$;
(b) C is closed in M;
(c) C is $\mathcal{F}$-saturated;
(d) C contains no proper subset with all of these properties.

For example, a closed leaf is a minimal set. Prove that, if M is compact, every closed, nonempty, $\mathcal{F}$-saturated subset of M contains at least one minimal set. (In particular, by Exercise 4.5.25, every leaf of $\mathcal{F}$ closes on at least one minimal set.) Show by an example that M itself may be a minimal set.

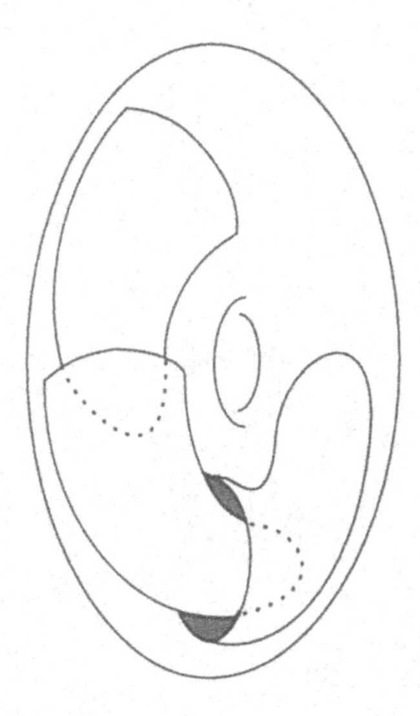

Figure 4.5.2. The Reeb foliation of $D^2 \times S^1$

Exercise 4.5.27. Let $L \subset M$ be a leaf of a foliation $\mathcal{F}$. The *limit set* of L is defined by

$$\lim(L) = \bigcap_{K \in \mathcal{K}} \overline{(L \smallsetminus K)},$$

where $\mathcal{K}$ is the family of compact subsets of the leaf L and the overline denotes closure in M. Prove the following.

(1) If M is compact, then $\lim(L)$ is a compact, $\mathcal{F}$-saturated set.
(2) If M is compact, $\lim(L) = \emptyset$ if and only if L is compact.
(3) If L is dense in M (but not equal to M) $\lim(L) = M$.
(4) If the leaf L is an imbedded submanifold of M, then $L \cap \lim(L) = \emptyset$.

CHAPTER 5

Lie Groups and Lie Algebras

Lie groups and their Lie algebras play a central role in geometry, topology, and analysis. Here we can only give a brief introduction to this fascinating topic.

5.1. Basic Definitions and Facts

A topological group is a topological space together with a group structure on that space such that the group operations are continuous. A Lie group is a differentiable manifold together with a group structure on that manifold such that the group operations are smooth. Lie groups are also topological groups, but not *vice versa.* Here are the precise definitions.

Definition 5.1.1. A topological group G is a topological space that is also a group such that the operations

$$\mu : G \times G \to G, \qquad \mu(x, y) = xy,$$
$$\iota : G \to G, \qquad \iota(x) = x^{-1}$$

are continuous maps. If G is a smooth manifold without boundary and these operations are smooth, G is called a Lie group.

Remark. In most of the literature, Lie groups are defined to be *real analytic.* That is, G is a manifold with a C^ω (real analytic) atlas and the group operations are real analytic. In fact, no generality is lost by this more restrictive definition. Smooth Lie groups always support an analytic group structure, and something even stronger is true. Hilbert's fifth problem was to show that if G is only assumed to be a topological manifold with continuous group operations, then it is, in fact, a real analytic Lie group. This was finally proven by the combined work of A. Gleason, D. Montgomery, and L. Zippin. The details are too deep to be discussed here.

Evidently, every finite dimensional vector space over $\mathbb{R}$ or $\mathbb{C}$ is a Lie group under vector addition. Here are some more interesting examples.

Example 5.1.2. The group $\mathrm{Gl}(n)$ is evidently a Lie group, the operations being given by rational functions of the coordinates. The subgroups $\mathrm{Sl}(n)$ and $\mathrm{O}(n)$ are smoothly imbedded submanifolds of $\mathrm{Gl}(n)$, hence the smoothness of the group operations on $\mathrm{Gl}(n)$ implies the smoothness of their restrictions to $\mathrm{Sl}(n)$ and $\mathrm{O}(n)$. It can be shown that $\mathrm{Gl}(n)$ and $\mathrm{O}(n)$ each have two connected components, distinguished by the sign of the determinant. The component of the identity I in each of these groups is itself a Lie group (Exercise 5.1.3), denoted by $\mathrm{Gl}_+(n)$ and $\mathrm{SO}(n)$ respectively. The group $\mathrm{SO}(n)$ is called the *special orthogonal group.* Recall that the orthogonal group $\mathrm{O}(n)$, hence also the special orthogonal group, is compact (Exercise 2.5.9).

Exercise 5.1.3. Prove that the component of the identity in a Lie group is itself a Lie group.

Example 5.1.4. The group $\mathrm{Gl}(k, n-k)$ consists of matrices

$$\begin{bmatrix} A & B \\ 0 & C \end{bmatrix}$$

where $A \in \mathrm{Gl}(k)$, $C \in \mathrm{Gl}(n-k)$, and B is an arbitrary $k \times (n-k)$ matrix. Thus, $\mathrm{Gl}(k, n-k)$ is a manifold diffeomorphic to $\mathrm{Gl}(k) \times \mathrm{Gl}(n-k) \times \mathbb{R}^{k(n-k)}$. This is an open subset of $\mathbb{R}^{k^2+(n-k)^2+k(n-k)}$ and the group operations are rational functions in the coordinates, so this is a Lie group.

Example 5.1.5. The group $\mathrm{Gl}(n, \mathbb{C})$ of nonsingular, $n \times n$ matrices over the complex field $\mathbb{C}$ is a Lie group, called the complex general linear group. Remark that $\mathrm{Gl}(1, \mathbb{C}) = \mathbb{C}^*$ is the multiplicative group of nonzero complex numbers. The unit circle $S^1 \subset \mathbb{C}^*$ is a subgroup and a smoothly imbedded submanifold, hence is also a Lie group.

Example 5.1.6. If G and H are Lie groups, then $G \times H$ is a Lie group under the usual Cartesian group operations and the smooth product structure. In particular, $T^n = S^1 \times \cdots \times S^1$ is a Lie group.

Example 5.1.7. Define $\varphi : \mathrm{Gl}(n, \mathbb{C}) \to \mathrm{Gl}(n, \mathbb{C})$ by $\varphi(A) = \overline{A}^{\mathrm{T}} A$. Here the overline indicates complex conjugation in each entry of the matrix. This map has constant rank n^2 and $\varphi(I) = I$. We define $\mathrm{U}(n) = \varphi^{-1}(I)$, a smoothly imbedded, nonempty submanifold of $\mathrm{Gl}(n, \mathbb{C})$ that has dimension $2n^2 - n^2 = n^2$. It is easy to check that $\mathrm{U}(n)$ is also a subgroup, hence it is a Lie group, called the unitary group. For the same reasons that $\mathrm{O}(n)$ is compact, the Lie group $\mathrm{U}(n)$ is compact. Since $A \in \mathrm{U}(n)$ if and only if $A^{-1} = \overline{A}^{\mathrm{T}}$, it follows that $|\det(A)| = 1$. Indeed, $\det : U(n) \to S^1$ is a group homomorphism and a submersion. One defines the *special unitary group* $\mathrm{SU}(n) = \det^{-1}(1) \subset \mathrm{U}(n)$, a compact Lie group of dimension $n^2 - 1$.

Exercise 5.1.8. Check the various assertions in Example 5.1.7, showing that $\mathrm{U}(n)$ and $\mathrm{SU}(n)$ are compact Lie groups of respective dimensions n^2 and $n^2 - 1$.

Example 5.1.9. Let $\mathbb{H}$ denote the division algebra of quaternions. The nonzero quaternions $\mathbb{H}^*$ form a multiplicative group and a manifold diffeomorphic to $\mathbb{R}^4 \setminus \{0\}$. It is clear that the group operations are smooth, so $\mathbb{H}^*$ is a Lie group. The 3-sphere $S^3 \subset \mathbb{H}^*$ consists of the unit length quaternions, hence it is closed under multiplication and passing to inverses. This gives a Lie group structure on S^3.

Usually, the identity element of a topological group or Lie group will be denoted by e. For matrix groups, however, the customary symbol for the identity is I. Because the Lie groups S^1 and S^3 are subgroups of the division algebras $\mathbb{C}$ and $\mathbb{H}$ respectively, the identity elements in these groups are denoted by 1, the unity of the respective algebras.

Definition 5.1.10. Let G be a topological group (respectively, a Lie group), $a \in G$. Left translation by a is the continuous (respectively, smooth) map $L_a : G \to G$ defined by $L_a(x) = ax$, $\forall\, x \in G$.

Remark that $L_{a^{-1}} = L_a^{-1}$, so $L_a \in \mathrm{Homeo}(G)$. Similarly, if G is a Lie group, $L_a \in \mathrm{Diff}(G)$. Also, the inversion map $\iota : G \to G$ is continuous and equal to its own inverse, so $\iota \in \mathrm{Homeo}(G)$ and, when G is a Lie group, $\iota \in \mathrm{Diff}(G)$.

The following discussion illustrates some of the striking ways in which algebra and topology interact in these structures.

Lemma 5.1.11. *If G is a topological group and $U \subseteq G$ is an open neighborhood of $e \in G$, then there is an open neighborhood $V \subseteq U$ of e with the property that $v \in V \Leftrightarrow v^{-1} \in V$. Such a neighborhood V of e is said to be symmetric.*

Proof. Indeed, $\iota(U)$ is also an open neighborhood of $e \in G$, so the neighborhood $V = U \cap \iota(U)$ is as desired. □

Lemma 5.1.12. *If $Z, W \subseteq G$ and if W is open in G, then the set*

$$ZW = \{zw \mid z \in Z \text{ and } w \in W\}$$

is open in G.

Proof. Indeed, $ZW = \bigcup_{z \in Z} L_z(W)$ is a union of open sets. □

Proposition 5.1.13. *Let G be a connected topological group and let $U \subseteq G$ be an arbitrary open neighborhood of the identity $e \in G$. Then U generates the group G.*

Proof. By Lemma 5.1.11, we lose no generality in assuming that U is a symmetric neighborhood of e. Using Lemma 5.1.12, we define open sets

$$U^n = UU^{n-1}$$

by induction on n. Since $e \in U$, these form an increasing nest

$$U \subseteq U^2 \subseteq \dots \subseteq U^n \subseteq \cdots$$

of open neighborhoods of the identity. By the symmetry of U and the formula for the inverse of a product, each U^n is symmetric. Thus, we obtain an open, symmetric neighborhood

$$U^\infty = \bigcup_{n=1}^{\infty} U^n$$

of the identity. Clearly, U^∞ is closed under group multiplication; hence, being a symmetric neighborhood of the identity, U^∞ is a subgroup of G. But the left cosets $\{aU^\infty\}_{a \in G}$ form a cover of G by disjoint open sets, hence the connectivity of G implies that there is only one coset. That is, $U^\infty = G$. □

We focus our attention on Lie groups G and their associated Lie algebras. It will be seen that the introduction of Lie algebras produces further remarkable interactions of algebra, topology and calculus.

Definition 5.1.14. A vector field $X \in \mathfrak{X}(G)$ is left-invariant if, for each $a \in G$, $L_{a*}(X) = X$. The set of left-invariant vector fields on G is denoted by $L(G)$.

The following is quite easy, but very important.

Proposition 5.1.15. *The subset $L(G) \subset \mathfrak{X}(G)$ is a Lie subalgebra.*

Proof. Indeed, the bracket of L_a-related fields is L_a-related to the bracket of these fields. It follows immediately that the bracket of left-invariant fields is a left-invariant field. □

Definition 5.1.16. If G is a Lie group, its Lie algebra is $L(G)$.

Example 5.1.17. We saw in Example 2.7.18 that $L(\mathrm{Gl}(n)) = \mathfrak{gl}(n)$ consists of the fields R_A, $A \in \mathfrak{M}(n)$, and that this defines a canonical isomorphism of Lie algebras $L(\mathrm{Gl}(n)) = \mathfrak{M}(n)$. Similarly, $L(\mathrm{Sl}(n)) = \mathfrak{sl}(n)$ is the Lie algebra of $n \times n$ matrices of trace 0 and $L(\mathrm{O}(n)) = \mathfrak{o}(n)$ is the Lie algebra of skew symmetric matrices (*cf.* Example 4.3.6).

Exercise 5.1.18. Identify $L(\mathrm{U}(n)) = \mathfrak{u}(n)$ and $L(\mathrm{SU}(n)) = \mathfrak{su}(n)$ as Lie algebras of complex matrices.

Proposition 5.1.19. *The evaluation map $\epsilon : L(G) \to T_e(G)$, defined by $\epsilon(X) = X_e$, is an isomorphism of vector spaces.*

Proof. This is clearly a linear map. The fact that it is injective follows immediately from

$$X_a = L_{a*}(X_e), \ \forall\, a \in G.$$

We prove that ϵ is surjective. Let $v \in T_e(G)$ and define

$$X_a = L_{a*}(v), \ \forall\, a \in G.$$

This defines $X : G \to T(G)$ carrying each $a \in G$ to $X_a \in T_a(G)$ and $X_e = v$. We must prove that X is smooth, hence a vector field, and that it is left-invariant.

For $f \in C^\infty(G)$, form the function $X(f) : G \to \mathbb{R}$ by setting

$$X(f)(x) = X_x(f), \ \forall\, x \in G.$$

If $X(f)$ is smooth, $\forall\, f \in C^\infty(G)$, it will follow that X is smooth. Indeed, in local coordinates, smoothness of $X(x^i) = f_i$ implies smoothness (locally) of X. Note that

$$X(f)(x) = X_x(f) = (L_x)_{*e}(X_e)(f) = X_e(f \circ L_x).$$

The function $g : G \times G \to \mathbb{R}$, defined by

$$g(x,y) = f(L_x(y)) = f(xy),$$

is smooth. Let $(U, x^1, \dots, x^n)$ be an arbitrary coordinate neighborhood in G. About $e \in G$, choose coordinates $(V, y^1, \dots, y^n)$ relative to which $e = (0, \dots, 0)$. Then $(x, e) \in U \times V$ has coordinates $(x^1, \dots, x^n, 0, \dots, 0)$. Write

$$X_e = \sum_{i=1}^n c^i \left.\frac{\partial}{\partial y^i}\right|_{(0,\dots,0)}.$$

Then

$$X_e(f \circ L_x) = \sum_{i=1}^n c^i \frac{\partial g}{\partial y^i}(x^1, \dots, x^n, 0, \dots, 0)$$

is smooth in the arbitrary coordinate neighborhood $(U, x^1, \dots, x^n)$, proving that $X \in \mathfrak{X}(G)$.

Finally, we prove that X is left-invariant. Indeed, if $a, b \in G$,

$$\begin{aligned}
(L_{a*}(X))_b &= (L_a)_{*a^{-1}b}(X_{a^{-1}b}) \\
&= (L_a)_{*a^{-1}b}((L_{a^{-1}b})_{*e}(X_e)) \\
&= (L_a \circ L_{a^{-1}b})_{*e}(X_e) \\
&= (L_b)_{*e}(X_e) \\
&= X_b.
\end{aligned}$$

Since b is arbitrary, $L_{a*}(X) = X$. Since a is arbitrary, $X \in L(G)$. □

Corollary 5.1.20. *If G is a Lie group, then* $\dim L(G) = \dim G$.

Corollary 5.1.21. *If G is a Lie group, there is a canonical trivialization of the tangent bundle $\pi : T(G) \to G$. In particular, every Lie group is parallelizable.*

Proof. Define $\varphi : G \times L(G) \to T(G)$ by $\varphi(a, X) = X_a$. Then the restriction $\varphi_a = \varphi|(\{a\} \times L(G))$ is an isomorphism

$$\varphi_a : \{a\} \times L(G) \to T_a(G)$$

of vector spaces, $\forall\, a \in G$. If we show that φ is smooth, it will be a bundle isomorphism. Fix a basis $X_1, \dots, X_n$ of $L(G)$. Coordinatize $L(G)$ via this basis, thereby defining an isomorphism $G \times L(G) \cong G \times \mathbb{R}^n$. Relative to this coordinatization, φ has the formula

$$\widetilde{\varphi}(a, (b^1, \dots, b^n)) = \sum_{i=1}^{n} b^i X_{ia}.$$

Since the fields X_i are smooth, so is $\widetilde{\varphi}$, hence φ. □

In particular, S^3 is parallelizable (*cf.* Theorem 1.2.13). This sphere is not, however, integrably parallelizable (*cf.* Exercise 4.4.15).

Definition 5.1.22. A 1-parameter subgroup of a Lie group G is a C^∞ map $s : \mathbb{R} \to G$ such that $s(0) = e$ and

$$s(t_1 + t_2) = s(t_1)s(t_2), \quad \forall\, t_1, t_2 \in \mathbb{R}.$$

Proposition 5.1.23. *If G is a Lie group and $X \in L(G)$, there is a unique 1-parameter subgroup $s_X : \mathbb{R} \to G$ such that $\dot{s}_X(0) = X_e$. Furthermore, X is a complete vector field, the flow that it generates being given by*

$$\Phi^X : \mathbb{R} \times G \to G, \qquad \Phi^X_t(a) = a s_X(t).$$

Proof. We first prove that X is complete. Indeed, if

$$\Phi : (-\epsilon, \epsilon) \times U \to W$$

is a local flow about $e \in G$ generated by $X|W$ and if $a \in G$, then the formula

$$\Phi^a_t(b) = L_a(\Phi_t(L_{a^{-1}}(b))), \quad \forall\, b \in L_a(U)$$

defines a local flow

$$\Phi^a : (-\epsilon, \epsilon) \times L_a(U) \to L_a(W)$$

about a having infinitesimal generator $L_{a*}(X|W) = X|L_a(W)$. These fit together to give

$$\Phi : (-\epsilon, \epsilon) \times G \to G$$

and the field X is complete by Lemma 4.1.10. Let Φ^X designate the global flow generated by X. Remark that we have also established the identity

$$L_a \circ \Phi^X_t \circ L_{a^{-1}} = \Phi^X_t, \quad \forall\, a \in G. \tag{5.1}$$

If σ is any 1-parameter subgroup with initial velocity $\dot{\sigma}(0) = X_e$, then the identity

$$\sigma(t + \tau) = \sigma(t)\sigma(\tau),$$

for fixed but arbitrary $t, \tau \in \mathbb{R}$, implies that

$$\dot{\sigma}(t + \tau) = L_{\sigma(t)*}(\dot{\sigma}(\tau)).$$

In particular, take $\tau = 0$ and conclude that

$$\dot{\sigma}(t) = L_{\sigma(t)*}(\dot{\sigma}(0)) = L_{\sigma(t)*}(X_e) = X_{\sigma(t)}.$$

That is, σ is the unique integral curve to X through e.

By the previous paragraph, we must define

$$s_X : \mathbb{R} \to G$$

by $s_X(t) = \Phi_t^X(e)$ and prove that this is a 1-parameter subgroup of G. Indeed, $s_X(0) = e$ and

$$\begin{aligned} s_X(t_1 + t_2) &= \Phi_{t_1+t_2}^X(e) \\ &= \Phi_{t_2}^X(\Phi_{t_1}^X(e)) \\ &= \Phi_{t_2}^X(s_X(t_1)) \\ &= s_X(t_1)s_X(t_1)^{-1}\Phi_{t_2}^X(s_X(t_1)e) \\ &= s_X(t_1)\Phi_{t_2}^X(e) \\ &= s_X(t_1)s_X(t_2), \end{aligned}$$

where the second-to-last equality is by (5.1). Evidently, $\dot{s}_X(0) = X_e$.

Finally, another application of the identity (5.1) yields

$$\Phi_t^X(a) = L_a \circ \Phi_t^X \circ L_{a^{-1}}(a) = as_X(t),$$

for arbitrary values of $t \in \mathbb{R}$ and $a \in G$. □

Example 5.1.24. Let $A \in \mathfrak{M}(n) = L(\mathrm{Gl}(n))$. The series

$$\exp(tA) = I + \frac{tA}{1!} + \frac{t^2A^2}{2!} + \cdots + \frac{t^nA^n}{n!} + \cdots$$

converges absolutely. Set $s(t) = \exp(tA)$. Clearly, $s(0) = I$ and, by basic properties of the exponential series, $s(t_1 + t_2) = s(t_1)s(t_2)$. In particular, $\exp(tA)\exp(-tA) = I$, so the matrix $\exp(tA)$ is invertible, $\forall t \in \mathbb{R}$. By these remarks, $s(t)$ is a 1-parameter subgroup of $\mathrm{Gl}(n)$. Finally, $\dot{s}(0) = A$ and, by Proposition 5.1.23, $\exp(tA)$ is the unique 1-parameter subgroup of $\mathrm{Gl}(n)$ with initial velocity vector A. This example provides the motivation for the following terminology.

Definition 5.1.25. If G is a Lie group, the exponential map

$$\exp : L(G) \to G$$

is defined by $\exp(X) = s_X(1)$, $\forall\, X \in L(G)$.

In turn, we obtain a perfect generalization of Example 5.1.24.

Lemma 5.1.26. *If G is a Lie group and $X \in L(G)$, then*

$$s_X(t) = \exp(tX), \quad -\infty < t < \infty.$$

Proof. Fix $\tau \in \mathbb{R}$ and set $\sigma(t) = s_X(\tau t)$. Then $\dot{\sigma}(t) = \tau X_{\sigma(t)}$ and $\sigma(0) = e$. By Proposition 5.1.23, $\sigma = s_{\tau X}$. In particular,

$$s_X(\tau) = \sigma(1) = s_{\tau X}(1) = \exp(\tau X),$$

for each $\tau \in \mathbb{R}$. □

Hereafter, the standard notation for the 1-parameter subgroup associated to $X \in L(G)$ will be $\exp(tX)$.

Via the identification $L(G) = T_e(G)$, we can view the exponential map as

$$\exp : T_e(G) \to G,$$

a map between smooth manifolds of the same dimension.

Proposition 5.1.27. *The map* $\exp : T_e(G) \to G$ *is smooth and carries some neighborhood of* 0 *in* $T_e(G)$ *diffeomorphically onto a neighborhood of* e *in* G.

Proof. Consider $G \times T_e(G) = G \times L(G)$ as a manifold and define the vector field $Y \in \mathfrak{X}(G \times L(G))$ by

$$Y_{(g,X)} = (X_g, 0), \quad \forall\, (g, X) \in G \times L(G).$$

Clearly, the curve

$$\sigma(t) = (g \cdot \exp(tX), X), \quad -\infty < t < \infty,$$

is integral to Y with $\sigma(0) = (g, X)$. Thus, Y is complete and its flow is

$$\Phi : \mathbb{R} \times (G \times L(G)) \to G \times L(G),$$
$$\Phi_t(g, X) = (g \cdot \exp(tX), X).$$

In particular, define a smooth map $\varphi : L(G) \to G \times L(G)$ by

$$\varphi = \Phi | (\{1\} \times \{e\} \times L(G)).$$

Then $\varphi(X) = (\exp(X), X)$ and the smoothness of exp follows. Under the canonical identity $T_0(T_e(G)) = T_e(G)$, we claim that

$$\exp_{*e} : T_e(G) \to T_e(G)$$

is the identity map. Indeed, represent the tangent vector X to $T_e(G)$ at e as the infinitesimal curve represented by $s(t) = tX$. Then $\exp \circ s$ is the curve $\exp(tX)$ which, as an infinitesimal curve, represents X as well as $\exp_{*e}(X)$. The inverse function theorem then gives the final assertion. □

Exercise 5.1.28. If G is a Lie group and $X, Y \in T_e(G)$, show that the curve $\sigma(t) = \exp(tX)\exp(tY)$ has initial velocity vector $\dot\sigma(0) = X + Y$.

Definition 5.1.29. If G and H are Lie groups, a Lie group homomorphism $\varphi : G \to H$ is a smooth map that is also a group homomorphism. If, in addition, φ is a diffeomorphism, it is called a Lie group isomorphism and G and H are said to be isomorphic Lie groups.

Remark that a 1-parameter subgroup $s : \mathbb{R} \to G$ is a Lie group homomorphism, where $\mathbb{R}$ is a Lie group under addition.

Exercise 5.1.30. A Lie algebra $\mathfrak{a}$ is said to be abelian if $[A, B] = 0$, $\forall\, A, B \in \mathfrak{a}$. Prove the following.

(1) A connected Lie group G is abelian if and only if its Lie algebra $L(G)$ is abelian.

(2) If the Lie group G is connected and abelian, the map

$$\exp : T_e(G) \to G$$

is a surjective Lie group homomorphism, where the vector space $T_e(G)$ is viewed as a Lie group under vector addition.

(3) Every connected, n-dimensional, abelian Lie group is Lie isomorphic to $T^k \times \mathbb{R}^{n-k}$ for some $k = 0, 1, \dots, n$.

Proposition 5.1.31. *If $\varphi : G \to H$ is a Lie group homomorphism, then there is a unique linear map $\varphi_* : L(G) \to L(H)$ such that X is φ-related to $\varphi_*(X)$, $\forall X \in L(G)$. Thus, φ_* is a Lie algebra homomorphism.*

Proof. The $\mathbb{R}$-linear map φ_* will be the composition

$$L(G) \xrightarrow{\epsilon} T_e(G) \xrightarrow{\varphi_{*e}} T_e(H) \xrightarrow{\epsilon^{-1}} L(H).$$

If $X \in L(G)$, $Y = \varphi_*(X)$, and $a \in G$, then

$$\begin{aligned}
\varphi_{*a}(X_a) &= \varphi_{*a}((L_a)_{*e}(X_e)) \\
&= (\varphi \circ L_a)_{*e}(X_e) \\
&= (L_{\varphi(a)} \circ \varphi)_{*e}(X_e) \\
&= (L_{\varphi(a)})_{*e}(\varphi_{*e}(X_e)) \\
&= (L_{\varphi(a)})_{*e}(Y_e) \\
&= Y_{\varphi(a)}.
\end{aligned}$$

Thus, X is φ-related to Y. The left-invariant field Y is uniquely determined by Y_e which, itself, is uniquely determined by the requirement that X be φ-related to Y. □

Remark. Using the canonical identifications $L(G) = T_e(G)$ and $L(H) = T_e(H)$, one obtains a Lie algebra structure on $T_e(G)$ and $T_e(H)$. Then the Lie algebra homomorphism φ_* becomes $\varphi_{*e} : T_e(G) \to T_e(H)$.

Definition 5.1.32. Let G be a topological group, $\Gamma \subseteq G$ a subgroup. If there is a neighborhood $U \subseteq G$ of the identity $e \in G$ such that $U \cap \Gamma = \{e\}$, then Γ is called a discrete subgroup of G.

By Theorem 4.4.9, a discrete subgroup of the additive Lie group $\mathbb{R}^n$ is exactly the same thing as a lattice Γ in $\mathbb{R}^n$. In this case, the projection map

$$p : \mathbb{R}^n \to \mathbb{R}^n/\Gamma = T^k \times \mathbb{R}^{n-k}$$

is readily seen to be the universal covering. We consider the generalization of this to arbitrary Lie groups.

Exercise 5.1.33. If G is a Lie group and $\Gamma \subseteq G$ is a discrete subgroup, prove that the space G/Γ has a canonical smooth manifold structure relative to which the quotient projection $p : G \to G/\Gamma$ is a regular covering map, the covering transformations being the right translations by elements of Γ. If Γ is also a normal subgroup, show that G/Γ is a Lie group and that $p : G \to G/\Gamma$ is a Lie group homomorphism. Finally, note that $p_* : L(G) \to L(G/\Gamma)$ is an isomorphism of Lie algebras.

Proposition 5.1.34. *Let G and H be Lie groups with H connected. Let $\varphi : G \to H$ be a Lie group homomorphism such that $\varphi_{*e} : T_e(G) \to T_e(H)$ is an isomorphism of vector spaces. Then there is a discrete normal subgroup $\Gamma \subseteq G$ such that H is isomorphic as a Lie group to G/Γ.*

Proof. By the inverse function theorem, φ carries some open neighborhood $U \subseteq G$ of $e \in G$ diffeomorphically onto an open neighborhood $\varphi(U) \subseteq H$ of $e \in H$. Since φ respects left translations in these groups, it follows that $\varphi(G)$ is an open neighborhood of the identity in H. Since $\varphi(G)$ is a subgroup of H, Proposition 5.1.13 implies that $\varphi(G) = H$. Finally, φ being one-to-one on $U \subseteq G$, the normal subgroup $\Gamma = \ker(\varphi)$ is a discrete subgroup of G and φ induces a Lie group isomorphism $\overline{\varphi} : G/\Gamma \to H$. □

Theorem 5.1.35. *Let G be a connected Lie group with identity e and let $\pi : (\widetilde{G}, \widetilde{e}) \to (G, e)$ be the universal covering space. Then $\widetilde{G}$ is canonically a Lie group in such a way that $\widetilde{e}$ is the identity element and π is a Lie group homomorphism.*

Exercise 5.1.36. Prove Theorem 5.1.35, proceeding as follows. First construct a commutative diagram

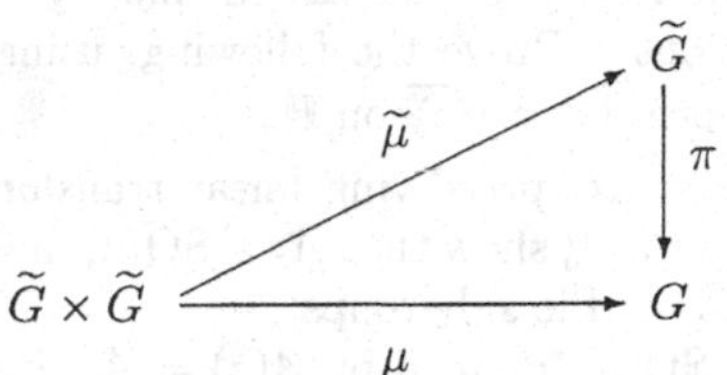

where $\mu(x, y) = \pi(x)\pi(y)^{-1}$ and $\widetilde{\mu}$ is the unique lift such that $\widetilde{\mu}(\widetilde{e}, \widetilde{e}) = \widetilde{e}$. The existence and uniqueness of this lift are guaranteed by Corollary 1.7.40 and the fact that $\widetilde{G}$, hence $\widetilde{G} \times \widetilde{G}$, is simply connected. Since π is a local diffeomorphism, it is elementary that $\widetilde{\mu}$ is smooth.

Given $x, y \in \widetilde{G}$, define y^{-1} to be $\widetilde{\mu}(\widetilde{e}, y)$ and xy to be $\widetilde{\mu}(x, y^{-1})$. Next, consider a commutative diagram

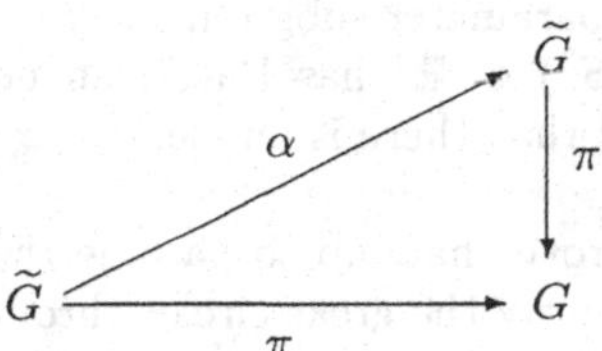

such that $\alpha(\widetilde{e}) = \widetilde{e}$. Since $\alpha = \mathrm{id}$ works, uniqueness of lifts shows that every α satisfying the condition is equal to the identity map. Using this, prove successively the identities

$$(x^{-1})^{-1} = x \text{ and } \widetilde{e}x = x = x\widetilde{e}.$$

Use similar arguments to prove that

$$x^{-1}x = \widetilde{e} = xx^{-1} \text{ and } (xy)z = x(yz).$$

Thus, $\widetilde{G}$ is a group with smooth operations, hence a Lie group. From the definitions, it is obvious that π is a group homomorphism.

Thus, combining Theorem 5.1.35 and Proposition 5.1.34, we see that every connected Lie group has the form G/Γ, where G is a simply connected Lie group and Γ is a discrete normal subgroup. All Lie groups sharing this same universal covering group G have canonically the same Lie algebra. Conversely, connected Lie groups with isomorphic Lie algebras have isomorphic universal covering groups, but we will not prove this.

Exercise 5.1.37. Let $\pi[M,N]$ denote the set of homotopy classes of maps $M \to N$. You may use the standard, but nontrivial fact that this set is canonically the same whether defined with continuous maps and continuous homotopies or smooth maps and smooth homotopies (*cf.* [**16**]).

(1) If G is a Lie group, show that $\pi[M,G]$ has a natural group structure.
(2) There is a natural map of sets $\iota : \pi_1(G,e) \to \pi[S^1,G]$. Define this map and show that it is actually an isomorphism of groups. Show simultaneously that this group is abelian.
(3) Using the above, we easily conclude that a discrete normal subgroup of a simply connected Lie group is abelian. In fact, give a direct proof that a discrete normal subgroup of any path-connected topological group G is a subgroup of the center of G.

Exercise 5.1.38. For each $z \in S^3 \subset \mathbb{H} = \mathbb{R}^4$, define $A_z : \mathbb{R}^4 \to \mathbb{R}^4$ by $A_z(w) = zwz^{-1}$ (quaternion operations). Prove the following, using standard facts about the skew field $\mathbb{H}$ and the norm $|z| = \sqrt{z\bar{z}}$ on $\mathbb{H}$.

(1) A_z is a nonsingular, norm-preserving linear transformation. That is, as a matrix, $A_z \in \mathrm{O}(4)$. In fact, show that $A_z \in \mathrm{SO}(3)$ under canonical inclusions $\mathrm{SO}(3) \subset \mathrm{O}(3) \subset \mathrm{O}(4)$ of Lie subgroups.
(2) The map $A : S^3 \to \mathrm{SO}(3)$, defined by $A(z) = A_z$, is a surjective homomorphism of Lie groups.
(3) The kernel of the homomorphism A is the normal subgroup
$$\mathbb{Z}_2 = \{\pm 1\} \subset S^3.$$
Thus, $\mathrm{SO}(3)$ is diffeomorphic to the projective space P^3 and S^3 is the universal covering group.
(4) Let s_1 and s_2 be 1-parameter subgroups of S^3 such that the initial velocity vector $\dot{s}_i(0) \in T_1(S^3) \subset \mathbb{R}^4$ has Euclidean norm 1, $i = 1,2$. Using the previous step, show that there is an element $z \in S^3$ such that $zs_1(t)z^{-1} = s_2(t)$, $\forall t \in \mathbb{R}$.
(5) Using the above, prove that, up to parametrization, the 1-parameter subgroups of S^3 are exactly the great circles through $1 \in S^3$.

5.2. Lie Subgroups and Subalgebras

We fix a choice of the Lie group G and discuss its Lie subgroups.

Definition 5.2.1. A subset $H \subseteq G$ is a Lie subgroup if H has a Lie group structure relative to which the inclusion map $i : H \hookrightarrow G$ is a one-to-one immersion and a group homomorphism.

In particular, the inclusion i of a Lie subgroup is a Lie group homomorphism. We emphasize that the topology of H as a Lie group may not coincide with its relative topology in G.

Example 5.2.2. A nontrivial 1-parameter subgroup $s : \mathbb{R} \to G$ is an immersion, generally not one-to-one. We will see that, if $n \geq 2$, uncountably many of the 1-parameter subgroups $s : \mathbb{R} \to T^n$ are one-to-one immersions, each having image dense in T^n (Example 5.3.9). By our definition, $s(\mathbb{R})$ with its manifold topology and additive group structure is a Lie subgroup with $i = s$, but the relative topology of this subgroup in T^n is wildly different from its manifold topology.

Definition 5.2.3. If $\mathfrak{g}$ is a Lie algebra, a vector subspace $\mathfrak{h} \subseteq \mathfrak{g}$ is a Lie subalgebra if $\mathfrak{h}$ is closed under the bracket.

Lemma 5.2.4. *Let $i : H \hookrightarrow G$ be a Lie subgroup. Then $i_* : L(H) \to L(G)$ imbeds $L(H)$ as a Lie subalgebra of $L(G)$.*

Indeed, $i_{*e} : T_e(H) \to T_e(G)$ is one-to-one, so the lemma follows from Proposition 5.1.31.

Theorem 5.2.5. *The correspondence between Lie subgroups $i : H \hookrightarrow G$ and their Lie subalgebras $i_* : L(H) \hookrightarrow L(G)$ induces a one-to-one correspondence between the set of connected Lie subgroups of G and the set of Lie subalgebras of $L(G)$.*

The proof of Theorem 5.2.5 is the main goal of this section. In light of the preceding lemma, what we have to prove is that, given a Lie subalgebra $\mathfrak{h} \subseteq L(G)$, there is a unique connected Lie subgroup $i : H \hookrightarrow G$ such that $\mathfrak{h} = i_*(L(H))$. The principal tool for this will be the Frobenius theorem.

The evaluation map $\epsilon : L(G) \to T_e(G)$ carries $\mathfrak{h}$ one-to-one onto a vector subspace $E_e \subseteq T_e(G)$. For each $a \in G$, define

$$E_a = (L_a)_{*e}(E_e) \subseteq T_a(G).$$

Then $E = \bigcup_{a\in G} E_a$ is a k-plane distribution on G, where $k = \dim \mathfrak{h}$. Indeed,

Lemma 5.2.6. *The subset $E \subseteq T(G)$ is an integrable k-plane distribution on G.*

Proof. Let $X_1, \dots, X_k$ be a basis of the vector space $\mathfrak{h}$. This is a set of everywhere linearly independent, left-invariant fields on G, proving that E is a k-plane distribution. Remark that $\mathfrak{h} \subseteq \Gamma(E)$ spans $\Gamma(E)$ as a $C^\infty(G)$-module and $[X_j, X_\ell] \in \mathfrak{h}$, so E is an integrable distribution by Theorem 4.5.5. □

Let H be the leaf through e of the corresponding foliation $\mathcal{H}$. Then H is a connected k-manifold together with a one-to-one immersion $i : H \hookrightarrow G$ with $e \in i(H)$ and $i_{*b}(T_b(H)) = E_b$, $\forall\, b \in H$. Our first goal will be to show that this leaf is a Lie subgroup of G with $i_*(L(H)) = \mathfrak{h}$. Secondly, we will show that this is the only connected Lie subgroup with this property.

Wherever explicit reference to the immersion i is not needed, we generally denote $i(H)$ by H. Similarly, if $a \in G$, aH will denote the immersed submanifold $L_a \circ i : H \hookrightarrow G$.

Lemma 5.2.7. *For each $a \in G$, aH is the leaf of $\mathcal{H}$ through a.*

Proof. We must show that aH is the maximal integral manifold to E through a.

To see that it is an integral manifold to E, note that $b \in H$ if and only if $ab \in aH$ and

$$\begin{aligned}
(L_a \circ i)_{*b}(T_b(H)) &= (L_a)_{*b}(i_{*b}(T_b(H))) \\
&= (L_a)_{*b}(E_b) \\
&= (L_a)_{*b}(L_b)_{*e}(E_e) \\
&= (L_a \circ L_b)_{*e}(E_e) \\
&= (L_{ab})_{*e}(E_e) \\
&= E_{ab}.
\end{aligned}$$

We deduce the maximality of aH from that of H. Let $j : K \hookrightarrow G$ be a connected integral manifold to E through $a = j(\alpha)$. Then, $L_{a^{-1}} \circ j : K \to G$ is a one-to-one immersion, integral to E and containing $L_{a^{-1}} \circ j(\alpha) = L_{a^{-1}}(a) = e$. By the maximality of H, $L_{a^{-1}} \circ j : K \to G$ carries K into H. Thus, the image of $j = L_a \circ L_{a^{-1}} \circ j$ is contained in aH. □

Thus, $\mathcal{H}$ is the foliation of G by the "left cosets" aH.

Corollary 5.2.8. *If $a \in H$, then $a^{-1} \in H$.*

Proof. Consider the leaf $a^{-1}H$ through a^{-1}. Since $a \in H$, $e \in a^{-1}H$, so this leaf coincides with H. □

Corollary 5.2.9. *If $a, b \in H$, then $ab \in H$.*

Proof. As above, $a^{-1} \in H$, so $e = aa^{-1} \in aH$, hence $aH = H$. This implies that $ab \in aH = H$. □

Thus, H is an abstract subgroup and a one-to-one immersed submanifold of G. It remains to be shown that the group operations are smooth in H. If H were an imbedded submanifold, this would be immediate, but we must allow i to be only an immersion.

Lemma 5.2.10. *The immersion $i : H \hookrightarrow G$ defines a connected Lie subgroup of G with $i_*(L(H)) = \mathfrak{h}$.*

Proof. The multiplication map $\mu_H : H \times H \to G$ is given by the composition

$$H \times H \xrightarrow{i \times i} G \times G \xrightarrow{\mu} G.$$

Since $\mu_H(H \times H) \subseteq i(H)$, the map

$$i^{-1} \circ \mu_H : H \times H \to H$$

is smooth by Exercise 4.5.17. This is the group multiplication in H. Similarly, the group inversion $\iota : H \to H$ is smooth. Finally, $i_{*e}(T_e(H)) = E_e$, so $i_*(L(H)) = \mathfrak{h}$. □

The following lemma completes the proof of Theorem 5.2.5.

Lemma 5.2.11. *The Lie subgroup $i : H \hookrightarrow G$ is the only connected one with the property that $i_*(L(H)) = \mathfrak{h}$.*

Proof. The Lie subgroup H is a leaf of the foliation $\mathcal{H}$ determined by the Lie subalgebra $\mathfrak{h}$. Suppose that $i' : H' \hookrightarrow G$ is also a connected Lie subgroup such that $i'_*(L(H')) = \mathfrak{h}$. Then H' must be a connected integral manifold through e to the distribution E determined by $\mathfrak{h}$. The maximal such integral manifold is H, and therefore $H' \subseteq H$. Indeed, H' is an open Lie subgroup of the connected Lie group H, so $H' = H$ by Proposition 5.1.13. □

Exercise 5.2.12. Let $i : H \hookrightarrow G$ be a Lie subgroup of G. Let

$$\exp_H : T_e(H) \to H,$$
$$\exp_G : T_e(G) \to G$$

be the respective exponential maps. Prove that the diagram

$$\begin{array}{ccc} T_e(H) & \xrightarrow{i_{*e}} & T_e(G) \\ {\scriptstyle \exp_H}\downarrow & & \downarrow{\scriptstyle \exp_G} \\ H & \xrightarrow[i]{} & G \end{array}$$

is commutative.

Exercise 5.2.13. Using the above exercise and the fact that

$$\exp : \mathfrak{M}(n, \mathbb{F}) \to \mathrm{Gl}(n, \mathbb{F})$$

is ordinary matrix exponentiation, $\mathbb{F} = \mathbb{R}$ or $\mathbb{C}$, give a new proof that $L(\mathrm{O}(n)) = \mathfrak{o}(n)$ is the algebra of skew symmetric matrices over $\mathbb{R}$ and determine the Lie subalgebra $L(U(n)) = \mathfrak{u}(n) \subset \mathfrak{M}(n, \mathbb{C})$.

Exercise 5.2.14. Let $\varphi : H \to G$ be a homomorphism of Lie groups and prove that $\varphi(H)$ is a Lie subgroup of G.

5.3. Closed Subgroups*

While a Lie subgroup $i : H \hookrightarrow G$ is generally only an immersed submanifold, we have seen a number of examples, such as $\mathrm{Sl}(n) \subset \mathrm{Gl}(n)$, $\mathrm{O}(n) \subset \mathrm{Gl}(n)$, and $\mathrm{U}(n) \subset \mathrm{Gl}(n, \mathbb{C})$, in which H is a properly imbedded submanifold. In this case, the Lie subgroup has the relative topology from G, making it easier to work with. Generally, imbedded submanifolds are not closed subsets, but this is true for Lie subgroups.

Proposition 5.3.1. *If the Lie subgroup $H \subset G$ is an imbedded submanifold, then H is closed as a subset of G. (That is, H is properly imbedded.)*

Proof. We use the foliation $\mathcal{H}$ from the previous section. The components of H are leaves of $\mathcal{H}$. Find a neighborhood U of e in which the foliation $\mathcal{H}|U$ becomes a foliation by plaques P_a. Since H is imbedded, this neighborhood can be chosen so that $H \cap U = P_0$, a single plaque. We can assume that $\overline{U} \subset \widetilde{U}$, a compact subset, where $\mathcal{H}|\widetilde{U}$ is also a foliation by plaques $\widetilde{P}_a$.

Let $\{h_n\}_{n=1}^{\infty} \subset H$ be a sequence that converges in G to an element g. We must prove that $g \in H$. Let $W = L_g(U)$, a neighborhood of g in which $\mathcal{H}|W$ is also a foliation by plaques $L_g(P_a)$. Similarly, let $\widetilde{W} = L_g(\widetilde{U})$. If we can show that all but finitely many h_n lie in a common plaque of $\mathcal{H}|\widetilde{W}$, then this plaque must be the one containing g *and* it must lie in the integral manifold H, so $g \in H$.

Since $\lim_{n\to\infty} h_n^{-1} h_{n+1} = e$, there is an integer $r \geq 1$ such that

$$h_n^{-1} h_{n+1} \in U, \quad \forall n \geq r.$$

That is, $h_n^{-1} h_{n+1} \in U \cap H = P_0$. Also, since $e \in P_0$, it follows that both $h_n \in L_{h_n}(P_0)$ and $h_{n+1} \in L_{h_n}(P_0)$. For r sufficiently large, $L_{h_n}(P_0) \subset \widetilde{P}$, where $\widetilde{P}$ is a plaque of $\mathcal{H}|\widetilde{W}$. That is, when $n \geq r$, h_n and h_{n+1} lie in a common plaque $\widetilde{P}$ of $\widetilde{W}$. Similarly, h_{n+1} and h_{n+2} lie in a common plaque $\widetilde{P}'$. Since $h_{n+1} \in \widetilde{P} \cap \widetilde{P}'$, it follows that $\widetilde{P} = \widetilde{P}'$. Proceeding in this way, we see that $\widetilde{P}$ contains h_m, $\forall m \geq r$. □

This proposition has a surprisingly strong converse.

Theorem 5.3.2 (Closed subgroup theorem). *If G is a Lie group and $H \subseteq G$ is a closed subset that is also an abstract subgroup, then H is a properly imbedded Lie subgroup.*

Exercise 5.3.3. Let G be a topological group whose underlying space is a topological manifold. Use Theorem 5.3.2 to prove that there is at most one differentiable structure on the topological manifold G making G into a Lie group. (The positive solution to Hilbert's fifth problem guarantees that a topological group-manifold does have a smooth (in fact, analytic) structure making it into a Lie group.)

Our proof will be modeled, in certain important ways, on the proof given in [**14**, pp. 105–106]. The problem with that proof is that it is based on a lemma [**14**, Lemma 1.8, p. 96] that assumes that Lie groups are real analytic groups. The fact that C^∞ groups are, in fact, real analytic, will not be used in our proof. For a somewhat different presentation, also carried out in the smooth category, the reader can consult [**49**, pp. 110–112].

Fix the assumptions in Theorem 5.3.2. Define

$$\mathfrak{h} = \{X \in L(G) \mid \exp(tX) \in H, -\infty < t < \infty\}.$$

This contains $0 \in L(G)$ and is closed under scalar multiplication. It is not evident that $\mathfrak{h}$ is a vector subspace, let alone a Lie subalgebra. It is also unclear that $\mathfrak{h} \neq 0$ if H is not discrete. In fact, $\mathfrak{h}$ will turn out to be the Lie algebra of an open Lie subgroup of H.

Let $V \subseteq L(G)$ be the vector space spanned by $\mathfrak{h}$. In the following proof, it will be convenient to use the notation $aX = L_{a*}(X)$ and $Xa = R_{a*}(X)$ (where R_a denotes right translation by a), $a \in G$, $X \in \mathfrak{X}(G)$. Thinking of vectors as infinitesimal curves makes this notation particularly natural.

Lemma 5.3.4. *The vector space V is a Lie subalgebra of $L(G)$.*

Proof. Since the Lie bracket is bilinear and V is spanned by $\mathfrak{h}$, it will be enough to prove that $[X,Y] \in V$, $\forall X, Y \in \mathfrak{h}$. By Theorem 4.3.2 and Proposition 5.1.23,

$$[X,Y] = \lim_{t\to 0} \frac{Y\exp(-tX) - Y}{t}.$$

Since Y is a left-invariant field,

$$Y\exp(-tX) = \exp(tX)\,Y\exp(-tX),$$

and, for a fixed t, this is a left-invariant field whose corresponding 1-parameter group is

$$\sigma(\tau) = \exp(tX)\exp(\tau Y)\exp(-tX).$$

Since $X, Y \in \mathfrak{h}$, $\sigma(\tau)$ is a product of elements of the subgroup H, hence $\sigma(\tau) \in H$, $-\infty < \tau < \infty$. It follows that, for each value of t, $Y\exp(-tX) \in \mathfrak{h} \subseteq V$. Thus, $[X,Y] \in V$, as desired. □

Let $H_0 \subset G$ be the connected Lie subgroup with $L(H_0) = V$.

Lemma 5.3.5. *There is an open neighborhood U of e in H_0 (in the manifold topology of H_0) that is contained in H.*

Proof. Let $\{Y_1, \dots, Y_q\} \subset \mathfrak{h}$ be a basis of V. The map $\varphi : V \to H_0$, defined by

$$\varphi\left(\sum_{i=1}^{q} t_i Y_i\right) = \exp(t_1Y_1)\exp(t_2Y_2)\cdots\exp(t_qY_q),$$

satisfies $\varphi_{*0}(Y_i) = Y_i$, $1 \le i \le q$, so the inverse function theorem implies that φ carries some neighborhood $U_0 \subseteq V$ of 0 diffeomorphically onto a neighborhood $U \subseteq H_0$ of e. But $\exp(t_iY_i) \in H$, $1 \le i \le q$, and H is a subgroup, so $U \subseteq H$. □

Corollary 5.3.6. *The Lie group H_0 is a subgroup of H.*

Proof. By Proposition 5.1.13, H_0 is generated by $U \subseteq H$, and so $H_0 \subseteq H$. □

Remark that, at this point, we know that $\mathfrak{h} = V$, hence $\mathfrak{h}$ is the Lie algebra of H_0.

Lemma 5.3.7. *The subgroup $H_0 \subseteq H$, with its manifold topology, is open in the relative topology of H.*

Proof. (Compare [**14**, p. 106].) It will be enough to prove that some open neighborhood U of e, in the manifold topology of H_0, is a neighborhood of e in the relative topology of H. The problem is that, for each such U, there might be a sequence $\{x_k\}_{k=1}^\infty \subset H \setminus U$ such that $x_k \to e$ in the topology of G. Assuming that this is so, we deduce a contradiction.

Find a direct sum decomposition $L(G) = \mathfrak{h} \oplus W$ and remark that, by the inverse function theorem, the map $\varphi : \mathfrak{h} \oplus W \to G$, defined by

$$\varphi(v, w) = \exp(v)\exp(w),$$

carries some neighborhood N of 0 in $L(G)$ diffeomorphically onto a neighborhood of e in G. Choose $U = \exp(\mathfrak{h} \cap N)$. Thus, for k sufficiently large, we can write

$$x_k = \exp(v_k)\exp(w_k) \in \varphi(N),$$

where $v_k \in \mathfrak{h} \cap N$ and $w_k \in W \cap N$. Since $x_k \notin U$, it is clear that $w_k \neq 0$, for all large values of k.

Select a bounded neighborhood $W_0 \subset W$ of 0 and positive integers n_k such that, for k sufficiently large, $n_k w_k \in W_0$, but $(n_k + 1)w_k \notin W_0$. Since W_0 is bounded, we can assume that $n_k w_k \to w \in W$. Since $w_k \to 0$ and $(n_k + 1)w_k \notin W_0$, we must have $w \neq 0$. For arbitrary $t \in \mathbb{R}$, we will show that $\exp(tw) \in H$. That is, $w \in \mathfrak{h}$, hence $0 \neq w \in \mathfrak{h} \cap W$, the desired contradiction.

Write $tn_k = s_k + t_k$, where $s_k \in \mathbb{Z}$ and $|t_k| < 1$. Thus, $t_k w_k \to 0$ and

$$\begin{aligned}
\exp(tw) &= \lim_{k\to\infty} \exp(tn_k w_k) \\
&= \lim_{k\to\infty} \exp(s_k w_k)\exp(t_k w_k) \\
&= \lim_{k\to\infty} \exp(s_k w_k) \\
&= \lim_{k\to\infty} \exp(w_k)^{s_k} \\
&= \lim_{k\to\infty} (\exp(-v_k)x_k)^{s_k}.
\end{aligned}$$

Since H is closed in G, it follows that $\exp(tw) \in H$. □

Proof of theorem 5.3.2. Let $i : H_0 \hookrightarrow H$ be the inclusion map. We have proven that i carries H_0, with its manifold topology, homeomorphically onto an open subset of H in the relative topology. In particular, the manifold topology of H_0 coincides with its relative topology, so H_0 is an imbedded Lie subgroup of G. By Proposition 5.3.1, H_0 is closed in G, so $H_0 = i(H_0)$ is a connected, open-closed subset of H. Thus, H_0 coincides with the component of the identity in H. The other components $L_a(H_0)$, $a \in H$, of H are also properly imbedded submanifolds of G, so H is a Lie subgroup. Since H has the relative topology, each of its components is relatively open in H, so H is a properly imbedded Lie subgroup. □

Corollary 5.3.8. *Let $\Gamma \subseteq G$ be an abstract subgroup. Then the closure $\overline{\Gamma}$ in G is a properly imbedded, Lie subgroup of G.*

Example 5.3.9. Let $v = (a^1, \dots, a^n) \in \mathbb{R}^n$ be a point such that, when $\mathbb{R}$ is viewed as a vector space over the rational number field $\mathbb{Q}$, the subset $\{a^1, \dots, a^n\} \subset \mathbb{R}$ is linearly independent. Let $p : \mathbb{R}^n \to T^n$ be the standard projection and let $\ell \subset \mathbb{R}^n$ be the line through v and 0. A classical theorem of Kronecker asserts that this line projects one-to-one to a 1-parameter subgroup $p(\ell) \subset T^n$ that is everywhere dense in T^n (for the case $n = 2$, *cf.* Example 4.1.8 and Exercise 4.4.12).

It is now fairly easy to prove Kronecker's theorem. Indeed, one proves (Exercise 5.3.10) that, if $v_1, \dots, v_k \in \mathbb{Z}^n$ and

$$v = \sum_{i=1}^{k} x^i v_i$$

for suitable coefficients $x^i \in \mathbb{R}$, then $k = n$. By Corollary 5.3.8, the closure $\overline{\ell} \subseteq T^n$ is a compact, connected, abelian Lie subgroup, hence a toroidal subgroup of dimension $r \leq n$ (Exercise 5.1.30). It follows that $v \in \ell \subset V$ where $V \subset \mathbb{R}^n$ is a subspace and $\overline{\ell} = p(V)$. In particular, V is spanned by $V \cap \mathbb{Z}^n$, so Exercise 5.3.10 implies that $\dim V = n$ and $\overline{\ell} = T^n$.

Exercise 5.3.10. Prove the assertion in Example 5.3.9 that the vector $v \in \mathbb{R}^n$, with rationally independent coefficients, cannot be expressed as a real linear combination of fewer than n elements of the integer lattice $\mathbb{Z}^n$.

Exercise 5.3.11. Let $v = (a^1, \dots, a^n) \in \mathbb{R}^n$ be a point such that the set of coefficients $\{1, a^1, \dots, a^n\}$ is linearly independent over $\mathbb{Q}$. Prove that the subgroup $A \subset T^n$, generated by $a = p(v)$, is everywhere dense. (Hint: Every point in the coset $qv + \mathbb{Z}^n$ has rationally independent coefficients, $\forall q \in \mathbb{Z}$. Prove this and use it to show that every 1-parameter subgroup of T^n meeting a nontrivial element of A is dense in T^n.)

Following Helgason [**14**, pp. 107–108], we deduce the following classical result as another corollary of Theorem 5.3.2. For a proof that does not depend on that theorem, see [**49**, p. 109].

Theorem 5.3.12. *If $\varphi : G \to H$ is a continuous group homomorphism between Lie groups, then φ is smooth.*

Proof. The product $G \times H$ is a Lie group and the projections

$$\begin{aligned} \pi_G : G \times H &\to G, \\ \pi_H : G \times H &\to H \end{aligned}$$

are smooth group homomorphisms. Let $\Gamma \subset G \times H$ be the graph of φ. That is, $\Gamma = \{(x, \varphi(x)) \mid x \in G\}$, clearly a closed subgroup of $G \times H$. Thus, Γ is a properly imbedded Lie subgroup. Also, $\pi_G|\Gamma = \psi : \Gamma \to G$ is a smooth group homomorphism and is bijective. If it can be shown that $\psi^{-1} : G \to \Gamma$ is smooth, then $\varphi = \pi_H \circ \psi^{-1}$ will be smooth and the assertion will be proven.

By the inverse function theorem, it will be enough to show that ψ_{*y} is bijective, $\forall y \in \Gamma$. Since Γ is a Lie group and ψ is a smooth homomorphism, it is enough to prove this at $y = (e, e)$.

Remark that the exponential maps for the groups $G \times H$, G, and H are related by

$$\exp_{G \times H} = \exp_G \times \exp_H .$$

This, together with the proof of Theorem 5.3.2, implies that

$$L(\Gamma) = \{(X,Y) \in L(G) \times L(H) \mid (\exp_G tX, \exp_H tY) \in \Gamma, \ \forall t \in \mathbb{R}\}.$$

Since $\psi_{*(e,e)}(X,Y) = X$, we must show that, for each $X \in L(G)$, there is a unique $Y \in L(H)$ such that $(X,Y) \in L(\Gamma)$.

We first show uniqueness of Y. If $(X,Y) \in L(\Gamma)$ and $(X,Z) \in L(\Gamma)$, then the difference is $(0, Y-Z) \in L(\Gamma)$, implying that $(e, \exp_H t(Y-Z)) \in \Gamma, \forall t \in \mathbb{R}$. Thus, $\exp_H t(Y-Z) = \varphi(e) = e$, $\forall t \in \mathbb{R}$, and $Y - Z = 0$.

Choose open neighborhoods $U_0 \subseteq L(G)$ and $V_0 \subseteq L(H)$ of the origin and $U_e \subseteq G$ and $V_e \subseteq H$ of the identity such that

1. $\exp_G : U_0 \to U_e$ is a diffeomorphism onto;
2. $\exp_H : V_0 \to V_e$ is a diffeomorphism onto;
3. $\varphi(U_e) \subseteq V_e$;
4. $\exp_{G\times H}$ carries $(U_0 \times V_0) \cap L(\Gamma)$ diffeomorphically onto $(U_e \times V_e) \cap \Gamma$.

Let $X \in L(G)$ and choose an integer $r > 0$ such that $(1/r)X \in U_0$. Thus, $\varphi(\exp_G(1/r)X) \in V_e$ and there is a unique $Y_r \in V_0$ such that $\exp_H Y_r = \varphi(\exp_G(1/r)X)$. There is also a unique $Z_r \in (U_0 \times V_0) \cap L(\Gamma)$ such that

$$\exp_{G\times H} Z_r = (\exp_G(1/r)X, \exp_H Y_r).$$

Since $\exp_{G\times H}$ is one-to-one on $U_0 \times V_0$, this implies that

$$((1/r)X, Y_r) = Z_r \in L(\Gamma).$$

Take $Y = rY_r$, obtaining $(X,Y) = rZ_r \in L(\Gamma)$. □

Corollary 5.3.13. *Let $\varphi : G \to H$ be a continuous homomorphism of Lie groups and let $K = \ker(\varphi)$. Then K is a properly imbedded, normal Lie subgroup of G, G/K is canonically a Lie group, and the induced map $\overline{\varphi} : G/K \to H$ is a one-to-one immersion of this Lie group as a Lie subgroup of H.*

Proof. Indeed, K is a normal subgroup by standard group theory and $K = \varphi^{-1}(e)$ is a closed subset of G, so Theorem 5.3.2 guarantees that K is a properly imbedded Lie subgroup of G. By Theorem 5.3.12, φ is a smooth homomorphism, so Exercise 5.2.14 guarantees that $\varphi(G)$ is a Lie subgroup of H. Obviously, $\overline{\varphi}$ is an isomorphism of the group G/K onto $\varphi(G)$, so $\overline{\varphi}$ can be used to transfer the Lie structure of $\varphi(G)$ back to G/K. □

Exercise 5.3.14. Maximal abelian subalgebras of Lie algebras play an important role in Lie theory, as do the maximal abelian subgroups of Lie groups. Prove the following.

(1) Show that every finite dimensional Lie algebra contains a nontrivial abelian subalgebra that is not itself contained properly in another such subalgebra. Similarly, show that every compact Lie group G contains a maximal subgroup that is Lie isomorphic to T^k, some $k \geq 1$. This is called a *maximal torus* of G.

(2) When G is compact, prove that the correspondence between Lie subalgebras and connected Lie subgroups sets up a one-to-one correspondence between the maximal abelian subalgebras of $L(G)$ and the maximal tori in G.

(3) Let G be compact and connected, $T \subseteq G$ a maximal torus. Prove that T is a maximal abelian subgroup.

(4) There are maximal abelian subgroups of a connected Lie group G that are not maximal tori. Find a finite subgroup of SO(3) that is maximal abelian.

Exercise 5.3.15. Let G be an n-dimensional Lie group and $\mathfrak{g} = L(G)$ its Lie algebra. Let $\mathrm{Gl}(\mathfrak{g})$ denote the group of nonsingular linear transformations of the vector space $\mathfrak{g}$ and let $\mathrm{Aut}(\mathfrak{g}) \subset \mathrm{Gl}(\mathfrak{g})$ be the subgroup of Lie algebra automorphisms of $\mathfrak{g}$. Prove the following.

(1) $\mathrm{Gl}(\mathfrak{g})$ has a canonical Lie group structure under which it is (non-canonically) isomorphic to $\mathrm{Gl}(n)$. Also, for use in Exercise 5.3.16, show that $L(\mathrm{Gl}(\mathfrak{g}))$ is canonically the space $\mathrm{End}(\mathfrak{g})$ of linear endomorphisms of the vector space $\mathfrak{g}$, the bracket in $\mathrm{End}(\mathfrak{g})$ being the commutator product of endomorphisms.
(2) $\mathrm{Aut}(\mathfrak{g})$ is a closed subgroup of $\mathrm{Gl}(\mathfrak{g})$, hence a properly imbedded Lie subgroup.
(3) Assume that G is connected and let $C \subseteq G$ denote the center of G, clearly a closed subgroup. Each element $a \in G$ determines an inner automorphism of G, denoted by $\mathrm{Ad}(a)$ and defined by

$$\mathrm{Ad}(a)(g) = aga^{-1}, \qquad \forall g \in G.$$

Prove that $\{\mathrm{Ad}(a)\}_{a\in G}$ is canonically a Lie subgroup of $\mathrm{Aut}(\mathfrak{g})$, isomorphic as a group to G/C. This subgroup is denoted by $\mathrm{Ad}(G)$ and called the *adjoint group* of G.

Exercise 5.3.16. Let G be a Lie group and again denote its Lie algebra by $\mathfrak{g}$. A *derivation* $D : \mathfrak{g} \to \mathfrak{g}$ is a linear transformation such that

$$D[X,Y] = [DX,Y] + [X,DY],$$

$\forall X, Y \in \mathfrak{g}$. Let $\mathfrak{D}(\mathfrak{g})$ be the space of derivations of $\mathfrak{g}$.

(1) Prove that, under the commutator product, $\mathfrak{D}(\mathfrak{g})$ is naturally identified as a Lie subalgebra of $L(\mathrm{Gl}(\mathfrak{g}))$ (*cf.* Exercise 5.3.15, part (a)).
(2) For each $X \in \mathfrak{g}$, define $\mathrm{ad}(X) : \mathfrak{g} \to \mathfrak{g}$ by

$$\mathrm{ad}(X)Y = [X,Y], \ \forall Y \in \mathfrak{g}$$

and prove that $\mathrm{ad}(X) \in \mathfrak{D}(\mathfrak{g})$.
(3) Prove that $\mathrm{ad} : \mathfrak{g} \to L(\mathrm{Gl}(\mathfrak{g}))$ is a homomorphism of Lie algebras. Thus, $\mathrm{ad}(\mathfrak{g}) \subseteq L(\mathrm{Gl}(\mathfrak{g}))$ is a Lie subalgebra.
(4) Assume that G is connected and prove that the connected Lie subgroup of $\mathrm{Gl}(\mathfrak{g})$ corresponding to the Lie subalgebra $\mathrm{ad}(\mathfrak{g})$ is exactly the adjoint group $\mathrm{Ad}(G)$. (Hint: Prove that $\mathrm{Ad}(\exp(tX)) = \exp(t\,\mathrm{ad}(X))$, $\forall X \in \mathfrak{g}$.)

5.4. Homogeneous Spaces*

Lie groups arise in many natural ways as transformation groups of differentiable manifolds. When the group action is transitive, the manifold is called a *homogeneous space* and one has considerable control over its structure.

Definition 5.4.1. Let M be a smooth manifold and G a Lie group. A smooth map

$$\mu : G \times M \to M,$$

written $\mu(g,x) = gx$, is said to be an action of G (from the left) on M, and G is called a Lie transformation group on M, if

(1) $g_1(g_2x) = (g_1g_2)x$, $\forall g_1, g_2 \in G$ and $\forall x \in M$;
(2) $ex = x$, $\forall x \in M$.

Remark. One can also define a *right* action

$$\mu : M \times G \to M$$

by making the obvious changes in the above definition.

Definition 5.4.2. An orbit of the action

$$G \times M \to M$$

is a set of points of the form $\{gx_0 \mid g \in G\}$, where $x_0 \in M$. The action is transitive if M itself is an orbit, in which case M is said to be a homogeneous space of G.

Remark. It is elementary that the orbits of a group action are equivalence classes, two points $x, y \in M$ being equivalent under the action if $\exists\, g \in G$ such that $gx = y$.

Example 5.4.3. The orthogonal group $\mathrm{O}(n)$ acts on $\mathbb{R}^n$ in the usual way, leaving invariant the unit sphere S^{n-1}. Note that, if $e_1 \in S^n$ is the column vector with first entry 1 and remaining entries 0, then Ae_1 is the first column of $A \in \mathrm{O}(n)$. Every unit vector appears as the first column of suitable orthogonal matrices, so the action

$$\mathrm{O}(n) \times S^{n-1} \to S^{n-1}$$

is transitive and S^{n-1} is a homogeneous space of $\mathrm{O}(n)$. In a completely similar way, there is a transitive action

$$\mathrm{U}(n) \times S^{2n-1} \to S^{2n-1},$$

where $S^{2n-1} \subset \mathbb{C}^n$ is the unit sphere in the standard Hermitian metric.

Definition 5.4.4. Let M be a homogeneous space of G and let $x_0 \in M$. The isotropy group of x_0 is the set $G_{x_0} = \{g \in G \mid gx_0 = x_0\}$.

Lemma 5.4.5. *The isotropy group G_{x_0} as above is a properly imbedded Lie subgroup of G.*

Proof. It is obvious that G_{x_0} is an abstract subgroup of G. If $\{g_n\}_{n=1}^{\infty}$ is a sequence in G_{x_0} converging to $g \in G$, then, by the continuity of the group action,

$$gx_0 = \lim_{n\to\infty} g_n x_0 = \lim_{n\to\infty} x_0 = x_0.$$

Thus, G_{x_0} is a closed subset of G. By Theorem 5.3.2, G_{x_0} is a properly imbedded Lie subgroup. □

Example 5.4.6. If $e_1 \in S^{n-1}$ is as in Example 5.4.3, the isotropy group $\mathrm{O}(n)_{e_1}$ is the set of matrices

$$\begin{bmatrix} 1 & 0 \\ 0 & A \end{bmatrix}$$

where $A \in \mathrm{O}(n-1)$. Similarly, for $e_1 \in S^{2n-1}$, $\mathrm{U}(n)_{e_1}$ is the set

$$\begin{bmatrix} 1 & 0 \\ 0 & A \end{bmatrix}$$

where $A \in \mathrm{U}(n-1)$.

We are going to show how to put a smooth structure on the quotient space G/G_{x_0} and prove that this manifold is diffeomorphic to the homogeneous space M. Under the identification $M = G/G_{x_0}$, the G-action on M becomes the action

$$G \times G/G_{x_0} \to G/G_{x_0}, \quad g(hG_{x_0}) = (gh)G_{x_0}.$$

In what follows, we consider an arbitrary properly imbedded Lie subgroup $H \subsetneq G$, put the quotient topology on G/H, and construct a natural smooth structure on this space. Throughout this discussion, we set $\mathfrak{h} = L(H)$. Decompose $L(G) = \mathfrak{m} \oplus \mathfrak{h}$, where $\mathfrak{m}$ is any fixed choice of complementary subspace. Let

$$\psi : \mathfrak{m} \oplus \mathfrak{h} \to G$$

be the map

$$\psi(A, B) = \exp(A)\exp(B)$$

and choose a neighborhood V of 0 in $\mathfrak{h}$ and a neighborhood W of 0 in $\mathfrak{m}$ such that ψ sends $W \times V$ diffeomorphically onto a neighborhood U of e in G. Choose a compact neighborhood $C \subset W$ of 0 with the property that $-C = C$ and $\exp(C)\exp(C) \subset U$.

We can assume that coordinates $x^1, \dots, x^k$ in $\mathfrak{h}$ define V by the inequalities $-1 < x^i < 1$, $1 \le i \le k$. Similarly, coordinates $y^1, \dots, y^q$ for $\mathfrak{m}$ define C by $-1 \le y^j \le 1$, $1 \le j \le q$. Coordinatize $Q = \psi(C \times V)$ by $\psi^{-1} : Q \to C \times V$. The foliation of G by the components of the left cosets of H is given in Q by plaques $\psi(\{c\} \times V)$ that are level sets $(y^1, \dots, y^q) = c$. Since H is properly imbedded in G, we can arrange that $H \cap U = \psi(\{0\} \times V)$.

Lemma 5.4.7. *Each coset aH meets Q in at most one plaque.*

Proof. Let $c_1, c_2 \in C$ be such that $\exp(c_1)\exp(V)$ and $\exp(c_2)\exp(V)$ lie in a common coset $\exp(c_1)H = \exp(c_2)H$. Then,

$$\exp(-c_1)\exp(c_2) \in H \cap U = \psi(\{0\} \times V).$$

This implies that

$$\psi(c_1, v) = \exp(c_1)\exp(v) = \exp(c_2) = \psi(c_2, 0),$$

for some $v \in V$. Since ψ is one-to-one, we conclude that $v = 0$ and $c_1 = c_2$. □

Exercise 5.4.8. Let $C_0 = \mathrm{int}(C)$ and $Q_0 = \mathrm{int}(Q)$. Prove that the map

$$\varphi : C_0 \times H \to Q_0 H,$$

given by

$$\varphi(c, h) = \exp(c)h,$$

is a diffeomorphism.

By this exercise, if $\pi_C : C \times H \to C$ denotes projection onto the first factor, we obtain a submersion

$$y = \pi_C \circ \varphi^{-1} : Q_0 H \to \mathfrak{m} = \mathbb{R}^q.$$

This smooth submersion assigns a coordinate q-tuple $(y^1(aH), \dots, y^q(aH))$ to each coset $aH \subset Q_0 H$, distinct cosets getting distinct coordinates.

Cover G by open sets of the form aQ_0H, $a \in G$. Such a set is also a union of cosets bH and we assign coordinates to each coset via the submersion $y_a = y \circ L_{a^{-1}}$. On overlaps $aQ_0H \cap bQ_0H$, the coordinates y_a and y_b are related by $y_b = y_a \circ L_{ab^{-1}}$. That is, the change of coordinates on overlaps is smooth.

Let $\pi : G \to G/H$ be the quotient map. This carries aQ_0H onto an open set $U_a \subset G/H$ and y_a induces $\tilde{y}_a : U_a \to \mathbb{R}^q$.

Lemma 5.4.9. *The map $\tilde{y}_a : U_a \to \mathbb{R}^q$ is a homeomorphism onto an open subset of $\mathbb{R}^q$.*

Proof. Indeed, we have coordinatized $\mathfrak{m}$ so that the image of $\tilde{y}_a$ is the open set C_0. It is clear that $\tilde{y}_a$ is one-to-one and continuous. We must prove that it is an open map. If $Z \subseteq U_a$ is open, then $\pi^{-1}(Z)$ is open and y_a, being a submersion, carries this open set onto an open set. But $\tilde{y}_a(Z) = y_a(\pi^{-1}(Z))$. □

We view $\{(U_a, \tilde{y}_a)\}_{a \in G}$ as a coordinate atlas on G/H. By the above remarks, this is a C^∞ atlas on the locally Euclidean space G/H.

The following exercise completes our analysis of G/H.

Exercise 5.4.10. With the above C^∞ atlas, prove that G/H is a smooth manifold, that the projection $\pi : G \to G/H$ is a submersion, and that the action

$$\mu : G \times G/H \to G/H,$$

defined by $\mu(a, bH) = abH$, is smooth.

Corollary 5.4.11. *If G is a Lie group and $H \subseteq G$ is a closed, normal subgroup, then the group G/H has a smooth structure in which it is a Lie group.*

We return to the smooth transitive action

$$\mu : G \times M \to M.$$

Let $x_0 \in M$ and let $H = G_{x_0}$ be the isotropy group. Define the map

$$\theta : G/H \to M$$

by $\theta(aH) = ax_0$. This is induced by the smooth map $\tilde{\theta} : G \to M$, $\tilde{\theta}(a) = ax_0$, so θ is continuous. Since $aH = bH$ if and only if $a^{-1}b \in H$, we see that $x_0 = a^{-1}bx_0$, hence $ax_0 = bx_0$, if and only if $aH = bH$. That is, θ is well defined, one-to-one, and continuous. Since the action of G is transitive, θ is a surjection. The following diagram is commutative:

$$\begin{array}{ccc} G \times G/H & \xrightarrow{\mu} & G/H \\ {\scriptstyle \mathrm{id} \times \theta}\downarrow & & \downarrow{\scriptstyle \theta} \\ G \times M & \xrightarrow[\mu]{} & M \end{array}.$$

Thus, if we prove that θ is a diffeomorphism, θ will be a canonical identification of G/H with M *as a homogeneous space of* G.

Proposition 5.4.12. *The map $\theta : G/H \to M$ is a diffeomorphism.*

Proof. Let L_a denote left translation by $a \in G$ on both G/H and M. That is,

$$L_a(bH) = abH, \qquad \forall\, bH \in G/H,$$
$$L_a(x) = ax, \qquad \forall\, x \in M.$$

Then

$$\theta = L_a \circ \theta \circ L_{a^{-1}}, \quad \forall\, a \in G.$$

Since $L_a : M \to M$ and $L_{a^{-1}} : G/H \to G/H$ are diffeomorphisms, it follows that θ will be smooth at aH if and only if it is smooth at eH. Furthermore, if smoothness has been established, then

$$\theta_{*aH} = (L_a)_{*x_0} \circ \theta_{*eH} \circ (L_{a^{-1}})_{*aH}$$

will be an isomorphism of $T_{aH}(G/H)$ onto $T_{ax_0}(M)$ if and only if

$$\theta_{*eH} : T_{eH}(G/H) \to T_{x_0}(M)$$

is an isomorphism.

We show smoothness at eH. Consider the commutative diagram

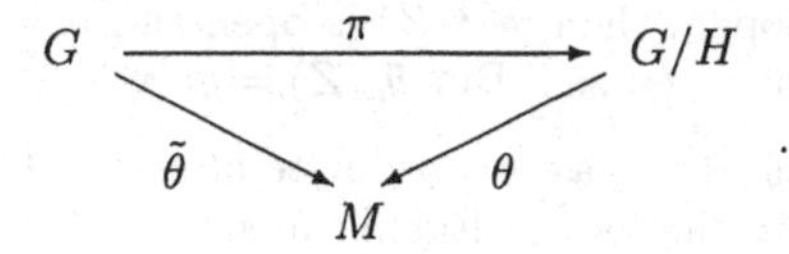

The map

$$\tilde{\theta}|\exp(C_0) : \exp(C_0) \to M$$

is smooth, and the map

$$\pi|\exp(C_0) : \exp(C_0) \to G/H$$

is a diffeomorphism onto the coordinate neighborhood U_e, so θ is smooth in a neighborhood of eH.

We show that $\theta_{*eH} : T_{eH}(G/H) \to T_{x_0}(M)$ is an isomorphism. Again, this translates, via the commutative triangle, to showing that $\tilde{\theta}_{*e}$ is an isomorphism of $T_e(\exp(C_0))$ onto $T_{x_0}(M)$. Let $v \in T_e(\exp(C_0)) = \mathfrak{m}$ and consider the curve $s(t) = \exp(tv)x_0$. Since $\tilde{\theta}_{*e}(v) = \dot{s}(0)$, we only need prove that $\dot{s}(0) = 0$ implies that $v = 0$. If $a = \exp(t_0 v)$, then $L_a(s(t)) = s(t + t_0)$, so $L_{*a}(\dot{s}(0)) = \dot{s}(t_0)$ and $\dot{s}(0) = 0$ implies that $\dot{s}(t) = 0$, $\forall t \in \mathbb{R}$. That is, $\exp(tv)x_0 = x_0$, $\forall t \in \mathbb{R}$, implying that $v \in \mathfrak{h} \cap \mathfrak{m} = \{0\}$. □

Example 5.4.13. Thus, as a homogeneous space of $\mathrm{O}(n)$,

$$S^{n-1} = \mathrm{O}(n)/\mathrm{O}(n-1),$$

where $\mathrm{O}(n-1)$ is properly imbedded as a Lie subgroup of $\mathrm{O}(n)$ as in Example 5.4.6. Similarly,

$$S^{2n-1} = \mathrm{U}(n)/\mathrm{U}(n-1).$$

Exercise 5.4.14. Let $G_{n,k}$ denote the set of k-dimensional vector subspaces of $\mathbb{R}^n$.

(1) Show how to make $G_{n,k}$ into a compact manifold that is a homogeneous space of $\mathrm{O}(n)$. This is called the (real) Grassmann manifold of k-planes in n-space.
(2) If $x_0 \in G_{n,k}$ is the standard $\mathbb{R}^k \subseteq \mathbb{R}^n$, identify the isotropy group $\mathrm{O}(n)_{x_0}$.
(3) Show that projective space P^{n-1} is the Grassmann manifold $G_{n,1}$ and identify the standard two-to-one map $S^{n-1} \to P^{n-1}$ as a map

$$\mathrm{O}(n)/\mathrm{O}(n-1) \to \mathrm{O}(n)/\mathrm{O}(n)_{x_0}.$$

(Remark: Using $\mathbb{C}^n$ instead of $\mathbb{R}^n$, one defines in a similar way the complex Grassmann manifolds $G_{n,k}(\mathbb{C})$ as homogeneous spaces of $\mathrm{U}(n)$. Complex projective space is defined to be $P^n(\mathbb{C}) = G_{n,1}(\mathbb{C})$. The real and complex Grassmann manifolds play an important role in differential geometry and topology.)

Exercise 5.4.15. A k-tuple $(v_1, \ldots, v_k)$ of orthonormal vectors in $\mathbb{R}^n$ will be called an orthonormal k-frame in $\mathbb{R}^n$. Let $V_{n,k}$ denote the set of all orthonormal k-frames in $\mathbb{R}^n$, identify this as a homogeneous space of $\mathrm{O}(n)$ (called the Stiefel manifold of k-frames in n-space). Remark that $V_{n,1} = S^{n-1}$ and that this case gives back Example 5.4.3. Using $\mathbb{C}^n$ and the standard positive definite Hermitian inner product on $\mathbb{C}^n$, one obtains the complex Stiefel manifolds $V_{n,k}(\mathbb{C})$VnkC@ and $V_{n,1}(\mathbb{C}) = S^{2n-1}$.

CHAPTER 6

Covectors and 1-Forms

An important analytic tool in our study of manifolds M has been the Lie algebra $\mathfrak{X}(M)$ of smooth vector fields. In this chapter, we begin the study of the dual object, the space $A^1(M)$ of differential 1-forms on M. One would expect this space of "covector fields" to be neither more nor less useful than $\mathfrak{X}(M)$, but for many purposes it is much more powerful. One reason for this is the "functoriality" of $A^1(M)$, as will be explained presently. Another is *exterior derivative* and *exterior multiplication*, operations that produce higher order objects, called *q-forms.* These q-forms can be integrated over suitable q-dimensional domains and differentiated. A version of the fundamental theorem of calculus, called Stokes' theorem, relates these operations and, in the global setting, leads to a remarkable tool (de Rham cohomology) for analyzing the topology of M. This, in fact, is the beginning of a major mathematical discipline called algebraic topology. Finally, theorems stated in terms of differential forms sometimes provide interesting and useful alternatives to equivalent vector field versions. An example of this will be a differential forms version of the Frobenius theorem.

These topics will require the next several chapters to do them justice. Here we deal only with 1-forms.

6.1. Dual Bundles

Let $\pi : E \to M$ be a k-plane bundle. In particular, E can be viewed as a parametrized family of k-dimensional vector spaces E_x, where the parameter x ranges over M. There is a general philosophy that linear algebra constructions which do not involve a choice of basis, being canonically defined on every E_x, can be extended smoothly to the entire bundle. We will see many examples of this, beginning here with the construction of the *dual* bundle $\pi : E^* \to M$.

Recall that a vector space V has a dual space V^*, this being the vector space

$$V^* = \mathrm{Hom}_{\mathbb{R}}(V, \mathbb{R})$$

of all linear functionals on V. Similarly, one constructs the dual bundle E^*, essentially by taking the vector space duals E_x^* of the fibers E_x $\forall\, x \in M$, and assembling them into a bundle by means of a suitable $\mathrm{Gl}(n)$-cocycle as in Section 3.4.

The dual space V^* of V is abstractly, but generally not canonically, isomorphic to V. (As we see below, there is a canonical choice of this isomorphism when $V = \mathbb{R}^k$.) If $\varphi : V_1 \to V_2$ is a linear map between vector spaces, the *adjoint*

$$\varphi^* : V_2^* \to V_1^*$$

is the linear map defined by

$$\varphi^*(f) = f \circ \varphi, \quad \forall\, f \in V_2^*.$$

Example 6.1.1. We represent elements $v \in \mathbb{R}^k$ by $k \times 1$ matrices and elements $f \in \mathbb{R}^{k*}$ by $1 \times k$ matrices. Then $f(v) = f \cdot v$ is just matrix multiplication. Thus, the transpose operation $v \mapsto v^{\mathrm{T}}$ defines a canonical isomorphism between $\mathbb{R}^k$ and $\mathbb{R}^{k*}$. If $\varphi : \mathbb{R}^k \to \mathbb{R}^m$ is a linear map, let A be the $m \times k$ matrix representing φ. Then, relative to the canonical identifications $\mathbb{R}^k = \mathbb{R}^{k*}$ and $\mathbb{R}^m = \mathbb{R}^{m*}$, the adjoint φ^* is represented by the $k \times m$ matrix A^{T}.

In the language of category theory, the associations $V \mapsto V^*$ and $\varphi \mapsto \varphi^*$ define a *contravariant* functor on the category of real, finite dimensional vector spaces and linear maps. That is, morphism arrows are reversed under $\varphi \mapsto \varphi^*$ and, consequently,

$$(\varphi \circ \psi)^* = \psi^* \circ \varphi^*.$$

This contravariance is a slight problem when we try to find a $\mathrm{Gl}(k)$-cocycle for the construction of the dual bundle E^*. The following saves the day.

Definition 6.1.2. If $\varphi : V_1 \to V_2$ is an isomorphism of vector spaces, then $\varphi' : V_1^* \to V_2^*$ is the isomorphism $\varphi' = (\varphi^*)^{-1}$.

Thus, if $V_1 = V_2 = \mathbb{R}^k$ and A is the $k \times k$ matrix representing φ, then $(A^{\mathrm{T}})^{-1} = (A^{-1})^{\mathrm{T}}$ represents φ'. One clearly has covariant functoriality

$$(\varphi \circ \psi)' = \varphi' \circ \psi',$$

corresponding to the matrix identity

$$((AB)^{\mathrm{T}})^{-1} = (A^{\mathrm{T}})^{-1}(B^{\mathrm{T}})^{-1}.$$

Let $\gamma = \{W_\alpha, \gamma_{\alpha\beta}\}_{\alpha,\beta\in\mathfrak{A}}$ be a $\mathrm{Gl}(k)$-cocycle arising from a family of local trivializations

$$\psi_\alpha : \pi^{-1}(W_\alpha) \xrightarrow{\cong} W_\alpha \times \mathbb{R}^k$$

of a k-plane bundle $\pi : E \to M$. We can recover that bundle from the disjoint union

$$\widetilde{E}_\gamma = \bigsqcup_{\alpha\in\mathfrak{A}} W_\alpha \times \mathbb{R}^k = \bigcup_{\alpha\in\mathfrak{A}} W_\alpha \times \mathbb{R}^k \times \{\alpha\}$$

by quotienting out an equivalence relation. (The standard device of reducing a disjoint union to an ordinary union by adding the index as a factor will be notationally useful.) The equivalence relation identifies an element $(x, v, \beta) \in W_\beta \times \mathbb{R}^k \times \{\beta\}$ with $(y, w, \alpha) \in W_\alpha \times \mathbb{R}^k \times \{\alpha\}$ whenever $x = y \in W_\alpha \cap W_\beta$ and $w = \gamma_{\alpha\beta}(x) \cdot v$. Indeed, it is straightforward to check that the quotient space E_γ has a canonical vector bundle structure with projection

$$\pi : E_\gamma \to M,$$
$$\pi([x, v, \beta]) = x.$$

One then checks that the local trivializations ψ_α fit together to define a *canonical* bundle isomorphism $\psi : E \to E_\gamma$. The reader who did not carefully think this through in Section 3.4 really should do so now.

By the above remarks, the $\mathrm{Gl}(k)$-cocycle $\gamma = \{W, \gamma_{\alpha\beta}\}_{\alpha,\beta\in\mathfrak{A}}$ gives rise to a $\mathrm{Gl}(k)$-cocycle γ',

$$\gamma'_{\alpha\beta}(x) = (\gamma_{\alpha\beta}(x)^{\mathrm{T}})^{-1}, \quad \forall x \in W_\alpha \cap W_\beta,$$

and the bundle $E_{\gamma'}$ constructed from this cocycle will be called the *dual bundle* E^*.

Proposition 6.1.3. *For each $x \in M$, the fiber $(E_{\gamma'})_x = E^*_x$ is* canonically isomorphic to the dual of the fiber $(E_\gamma)_x = E_x$.

Proof. Let the equivalence classes in $(E_\gamma)_x$ of elements (x, v, α) of $W_\alpha \times \mathbb{R}^k \times \{\alpha\}$ be denoted by $[x, v, \alpha]$ and those in $(E_{\gamma'})_x = E^*_x$ by $[x, v, \alpha]^*$. We attempt to evaluate $[x, w, \alpha]^*$ on $[x, v, \alpha]$ by the formula

$$[x, w, \alpha]^* \cdot [x, v, \alpha] = w^{\mathrm{T}} \cdot v.$$

We show that this is well defined. Indeed, if $x \in W_\alpha \cap W_\beta$,

$$\begin{aligned}[x, w, \alpha]^* &= [x, \gamma'_{\beta\alpha}(x) \cdot w, \beta]^*, \\ [x, v, \alpha] &= [x, \gamma_{\beta\alpha}(x) \cdot v, \beta],\end{aligned}$$

and

$$\begin{aligned}(\gamma'_{\beta\alpha}(x) \cdot w)^{\mathrm{T}} \cdot \gamma_{\beta\alpha}(x) \cdot v &= ((\gamma_{\beta\alpha}(x)^{-1})^{\mathrm{T}} \cdot w)^{\mathrm{T}} \cdot \gamma_{\beta\alpha}(x) \cdot v \\ &= w^{\mathrm{T}} \cdot \gamma_{\beta\alpha}(x)^{-1} \cdot \gamma_{\beta\alpha}(x) \cdot v \\ &= w^{\mathrm{T}} \cdot v.\end{aligned}$$

This action of $[x, w, \alpha]^*$ on E_x is clearly linear, and defines E^*_x as a vector subspace of the dual space $(E_x)^*$. Since these spaces have the same dimension, $E^*_x = (E_x)^*$ canonically. □

Exercise 6.1.4. Let $\sigma : M \to E^*$ be a section, not necessarily smooth or continuous. Show that σ is continuous (respectively, smooth) if the map

$$\begin{aligned}M &\to \mathbb{R}, \\ x &\mapsto \sigma_x(\tau_x)\end{aligned}$$

is continuous (respectively, smooth) for every continuous (respectively, for every smooth) section τ of E.

Remark. The double dual of an n-plane bundle $\pi : E \to M$ is canonically isomorphic to the original bundle. That is, $(E^*)^* = E$. Indeed, it is immediate that $(\gamma'_{\alpha\beta})' = \gamma_{\alpha\beta}$.

6.2. The space of 1-forms

We apply the construction of the previous section to the tangent bundle $T(M)$.

Definition 6.2.1. Let M be a differentiable manifold, $x \in M$. The dual space $(T_x(M))^*$ is called the cotangent space of M at the point x and will be denoted by $T^*_x(M)$. Each element $\alpha \in T^*_x(M)$ is called a cotangent vector to M at x. The dual bundle $T^*(M)$ to the tangent bundle is called the cotangent bundle of M.

Exercise 6.2.2. Show that a choice of Riemannian metric on M induces a bundle isomorphism $T(M) \cong T^*(M)$. Thus, in the case that M is an open subset of Euclidean space, the standard Euclidean inner product defines a canonical choice of isomorphism. (In classical physics and advanced calculus courses, it is quite common not to distinguish tangent vectors and cotangent vectors. It is precisely because these treatments are carried out in Euclidean domains that this identification is legitimate. In general, we do *not* identify $T(M)$ and $T^*(M)$.)

A typical cotangent vector is the differential of a map. Let $U \subseteq M$ be open, $x \in U$, and let $f \in C^\infty(U)$. Since $T_{f(x)}(\mathbb{R}) = \mathbb{R}$ canonically, we obtain a linear functional

$$df_x : T_x(M) \to \mathbb{R},$$

so $df_x \in T_x^*(M)$. It is evident that df_x depends only on the germ $[f]_x \in \mathfrak{G}_x$, so we obtain an $\mathbb{R}$-linear map

$$d : \mathfrak{G}_x \to T_x^*(M).$$

Lemma 6.2.3. *For each $X_x \in T_x(M)$, $df_x(X_x) = X_x(f)$.*

Proof. Let $(U, x^1, \dots, x^n)$ be a coordinate chart about x. Then

$$df_x = Jf_x = \left[\frac{\partial f}{\partial x^1}(x), \dots, \frac{\partial f}{\partial x^n}(x)\right].$$

If

$$X_x = \sum_{i=1}^n a^i \left.\frac{\partial}{\partial x^i}\right|_x \in T_x(M),$$

then

$$df_x(X_x) = Jf_x \cdot \begin{bmatrix} a^1 \\ \vdots \\ a^n \end{bmatrix} = \sum_{i=1}^n a^i \frac{\partial f}{\partial x^i}(x) = X_x(f).$$

□

Corollary 6.2.4. *Relative to local coordinates $x^1, \dots, x^n$ about $x \in M$, the covectors $dx_x^1, \dots, dx_x^n$ form a basis of $T_x^*(M)$.*

Proof. Since $\dim T_x^*(M) = n$, it will be enough to show that this set of covectors is linearly independent. By the lemma,

$$dx_x^i\left(\left.\frac{\partial}{\partial x^j}\right|_x\right) = \frac{\partial x^i}{\partial x^j}(x) = \delta_{ij}.$$

Thus

$$\begin{aligned} \sum_{i=1}^n b_i dx_x^i = 0 \Leftrightarrow 0 &= \sum_{i=1}^n b_i dx_x^i \left(\left.\frac{\partial}{\partial x^j}\right|_x\right), \qquad 1 \le j \le n, \\ \Leftrightarrow 0 &= b_j, \qquad 1 \le j \le n. \end{aligned}$$

□

Corollary 6.2.5. *The linear map $d : \mathfrak{G}_x \to T_x^*(M)$ is surjective.*

That is, every covector is the differential of a function.

Let $U \subseteq M$ be an open subset, $f \in C^\infty(U)$, and consider the assignment

$$x \mapsto df_x \in T_x^*(U), \quad \forall x \in U.$$

In local coordinates,

$$df_x = \sum_{i=1}^n \frac{\partial f}{\partial x^i}(x)\, dx_x^i.$$

Since $dx_x^1(\partial/\partial x^j|_x) = \delta_{ij}$, it is clear that, for each smooth vector field $X \in \mathfrak{X}(U)$, the map

$$x \mapsto df_x(X_x)$$

defines a smooth function on U. By Exercise 6.1.4, it follows that $x \mapsto df_x$ defines a smooth section df of $T^*(U)$. More generally, smooth sections ω of $T^*(U)$ have local coordinate formulas

$$\omega_x = \sum_{i=1}^{n} f_i dx^i,$$

where $f_i \in C^\infty(U)$, $1 \le i \le n$.

Definition 6.2.6. The $C^\infty(M)$-module $\Gamma(T^*(M))$ of smooth sections of the cotangent bundle is denoted by

$$A^1(M) = \Gamma(T^*(M)).$$

The elements of $A^1(M)$ are called *covector fields* or (more commonly) 1-forms on M. If $\omega \in A^1(M)$, then its value at $x \in M$ is denoted by $\omega_x \in T^*_x(M)$.

Denote by $\mathrm{Hom}_{C^\infty(M)}(\mathfrak{X}(M), C^\infty(M))$ the $C^\infty(M)$-module of all maps $\mathfrak{X}(M) \to C^\infty(M)$ that are $C^\infty(M)$-linear. If $\omega \in A^1(M)$ and $X \in \mathfrak{X}(M)$, we obtain $\omega(X) \in C^\infty(M)$ by setting

$$\omega(X)(x) = \omega_x(X_x), \quad \forall x \in M.$$

It is clear that

$$\omega(fX) = f\omega(X), \quad \forall f \in C^\infty(M),$$

so we can view this 1-form as an element

$$\omega \in \mathrm{Hom}_{C^\infty(M)}(\mathfrak{X}(M), C^\infty(M)).$$

This defines an injective homomorphism

$$A^1(M) \hookrightarrow \mathrm{Hom}_{C^\infty(M)}(\mathfrak{X}(M), C^\infty(M))$$

of $C^\infty(M)$-modules. We will now show that this is also a surjection, proving that these $C^\infty(M)$-modules are canonically isomorphic.

Let $\alpha \in \mathrm{Hom}_{C^\infty(M)}(\mathfrak{X}(M), C^\infty(M))$ and let $U \subseteq M$ be an open subset.

Lemma 6.2.7. *If $X \in \mathfrak{X}(M)$ and $X|U \equiv 0$, then $\alpha(X)|U \equiv 0$.*

Proof. Let $x \in U$ and choose $f \in C^\infty(M)$, vanishing at x and identically equal to 1 on $M \smallsetminus U$. Then $fX = X$ and

$$\alpha(X) = \alpha(fX) = f\alpha(X).$$

This shows that

$$\alpha(X)(x) = f(x)\alpha(X)(x) = 0.$$

Since this is true for arbitrary $x \in U$, it follows that $\alpha(X)|U \equiv 0$. □

Lemma 6.2.8. *There is a canonical*

$$\widetilde{\alpha} \in \mathrm{Hom}_{C^\infty(U)}(\mathfrak{X}(U), C^\infty(U))$$

such that

$$\widetilde{\alpha}(X|U) = \alpha(X)|U, \quad \forall X \in \mathfrak{X}(M).$$

Proof. If $Y \in \mathfrak{X}(U)$, define $\widetilde{\alpha}(Y) \in C^\infty(U)$ as follows. For arbitrary $y \in U$, choose $f \in C^\infty(M)$ such that $f \equiv 1$ on some open neighborhood $V \subset U$ of y and $f|(M \smallsetminus U) \equiv 0$. Then we can interpret fY as a field defined on all of M ($\equiv 0$ outside of U) and $fY|V = Y|V$. Define

$$\widetilde{\alpha}(Y)(y) = \alpha(fY)(y).$$

If $\widehat{f}$ and $\widehat{V}$ are different choices, Lemma 6.2.7 implies that the two definitions of $\widetilde{\alpha}(Y)$ agree at y (and, indeed, on the neighborhood $V \cap \widehat{V}$ of y in U). It is clear that this defines $\widetilde{\alpha} \in \operatorname{Hom}_{C^\infty(U)}(\mathfrak{X}(U), C^\infty(U))$ and that $\widetilde{\alpha}(X|U) = \alpha(X)|U$, $\forall X \in \mathfrak{X}(M)$. □

By this lemma, we can define $\alpha|U = \widetilde{\alpha}$, calling this the *restriction* of α to U.

Corollary 6.2.9. *If $\alpha \in \operatorname{Hom}_{C^\infty(M)}(\mathfrak{X}(M), C^\infty(M))$ then $\alpha(X)(x)$ depends on X_x but not otherwise on X, $\forall x \in M$, $\forall X \in \mathfrak{X}(M)$.*

Proof. Let $x \in M$. Choose a neighborhood U of x in M over which $T(M)$ is trivial and let $Y^1, \dots, Y^n \in \mathfrak{X}(U)$ give a basis of the tangent space at each point of U. Then arbitrary $X \in \mathfrak{X}(M)$ can be written on U as

$$X|U = \sum_{i=1}^{n} f_i Y^i$$

and

$$\alpha(X)(x) = (\alpha|U)(X|U)(x) = \sum_{i=1}^{n} f_i(x)(\alpha|U)(Y^i)(x).$$

On the right-hand side of this equation, the only dependence on X is in the values $f_i(x)$, $1 \le i \le n$. □

The property of α in the above corollary is called the *tensor* property.

Lemma 6.2.10. *If $\eta : \mathfrak{X}(M) \to C^\infty(M)$ is an $\mathbb{R}$-linear map, then η has the tensor property if and only if $\eta \in \operatorname{Hom}_{C^\infty(M)}(\mathfrak{X}(M), C^\infty(M))$.*

Proof. We have proven the "if" part. For the converse, assume that η has the tensor property and let $f \in C^\infty(M)$, $X \in \mathfrak{X}(M)$. For each $x \in M$, $\eta(fX)(x)$ depends only on $f(x)X_x$, hence

$$\eta(fX)(x) = \eta(f(x)X)(x) = f(x)\eta(X)(x)$$

by $\mathbb{R}$-linearity. Since $x \in M$ is arbitrary, $\eta(fX) = f\eta(X)$. □

This equivalence between $C^\infty(M)$-linearity and the tensor property will recur in the broader context of $C^\infty(M)$-*multi*linearity later in this book.

For $x \in M$ and $\alpha \in \operatorname{Hom}_{C^\infty(M)}(\mathfrak{X}(M), C^\infty(M))$, we define $\alpha_x \in T_x^*(M)$ as follows. Given $v \in T_x(M)$, let $X \in \mathfrak{X}(M)$ be any vector field such that $X_x = v$. Define $\alpha_x(v) = \alpha(X)(x)$. By the above, this depends only on v, not on the choice of extension X, so we get $\alpha_x \in T_x^*(M)$, $\forall x \in M$. An application of Exercise 6.1.4 proves that the map $x \mapsto \alpha_x$ defines a smooth section of $T^*(M)$. This identifies α as an element of $A^1(M)$, completing the proof of the following.

Proposition 6.2.11. *There is a canonical isomorphism,*

$$A^1(M) = \operatorname{Hom}_{C^\infty(M)}(\mathfrak{X}(M), C^\infty(M))$$

of $C^\infty(M)$-modules.

Remark. Similarly, $\mathfrak{X}(M) = \operatorname{Hom}_{C^\infty(M)}(A^1(M), C^\infty(M))$.

Let $\varphi : M \to N$ be smooth. If $\omega \in A^1(N)$, define $\varphi^*(\omega) : M \to T^*(M)$ by

$$\varphi^*(\omega)_x = \varphi_x^*(\omega_{\varphi(x)}), \quad \forall x \in M.$$

If $f \in C^\infty(N)$, define $\varphi^*(f) = f \circ \varphi \in C^\infty(M)$. The proof of the following two lemmas will be left to the reader.

Lemma 6.2.12. *If $\varphi : M \to N$ is a smooth map of manifolds and if $\omega \in A^1(N)$, then $\varphi^*(\omega) \in A^1(M)$ and this defines a linear map*

$$\varphi^* : A^1(N) \to A^1(M)$$

of vector spaces over $\mathbb{R}$. Furthermore, if $f \in C^\infty(M)$,

$$\varphi^*(f\omega) = \varphi^*(f)\varphi^*(\omega).$$

Lemma 6.2.13. *If*

$$M \xrightarrow{\varphi} N \xrightarrow{\psi} P$$

are smooth maps of manifolds, then $(\psi \circ \varphi)^* = \varphi^* \circ \psi^*$ on both $C^\infty(P)$ and $A^1(P)$.

Thus, A^1 is a *contravariant functor* (an *anti-homomorphism* of categories) from the category of differentiable manifolds and smooth maps to the category of real vector spaces and linear maps.

Definition 6.2.14. If $f \in C^\infty(M)$, then df is called the exterior derivative of f. The $\mathbb{R}$-linear map $d : C^\infty(M) \to A^1(M)$ is called exterior differentiation.

Lemma 6.2.15. *If $f, g \in C^\infty(M)$, then $d(fg) = f\,dg + g\,df$.*

Proof. Indeed, if $X \in \mathfrak{X}(M)$, then

$$\begin{aligned} d(fg)(X) &= X(fg) \\ &= X(f)g + fX(g) \\ &= g\,df(X) + f\,dg(X) \\ &= (f\,dg + g\,df)(X). \end{aligned}$$

Since $X \in \mathfrak{X}(M)$ is arbitrary, the assertion follows. □

This lemma is a *Leibnitz rule* for exterior differentiation.

Exercise 6.2.16. If $\varphi : M \to N$ is smooth, prove that the diagram

$$\begin{array}{ccc} C^\infty(N) & \xrightarrow{\varphi^*} & C^\infty(M) \\ {\scriptstyle d}\downarrow & & \downarrow{\scriptstyle d} \\ A^1(N) & \xrightarrow[\varphi^*]{} & A^1(M) \end{array}$$

is commutative. That is, $d(\varphi^*(f)) = \varphi^*(df)$, $\forall f \in C^\infty(N)$.

The property of d in this exercise is called the *naturality* of the exterior derivative.

Exercise 6.2.17. Let $X \in \mathfrak{X}(M)$ and let Φ denote the local flow generated by X. One defines the Lie derivative

$$\mathfrak{L}_X : A^1(M) \to A^1(M)$$

by

$$(*) \qquad \mathfrak{L}_X(\omega) = \lim_{t \to 0} \frac{\Phi_t^*(\omega) - \omega}{t}, \quad \forall \omega \in A^1(M),$$

taken pointwise on M. Recall from Section 2.8 the analogous definitions of $\mathfrak{L}_X(f)$ and $\mathfrak{L}_X(Y)$ for $f \in C^\infty(M)$ and $Y \in \mathfrak{X}(M)$. Prove that $(*)$ is defined and satisfies the following identities for arbitrary $f \in C^\infty(M)$, $\omega \in A^1(M)$, and $Y \in \mathfrak{X}(M)$.

(1) $\mathcal{L}_X(df) = d\mathcal{L}_X(f)$.
(2) $\mathcal{L}_X(f\omega) = \mathcal{L}_X(f)\omega + f\mathcal{L}_X(\omega)$.
(3) $\mathcal{L}_X(\omega(Y)) = \mathcal{L}_X(\omega)(Y) + \omega(\mathcal{L}_X(Y))$.

6.3. Line Integrals

If $\omega \in A^1(M)$ and $s : [a,b] \to M$ is a smooth curve, then $s^*(\omega) \in A^1([a,b])$. We can write $s^*(\omega) = f\,dt$.

Definition 6.3.1. The line integral of $\omega \in A^1(M)$ along a smooth curve $s : [a,b] \to M$ is

$$\int_s \omega = \int_a^b s^*(\omega) = \int_a^b f(t)\,dt.$$

Line integrals are insensitive to orientation preserving changes of parameter and experience a sign change only under an orientation reversing reparametrization. It is not even necessary to require that the change of parameter be nonsingular or monotonic.

Lemma 6.3.2. *Let $s : [a,b] \to M$ and $u : [c,d] \to [a,b]$ be smooth. Set $\sigma = s \circ u$. Then,*

(1) *if $u(c) = a$ and $u(d) = b$, $\int_s \omega = \int_\sigma \omega$, $\forall \omega \in A^1(M)$;*
(2) *if $u(c) = b$ and $u(d) = a$, $-\int_s \omega = \int_\sigma \omega$, $\forall \omega \in A^1(M)$.*

Proof. Let t denote the coordinate of $[a,b]$ and τ the coordinate of $[c,d]$. Then

$$\begin{aligned}\int_\sigma \omega &= \int_c^d \sigma^*(\omega)\\ &= \int_c^d u^*(s^*(\omega))\\ &= \int_c^d u^*(f\,dt)\\ &= \int_c^d (f\circ u)\frac{du}{d\tau}\,d\tau.\end{aligned}$$

In case (1), the rule for change of variable in integrals gives

$$\int_\sigma \omega = \int_a^b f\,dt = \int_s \omega.$$

In case (2), the same rule gives

$$\int_\sigma \omega = \int_b^a f\,dt = -\int_s \omega.$$

□

Lemma 6.3.3. *Let $s_1 : [a,b] \to M$ and $s_2 : [c,d] \to M$ be smooth paths with the same initial point and the same terminal point. That is,*

$$s_1(a) = s_2(c) = x \text{ and } s_1(b) = s_2(d) = y.$$

If $f \in C^\infty(M)$, then $\int_{s_1} df = \int_{s_2} df = f(y) - f(x)$.

Proof. Appealing to Lemma 6.3.2, we assume, without loss of generality, that $[a,b] = [c,d]$. Then,

$$\begin{aligned}\int_{s_1} df &= \int_a^b s_1^*(df)\\ &= \int_a^b d(f \circ s_1)\\ &= \int_a^b \frac{d}{dt} f(s_1(t))\,dt\\ &= f(s_1(b)) - f(s_1(a))\\ &= f(s_2(b)) - f(s_2(a))\\ &= \cdots\\ &= \int_{s_2} df.\end{aligned}$$

□

This lemma is a 1-dimensional version of *Stokes' Theorem.* As the proof makes clear, it is just the fundamental theorem of calculus.

Definition 6.3.4. A form $\omega \in A^1(M)$ is said to be exact if $\omega = df$, for some $f \in C^\infty(M)$.

Exercise 6.3.5. Show that every 1-form on $\mathbb{R}$ is exact, but exhibit a 1-form on $\mathbb{R}^2$ that is not exact.

Lemma 6.3.3 says that the line integral $\int_s \omega$ of an exact 1-form ω depends only on the endpoints of the path s, not otherwise on s. In physics, the law of conservation of energy is a special case of this result. Lemma 6.3.3 is a part of Theorem 6.3.10, which will be stated and proven shortly.

The notion of a line integral can be extended to allow integration of 1-forms along paths $s : [a,b] \to M$ that are only *piecewise* smooth. That is, s is continuous and there exists a partition $a = t_0 < t_1 < \cdots < t_q = b$ such that $s_i = s|[t_{i-1}, t_i]$ is smooth, $1 \le i \le q$. We write $s = s_1 + s_2 + \cdots + s_q$ and define

$$\int_s \omega = \sum_{i=1}^q \int_{s_i} \omega.$$

Since it is not assumed that the partition contains only points at which s is not smooth, it is necessary to observe that this definition is independent of the choice of allowable partition. This is elementary and is left to the reader. The proof of the following consequence of Lemma 6.3.3 is also left to the reader.

Corollary 6.3.6. *Let $s_1 : [a,b] \to M$ and $s_2 : [c,d] \to M$ be piecewise smooth paths with the same initial point x and the same terminal point y. Then, if $f \in C^\infty(M)$,*

$$\int_{s_1} df = \int_{s_2} df = f(y) - f(x).$$

Lemma 6.3.7. *If $\omega \in A^1(M)$ and if, for every piecewise smooth path s, the integral $\int_s \omega = 0$, then $\omega = 0$.*

Proof. Otherwise, there is a point $z \in M$ and a vector $v \in T_z(M)$ such that $\omega_z(v) > 0$. Let $s : [-\epsilon, \epsilon] \to M$ be smooth such that $s(0) = z$ and $\dot{s}(0) = v$. Choosing $\epsilon > 0$ smaller, if necessary, we can assume that

$$\omega_{s(t)}(\dot{s}(t)) > 0, \quad -\epsilon \leq t \leq \epsilon.$$

An elementary computation shows that $s^*(\omega)_t = \omega_{s(t)}(\dot{s}(t))\,dt$, so

$$\int_s \omega = \int_{-\epsilon}^{\epsilon} s^*(\omega) = \int_{-\epsilon}^{\epsilon} \omega_{s(t)}(\dot{s}(t))\,dt > 0,$$

contradicting the hypothesis. □

Definition 6.3.8. We say that $\omega \in A^1(M)$ has path-independent line integrals if, for every piecewise smooth path $s : [a,b] \to M$, $\int_s \omega$ depends only on $s(a)$ and $s(b)$ and not otherwise on s.

Definition 6.3.9. A piecewise smooth path $s : [a,b] \to M$ is a loop if $s(a) = s(b)$.

Theorem 6.3.10. *For $\omega \in A^1(M)$, the following are equivalent.*

(1) *ω is an exact form.*
(2) *$\int_s \omega = 0$, for all piecewise smooth loops s.*
(3) *ω has path-independent line integrals.*

Proof. We prove that (1) $\Rightarrow$ (2). If $\omega = df$ is exact and $s : [a,b] \to M$ is a piecewise smooth loop, $s(a) = q = s(b)$, then Corollary 6.3.6 implies that

$$\int_s \omega = \int_q \omega = 0,$$

where q denotes the constant path $q(t) = q$, $a \leq t \leq b$

We prove that (2) $\Rightarrow$ (3). Let s_1 and s_2 be piecewise smooth curves starting at the same point x and ending at the same point y. Without loss of generality, assume that s_1 is parametrized on $[-1,0]$ and s_2 on $[0,1]$. Let $u : [0,1] \to [0,1]$ be defined by $u(t) = 1 - t$. Then $s_2 \circ u$ starts at y and ends at x and

$$s_1 + s_2 \circ u = s : [-1,1] \to M$$

is a piecewise smooth loop. By our assumption,

$$0 = \int_s \omega = \int_{s_1} \omega - \int_{s_2} \omega,$$

where we have used part (2) of Lemma 6.3.2 to write

$$\int_{s_2 \circ u} \omega = -\int_{s_2} \omega.$$

We prove that (3) $\Rightarrow$ (1) by using (3) to construct $f \in C^\infty(M)$ such that $\omega = df$. Without loss of generality, we assume that M is connected (otherwise, carry out the construction of f on each component individually). Fix a basepoint $x_0 \in M$. Given any point $x \in M$, use connectivity to find a piecewise smooth path $s : [a,b] \to M$ such that $s(a) = x_0$ and $s(b) = x$. (In fact, the homogeneity lemma, Theorem 3.8.7, implies the existence of a smooth path, but the present claim is more elementary and is left to the reader.) Set

$$f(x) = \int_s \omega.$$

By the assumption of path-independence, this is independent of the choice of piecewise smooth path s from x_0 to x. Remark that $f(x_0) = 0$.

We prove first that $f : M \to \mathbb{R}$ is smooth. Let $q \in M$ be arbitrary and choose a coordinate chart $(U, x^1, \dots, x^n)$ about q in which q is the origin and $U = \operatorname{int} D^n$, where D^n is the unit ball in $\mathbb{R}^n$. In these coordinates, we can write

$$\omega|U = \sum_{i=1}^{n} g_i \, dx^i.$$

For each $x \in U$, let $s_x : [0,1] \to U$ be defined by $s_x(t) = tx$, $0 \le t \le 1$, and express $f|U$ by the formula

$$f(x) = f(q) + \int_{s_x} \omega.$$

Thus, on U,

$$\begin{aligned} f(x^1, \dots, x^n) &= f(0) + \sum_{i=1}^{n} \int_0^1 g_i(tx^1, \dots, tx^n) \frac{d}{dt}(tx^i)\, dt \\ &= f(0) + \sum_{i=1}^{n} x^i \int_0^1 g_i(tx^1, \dots, tx^n)\, dt. \end{aligned}$$

This is clearly smooth. Since $q \in M$ is arbitrary, $f \in C^\infty(M)$.

Next, we prove that $\omega = df$. Let $s : [a,b] \to M$ be an arbitrary piecewise smooth path. Let $c < a$ and let $s_0 : [c,a] \to M$ be piecewise smooth such that $s_0(c) = x_0$ and $s_0(a) = s(a)$. Then

$$f(s(b)) = \int_{s_0 + s} \omega = \int_{s_0} \omega + \int_s \omega = f(s(a)) + \int_s \omega.$$

That is

$$\int_s \omega = f(s(b)) - f(s(a)) = \int_s df.$$

It follows that the form $\widetilde{\omega} = \omega - df$ satisfies

$$\int_s \widetilde{\omega} = 0,$$

for all piecewise smooth paths s. By Lemma 6.3.7, $\widetilde{\omega} = 0$, so $\omega = df$. □

Definition 6.3.11. A 1-form $\omega \in A^1(M)$ is locally exact if, for each $x \in M$, there is an open neighborhood U of x such that $\omega|U \in A^1(U)$ is exact.

Example 6.3.12. On the manifold $M = \mathbb{R}^2 \smallsetminus \{(0,0)\}$, define the 1-form

$$\eta = \frac{-y}{x^2 + y^2}\, dx + \frac{x}{x^2 + y^2}\, dy.$$

We claim that η is locally exact. Indeed, if $q \in M$ is not on the y-axis, a branch of $\theta = \arctan(y/x)$ is defined and smooth on a neighborhood of q. A direct computation gives $d\theta = \eta$. Similarly, if $q \in M$ is not on the x-axis, select a branch of $\theta = -\arctan(x/y)$ and check that $d\theta = \eta$. Since no point of M is on both axes, this proves that η is locally exact.

We claim, however, that η is not exact. Indeed, consider the smooth loop $s : [0,1] \to M$ defined by $s(t) = (\cos 2\pi t, \sin 2\pi t)$. Clearly,

$$\eta_{s(t)} = -\sin 2\pi t \, dx_{s(t)} + \cos 2\pi t \, dy_{s(t)},$$

and

$$s^*(dx) = -2\pi \sin 2\pi t \, dt,$$
$$s^*(dy) = 2\pi \cos 2\pi t \, dt,$$

so

$$s^*(\eta) = 2\pi(\sin^2 2\pi t + \cos^2 2\pi t) \, dt = 2\pi \, dt.$$

Thus

$$\int_s \eta = 2\pi \int_0^1 dt = 2\pi \neq 0.$$

Remark that the form η in the above example cannot be extended to a 1-form on $\mathbb{R}^2$. We are going to see shortly (Corollary 6.3.15) that, on $\mathbb{R}^2$, every locally exact 1-form is, in fact, exact. The above example reflects a topological feature of $\mathbb{R}^2 \setminus \{(0,0)\}$, the missing point, that distinguishes that space from $\mathbb{R}^2$.

The notion of smooth homotopy extends to a notion of *piecewise smooth* homotopy in a fairly obvious way. Here is the formal definition.

Definition 6.3.13. Let $s_0, s_1 : [a,b] \to M$ be piecewise smooth loops. We say that s_0 is (piecewise smoothly) homotopic to s_1, and write $s_0 \sim s_1$, if there is a continuous map

$$H : [a,b] \times [0,1] \to M$$

and a partition $a = t_0 < t_1 < \cdots < t_r = b$ such that

(1) $H|([t_{i-1}, t_i] \times [0,1])$ is smooth, $1 \leq i \leq r$;
(2) $H(t,0) = s_0(t)$ and $H(t,1) = s_1(t)$, $a \leq t \leq b$;
(3) $H(a,\tau) = H(b,\tau)$, $0 \leq \tau \leq 1$.

As usual, (piecewise smooth) homotopy is an equivalence relation.

Exercise 6.3.14. If $\omega \in A^1(M)$ is locally exact and if the loops s_1 and s_2 are piecewise smooth and homotopic, show that

$$\int_{s_1} \omega = \int_{s_2} \omega.$$

Proceed as follows.

(1) Let $U \subseteq \mathbb{R}^2$ be open, let $R = [a,b] \times [c,d] \subset U$, and let $\omega \in A^1(U)$ be locally exact. For $\epsilon > 0$, let $R_\epsilon = (a-\epsilon, b+\epsilon) \times (c-\epsilon, d+\epsilon)$. Show that there is $\epsilon > 0$ such that $R_\epsilon \subseteq U$ and $\omega|R_\epsilon$ is exact.
(2) Let $\varphi : M \to N$ be a smooth map between manifolds and let $\omega \in A^1(N)$ be locally exact. Prove that $\varphi^*(\omega) \in A^1(M)$ is locally exact.
(3) Use these two results to prove the proposition.

Corollary 6.3.15. *Every locally exact 1-form on $\mathbb{R}^n$ is exact.*

Proof. Let $s : [a,b] \to \mathbb{R}^n$ be a piecewise smooth loop and define

$$H : [a,b] \times [0,1] \to \mathbb{R}^n$$

by $H(t,\tau) = \tau s(t)$. Then H is a homotopy of the constant loop 0 to the loop s. If ω is locally exact, Exercise 6.3.14 implies that

$$\int_s \omega = \int_0 \omega = 0.$$

Since the loop s is arbitrary, Theorem 6.3.10 implies that ω is an exact form. □

Corollary 6.3.15 is a special case of one version of the *Poincaré Lemma*, to be treated later.

6.4. The First Cohomology Space

In Example 6.3.12, we saw that a locally exact form on a manifold can fail to be exact and that this seems to be related to the topology of the manifold. This insight is formalized and exploited by the de Rham cohomology $H^1(M)$, a vector space associated to the manifold M which measures, in some sense, how much the notions of "locally exact" and "exact" differ on M.

Definition 6.4.1. The space of (de Rham) 1-cocycles on M is

$$Z^1(M) = \{\omega \in A^1(M) \mid \omega \text{ is locally exact}\}.$$

The space of (de Rham) 1-coboundaries is

$$B^1(M) = \{\omega \in A^1(M) \mid \omega \text{ is exact}\}.$$

Remark that, if we regard $A^1(M)$ as a vector space over $\mathbb{R}$, then $Z^1(M)$ and $B^1(M)$ are vector subspaces. They are *not* $C^\infty(M)$-submodules. It is also clear that $B^1(M) \subseteq Z^1(M)$.

Definition 6.4.2. The vector space

$$H^1(M) = Z^1(M)/B^1(M)$$

is called the first (de Rham) cohomology space of the manifold M. If ω is a 1-cocycle, its cohomology class is $[\omega] = \omega + B^1(M) \in H^1(M)$.

Although, whenever $\dim M > 0$, the vector spaces $Z^1(M)$ and $B^1(M)$ are infinite dimensional, it frequently happens that $H^1(M)$ is finite dimensional. We will see, for instance, that this is the case whenever M is compact.

Cohomology is a contravariant functor from the category of differentiable manifolds (smooth maps are the morphisms) to the category of real vector spaces (and linear maps). Indeed, by Exercise 6.2.16 an arbitrary smooth map $\varphi : M \to N$ induces a linear map $\varphi^* : Z^1(N) \to Z^1(M)$ and $\varphi^*(B^1(N)) \subseteq B^1(M)$, so φ^* passes to a well-defined linear map (of the same name)

$$\varphi^* : H^1(N) \to H^1(M).$$

It is trivial to check that $(\varphi \circ \psi)^* = \psi^* \circ \varphi^*$ and $(\mathrm{id}_M)^* = \mathrm{id}_{H^1(M)}$.

Proposition 6.4.3. *Let $\omega, \widetilde{\omega} \in Z^1(M)$. Then $[\omega] = [\widetilde{\omega}] \in H^1(M)$ if and only if $\int_s \omega = \int_s \widetilde{\omega}$ as s varies over all piecewise smooth loops in M. These numbers are called the periods of ω and of the cohomology class $[\omega]$.*

Proof. The locally exact forms $\omega, \widetilde{\omega} \in Z^1(M)$ have the same periods $\int_s \omega = \int_s \widetilde{\omega}$, for every piecewise smooth loop s, if and only if $\int_s (\omega - \widetilde{\omega}) = 0$ for all such loops. By Theorem 6.3.10, this holds precisely when $\omega - \widetilde{\omega}$ belongs to $B^1(M)$. Equivalently, $[\omega] = [\widetilde{\omega}]$. □

Example 6.4.4. Consider the sphere S^n, $n \geq 2$. By stereographic projection, we know that the complement of a point in S^n is diffeomorphic to $\mathbb{R}^n$, so every piecewise smooth loop in S^n that misses a point is homotopic to a constant loop. By Sard's theorem, no piecewise smooth curve in S^n can be space-filling if $n \geq 2$ (see

Example 2.9.4), so all piecewise smooth loops σ on this sphere are homotopically trivial. By Exercise 6.3.14,

$$\int_\sigma \omega = 0,$$

for all locally exact 1-forms ω. By Proposition 6.4.3, we conclude that

$$H^1(S^n) = 0, \quad \text{whenever } n \geq 2.$$

Proposition 6.4.5. *If $f_0, f_1 : M \to N$ are smooth and homotopic, then*

$$f_0^* = f_1^* : H^1(N) \to H^1(M).$$

Proof. Let $[\omega] \in H^1(N)$. If $s : [a,b] \to M$ is a piecewise smooth loop, then $s_i = f_i \circ s : [a,b] \to N$ is also a piecewise smooth loop, $i = 0, 1$. Let $H : M \times \mathbb{R} \to M$ be a homotopy of f_0 to f_1. Then, the composition

$$[a,b] \times [0,1] \xrightarrow{s \times \mathrm{id}} M \times \mathbb{R} \xrightarrow{H} N$$

is a homotopy of s_0 to s_1, so

$$\begin{aligned} \int_s f_0^*(\omega) &= \int_a^b s^* f_0^*(\omega) \\ &= \int_a^b (f_0 \circ s)^*(\omega) \\ &= \int_{s_0} \omega \\ &= \int_{s_1} \omega \quad \text{(Exercise 6.3.14)} \\ &= \cdots \\ &= \int_s f_1^*(\omega). \end{aligned}$$

Since s is an arbitrary piecewise smooth loop, Proposition 6.4.3 implies that

$$f_1^*[\omega] = [f_1^*(\omega)] = [f_0^*(\omega)] = f_0^*[\omega].$$

Since $[\omega] \in H^1(M)$ is arbitrary, $f_1^* = f_0^*$ at the cohomology level, as desired. □

Definition 6.4.6. A smooth map $f : M \to N$ is a homotopy equivalence if there exists a smooth map $g : N \to M$ such that $f \circ g \sim \mathrm{id}_N$ and $g \circ f \sim \mathrm{id}_M$.

Corollary 6.4.7. *A homotopy equivalence $f : M \to N$ induces a linear isomorphism $f^* : H^1(N) \to H^1(M)$.*

Proof. Since $f \circ g \sim \mathrm{id}_N$, it follows by the (contravariant) functoriality of cohomology and Proposition 6.4.5 that

$$g^* \circ f^* = (f \circ g)^* = \mathrm{id}_N^* = \mathrm{id}_{H^1(N)}.$$

Similarly, $f^* \circ g^* = \mathrm{id}_{H^1(M)}$, so f^* and g^* are mutually inverse isomorphisms on cohomology. □

Example 6.4.8. Let $f : \{0\} \hookrightarrow D^n$ be the inclusion. Let $g : D^n \to \{0\}$ be the only map. These maps are smooth and $g \circ f = \mathrm{id}_{\{0\}}$. Consider the map $f \circ g : D^n \to D^n$ having image $\{0\}$. We claim that this is homotopic to id_{D^n}.

Indeed, let $\varphi : \mathbb{R} \to [0,1]$ be smooth such that $\varphi(0) = 0$ and $\varphi(1) = 1$. Define $H : D^n \times \mathbb{R} \to D^n$ by

$$H(x,t) = \varphi(t)x.$$

Then

$$H(x,0) = 0 = f(g(x)), \quad \forall x \in D^n,$$

and

$$H(x,1) = x = \mathrm{id}_{D^n}(x), \quad \forall x \in D^n.$$

This establishes the desired homotopy and completes the proof that f is a homotopy equivalence, so

$$f^* : H^1(D^n) \to H^1(\{0\}) = 0$$

is an isomorphism. That is,

$$H^1(D^n) = 0$$

or, equivalently, every locally exact 1-form on D^n is exact. A similar proof shows that $\mathbb{R}^n$ is homotopically equivalent to a point, and we recover Corollary 6.3.15.

Example 6.4.9. Let

$$i : S^{n-1} \hookrightarrow \mathbb{R}^n \smallsetminus \{0\}$$

be the inclusion. Let

$$g : \mathbb{R}^n \smallsetminus \{0\} \to S^{n-1}$$

be the map defined by

$$g(v) = \frac{v}{\|v\|}.$$

These maps are smooth, and

$$g \circ i = \mathrm{id}_{S^{n-1}}.$$

We claim that

$$i \circ g \sim \mathrm{id}_{\mathbb{R}^n \smallsetminus \{0\}},$$

hence that i is a homotopy equivalence. Indeed, define

$$H : (\mathbb{R}^n \smallsetminus \{0\}) \times [0,1] \to \mathbb{R}^n \smallsetminus \{0\}$$

by the formula

$$H(v,t) = \frac{v}{t + (1-t)\|v\|}.$$

This is smooth since $\|v\| > 0$ implies that $t + (1-t)\|v\| > 0$, $0 \le t \le 1$. Then

$$H(v,1) = v, \quad \forall v \in \mathbb{R}^n \smallsetminus \{0\},$$

and

$$H(v,0) = \frac{v}{\|v\|} = i(g(v)), \quad \forall v \in \mathbb{R}^n \smallsetminus \{0\}.$$

It follows that

$$H^1(\mathbb{R}^n \smallsetminus \{0\}) = H^1(S^{n-1}).$$

In particular, together with Example 6.4.4, this proves that $H^1(\mathbb{R}^n \smallsetminus \{0\})$ is trivial, whenever $n \ge 3$.

Proposition 6.4.10. *There is a canonical isomorphism* $H^1(S^1) = \mathbb{R}$.

We will prove this via three lemmas. Recall the universal covering map

$$p : \mathbb{R} \to S^1,$$
$$p(t) = (\cos 2\pi t, \sin 2\pi t).$$

This is the map that induces the standard diffeomorphism $\mathbb{R}/\mathbb{Z} = S^1$. Define

$$\alpha : Z^1(S^1) \to \mathbb{R}$$

by

$$\alpha(\omega) = \int_0^1 p^*(\omega),$$

a linear map.

Lemma 6.4.11. *The linear map α passes to a well-defined linear map*

$$\alpha : H^1(S^1) \to \mathbb{R}.$$

Proof. Indeed, $\sigma = p|[0,1]$ is a smooth loop and $\omega \in B^1(S^1)$ implies that

$$\alpha(\omega) = \int_\sigma \omega = 0,$$

by Theorem 6.3.10. □

Lemma 6.4.12. *The linear map $\alpha : H^1(S^1) \to \mathbb{R}$ is injective.*

Proof. Let $\omega \in Z^1(S^1)$ be such that $\alpha(\omega) = 0$. We must prove that $\omega \in B^1(S^1)$.

For $n \in \mathbb{Z}$, let $\tau_n : \mathbb{R} \to \mathbb{R}$ be the translation $\tau_n(t) = t + n$. Then, $p \circ \tau_n = p$, so

$$\tau_n^* \circ p^* = p^* : A^1(S^1) \to A^1(\mathbb{R}).$$

This and the change of variable formula for the integral gives

$$\int_n^{t+n} p^*(\omega) = \int_0^t \tau_n^*(p^*(\omega)) = \int_0^t p^*(\omega), \quad \forall n \in \mathbb{Z}, \quad 0 \le t \le 1.$$

In particular, since $\alpha(\omega) = 0$, we obtain

$$\int_n^{n+1} p^*(\omega) = 0 = \int_{n+1}^n p^*(\omega).$$

Define $f_\omega \in C^\infty(S^1)$ by

$$f_\omega(\cos 2\pi t, \sin 2\pi t) = \int_0^t p^*(\omega).$$

If this is well defined, it will be smooth. It will be well defined precisely if

$$\int_0^{t+n} p^*(\omega) = \int_0^t p^*(\omega), \quad \forall n \in \mathbb{Z}, \quad \forall t \in \mathbb{R}.$$

If $n = 0$, this is obvious. If $n > 0$,

$$\begin{aligned}
\int_0^{t+n} p^*(\omega) &= \int_n^{t+n} p^*(\omega) + \sum_{i=0}^{n-1} \int_i^{i+1} p^*(\omega) \\
&= \int_n^{t+n} p^*(\omega) \\
&= \int_0^t p^*(\omega).
\end{aligned}$$

A similar computation for the case $n < 0$ is left to the reader. For the lift $\widetilde{f}_\omega = p^*(f_\omega) \in C^\infty(\mathbb{R})$, we get

$$\widetilde{f}_\omega(t) = \int_0^t p^*(\omega),$$

so the fundamental theorem of calculus and Exercise 6.2.16 give

$$p^*(\omega) = d\widetilde{f}_\omega = d(p^*(f_\omega)) = p^*(df_\omega).$$

But $p : \mathbb{R} \to S^1$ is a local diffeomorphism, so ω and df_ω are equal locally, hence globally. That is, $\omega \in B^1(S^1)$ as desired. □

Recall the locally exact form

$$\eta = \frac{-y}{x^2 + y^2}\, dx + \frac{x}{x^2 + y^2}\, dy,$$

of Example 6.3.12 and let

$$\widetilde{\eta} = i^*(\eta) \in Z^1(S^1),$$

where i is the inclusion map of S^1 into $\mathbb{R}^2 \smallsetminus \{(0,0)\}$. For the loop $\sigma = p|[0,1]$, $s = i \circ \sigma$ is as in Example 6.3.12, and we showed that

$$\sigma^*(\widetilde{\eta}) = \sigma^*(i^*(\eta)) = (i \circ \sigma)^*(\eta) = s^*(\eta) = 2\pi dt.$$

Lemma 6.4.13. *The linear map $\alpha : H^1(S^1) \to \mathbb{R}$ is surjective.*

Proof. It is enough to show that α is nontrivial. But $[\widetilde{\eta}] \in H^1(S^1)$ and

$$\alpha[\widetilde{\eta}] = \int_0^1 p^*(\widetilde{\eta}) = \int_0^1 \sigma^*(\widetilde{\eta}) = 2\pi.$$

□

The proof of Proposition 6.4.10 is complete.

Corollary 6.4.14. $H^1(\mathbb{R}^2 \smallsetminus \{0\}) = \mathbb{R}$.

Example 6.4.15. Recall that the Brouwer fixed point theorem for arbitrary smooth (in fact, continuous) maps $f : D^2 \to D^2$ follows from the nonexistence of a smooth retraction $\rho : D^2 \to \partial D^2 = S^1$. The proof we gave using the fundamental group can be mimicked using cohomology instead.

Recall that, for ρ to be a retraction, it is required that the diagram

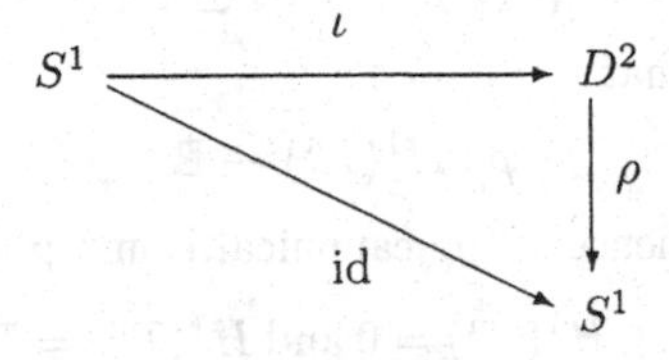

commute, where ι is the inclusion of the boundary circle. By the functoriality of cohomology, this produces a commutative diagram

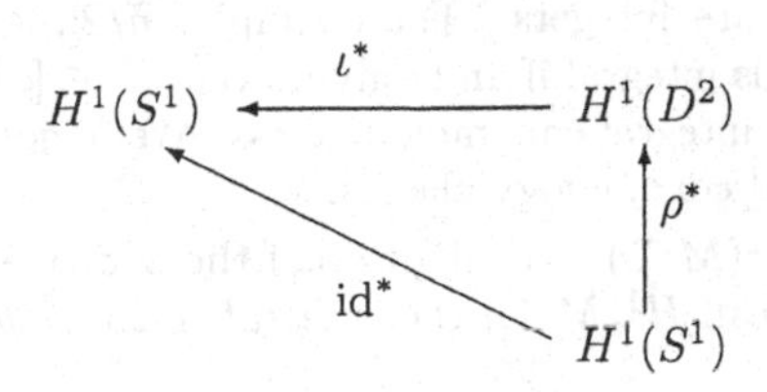

That is,

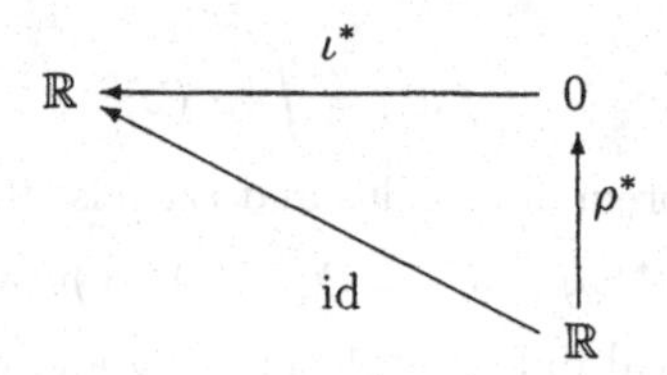

commutes, which is absurd.

Our computation of $H^1(S^1)$ generalizes to the higher dimensional tori.

Exercise 6.4.16. Let $\exp : \mathbb{R}^n \to T^n$ be the homomorphism of abelian Lie groups defined by

$$\exp(x^1, \dots, x^n) = \left(e^{2\pi i x^1}, \dots, e^{2\pi i x^n}\right).$$

If $\sigma : [a,b] \to \mathbb{R}^n$ is piecewise smooth such that $\vec{v}_\sigma = \sigma(b) - \sigma(a) \in \mathbb{Z}^n$, then $\exp \circ \sigma$ is a piecewise smooth loop on T^n. Using this observation, prove that $H^1(T^n) = \mathbb{R}^n$, proceeding as follows.

(1) Prove that every piecewise smooth loop on T^n is of the form $\exp \circ \sigma$ as above and that the homotopy class $[\exp \circ \sigma]$ is completely determined by $\vec{v}_\sigma$.

(2) If $\omega \in Z^1(T^n)$, define $\varphi_\omega : \mathbb{Z}^n \to \mathbb{R}$ as follows. Given $\vec{v} \in \mathbb{Z}^n$, choose a piecewise smooth path $\sigma : [a,b] \to \mathbb{R}^n$ such that $\vec{v} = \vec{v}_\sigma$ and set

$$\varphi_\omega(\vec{v}) = \int_{\exp \circ \sigma} \omega.$$

Show that this is well defined and that $\varphi_\omega : \mathbb{Z}^n \to \mathbb{R}$ extends uniquely to a linear functional of the same name

$$\varphi_\omega : \mathbb{R}^n \to \mathbb{R}.$$

(3) Prove that the assignment $\omega \mapsto \varphi_\omega$ passes to a well-defined linear *injection*

$$\varphi : H^1(T^n) \to (\mathbb{R}^n)^* = \mathbb{R}^n.$$

(4) Show that there are forms $\theta^1, \dots, \theta^n \in Z^1(T^n)$ such that

$$\exp^*(\theta^i) = dx^i, \ 1 \le i \le n.$$

Use this to show that

$$\varphi : H^1(T^n) \to \mathbb{R}^n$$

is also *surjective*, hence is the canonical isomorphism we seek.

Example 6.4.17. If $n \ge 2$, $H^1(S^n) = 0$ and $H^1(T^n) = \mathbb{R}^n$, proving that S^n and T^n are not homotopically equivalent. Of course, $S^1 = T^1$.

Exercise 6.4.18. We will say that a locally exact form $\omega \in Z^1(M)$ is *integral* if all of its periods are integers. For example, $\tilde{\eta}/2\pi \in Z^1(S^1)$ is integral. By Proposition 6.4.3, ω is integral if and only if every $\omega' \in [\omega]$ is integral, in which case we say that $[\omega]$ is an integral cohomology class. We denote by $H^1(M;\mathbb{Z}) \subset H^1(M)$ the subset of integral cohomology classes.

(1) Prove that $H^1(M;\mathbb{Z})$ is a subgroup of the additive group of the vector space $H^1(M)$. We call $H^1(M;\mathbb{Z})$ the *integral cohomology* of M.

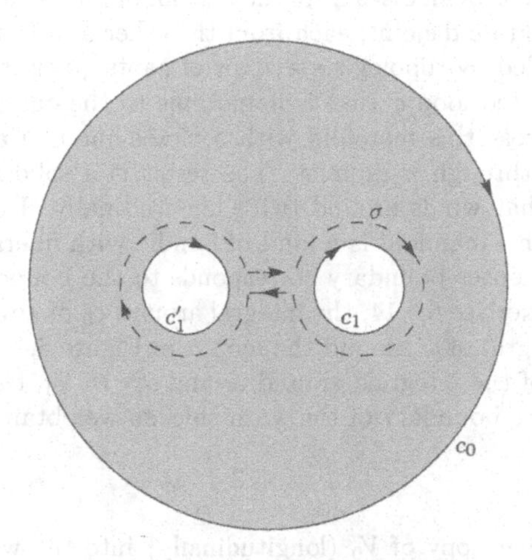

Figure 6.4.1. The pair of pants P

(2) If $f : M \to N$ is smooth and $\omega \in Z^1(N)$ is integral, prove that $f^*(\omega) \in Z^1(M)$ is also integral. Using this, show that integral cohomology is a contravariant functor from the category $\mathcal{M}$ of smooth manifolds and smooth maps to the category $\mathcal{G}$ of abelian groups and group homomorphisms.

(3) Referring to Exercise 6.4.16, prove that $H^1(T^n;\mathbb{Z})$ is canonically the integer lattice
$$\mathbb{Z}^n \subset \mathbb{R}^n = H^1(T^n).$$

(4) To each smooth map $f : T^n \to T^n$, show how to assign canonically an $n \times n$ matrix A_f of integers, depending only on the homotopy class of f, such that $A_{f \circ g} = A_g A_f$ (matrix multiplication). If f is a diffeomorphism of T^n onto itself, prove that A_f is unimodular (*i.e.*, has determinant ± 1).

(5) Prove that every $n \times n$ unimodular matrix of integers occurs as the matrix A_f assigned to some diffeomorphism $f : T^n \to T^n$.

Exercise 6.4.19. If the vector space $H^1(M)$ has finite dimension k, prove that there is a set of piecewise smooth loops $\{\sigma_1, \dots, \sigma_k\}$ on M such that the map
$$H^1(M) \to \mathbb{R}^k$$
defined by
$$[\omega] \mapsto \left(\int_{\sigma_1} \omega, \dots, \int_{\sigma_k} \omega\right)$$
is an isomorphism of vector spaces.

In light of Exercise 6.4.19, one might expect to generalize part (3) of Exercise 6.4.18 to all manifolds with finite dimensional first cohomology. That is, one asks whether the loops in Exercise 6.4.19 can be chosen so as to carry $H^1(M;\mathbb{Z})$ isomorphically onto $\mathbb{Z}^k$. In fact, this can be done if M is compact, but we sketch an example that shows what can go wrong in general.

Example 6.4.20. Let P denote the 2-manifold with boundary obtained by removing two small, disjoint, open disks from the interior of D^2. The boundaries c_1 and c_1' of these disks should be disjoint, each from the other and from $c_0 = \partial D^2$. The resulting surface, called by topologists a "pair of pants", is pictured in Figure 6.4.1, together with a dotted loop σ that is homotopic to the outer boundary circle. In Figure 6.4.2, we cross this manifold with a closed interval and identify opposite ends with a twist through π radians. The result is a solid torus with a "wormhole" drilled out that winds around twice longitudinally. Denote this 3-manifold by V_0. Note that this manifold is a kind of bundle with fibers diffeomorphic to P. A meridian on the outer boundary corresponds to the boundary curve c_0 of P in Figure 6.4.1. By Exercise 6.3.14, the integral around c_0 of any locally exact form ω is equal to the integral of ω around the loop σ in Figure 6.4.1 and this, in turn, is equal to the sum of the integrals around c_1 and c_1'. In V_0, the loops c_1 and c_1' are homotopic along the boundary of the wormhole, so we obtain

$$\int_{c_0} \omega = 2 \int_{c_1} \omega.$$

We now glue another copy of V_0 (longitudinally) into the wormhole, obtaining a manifold V_1 containing a loop c_2 such that

$$\int_{c_0} \omega = 2 \int_{c_1} \omega = 4 \int_{c_2} \omega.$$

Inductively, a manifold V_n is obtained by gluing a copy of V_0 longitudinally into the wormhole of V_{n-1} and V_n contains a new loop c_n such that

$$\int_{c_0} \omega = 2^n \int_{c_n} \omega.$$

Proceeding *ad infinitum*, we obtain a limit manifold V_∞. This is the complement in the solid torus of a very complicated compact subspace Σ called the *solenoid.* If one first imbeds the solid torus in S^3 in the standard unknotted fashion and then removes the solenoid, the noncompact manifold $M = S^3 \smallsetminus \Sigma$ that results can be shown to have first de Rham cohomology $H^1(M) = \mathbb{R}$ and the set of loops chosen in Exercise 6.4.19 can be taken to be the singleton $\{c_0\}$. In fact, one can show that all periods of any locally exact form ω are sums of periods of ω corresponding to loops c_i, $i \geq 0$. If ω is a locally exact 1-form that is integral, we obtain integers

$$\int_{c_i} \omega = n_i, \quad i \geq 0,$$

and $n_0 = 2n_1 = \cdots = 2^i n_i = \cdots$. This can only happen if $n_0 = 0$, in which case every $n_i = 0$ and ω has all periods 0. That is, the isomorphism in Exercise 6.4.19 identifies $H^1(M;\mathbb{Z}) = 0$.

6.5. Degree Theory on S^1*

Recall from Example 1.7.33 that $\pi_1(S^1, 1) = \mathbb{Z}$. This is a *canonical* isomorphism, produced by lifting a loop σ based at 1 in S^1 to a path $\widetilde{\sigma}$ in the universal cover $\mathbb{R}$ starting at 0. The endpoint of this path is the integer corresponding to $[\sigma]$. By Exercise 5.1.37 and the fact that S^1 is a Lie group, the group $\pi[S^1, S^1]$ is canonically isomorphic to $\pi_1(S^1, 1)$. This isomorphism, denoted by

$$\deg : \pi[S^1, S^1] \to \mathbb{Z},$$

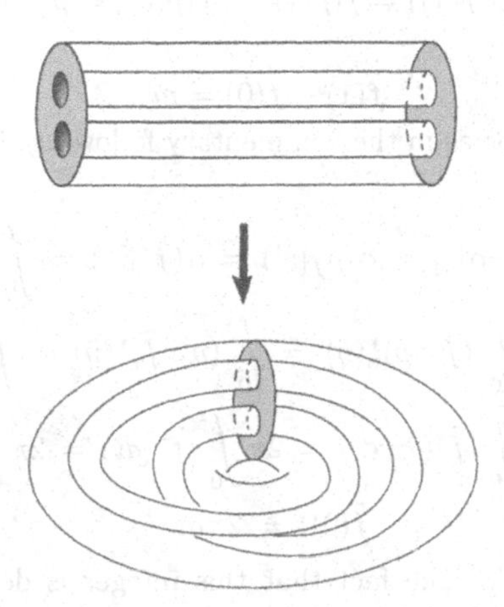

Figure 6.4.2. Forming the manifold V_0

is called the *degree* map and $\deg([f])$ is also called the degree of any $f \in [f]$ and denoted by $\deg(f)$.

We are going to give two equivalent definitions of this degree, one using cohomology and one in terms of regular values.

The first remark is that, since $H^1(S^1) = \mathbb{R}$, f induces a linear map

$$f^* : \mathbb{R} \to \mathbb{R}$$

depending only on the homotopy class of f. Thus, f^* is just multiplication by a certain constant $a_f \in \mathbb{R}$ and a_f depends only on the homotopy class of f. By part (4) of Exercise 6.4.18, a_f is an integer, being the sole entry in the 1×1 integer matrix A_f. We are going to give another way to see that $a_f \in \mathbb{Z}$.

By an application of Theorem 1.7.39, if $f : S^1 \to S^1$ is smooth, we can lift the map $f \circ p : \mathbb{R} \to S^1$ to a smooth map $\widetilde{f} : \mathbb{R} \to \mathbb{R}$. That is, the diagram

$$\begin{array}{ccc} \mathbb{R} & \xrightarrow{\widetilde{f}} & \mathbb{R} \\ {\scriptstyle p}\big\downarrow & & \big\downarrow{\scriptstyle p} \\ S^1 & \xrightarrow[f]{} & S^1 \end{array}$$

is commutative. Also, since the group of covering transformations consists of translations by integers, $\widehat{f} : \mathbb{R} \to \mathbb{R}$ will be a lift of $f \circ p$ if and only if $\widehat{f} = \widetilde{f} + k$, for some integer k.

Proposition 6.5.1. *If $f : S^1 \to S^1$ is smooth, then*

$$a_f = \widetilde{f}(1) - \widetilde{f}(0) = \deg(f) \in \mathbb{Z},$$

where $\widetilde{f}$ is any lift of $f \circ p$.

Proof. Remark that

$$p(\widetilde{f}(1)) = f(p(1)) = f(p(0)) = p(\widetilde{f}(0)),$$

so

$$\widetilde{f}(1) - \widetilde{f}(0) = m \in \mathbb{Z}.$$

Thus, let $\widetilde{\eta} \in Z^1(S^1)$ be as in the commentary following the proof of Lemma 6.4.12 and compute

$$\begin{aligned}
2\pi a_f = a_f \alpha[\widetilde{\eta}] &= \alpha(a_f[\widetilde{\eta}]) = \alpha(f^*[\widetilde{\eta}]) = \int_0^1 p^*(f^*(\widetilde{\eta})) \\
&= \int_0^1 (f \circ p)^*(\widetilde{\eta}) = \int_0^1 (p \circ \widetilde{f})^*(\widetilde{\eta}) = \int_0^1 \widetilde{f}^*(p^*(\widetilde{\eta})) \\
&= \int_0^1 \widetilde{f}^*(2\pi\, dt) = 2\pi \int_0^1 \widetilde{f}^*(dt) = 2\pi \int_0^1 \widetilde{f}'(t)\, dt \\
&= 2\pi(\widetilde{f}(1) - \widetilde{f}(0)) = 2\pi m.
\end{aligned}$$

That is, $a_f = m \in \mathbb{Z}$. The fact that this integer is $\deg(f)$ as defined above is elementary and left to the reader. □

Remark. In Section 3.9, we defined $\deg_2(f) \in \mathbb{Z}_2$ for smooth maps between manifolds (without boundary) of the same dimension. We will see that, for smooth maps of the circle to itself, $\deg_2(f)$ is just the residue class modulo 2 of $\deg(f)$ (Corollary 6.5.4).

Corollary 6.5.2. *If $f : S^1 \to S^1$ is smooth, if $\widetilde{f}$ is a lift of $f \circ p$, and if $t \in \mathbb{R}$ is arbitrary, then* $\deg(f) = \widetilde{f}(t+1) - \widetilde{f}(t)$.

Proof. View $p : \mathbb{R} \to S^1 \subset \mathbb{C}$ as a group homomorphism. Then

$$\begin{aligned}
p(\widetilde{f}(t+1) - \widetilde{f}(t)) &= \frac{p(\widetilde{f}(t+1))}{p(\widetilde{f}(t))} \\
&= \frac{f(p(t+1))}{f(p(t))} \\
&= 1,
\end{aligned}$$

so $\widetilde{f}(t+1) - \widetilde{f}(t) \in \mathbb{Z}$, $\forall t \in \mathbb{R}$. This function of t, being continuous and integer-valued, is constant on $\mathbb{R}$, hence equal to $\widetilde{f}(1) - \widetilde{f}(0) = \deg(f)$. □

We turn to the description of $\deg(f)$ in terms of regular values. Let $z_0 \in S^1$ be a regular value of $f : S^1 \to S^1$. Then $f^{-1}(z_0) = \{z_1, \dots, z_r\}$. Here, if $f^{-1}(z_0) = \emptyset$, we take $r = 0$. Recall that $\deg_2(f) = r \pmod 2$.

Choose $\widetilde{z}_i \in \mathbb{R}$ such that $p(\widetilde{z}_i) = z_i$, $1 \le i \le r$. The smooth map $f : S^1 \to S^1$ preserves orientation at z_i if $\widetilde{f}'(\widetilde{z}_i) > 0$ and reverses orientation at z_i if $\widetilde{f}'(\widetilde{z}_i) < 0$. Let $\epsilon_i = \widetilde{f}'(\widetilde{z}_i)/|\widetilde{f}'(\widetilde{z}_i)| \in \{-1, 1\}$ and remark that this depends only on f and z_i.

Proposition 6.5.3. *With the above conventions,* $\deg(f) = \sum_{i=1}^r \epsilon_i$.

Proof. Choose $a \in \mathbb{R} \smallsetminus \{p^{-1}\{z_1, \dots, z_r\}\}$. Then $p^{-1}(z_i) \cap (a, a+1)$ is a singleton and we choose this point as our $\widetilde{z}_i$, $1 \le i \le r$. We will also use the fact that $\widetilde{f}(a+1) = \widetilde{f}(a) + \deg(f)$.

Let $p^{-1}(z_0) = \{b + k\}_{k \in \mathbb{Z}}$ and consider the graph of $s = \widetilde{f}(t)$ over the open interval $(a, a+1)$, together with the horizontal lines $s = b + k$, $k \in \mathbb{Z}$. Each time

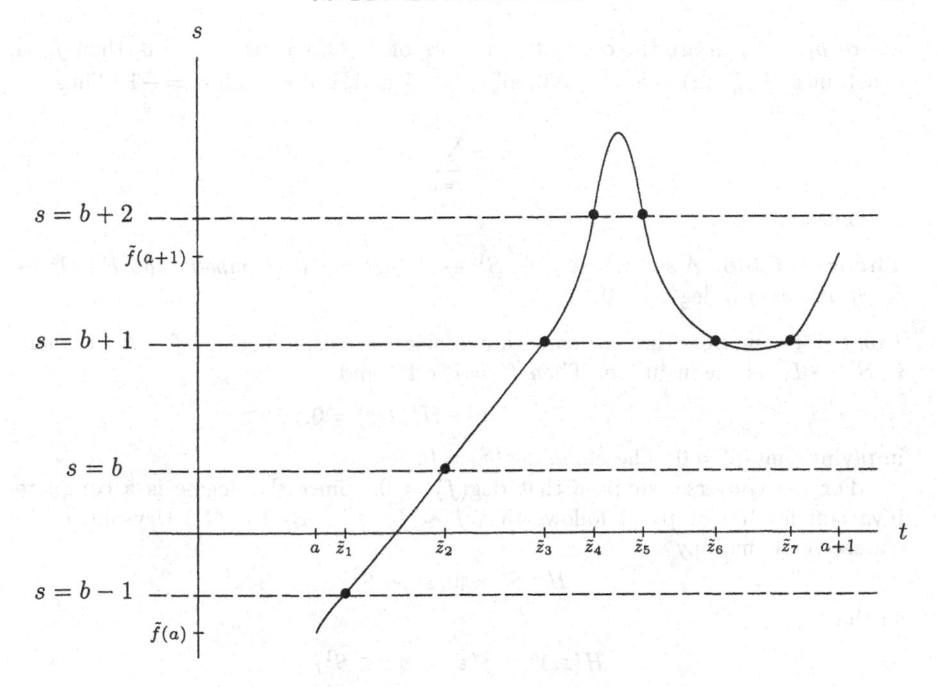

Figure 6.5.1. Graph of $\widetilde{f}$

the graph crosses a line $s = b + k$, the parameter t is equal to one of the $\widetilde{z}_i$ and ϵ_i records whether the graph crosses this line while increasing ($\epsilon_i = 1$) or decreasing ($\epsilon_i = -1$). Figure 6.5.1 illustrates a case in which $r = 7$, $\epsilon_1 = \epsilon_2 = \epsilon_3 = \epsilon_4 = \epsilon_7 = 1$, and $\epsilon_5 = \epsilon_6 = -1$.

The sum of the ϵ_is pertaining to a single line $s = b + k$ is 1, -1, or 0, the net number of directed crossings. Clearly, the sum of all these net numbers is

$$\sum_{i=1}^{r} \epsilon_i = \widetilde{f}(a+1) - \widetilde{f}(a) = \deg(f).$$

(In Figure 6.5.1, the degree is 3.) □

Corollary 6.5.4. $\deg_2(f) = \deg(f) \pmod 2$.

Example 6.5.5. For each $n \in \mathbb{Z}$, define

$$f_n : S^1 \to S^1,$$
$$f_n(z) = z^n.$$

Here, of course, we view $S^1 \subset \mathbb{C}$. We can choose the lift $\widetilde{f}_n : \mathbb{R} \to \mathbb{R}$ to be $\widetilde{f}_n(t) = nt$, so

$$\deg(f_n) = \widetilde{f}_n(1) - \widetilde{f}_n(0) = n.$$

If $z \in S^1$ is a regular value of f_n, then

$$f_n^{-1}(z) = \{\rho_1, \dots, \rho_{|n|}\},$$

where $\rho_1, \dots, \rho_{|n|}$ are the distinct nth roots of z. Of course, if $n = 0$, then f_0 is constant and $f_0^{-1}(z) = \emptyset$. If $n > 0$, all $\epsilon_i = +1$ and, if $n < 0$, all $\epsilon_i = -1$. Thus,

$$n = \sum_{i=1}^{|n|} \epsilon_i$$

in all cases.

Theorem 6.5.6. *A smooth map $f : S^1 \to S^1$ extends to a smooth map $F : D^2 \to S^1$ if and only if* $\deg(f) = 0$.

Proof. First suppose that the smooth extension F exists. That is, $f = F \circ i$ where $i : S^1 \hookrightarrow D^2$ is the inclusion. Then $f^* = i^* \circ F^*$ and

$$F^* : H^1(S^1) \to H^1(D^2) = 0,$$

implying that $f^* = 0$. Therefore, $\deg(f) = 0$.

For the converse, suppose that $\deg(f) = 0$. Since the degree is a complete invariant for homotopy, it follows that $f \sim f_0 \equiv 1$. By the C^∞ Urysohn trick, choose the homotopy

$$H : S^1 \times [0,1] \to S^1$$

so that

$$H(z,1) = f(z), \quad \forall\, z \in S^1,$$
$$H(z,t) \equiv 1, \quad 0 \le t \le 1/2.$$

Then H induces a smooth map of the disk to the circle as follows. Define a smooth surjection

$$\varphi : S^1 \times [0,1] \to D^2 \subset \mathbb{C}$$

by

$$\varphi(z,t) = tz.$$

Then φ carries $S^1 \times (0,1]$ diffeomorphically onto $D^2 \setminus \{0\}$. Define

$$F : D^2 \to S^1$$

by

$$F(\varphi(z,t)) = H(z,t).$$

This is well defined. It is smooth on $D^2 \setminus \{0\}$ and, on $\{w \in D^2 \mid |w| \le \frac{1}{2}\}$ it is constant, so F is smooth. Evidently, $F(z) = H(z,1) = f(z)$, $\forall\, z \in S^1$, so F extends f. □

Recall that the mod 2 degree allowed us to prove the fundamental theorem of algebra for polynomials of odd degree (Theorem 3.9.14). The integer degree makes it possible to carry out essentially the same argument for all positive degrees.

Theorem 6.5.7 (Fundamental Theorem of Algebra). *Let $f : \mathbb{C} \to \mathbb{C}$ be a polynomial of degree $n \ge 1$. Then there is $z_0 \in \mathbb{C}$ such that $f(z_0) = 0$.*

Proof. We can assume that the leading coefficient is 1 and write

$$f(z) = z^n + a_1 z^{n-1} + \cdots + a_{n-1} z + a_n.$$

Suppose this has no root. For each positive real number r, define a smooth function

$$F_r : D^2 \to S^1$$

by the formula

$$F_r(z) = \frac{f(rz)}{|f(rz)|}.$$

This is where we use the hypothesis that f has no roots. Let $g_r = F_r|S^1$. Then set

$$\widehat{H}_r(z,t) = (rz)^n + t(a_1(rz)^{n-1} + \cdots + a_n)$$

and note that, if r is large enough, this vanishes nowhere on $S^1 \times [0,1]$. Indeed,

$$\frac{\widehat{H}_r(z,t)}{(rz)^n} = 1 + t\left(\frac{a_1}{rz} + \cdots + \frac{a_n}{(rz)^n}\right)$$

approaches 1 as $r \to \infty$, uniformly on $S^1 \times [0,1]$. Thus, fix a large enough value of r and define

$$H : S^1 \times [0,1] \to S^1$$

by the formula

$$H(z,t) = \frac{\widehat{H}_r(z,t)}{|\widehat{H}_r(z,t)|}.$$

Then $H(z,1) = g_r(z)$ and $H(z,0) = z^n, \forall\, z \in S^1$. Thus, $g_r \sim f_n$ and $\deg(g_r) = n > 0$. But g_r extends smoothly to $F_r : D^2 \to S^1$, a contradiction to Theorem 6.5.6. □

Lemma 6.5.8. *If $f, g : S^1 \to S^1$ are smooth, then*

$$\deg(f \circ g) = \deg(f)\deg(g).$$

Indeed, functoriality of cohomology implies that $a_{f\circ g} = a_f a_g$, so the lemma is immediate.

Corollary 6.5.9. *If $f, g : S^1 \to S^1$ are smooth, then $f \circ g$ and $g \circ f$ are homotopic.*

By Exercise 5.1.37, the Lie group structure on S^1 makes $\pi[M, S^1]$ into an abelian group and one obtains the following.

Theorem 6.5.10. *The map*

$$\chi : \pi[M, S^1] \to H^1(M; \mathbb{Z}),$$

defined by

$$\chi[f] = f^*[\widetilde{\eta}/2\pi],$$

is an isomorphism of groups.

Here, the integral cohomology $H^1(M;\mathbb{Z})$ is defined as in Exercise 6.4.18. The fact that $[\widetilde{\eta}/2\pi]$ is an integral class implies that $f^*[\widetilde{\eta}/2\pi]$ is also integral by that same exercise.

Exercise 6.5.11. Prove Theorem 6.5.10. Proceed as follows.

(1) Show that χ is a group homomorphism.
(2) Let $\omega \in Z^1(M)$ be an integral form. You are going to define a smooth map $f_\omega : M \to S^1$ such that $f_\omega^*(\widetilde{\eta}/2\pi) = \omega$. For this, no generality will be lost in assuming that M is connected (why?), so make that assumption and fix a basepoint $x_0 \in M$. For each $x \in M$, choose any piecewise smooth path

$$s : [a,b] \to M$$

such that $s(a) = x_0$ and $s(b) = x$, and show that

$$f_\omega(x) = \left(\int_s \omega \pmod{\mathbb{Z}}\right) \in \mathbb{R}/\mathbb{Z} = S^1$$

depends only on x (and x_0), not on the choice of path s.

(3) Prove that $f_\omega : M \to S^1$ is smooth and that its homotopy class is independent of the choice of basepoint x_0.

(4) Prove that $f_\omega^*(\widetilde{\eta}/2\pi) = \omega$. In particular, conclude that χ is surjective.

(5) Let $f : M \to S^1$ be such that $\omega = f^*(\widetilde{\eta}/2\pi)$ is an exact form. You are to prove that $f \sim 1$, so note that, again, no generality is lost in assuming that M is connected. In this case, show that f_ω, as defined in step (b), is actually well defined as a map $f_\omega : M \to \mathbb{R}$ and that there is a constant c such that $f = p \circ (f_\omega + c)$. Conclude that $f \sim 1$, hence that χ is one-to-one.

Exercise 6.5.12. Use degree theory to show that the group $\mathrm{Diff}(S^1)$ has exactly two isotopy classes. (Hint. An easy application of degree theory will show that there are *at least* two isotopy classes. The hard step is to show that, if $f \in \mathrm{Diff}(S^1)$ and $\deg(f) = 1$, then f is isotopic to f_1. It then follows fairly easily that there are *at most* two isotopy classes.)

CHAPTER 7

Multilinear Algebra and Tensors

Smooth functions, vector fields and 1-forms are *tensors* of fairly simple types. In order to handle higher order tensors, we will need some rather sophisticated multilinear algebra. The reader who is well grounded in the multilinear algebra of R–modules can skip ahead to Section 7.4, referring to the first three sections only as needed.

7.1. Tensor Algebra

We will be working in the category $\mathcal{M}(R)$ of R-modules and R–linear maps, where R is a fixed commutative ring with unity 1. In order to study R-multilinear maps, we build a *universal model* of multilinear objects called the *tensor algebra* over R. In the typical applications in this book, R will be either the real field $\mathbb{R}$ or the ring $C^\infty(M)$.

Definition 7.1.1. An R-module V is free if there is a subset $B \subset V$ such that every nonzero element $v \in V$ can be written uniquely as a finite R–linear combination of elements of B (terms with coefficient 0 being suppressed). The set B will be called a (free) basis of V.

If R is a field, every R module is free. Another example is the integer lattice $\mathbb{Z}^k$, a free $\mathbb{Z}$-module. At the other extreme, the abelian group $\mathbb{Z}_2$, when viewed as a $\mathbb{Z}$-module, is not free. A basis would have to contain $1 \in \mathbb{Z}_2$, but $0 \in \mathbb{Z}_2$ would then have infinitely many representations $a \cdot 1$, $a \in 2\mathbb{Z}$. The following example will be very important.

Example 7.1.2. Let $\pi : E \to M$ be an n-plane bundle. Then $\Gamma(E)$ is a free $C^\infty(M)$-module on a basis of n elements if E is trivial. Indeed, if $E \cong M \times \mathbb{R}^n$, let $\{e_1, \dots, e_n\}$ be the standard basis of $\mathbb{R}^n$, and define $s_i \in \Gamma(E)$ by the formula $s_i(x) = (x, e_i)$, $1 \le i \le n$. An arbitrary section $s(x) = (x, f_1(x), \dots, f_n(x))$ has the unique expression $s = \sum_{i=1}^n f_i s_i$.

Exercise 7.1.3. Suppose that $\pi : E \to M$ is an n-plane bundle and that $\Gamma(E)$ is a free $C^\infty(M)$-module with basis B. One easily checks that B must contain at least n elements. Using local triviality and the C^∞ Urysohn lemma, show that B has exactly n elements $s_1, \dots, s_n$ and that $s_1(x), \dots, s_n(x)$ form a basis of E_x, for each $x \in M$. Thus, E must be trivial.

Remarks. There are strong but limited analogies between vector spaces over a field and free R-modules. Here are some of the facts.

(1) If V is free on the basis B, then R–linear maps $\varphi : V \to W$ into arbitrary R-modules W correspond one-to-one to maps $\overline{\varphi} : B \to W$ of sets, the correspondence being $\overline{\varphi} = \varphi | B$.

(2) If V is a free R-module, it can be shown that any two bases of V have the same cardinality, called $\dim_R V$. For example, $\dim_{\mathbb{Z}} \mathbb{Z}^k = k$.
(3) On the other hand, there are important dissimilarities. A submodule $W \subset V$ of a free R-module can fail to be free and, even when the submodule W is free, it may have no basis that extends to a basis of V.

We give two examples illustrating Remark (3).

Example 7.1.4. If $M \subset \mathbb{R}^n$ is a nonparallelizable submanifold, then we have the canonical inclusion $T(M) \hookrightarrow M \times \mathbb{R}^n$ of the tangent bundle as a subbundle of the trivial bundle. Thus, $\mathfrak{X}(M) \subset \Gamma(M \times \mathbb{R}^n)$ is a $C^\infty(M)$-submodule. By Exercise 7.1.3, $\mathfrak{X}(M)$ is not free, but by Example 7.1.2, $\Gamma(M \times \mathbb{R}^n)$ is free.

Example 7.1.5. The submodule $2\mathbb{Z} \subset \mathbb{Z}$ is a free $\mathbb{Z}$-submodule of a free $\mathbb{Z}$-module. But there are two bases $\{2\}$ and $\{-2\}$ of $2\mathbb{Z}$, neither of which extends to a basis of $\mathbb{Z}$.

Modules will not be assumed free unless that is explicitly stated.

Definition 7.1.6. If V_1, V_2, V_3 are objects in $\mathcal{M}(R)$, a map

$$\varphi : V_1 \times V_2 \to V_3$$

is R–bilinear if

$$\varphi(\cdot, v_2) : V_1 \to V_3$$
$$\varphi(v_1, \cdot) : V_2 \to V_3$$

are R–linear, $\forall\, v_i \in V_i$, $i = 1, 2$.

Remark. For fixed choices of $V_1, V_2, V_3 \in \mathcal{M}(R)$, the set of R–bilinear maps $\varphi : V_1 \times V_2 \to V_3$ is itself an R-module under the pointwise operations.

Definition 7.1.7. If V_1 and V_2 are R-modules, their tensor product is an R–module $V_1 \otimes V_2$, together with an R–bilinear map

$$\otimes : V_1 \times V_2 \to V_1 \otimes V_2$$

with the following "universal property": given any R-module V_3 and any R–bilinear map $\varphi : V_1 \times V_2 \to V_3$, there is a unique R–linear map $\widetilde{\varphi}$ such that the diagram

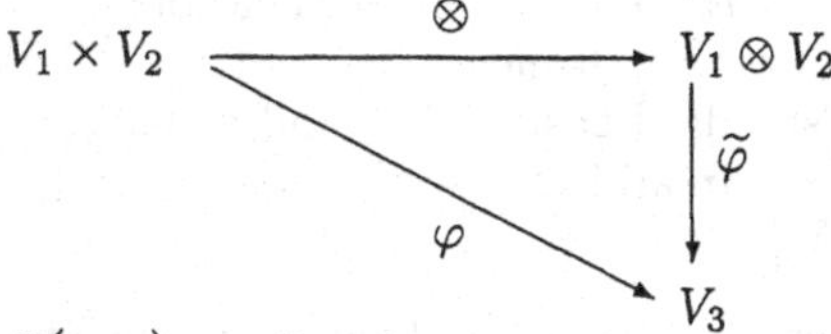

commutes. We write $\otimes(v, w) = v \otimes w$.

Thus, the R-module of R–bilinear maps $V_1 \times V_2 \to V_3$ is canonically isomorphic to the R-module $\mathrm{Hom}_R(V_1 \otimes V_2, V_3)$ of R–linear maps

$$V_1 \otimes V_2 \to V_3.$$

Theorem 7.1.8. *Given V_1, V_2 as above, a tensor product $V_1 \otimes V_2$ exists and is unique up to a unique isomorphism. That is, if*

$$\otimes : V_1 \times V_2 \to V_1 \otimes V_2,$$
$$\widetilde{\otimes} : V_1 \times V_2 \to V_1 \widetilde{\otimes} V_2$$

are two such tensor products, there is a unique isomorphism

$$\theta : V_1 \otimes V_2 \to V_1 \tilde{\otimes} V_2$$

of R-modules such that the diagram

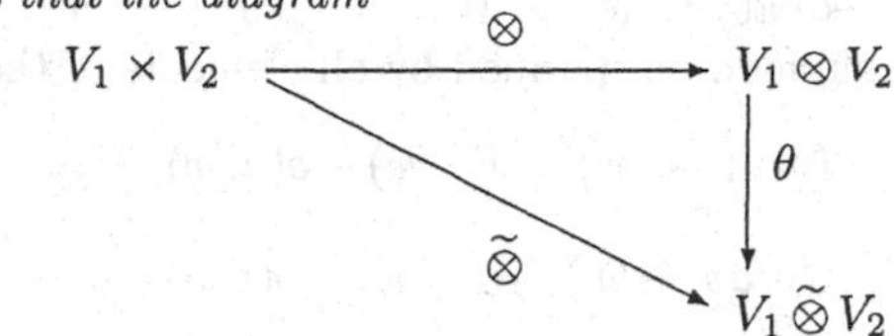

commutes.

Proof. First we prove uniqueness. If $\otimes$ and $\tilde{\otimes}$ are two tensor products, the universal property gives unique R-linear maps θ_1 and θ_2 making the following diagrams commute:

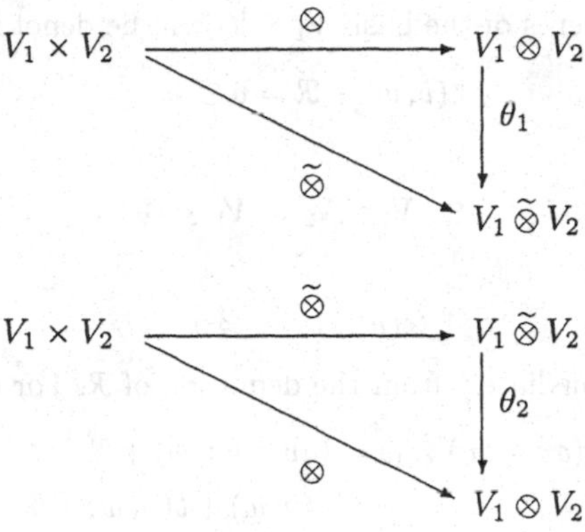

Then the diagram

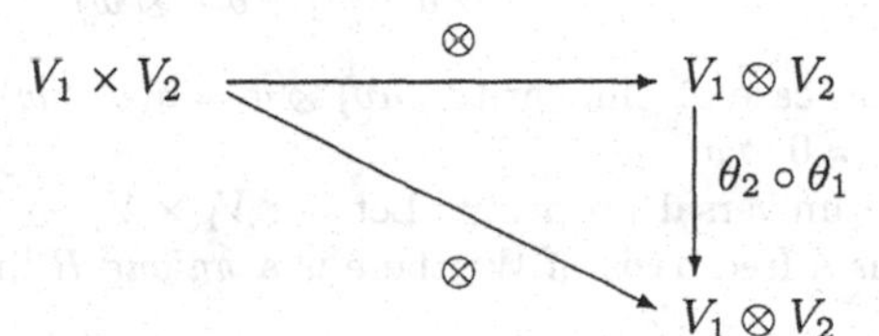

also commutes, as does

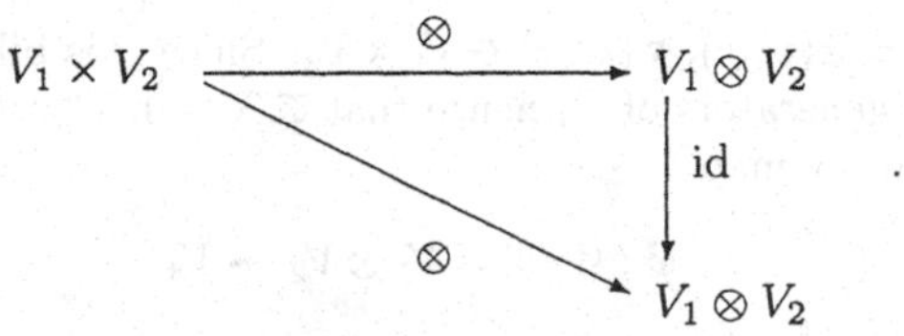

$$\begin{array}{ccc} V_1 \times V_2 & \xrightarrow{\ \otimes\ } & V_1 \otimes V_2 \\ & \searrow^{\otimes} & \downarrow \mathrm{id} \\ & & V_1 \otimes V_2 \end{array} .$$

By the universal property, we conclude that $\theta_2 \circ \theta_1 = \mathrm{id}$ and, similarly, that $\theta_1 \circ \theta_2 = \mathrm{id}$, so θ_1 and θ_2 are mutually inverse R-linear isomorphisms. Since θ_1 is unique, we are done.

The existence proof, though elementary, is a bit more long winded. Let W be the free R-module spanned by the *set* $V_1 \times V_2$. The module W is just the set of all formal linear combinations

$$\sum_{i=1}^{k} a_i(v_i, w_i)$$

where $a_i \in R$ and $(v_i, w_i) \in V_1 \times V_2$. This is an R-module under the obvious operations and each element $0 \neq w \in W$ is uniquely expressed as an R-linear combination of finitely many members of the basis $V_1 \times V_2$. Any linear combination with all coefficients 0 is equal to the $0 \in W$.

Let $\mathcal{R} \subseteq W$ be the submodule spanned by all elements of the form

$$(av + bu, w) - a(v, w) - b(u, w),$$

$$(v, aw + bu) - a(v, w) - b(v, u)$$

where $a, b \in R$ and u, v, w are in V_1 or V_2 appropriately. We think of $\mathcal{R}$ as the submodule of bilinear relations and set

$$V_1 \otimes V_2 = W/\mathcal{R}.$$

The cosets of the elements of the basis $V_1 \times V_2$ will be denoted by

$$(v, w) + \mathcal{R} = v \otimes w,$$

and we define

$$\otimes : V_1 \times V_2 \to V_1 \otimes V_2$$

by

$$\otimes(v, w) = v \otimes w.$$

Bilinearity follows immediately from the definition of $\mathcal{R}$. For example,

$$\begin{aligned}(av + bu) \otimes w &= (av + bu, w) + \mathcal{R} \\ &= a(v, w) + b(u, w) + \mathcal{R} \\ &= a(v \otimes w) + b(u \otimes w).\end{aligned}$$

Note that, as a special case of bilinearity, $(av) \otimes w = a(v \otimes w) = v \otimes (aw)$ and, in particular, $v \otimes 0 = 0 = 0 \otimes v$.

We establish the universal property. Let $\varphi : V_1 \times V_2 \to V_3$ be an R-bilinear map. Since $V_1 \times V_2$ is a free basis of W, there is a *unique* R-linear map

$$\overline{\varphi} : W \to V_3$$

such that $\overline{\varphi}(v, w) = \varphi(v, w)$, $\forall\, (v, w) \in V_1 \times V_2$. Since φ is bilinear, it follows that $\overline{\varphi}$ vanishes on the generators of $\mathcal{R}$, hence that $\overline{\varphi}|\mathcal{R} \equiv 0$. Consequently, $\overline{\varphi}$ passes to a well-defined R-linear map

$$\widetilde{\varphi} : W/\mathcal{R} = V_1 \otimes V_2 \to V_3$$

such that the diagram

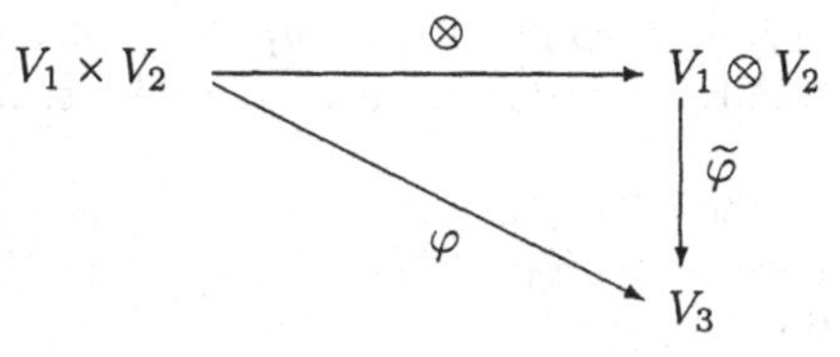

$$\begin{array}{ccc} V_1 \times V_2 & \xrightarrow{\otimes} & V_1 \otimes V_2 \\ & \searrow^{\varphi} & \downarrow \widetilde{\varphi} \\ & & V_3 \end{array}$$

commutes. Since $V_1 \otimes V_2$ is spanned by elements of the form $v \otimes w$, $\widetilde{\varphi}$ is unique. □

In a completely parallel way, one can consider R-trilinear maps and prove the existence and uniqueness of a universal R-trilinear map

$$V_1 \times V_2 \times V_3 \xrightarrow{\otimes} V_1 \otimes V_2 \otimes V_3$$

sending

$$(v_1, v_2, v_3) \mapsto v_1 \otimes v_2 \otimes v_3.$$

It is a trivial exercise to check that the composition

$$(V_1 \times V_2) \times V_3 \xrightarrow{\otimes \times \mathrm{id}_{V_3}} (V_1 \otimes V_2) \times V_3 \xrightarrow{\otimes} (V_1 \otimes V_2) \otimes V_3$$

also has the universal property, as does

$$V_1 \times (V_2 \times V_3) \xrightarrow{\mathrm{id}_{V_1} \times \otimes} V_1 \times (V_2 \otimes V_3) \xrightarrow{\otimes} V_1 \otimes (V_2 \otimes V_3).$$

Corollary 7.1.9. *If V_i is an R-module, $i = 1, 2, 3$, there are unique R-linear isomorphisms $V_1 \otimes (V_2 \otimes V_3) = (V_1 \otimes V_2) \otimes V_3 = V_1 \otimes V_2 \otimes V_3$ identifying $v_1 \otimes (v_2 \otimes v_3) = (v_1 \otimes v_2) \otimes v_3 = v_1 \otimes v_2 \otimes v_3$, $\forall v_i \in V_i$, $i = 1, 2, 3$.*

More generally, for each integer $k \geq 2$, there is a unique universal, k-linear map (over R)

$$V_1 \times V_2 \times \cdots V_k \xrightarrow{\otimes} V_1 \otimes V_2 \otimes \cdots \otimes V_k$$

and canonical identifications

$$V_1 \otimes (V_2 \otimes \cdots \otimes V_k) = (V_1 \otimes \cdots \otimes V_{k-1}) \otimes V_k = V_1 \otimes V_2 \otimes \cdots \otimes V_k.$$

An obvious induction shows that all groupings by parentheses are equivalent, so parentheses can be dropped or used selectively as desired.

Definition 7.1.10. An element $v \in V_1 \otimes \cdots \otimes V_k$ is decomposable if it can be written as a monomial $v = v_1 \otimes \cdots \otimes v_k$, for suitable elements $v_i \in V_i$, $1 \leq i \leq k$. Otherwise, v is indecomposable.

By the construction of the tensor product in the proof of Theorem 7.1.8, the decomposable elements span.

Lemma 7.1.11. *If V and W are free R-modules with respective bases A and B, then $V \otimes W$ is free with basis $C = \{a \otimes b \mid a \in A, b \in B\}$.*

Proof. An arbitrary element $v \in A \otimes B$ can be written as a linear combination of decomposables. A decomposable element $v \otimes w$ can be expanded, via the multilinearity of tensor product, to a linear combination of elements of C, proving that C spans $V \otimes W$. It remains for us to show that, if

$$\sum_{i,j=1}^{p,q} c_{ij} a_i \otimes b_j = \sum_{i,j=1}^{p,q} d_{ij} a_i \otimes b_j,$$

where $a_i \in A$ and $b_j \in B$, $1 \leq i \leq p$, $1 \leq j \leq q$ then all $c_{ij} = d_{ij}$. Subtracting one expression from the other, we only need to prove that

$$(*) \qquad \sum_{i,j=1}^{p,q} c_{ij} a_i \otimes b_j = 0$$

implies that all $c_{ij} = 0$.

The bilinear functionals $\varphi : V \times W \to R$ correspond one-to-one to arbitrary functions $f : A \times B \to R$. The correspondence is $\varphi \leftrightarrow \varphi|(A \times B)$. Thus, the linear functionals $\widetilde{\varphi} : V \otimes W \to R$ also correspond one-to-one to these functions

$f : A \times B \to R$. If $(a, b) \in A \times B$, let $f_{a,b} : A \times B \to R$ be the function taking the value 1 on (a, b) and the value 0 on every other element of $A \times B$. The corresponding linear functional will be denoted by $\widetilde{\varphi}_{a,b}$. Applying $\widetilde{\varphi}_{a_i,b_j}$ to equation $(*)$, we see that all $c_{ij} = 0$ as desired. □

By an obvious induction on the number of factors, this lemma generalizes to the following.

Corollary 7.1.12. *If $V_1, \dots, V_k$ are free R-modules having respective bases $B_1, \dots, B_k$, then $V_1 \otimes \cdots \otimes V_k$ is a free R-module with basis*

$$B = \{v_1 \otimes \cdots \otimes v_k \mid v_i \in B_i,\ 1 \le i \le k\}.$$

Proposition 7.1.13. *If $\lambda_i : V_i \to W_i$ is an R-linear map, $1 \le i \le k$, there is a unique R-linear map*

$$\lambda_1 \otimes \cdots \otimes \lambda_k : V_1 \otimes \cdots \otimes V_k \to W_1 \otimes \cdots \otimes W_k$$

that, on decomposable elements, has the formula

$$(\lambda_1 \otimes \cdots \otimes \lambda_k)(v_1 \otimes \cdots \otimes v_k) = \lambda_1(v_1) \otimes \cdots \otimes \lambda_k(v_k).$$

Proof. Since the decomposables span, uniqueness is immediate. For existence, define the multilinear map

$$\lambda : V_1 \times \cdots \times V_k \to W_1 \otimes \cdots \otimes W_k$$

by

$$\lambda(v_1, \dots, v_k) = \lambda_1(v_1) \otimes \cdots \otimes \lambda_k(v_k).$$

Then $\lambda_1 \otimes \cdots \otimes \lambda_k$ is defined to be the unique associated linear map. □

Definition 7.1.14. The dual V^* of an R-module V is $\mathrm{Hom}_R(V, R)$, the module of R-linear functionals.

Lemma 7.1.15. *If V has a finite free basis $\{v_1, \dots, v_n\}$, then V^* has a finite free basis $\{v_1^*, \dots, v_n^*\}$, called the dual basis and defined by*

$$v_i^*(v_j) = \delta_j^i, \quad 1 \le i, j \le n.$$

Proposition 7.1.16. *There is a unique R-linear map*

$$\iota : V_1^* \otimes \cdots \otimes V_k^* \to (V_1 \otimes \cdots \otimes V_k)^*$$

that, on decomposable elements, has the formula

$$\iota(\eta_1 \otimes \cdots \otimes \eta_k)(v_1 \otimes \cdots \otimes v_k) = \eta_1(v_1)\eta_2(v_2) \dots \eta_k(v_k).$$

If the R-modules V_i are all free on finite bases, then ι is a canonical isomorphism.

Proof. Uniqueness is immediate by the fact that decomposables span. For existence, define the multilinear functional

$$\theta : V_1^* \times \cdots \times V_k^* \times V_1 \times \cdots \times V_k \to R$$

by

$$\theta(\eta_1, \dots, \eta_k, v_1, \dots, v_k) = \eta_1(v_1)\eta_2(v_2) \dots \eta_k(v_k).$$

By the universal property, this gives the associated linear functional

$$\widetilde{\theta} : V_1^* \otimes \cdots \otimes V_k^* \otimes V_1 \otimes \cdots \otimes V_k \to R,$$

and we define

$$\iota : V_1^* \otimes \cdots \otimes V_k^* \to (V_1 \otimes \cdots \otimes V_k)^*$$

by

$$\iota(\eta)(v) = \widetilde{\theta}(\eta \otimes v).$$

If $\{v_{i,1}, \dots, v_{i,m_i}\}$ is a free basis of V_i, $1 \le i \le k$, let $\{v^*_{i,1}, \dots, v^*_{i,m_i}\}$ be the dual basis. Let B and B^* be the respective bases of $V_1 \otimes \cdots \otimes V_k$ and $V_1^* \otimes \cdots \otimes V_k^*$ given by Corollary 7.1.12. The formula

$$\iota(v^*_{1,j_1} \otimes \cdots \otimes v^*_{k,j_k})(v_{1,i_1} \otimes \cdots \otimes v_{k,i_k}) = \delta^{j_1}_{i_1} \dots \delta^{j_k}_{i_k} = \delta^{j_1 \dots j_k}_{i_1 \dots i_k}$$

shows that ι carries the basis B^* one-to-one onto the basis dual to B, so ι is an isomorphism. □

Let V be an R-module and view R as a module over itself.

Lemma 7.1.17. *Scalar multiplication*

$$R \times V \to V,$$
$$V \times R \to V$$

induces canonical isomorphisms $R \otimes V = V \otimes R = V$ *relative to which* $1 \otimes v = v \otimes 1 = v$.

Indeed, scalar multiplication is R-bilinear, so there are canonical R-linear maps

$$R \otimes V, \to V$$
$$V \otimes R \to V.$$

These are inverted by the R-linear maps

$$v \mapsto 1 \otimes v$$
$$v \mapsto v \otimes 1$$

respectively.

Definition 7.1.18. Let V be an R-module. For each integer $r \ge 0$, the rth tensor power of V is

$$\mathcal{T}^r(V) = \begin{cases} R, & r = 0, \\ V, & r = 1, \\ \underbrace{V \otimes \cdots \otimes V}_{r}, & r \ge 2. \end{cases}$$

Remark. By Lemma 7.1.17, $\mathcal{T}^0(V) \otimes \mathcal{T}^n(V) = \mathcal{T}^n(V) = \mathcal{T}^n(V) \otimes \mathcal{T}^0(V)$. When n and m are both positive, the identity $\mathcal{T}^n(V) \otimes \mathcal{T}^m(V) = \mathcal{T}^{n+m}(V)$ is given by the associativity of the tensor product.

Set $\mathcal{T}(V) = \{\mathcal{T}^r(V)\}_{r=0}^\infty$ and note that $\otimes$ defines an R-bilinear map

$$\mathcal{T}^n(V) \times \mathcal{T}^m(V) \xrightarrow{\otimes} \mathcal{T}^n(V) \otimes \mathcal{T}^m(V) = \mathcal{T}^{n+m}(V).$$

This makes $\mathcal{T}(V)$ into a *graded algebra* over R in the following sense.

Definition 7.1.19. A graded (associative) algebra A over R is a sequence $\{A^r\}_{r=0}^\infty$ of R-modules, together with R-bilinear maps (multiplication)

$$A^n \times A^m \xrightarrow{\cdot} A^{n+m}, \quad \forall n, m \ge 0,$$

(written $(a,b) \mapsto a \cdot b$ or, sometimes, $(a,b) \mapsto ab$) that is associative in the sense that the compositions

$$(A^n \times A^m) \times A^k \xrightarrow{\cdot \times \mathrm{id}} A^{n+m} \times A^k \xrightarrow{\cdot} A^{n+m+k}$$

$$A^n \times (A^m \times A^k) \xrightarrow{\mathrm{id} \times \cdot} A^n \times A^{m+k} \xrightarrow{\cdot} A^{n+m+k}$$

are equal, $\forall n, m, k \geq 0$.

Definition 7.1.20. The graded algebra A is connected if $A^0 = R$ and

$$A^0 \times A^m \xrightarrow{\cdot} A^m \xleftarrow{\cdot} A^m \times A^0$$

are equal to scalar multiplication, $\forall m \geq 0$.

Remark that a connected graded algebra has unity $1 \in R = A^0$.

Definition 7.1.21. If V is an R-module, then $\mathcal{T}(V)$, with multiplication $\otimes$, is called the tensor algebra of V.

It is clear that the tensor algebra $\mathcal{T}(V)$ is connected.

Definition 7.1.22. A homomorphism $\varphi : A \to B$ of graded R–algebras is a collection of R-linear maps $\varphi^n : A^n \to B^n$, $\forall n \geq 0$, such that the diagrams

$$\begin{array}{ccc} A^n \times A^m & \xrightarrow{\cdot} & A^{n+m} \\ {\scriptstyle \varphi^n \times \varphi^m}\downarrow & & \downarrow{\scriptstyle \varphi^{n+m}} \\ B^n \times B^m & \xrightarrow{\cdot} & B^{n+m} \end{array}$$

commute, $\forall n, m \geq 0$. The homomorphism φ is an isomorphism if φ^n is bijective, $\forall n \geq 0$.

Theorem 7.1.23. *If $\lambda : V \to W$ is an R-linear map, then there is a unique induced homomorphism $\mathcal{T}(\lambda) : \mathcal{T}(V) \to \mathcal{T}(W)$ of graded R-algebras such that $\mathcal{T}^0(\lambda) = \mathrm{id}_R$ and $\mathcal{T}^1(\lambda) = \lambda$. This homomorphism satisfies*

$$\mathcal{T}^n(\lambda)(v_1 \otimes v_2 \otimes \cdots \otimes v_n) = \lambda(v_1) \otimes \lambda(v_2) \otimes \cdots \otimes \lambda(v_n),$$

$\forall n \geq 2$, $\forall v_i \in V$, $1 \leq i \leq n$. Finally, this induced homomorphism makes $\mathcal{T}$ a covariant functor from the category of R-modules and R-linear maps to the category of graded algebras over R and graded algebra homomorphisms.

Proof. The formula on decomposable tensors is imposed by the requirement that $\mathcal{T}(\lambda)$ be a homomorphism of graded algebras, together with the stipulation that $\mathcal{T}^1(\lambda) = \lambda$. Existence and uniqueness of the linear maps $\mathcal{T}^n(\lambda)$ are given by Proposition 7.1.13. The fact that $\mathcal{T}(\lambda)$ preserves $\otimes$ multiplication is immediate. The final assertion amounts to the obvious identities

$$\mathcal{T}(\lambda \circ \mu) = \mathcal{T}(\lambda) \circ \mathcal{T}(\mu),$$
$$\mathcal{T}(\mathrm{id}_V) = \mathrm{id}_{\mathcal{T}(V)}.$$

□

The following is an elementary consequence of Corollary 7.1.12 and Proposition 7.1.16.

Theorem 7.1.24. *If V is a free R-module with basis $\{e_1, \ldots, e_m\}$ then*

$$\{e_{i_1} \otimes \cdots \otimes e_{i_k}\}_{1 \leq i_1, \ldots, i_k \leq m}$$

is a free basis of $\mathcal{T}^k(V)$ and $\mathcal{T}^k(V^*) = \mathcal{T}^k(V)^*$.

Remark. In particular, if V is a free R-module with $\dim_R V = m$, then

$$\dim_R \mathcal{T}^k(V) = \dim_R \mathcal{T}^k(V^*) = m^k.$$

Terminology. The established terminology about covariance and contravariance of tensors in geometry is inconsistent with the usage of "covariant" and "contravariant" in category theory. For later reference, here are the geometer's definitions.

Let V be a finite dimensional vector space over the field $\mathbb{F}$.

Definition 7.1.25. For each integer $r \geq 0$, $\mathcal{T}^r(V^*)$, viewed as the space of r-linear maps $V^r \to \mathbb{F}$, is called the space of covariant tensors on V of degree r and is denoted by $\mathcal{T}^r_0(V)$.

Definition 7.1.26. For each integer $s \geq 0$, $\mathcal{T}^s(V)$, viewed as the space of s-linear maps $(V^*)^s \to \mathbb{F}$, is called the space of contravariant tensors on V of degree s and is denoted by $\mathcal{T}^0_s(V)$.

Definition 7.1.27. The space of tensors on V of type (r, s) is the tensor product

$$\mathcal{T}^r_s(V) = \mathcal{T}^r_0(V) \otimes \mathcal{T}^0_s(V).$$

A tensor $\alpha \in \mathcal{T}^r_s(V)$ is said to have covariant degree r and contravariant degree s.

Obviously, $\mathcal{T}^r_s(V)$ is the space of $(r+s)$-linear maps

$$\underbrace{V \times \cdots \times V}_{r} \times \underbrace{V^* \times \cdots \times V^*}_{s} \to \mathbb{F}.$$

Exercise 7.1.28. Let V be an R-module, V^* its dual.

(1) Exhibit a canonical R-linear map $\alpha : V^* \otimes V \to R$.
(2) If V is free, prove that α is a surjection. If, in addition, V has a basis with one element, prove that α is a bijection.
(3) If V and W are R-modules (not necessarily free), exhibit a canonical R-linear map $\beta : V^* \otimes W \to \operatorname{Hom}_R(V, W)$.
(4) If R is a field, prove that β is injective. Do not assume that V and W are finite dimensional.
(5) If R is a field, prove that β is surjective if and only if either $\dim_R V < \infty$ or $\dim_R W < \infty$.

Exercise 7.1.29. Let A be a connected, graded R-algebra.

(1) Show that there is a unique homomorphism
$$\varphi : \mathcal{T}(A^1) \to A$$
of graded algebras such that $\varphi^0 = \mathrm{id}_R$ and $\varphi^1 = \mathrm{id}_{A^1}$.
(2) Define a suitable notion of graded 2-sided ideal $I \subseteq A$ so that $A/I = \{A^n/I^n\}_{n=0}^{\infty}$ is again a graded R-algebra.
(3) If A is generated, as a graded algebra, by A^1, show that there is a canonical ideal $I \subset \mathcal{T}(A^1)$, with $I^0 = \{0\} = I^1$, and a canonical isomorphism
$$\gamma : \mathcal{T}(A^1)/I \to A$$
such that $\gamma^0 = \mathrm{id}_R$ and $\gamma^1 = \mathrm{id}_{A^1}$.

7.2. Exterior Algebra

Let R be any commutative ring with unity 1 such that $\frac{1}{2} \in R$. That is, if $2 = 1 + 1 \in R$, then $\frac{1}{2} \in R$ has the property that $\frac{1}{2} \cdot 2 = 1$. In the case that $R = \mathbb{F}$ is a field, this means that the characteristic of $\mathbb{F}$ is not 2.

Lemma 7.2.1. *Let V be an R-module, $v \in V$. Then $v = -v \Leftrightarrow v = 0$.*

Proof. Evidently, $v = 0 \Rightarrow v = -v$. For the converse,

$$v = -v \Rightarrow 2v = 0 \Rightarrow v = \frac{1}{2}(2v) = \frac{1}{2}(0) = 0.$$

□

Let Σ_k be the group of permutations of $\{1, 2, \ldots, k\}$, a group of order $k!$.

Definition 7.2.2. The sign of $\sigma \in \Sigma_k$ is

$$(-1)^\sigma = \begin{cases} 1, & \sigma \text{ an even permutation,} \\ -1, & \sigma \text{ an odd permutation.} \end{cases}$$

Definition 7.2.3. Let V and W be R-modules. An antisymmetric k-linear map $\varphi : V^k \to W$ is a k-linear map such that

$$\varphi(v_{\sigma(1)}, \ldots, v_{\sigma(k)}) = (-1)^\sigma \varphi(v_1, \ldots, v_k),$$

$\forall v_1, \ldots, v_k \in V$, $\forall \sigma \in \Sigma_k$.

Remark that this definition will be useful only because $1 \neq -1$ in R.

As in the definition of tensor product, for each $k \geq 2$, define a universal antisymmetric k-linear map

$$\underbrace{V \times \cdots \times V}_{k} \xrightarrow{\wedge} \Lambda^k(V),$$

written

$$\wedge(v_1, \ldots, v_k) = v_1 \wedge \cdots \wedge v_k.$$

Here, the subspace $\mathcal{R}$ of relations is generated by the k-linear relations and all elements

$$(v_1, \ldots, v_k) - (-1)^\sigma (v_{\sigma(1)}, \ldots, v_{\sigma(k)}), \quad \sigma \in \Sigma_k.$$

Existence and uniqueness, up to unique isomorphism, are established exactly as for tensor product. One sets $\Lambda^0(V) = R$ and $\Lambda^1(V) = V$.

Definition 7.2.4. The R-module $\Lambda^k(V)$ is called the kth exterior power of V. We set $\Lambda(V) = \{\Lambda^k(V)\}_{k=0}^\infty$.

Definition 7.2.5. An element $\omega \in \Lambda^k(V)$ that can be expressed in the form $v_1 \wedge v_2 \wedge \cdots \wedge v_k$, where $v_i \in V$, $1 \leq i \leq k$, is said to be decomposable. Otherwise, ω is indecomposable.

It is clear that $\Lambda^k(V)$ is spanned by decomposable elements, but generally there are plenty of indecomposable elements as well. As before, R-linear maps $\lambda : V \to W$ will induce canonical R-linear maps

$$\Lambda^k(\lambda) : \Lambda^k(V) \to \Lambda^k(W)$$

such that $\Lambda^0(\lambda) = \mathrm{id}_R$, $\Lambda^1(\lambda) = \lambda$, and

$$\Lambda^k(\lambda)(v_1 \wedge \cdots \wedge v_k) = \lambda(v_1) \wedge \cdots \wedge \lambda(v_k).$$

We are going to define a bilinear, associative multiplication

$$\Lambda^p(V) \times \Lambda^q(V) \xrightarrow{\wedge} \Lambda^{p+q}(V)$$

that will satisfy

$$(v_1 \wedge \cdots \wedge v_p) \wedge (v_{p+1} \wedge \cdots \wedge v_{p+q}) = v_1 \wedge \cdots \wedge v_{p+q}.$$

By the above remarks, this will make Λ a covariant functor from the category of R-modules and R-linear maps to the category of graded algebras over R and graded algebra homomorphisms.

To define the algebra structure on $\Lambda(V)$, we must relate $\Lambda(V)$ more directly to $\mathfrak{T}(V)$.

Definition 7.2.6. The ideal $\mathfrak{A}(V) \subset \mathfrak{T}(V)$ is the 2-sided ideal generated by all elements in $\mathfrak{T}^2(V)$ of the form $v_1 \otimes v_2 + v_2 \otimes v_1$, $v_1, v_2 \in V$.

Thus,

$$\begin{aligned}
\mathfrak{A}^0(V) &= 0, \\
\mathfrak{A}^1(V) &= 0, \\
\mathfrak{A}^2(V) &= \operatorname{span}\{v_1 \otimes v_2 + v_2 \otimes v_1 \mid v_1, v_2 \in V\}, \\
&\vdots \\
\mathfrak{A}^k(V) &= \operatorname{span} \bigcup_{p+q=k-2} \mathfrak{T}^p(V) \otimes \mathfrak{A}^2(V) \otimes \mathfrak{T}^q(V), \\
&\vdots
\end{aligned}$$

Let $\varphi : V^k \to W$ be an antisymmetric k-linear map. As a k-linear map, φ can be interpreted as a linear map which, for clarity, we denote by $\widetilde{\varphi} : \mathfrak{T}^k(V) \to W$.

Lemma 7.2.7. *If $\varphi : V^k \to W$ is antisymmetric, then $\widetilde{\varphi}(\mathfrak{A}^k(V)) = \{0\}$.*

Proof. It will be enough to show that $\widetilde{\varphi}$ vanishes on a set spanning $\mathfrak{A}^k(V)$. Thus, if $w \in \mathfrak{T}^p(V)$, $u \in \mathfrak{T}^q(V)$, $p + q = k - 2$, and $v_1, v_2 \in V$, we will show that

$$\widetilde{\varphi}(w \otimes (v_1 \otimes v_2 + v_2 \otimes v_1) \otimes u) = 0.$$

But the antisymmetry of φ implies that

$$\widetilde{\varphi}(w \otimes v_1 \otimes v_2 \otimes u) = -\widetilde{\varphi}(w \otimes v_2 \otimes v_1 \otimes u),$$

and the assertion follows from linearity. □

Corollary 7.2.8. *There is a canonical isomorphism*

$$\Lambda^k(V) = \mathfrak{T}^k(V)/\mathfrak{A}^k(V)$$

of R-modules, $\forall\, k \geq 0$. Thus, $\Lambda(V)$ is a graded algebra over R (the exterior algebra *of V) and Λ is a covariant functor from R-modules to graded R-algebras.*

Proof. For $k = 0, 1$, it is clear that $\Lambda^k(V) = \mathfrak{T}^k(V)/\mathfrak{A}^k(V)$. For each $k \geq 2$, consider the k-linear map

$$V^k \xrightarrow{\otimes} \mathfrak{T}^k(V) \xrightarrow{\pi} \mathfrak{T}^k(V)/\mathfrak{A}^k(V),$$

where π is the quotient projection. The reader will check easily that this is antisymmetric. Given an arbitrary antisymmetric k-linear map $\varphi : V^k \to W$, we obtain the commutative triangle

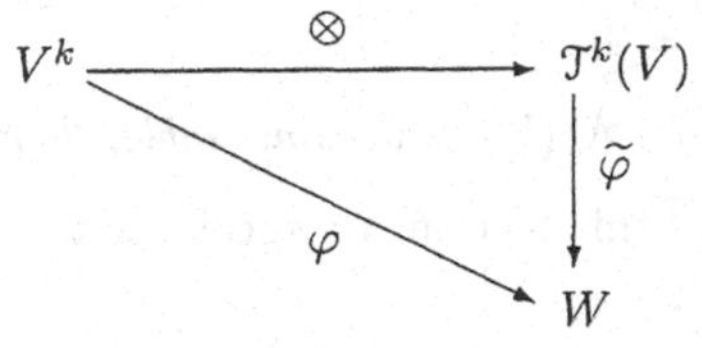

and $\widetilde{\varphi}|\mathfrak{A}^k(V) \equiv 0$. Thus, $\widetilde{\varphi}$ induces $\overline{\varphi} : \mathfrak{T}^k(V)/\mathfrak{A}^k(V) \to W$ making the following diagram commutative:

$$\begin{array}{ccccc} V^k & \xrightarrow{\otimes} & \mathfrak{T}^k(V) & \xrightarrow{\pi} & \mathfrak{T}^k(V)/\mathfrak{A}^k(V) \\ {\scriptstyle\varphi}\downarrow & & \downarrow{\scriptstyle\widetilde{\varphi}} & & \downarrow{\scriptstyle\overline{\varphi}} \\ W & \xrightarrow[\text{id}]{} & W & \xrightarrow[\text{id}]{} & W \end{array} .$$

That is, the triangle

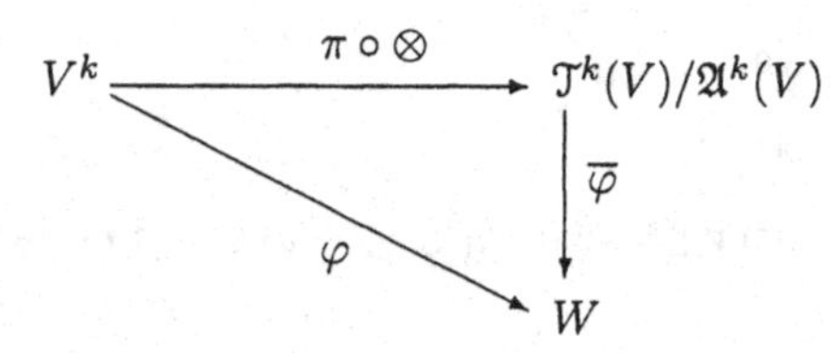

commutes. Since $\mathfrak{T}^k(V)/\mathfrak{A}^k(V)$ is spanned by

$$\{\pi(v_1 \otimes \cdots \otimes v_k) \mid v_1, \dots, v_k \in V\}$$

and commutativity of the diagrams forces

$$\overline{\varphi}(\pi(v_1 \otimes \cdots \otimes v_k)) = \varphi(v_1, \dots, v_k),$$

we see that $\overline{\varphi}$ is the *only* linear map making the triangle commute. That is, $\pi \circ \otimes : V^k \to \mathfrak{T}^k(V)/\mathfrak{A}^k(V)$ has the universal property for antisymmetric, k-linear maps, hence is uniquely identified with $\wedge : V^k \to \Lambda^k(V)$. Finally, this identification makes $v_1 \wedge \cdots \wedge v_k = v_1 \otimes \cdots \otimes v_k + \mathfrak{A}^k(V)$, and all assertions follow. □

Definition 7.2.9. A graded algebra A is anticommutative if

$$\alpha \in A^k \text{ and } \beta \in A^r \Rightarrow \alpha\beta = (-1)^{kr}\beta\alpha.$$

Corollary 7.2.10. *The graded algebra $\Lambda(V)$ is anticommutative.*

Proof. It is enough to verify Definition 7.2.9 for decomposable elements of $\Lambda^k(V)$ and $\Lambda^r(V)$. But that case is an elementary consequence of the case $k = r = 1$, and this latter case is given by

$$\begin{aligned} v \wedge w &= v \otimes w + \mathfrak{A}^2(V) \\ &= -w \otimes v + \mathfrak{A}^2(V) \\ &= -w \wedge v, \end{aligned}$$

$\forall\, v, w \in V$. □

Corollary 7.2.11. *If $\omega \in \Lambda^{2r+1}(V)$, then $\omega \wedge \omega = 0$.*

Proof. Indeed,

$$\omega \wedge \omega = (-1)^{(2r+1)(2r+1)}\omega \wedge \omega = -\omega \wedge \omega.$$

By Lemma 7.2.1, $\omega \wedge \omega = 0$. □

Corollary 7.2.12. *If $\omega \in \Lambda^k(V)$ is decomposable, then $\omega \wedge \omega = 0$.*

For the remainder of this section, we specialize to the case in which V is a free R-module on a basis $\{e_1, \dots, e_m\}$.

Lemma 7.2.13. *If V is as above, then*

$$\{e_{i_1} \wedge e_{i_2} \wedge \cdots \wedge e_{i_k}\}_{1 \le i_1 < i_2 < \cdots < i_k \le m}$$

is a free basis of $\Lambda^k(V)$, $k \ge 2$. In particular,

$$\dim_R \Lambda^k(V) = \binom{m}{k} = \frac{m!}{k!(m-k)!}$$

and, if $k > m$, $\Lambda^k(V) = \{0\}$.

Proof. The basis

$$\{e_{j_1} \otimes \cdots \otimes e_{j_k}\}_{1 \le j_1, \dots, j_k \le m}$$

of $\mathfrak{T}^k(V)$ projects to a spanning set

$$\{e_{j_1} \wedge \cdots \wedge e_{j_k}\}_{1 \le j_1, \dots, j_k \le m}$$

of $\Lambda^k(V)$. If some $j_r = j_q$, $r \ne q$, then $e_{j_1} \wedge \cdots \wedge e_{j_k} = 0$. If all j_i are distinct, $1 \le i \le k$, choose $\sigma \in \Sigma_k$ such that

$$j_{\sigma(1)} < j_{\sigma(2)} < \cdots < j_{\sigma(k)}.$$

Letting $i_r = j_{\sigma(r)}$, $1 \le r \le k$, we get

$$e_{i_1} \wedge \cdots \wedge e_{i_k} = (-1)^\sigma e_{j_1} \wedge \cdots \wedge e_{j_k}.$$

Thus, the set

$$B = \{e_{i_1} \wedge \cdots \wedge e_{i_k}\}_{1 \le i_1 < i_2 < \cdots < i_k \le m}$$

spans $\Lambda^k(V)$. We must prove that B is free.

For $1 \le i_1 < i_2 < \cdots < i_k \le m$, define an antisymmetric k-linear map $\varphi_{i_1 i_2 \cdots i_k} : V^k \to R$ by stipulating that

$$\varphi_{i_1 i_2 \cdots i_k}(e_{j_1}, e_{j_2}, \dots, e_{j_k}) = \delta^{j_1 j_2 \cdots j_k}_{i_1 i_2 \cdots i_k}$$

whenever $1 \le j_1 < j_2 < \cdots < j_k \le m$. By antisymmetry and k–multilinearity, this uniquely determines the desired map. As a linear map,

$$\varphi_{i_1 i_2 \cdots i_k} : \Lambda^k(V) \to R$$

has the property that

$$\varphi_{i_1 i_2 \cdots i_k}(e_{j_1} \wedge e_{j_2} \wedge \cdots \wedge e_{j_k}) = \delta^{j_1 j_2 \cdots j_k}_{i_1 i_2 \cdots i_k}.$$

This proves that the spanning set B is free and provides, simultaneously, the dual basis

$$B^* = \{\varphi_{i_1 i_2 \cdots i_k}\}_{1 \le i_1 < i_2 < \cdots < i_k \le m}$$

of $\Lambda^k(V)^*$. □

Corollary 7.2.14. $\dim_R \Lambda^k(V) = \dim_R \Lambda^{m-k}(V)$, $0 \le k \le m$.

We emphasize that

$$\dim_R \Lambda^m(V) = \dim_R \Lambda^0(V) = 1$$

and, if $\{e_1, \dots, e_m\}$ is a basis of V, then $e_1 \wedge \cdots \wedge e_m$ constitutes a basis of $\Lambda^m(V)$.

Exercise 7.2.15. If R is a field and $\dim_R V = n$, prove that every element of $\Lambda^{n-1}(V)$ is decomposable. (Hint: Given $0 \ne \alpha \in \Lambda^{n-1}(V)$, consider the linear map $\alpha\wedge : V \to \Lambda^n(V)$.)

Remark. If $\lambda : V \to V$ is linear, then $\det(\lambda) \in R$ is well defined. Indeed, if A, B are $m \times m$ matrices over R that represent λ relative to two choices of basis, then there is an invertible matrix P over R such that $A = PBP^{-1}$, so $\det(A) = \det(B)$.

Lemma 7.2.16. *If $\lambda : V \to V$ is linear, then $\Lambda^m(\lambda) : \Lambda^m(V) \to \Lambda^m(V)$ is multiplication by* $\det(\lambda)$.

Proof. Relative to a basis $\{e_1, \dots, e_m\}$ of V, write

$$\lambda(e_i) = \sum_{j=1}^{m} a_i^j e_j, \quad 1 \le i \le m.$$

Then,

$$\begin{aligned}
\Lambda^m(\lambda)(e_1 \wedge \cdots \wedge e_m) &= \lambda(e_1) \wedge \cdots \wedge \lambda(e_m) \\
&= \left(\sum_{j=1}^{m} a_1^j e_j\right) \wedge \cdots \wedge \left(\sum_{j=1}^{m} a_m^j e_j\right) \\
&= \sum_{1 \le j_1, \dots, j_m \le m} a_1^{j_1} \cdots a_m^{j_m} e_{j_1} \wedge \cdots \wedge e_{j_m}.
\end{aligned}$$

Any term with a repeated j index vanishes. If $J = (j_1, j_2, \dots, j_m)$ contains no repetitions, there is a unique permutation $\sigma_J \in \Sigma_m$ such that

$$j_{\sigma_J(r)} = r, \quad 1 \le r \le m.$$

Thus,

$$\begin{aligned}
\Lambda^m(\lambda)(e_1 \wedge \cdots \wedge e_m) &= \left(\sum_{\sigma \in \Sigma_m} (-1)^\sigma a^1_{\sigma(1)} \cdots a^m_{\sigma(m)}\right) e_1 \wedge \cdots \wedge e_m \\
&= \det(\lambda) e_1 \wedge \cdots \wedge e_m.
\end{aligned}$$

□

Lemma 7.2.17. *If R is a field, a set of vectors $w_1, \dots, w_k \in V$, $k \ge 2$, is linearly independent if and only if $w_1 \wedge \cdots \wedge w_k \ne 0$.*

Proof. If the set is dependent, the existence of inverses in R allows us to assume, without loss of generality, that

$$w_1 = \sum_{i=2}^{k} a_i w_i.$$

Then,

$$w_1 \wedge \cdots \wedge w_k = \sum_{i=2}^{k} a_i w_i \wedge w_2 \wedge \cdots \wedge w_k = 0.$$

Conversely, if the set is linearly independent, extend it to a basis by suitable choices of $w_{k+1}, \dots, w_m \in V$. Then, $w_1 \wedge \cdots \wedge w_k \wedge \cdots \wedge w_m$ is a basis of the one-dimensional space $\Lambda^m(V)$, hence is not 0. □

In an obvious sense, one can view $\Lambda(V)$ and $\mathcal{T}(V)$ as *graded R-modules* by ignoring multiplication. It will be convenient to construct a (graded) R-linear map $A : \Lambda(V) \to \mathcal{T}(V)$. That is,

$$A = \{A^k : \Lambda^k(V) \to \mathcal{T}^k(V)\}_{k=0}^{\infty}.$$

We emphasize that this will not be a homomorphism of graded algebras.

Define an antisymmetric, k-linear map $A^k : V^k \to \mathcal{T}^k(V)$ by

$$A^k(v_1, \dots, v_k) = \sum_{\sigma \in \Sigma_k} (-1)^\sigma v_{\sigma(1)} \otimes \cdots \otimes v_{\sigma(k)}.$$

By the universal property of the exterior powers, we view this as a linear map

$$A^k : \Lambda^k(V) \to \mathcal{T}^k(V).$$

The sequence $A = \{A^k\}_{0 \le k < \infty}$ is a linear map of graded R-modules

$$A : \Lambda(V) \to \mathcal{T}(V).$$

Generally, A^k is not injective, but the next two lemmas give conditions that insure injectivity.

Lemma 7.2.18. *If V is a free R-module on a finite basis, then each A^k is one-to-one, hence $A : \Lambda(V) \hookrightarrow \mathcal{T}(V)$ is a canonical graded linear imbedding.*

Proof. Let $\{e_1, \dots, e_m\} \subset V$ be a basis and consider the basis

$$\{e_{i_1} \wedge \cdots \wedge e_{i_k}\}_{1 \le i_1 < \cdots < i_k \le m}$$

of $\Lambda^k(V)$. Let $\{e_1^*, \dots, e_k^*\} \subset V^*$ be the dual basis. Since $\mathcal{T}^k(V^*) = \mathcal{T}^k(V)^*$ (Theorem 7.1.24), we obtain a subset

$$\{e_{j_1}^* \otimes \cdots \otimes e_{j_k}^*\}_{1 \le j_1 < \cdots < j_k \le m} \subset \mathcal{T}^k(V)^*,$$

that is part of a free basis. Then, since $j_1 < \cdots < j_k$ and $i_1 < \cdots < i_k$,

$$\begin{aligned}
&(e_{j_1}^* \otimes \cdots \otimes e_{j_k}^*)(A^k(e_{i_1} \wedge \cdots \wedge e_{i_k})) \\
&= (e_{j_1}^* \otimes \cdots \otimes e_{j_k}^*)\left(\sum_{\sigma \in \Sigma_k} (-1)^\sigma e_{i_{\sigma(1)}} \otimes \cdots \otimes e_{i_{\sigma(k)}}\right) \\
&= (e_{j_1}^* \otimes \cdots \otimes e_{j_k}^*)(e_{i_1} \otimes \cdots \otimes e_{i_k}) \\
&= \delta_{i_1 \cdots i_k}^{j_1 \cdots j_k}
\end{aligned}$$

and the assertion follows. □

Lemma 7.2.19. *If $1/k! \in R$, then A^k is one-to-one for every R-module V. In particular, if the rational field is imbedded as a subring $\mathbb{Q} \subseteq R$ containing the unity, then*

$$A : \Lambda(V) \hookrightarrow \mathcal{T}(V)$$

is a canonical graded linear imbedding.

Proof. First, consider A^k as a multilinear map on V^k and define $\alpha = (1/k!)A^k$. Denote by $\mathcal{A}^k(V)$ the submodule of $\mathcal{T}^k(V)$ spanned by the image of α. Of course, this submodule is also spanned by the image of A^k. We will show that the antisymmetric, k-linear map $\alpha : V^k \to \mathcal{A}^k(V)$ has the universal property, hence is canonically the same as $\wedge : V^k \to \Lambda^k(V)$. By universality, the corresponding linear map $\alpha : \Lambda^k(V) \to \mathcal{A}^k(V) = \Lambda^k(V)$ is the identity. In particular, as a linear map, $A^k = k!\alpha$ is injective.

We verify the universal property. Let $\varphi : V^k \to W$ be an arbitrary antisymmetric, k-linear map. Since it is k-linear, φ induces a unique linear map $\widetilde{\varphi} : \mathcal{T}^k(V) \to W$ and we restrict this linear map to the submodule $\mathcal{A}^k(V)$. The diagram

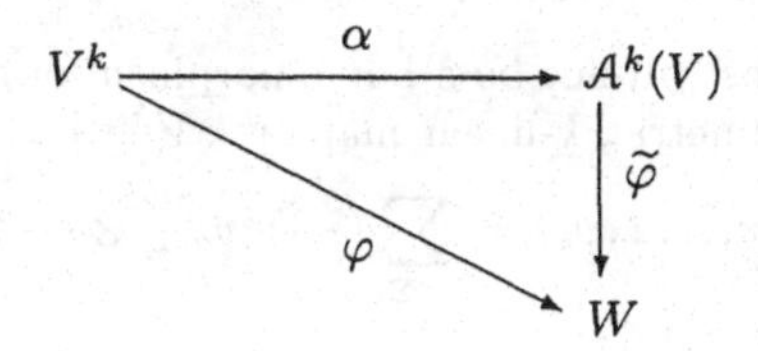

commutes. Indeed,

$$\begin{aligned}\widetilde{\varphi}(\alpha(v_1,\dots,v_k)) &= \widetilde{\varphi}\Big(\frac{1}{k!}\sum_{\sigma\in\Sigma_k}(-1)^\sigma v_{\sigma(1)}\otimes\cdots\otimes v_{\sigma(k)}\Big)\\ &= \frac{1}{k!}\sum_{\sigma\in\Sigma_k}(-1)^\sigma\varphi(v_{\sigma(1)},\dots,v_{\sigma(k)})\\ &= \frac{1}{k!}\sum_{\sigma\in\Sigma_k}(-1)^\sigma(-1)^\sigma\varphi(v_1,\dots,v_k)\\ &= \varphi(v_1,\dots,v_k),\end{aligned}$$

where the second equality is by the universal property of $\mathcal{T}^k(V)$ and the third by the antisymmetry of φ. This commutativity forces the definition of $\widetilde{\varphi}$ on the spanning set $\operatorname{im}(\alpha)$, so $\widetilde{\varphi}$ is the unique linear map making the diagram commute. □

Since A is canonical, we will generally suppress it from the notation, writing

$$\Lambda(V)\subset\mathcal{T}(V)$$

$$v_1\wedge\cdots\wedge v_k = \sum_{\sigma\in\Sigma_k}(-1)^\sigma v_{\sigma(1)}\otimes\cdots\otimes v_{\sigma(k)},$$

provided either that V is free and finite dimensional or $\mathbb{Q}\subseteq R$.

Theorem 7.2.20. *If V is a free R-module on a finite basis, there is a canonical isomorphism $\Lambda^k(V^*) = \Lambda^k(V)^*$.*

Proof. By Lemma 7.2.18 and Theorem 7.1.24,

$$\Lambda^k(V^*)\subset\mathcal{T}^k(V^*) = \mathcal{T}^k(V)^*,$$

so $\Lambda^k(V^*)$ can be viewed as a space of k-linear maps $V^k\to R$.

We prove first that each $\omega\in\Lambda^k(V^*)$, interpreted as $\omega : V^k\to R$, is antisymmetric. Indeed, let $\{e_1^*,\dots,e_m^*\}\subset V^*$ be a basis and suppose that $\omega = e_{i_1}^*\wedge\cdots\wedge e_{i_k}^*$, $1\le i_1<\cdots<i_k\le m$. Then,

$$\begin{aligned}\omega(v_1,\dots,v_k) &= \sum_{\sigma\in\Sigma_k}(-1)^\sigma(e_{i_{\sigma(1)}}^*\otimes\cdots\otimes e_{i_{\sigma(k)}}^*)(v_1,\dots,v_k)\\ &= \sum_{\sigma\in\Sigma_k}(-1)^\sigma e_{i_{\sigma(1)}}^*(v_1)\cdots e_{i_{\sigma(k)}}^*(v_k).\end{aligned}$$

That is,

$$\omega(v_1,\dots,v_k) = \det[e_{i_j}^*(v_\ell)].$$

Thus, if $\tau\in\Sigma_k$,

$$\omega(v_{\tau(1)},\dots,v_{\tau(k)}) = (-1)^\tau\omega(v_1,\dots,v_k),$$

proving that $\omega = e_{i_1}^*\wedge\cdots\wedge e_{i_k}^*$ is antisymmetric, $1\le i_1<\cdots<i_k\le m$. These monomials range over a basis of $\Lambda^k(V^*)$, so every element ω of this space is antisymmetric as a k-linear map $\omega : V^k\to R$.

Consequently, we can view $\omega \in \Lambda^k(V^*)$ as defining a linear map

$$\overline{\omega} : \Lambda^k(V) \to R.$$

That is, $\overline{\omega} \in \Lambda^k(V)^*$ and, if

$$1 \leq i_1 < \cdots < i_k \leq m,$$
$$1 \leq j_1 < \cdots < j_k \leq m,$$

then

$$\overline{e^*_{i_1} \wedge \cdots \wedge e^*_{i_k}}(e_{j_1} \wedge \cdots \wedge e_{j_k}) = \delta^{i_1 \cdots i_k}_{j_1 \cdots j_k}.$$

It follows that

$$\overline{e^*_{i_1} \wedge \cdots \wedge e^*_{i_k}} = (e_{i_1} \wedge \cdots \wedge e_{i_k})^*,$$

hence that the map $\omega \mapsto \overline{\omega}$ carries $\Lambda^k(V^*)$ isomorphically onto $\Lambda^k(V)^*$. □

Exercise 7.2.21. If R is a field, $\dim_R V = n$, and $0 \neq \alpha \in \Lambda^2(V)$, let $r \geq 1$ be the integer such that the r–fold exterior power $\alpha \wedge \cdots \wedge \alpha \neq 0$, but the $(r+1)$–fold power $= 0$. This integer is called the *rank* of α. Show that there exists a basis $\{v_1, \ldots, v_n\}$ of V so that

$$\alpha = v_1 \wedge v_2 + v_3 \wedge v_4 + \cdots + v_{2r-1} \wedge v_{2r}.$$

(Hint: Proceed by induction on r. For the inductive step, show that there is always a basis $\{w_1, \ldots, w_n\}$ such that $\alpha = w_1 \wedge w_2 + \alpha'$ where α' is a linear combination of terms $w_i \wedge w_j$ with $3 \leq i < j \leq n$.)

Exercise 7.2.22. Recall the Grassmann manifold $G_{n,k}$ of k-planes in $\mathbb{R}^n$ (Exercise 5.4.14). If V is a vector space, one defines similarly the Grassmann manifold $G_k(V)$ of k-dimensional subspaces of V. We use this notation in what follows.

(1) Using exterior algebra, define a canonical imbedding (of sets)

$$i^m_k : G_k(R^m) \hookrightarrow G_1(\Lambda^k(\mathbb{R}^m)).$$

(Hint: Consider the decomposable elements of $\Lambda^k(\mathbb{R}^m)$).

(2) Exhibit a natural linear (hence, smooth) group action

$$\mathrm{Gl}(m) \times \Lambda^k(\mathbb{R}^m) \to \Lambda^k(\mathbb{R}^m).$$

(3) Using the above, exhibit a smooth action

$$\mathrm{Gl}(m) \times G_1(\Lambda^k(\mathbb{R}^m)) \to G_1(\Lambda^k(\mathbb{R}^m)).$$

(4) Let $x_0 = \mathrm{span}\{e_1 \wedge e_2 \wedge \cdots \wedge e_k\} \in G_1(\Lambda^k(\mathbb{R}^m))$ and show that, relative to the above action, the isotropy group of x_0 is

$$\mathrm{Gl}(m)_{x_0} = \mathrm{Gl}(k, m-k).$$

(5) Show that $i^m_k(G_k(\mathbb{R}^m)) = \mathrm{Gl}(m) \cdot x_0$, hence $G_k(\mathbb{R}^m)$ is expressed as the homogeneous space $\mathrm{Gl}(m)/\mathrm{Gl}(k, m-k)$.

(6) Finally, restrict the above action to the orthogonal group

$$\mathrm{O}(m) \times G_1(\Lambda^k(\mathbb{R}^m)) \to G_1(\Lambda^k(\mathbb{R}^m))$$

and prove that this has exactly the same orbits as $\mathrm{Gl}(m)$. This gives $G_k(\mathbb{R}^m)$ as a homogeneous space of $\mathrm{O}(m)$.

Exercise 7.2.23. If V is a free R-module on a finite basis and $v \in V$, define the *interior product*

$$i_v : \Lambda^k(V^*) \to \Lambda^{k-1}(V^*)$$

as follows. Viewing $\omega \in \Lambda^k(V^*)$ as an antisymmetric k-linear map

$$\omega : V^k \to R,$$

let $i_v(\omega)$ be the antisymmetric $(k-1)$– linear map defined by the formula

$$i_v(\omega)(v_1, \dots, v_{k-1}) = \omega(v, v_1, \dots, v_{k-1}).$$

If $\omega \in \Lambda^p(V^*)$ and $\eta \in \Lambda^q(V^*)$, prove that

$$i_v(\omega \wedge \eta) = i_v(\omega) \wedge \eta + (-1)^p \omega \wedge i_v(\eta).$$

7.3. Symmetric Algebra

This will be an abbreviated treatment, not because the subject is unimportant, but because the ideas and proofs are so analogous to those for exterior algebra. There are, however, notable differences.

Again, our initial hypothesis is that V is a module over a commutative ring R with unity.

Definition 7.3.1. A k-linear map $\varphi : V^k \to W$ is symmetric if, for each $\sigma \in \Sigma_k$,

$$\varphi(v_{\sigma(1)}, \dots, v_{\sigma(k)}) = \varphi(v_1, \dots, v_k),$$

$\forall v_1, \dots, v_k \in V$.

In the usual way, we build a universal, symmetric, k-linear map

$$V^k \xrightarrow{\cdot} \mathcal{S}^k(V),$$

usually written with the dots suppressed:

$$(v_1, v_2, \dots, v_k) \mapsto v_1 v_2 \cdots v_k.$$

Definition 7.3.2. The space $\mathcal{S}^k(V)$ is called the kth symmetric power of V, where, as usual, $\mathcal{S}^0(V) = R$ and $\mathcal{S}^1(V) = V$.

We define the graded, 2-sided ideal $\mathfrak{S}(V) \subset \mathcal{T}(V)$, generated by all

$$v_1 \otimes v_2 - v_2 \otimes v_1 \in \mathcal{T}^2(V)$$

and obtain

Proposition 7.3.3. *There is a canonical isomorphism*

$$\mathcal{S}^k(V) = \mathcal{T}^k(V)/\mathfrak{S}^k(V)$$

of graded R-modules. The connected, graded algebra $\mathcal{S}(V) = \{\mathcal{S}^k(V)\}_{k=0}^{\infty}$, with multiplication "$\cdot$", is called the symmetric algebra of V.

Remark that, if $\alpha \in \mathcal{S}^p(V)$ and $\beta \in \mathcal{S}^q(V)$, then

$$\alpha\beta = \beta\alpha \in \mathcal{S}^{p+q}(V).$$

If $\lambda : V \to W$ is R-linear, there is induced a homomorphism

$$\mathcal{S}(\lambda) : \mathcal{S}(V) \to \mathcal{S}(W)$$

of graded algebras such that

$$\mathcal{S}^k(\lambda)(v_1 v_2 \cdots v_k) = \lambda(v_1)\lambda(v_2) \cdots \lambda(v_k).$$

Once again, $\mathcal{S}$ is a covariant functor.

Specializing to the case in which V is a free R-module on a finite basis, we obtain

Lemma 7.3.4. *If $\{e_1, \dots, e_m\}$ is a basis of V, then*

$$\{e_{i_1}e_{i_2}\cdots e_{i_k}\}_{1\le i_1\le i_2\le\cdots\le i_k\le m}$$

is a basis of $\mathcal{S}^k(V)$, $k \geq 2$.

Remark. If the space V is nontrivial, so is $\mathcal{S}^k(V)$, $\forall k \geq 0$, in strong contradistinction to the fact that $\Lambda^k(V) = \{0\}$, $\forall k > m$.

As for exterior algebras, we define a canonical map

$$S^k : \mathcal{S}^k(V) \to \mathcal{T}^k(V),$$

this being the linear map defined by the symmetric, k-linear map

$$(v_1, \dots, v_k) \mapsto \sum_{\sigma\in\Sigma_k} v_{\sigma(1)} \otimes \cdots \otimes v_{\sigma_k}.$$

Exercise 7.3.5. If V is free on a finite basis or if $\mathbb{Q} \subseteq R$ is imbedded as a subring containing the unity, show that $S^k : \mathcal{S}(V) \hookrightarrow \mathcal{T}(V)$ is an inclusion of graded R-modules. In the first case, use this to prove that

$$\mathcal{S}^k(V^*) = \mathcal{S}^k(V)^*, \quad \forall k \geq 2.$$

Show that, if $\{e_1, \dots, e_m\}$ is a basis of V, this identifies the monomial $e^*_{i_1}e^*_{i_2}\cdots e^*_{i_k}$ with $(e_{i_1}e_{i_2}\cdots e_{i_k})^*$.

Definition 7.3.6. Let V be a finite dimensional vector space over a field $\mathbb{F}$ of characteristic zero. A function $f : V \to \mathbb{F}$ is a homogeneous polynomial of degree k on V if, relative to some (hence, every) basis $\{e_1, \dots, e_m\}$ of V,

$$f\left(\sum_{i=1}^m x_i e_i\right) = P(x_1, \dots, x_m)$$

is a homogeneous polynomial of degree k in the variables $x_1, \dots, x_m$. The vector space of all homogeneous polynomials of degree k on V will be denoted by $\mathcal{P}^k(V)$.

Exercise 7.3.7. For all $k \geq 0$, establish a canonical isomorphism

$$\theta : \mathcal{S}^k(V^*) \to \mathcal{P}^k(V)$$

of vector spaces. For the case $k = 2$, construct θ^{-1} explicitly. ($\mathcal{P}^2(V)$ is called the space of *quadratic forms* on V and the process θ^{-1} of recovering the symmetric bilinear form from its associated quadratic form is called *polarization.*)

7.4. Multilinear Bundle Theory

Just as the linear construction of dualizing a vector space passes to the construction of dualizing a vector bundle, so the multilinear constructions of the previous three sections pass to corresponding constructions on vector bundles.

Let $\pi_i : E_i \to M$ be a k_i-plane bundle, $1 \leq i \leq m$. We want to define a bundle

$$\pi : E_1 \otimes \cdots \otimes E_m \to M$$

with fiber over $x \in M$ canonically equal to $E_{1x} \otimes \cdots \otimes E_{mx}$. The fiber dimension will be $k_1 k_2 \cdots k_m$.

Let $\{U_\alpha, \psi^i_\alpha\}_{\alpha \in \mathfrak{A}, 1 \le i \le m}$ be the maximal family of simultaneous trivializations

$$\begin{array}{ccc} E_i|U_\alpha & \xrightarrow{\psi^i_\alpha} & U_\alpha \times \mathbb{R}^{k_i} \\ {\scriptstyle \pi_i}\downarrow & & \downarrow{\scriptstyle p_1} \\ U_\alpha & \xrightarrow[\text{id}]{} & U_\alpha. \end{array}$$

Note that, if $A_j \in \mathrm{Gl}(k_j)$ is viewed as a nonsingular linear transformation of $\mathbb{R}^{k_j}$, $1 \le j \le m$, then $A_1 \otimes A_2 \otimes \cdots \otimes A_m$ is a nonsingular linear transformation of $\mathbb{R}^{k_1} \otimes \mathbb{R}^{k_2} \otimes \cdots \otimes \mathbb{R}^{k_m}$, hence

$$A_1 \otimes A_2 \otimes \cdots \otimes A_m \in \mathrm{Gl}(k_1 k_2 \cdots k_m).$$

Here, we identify

$$\mathbb{R}^{k_1} \otimes \cdots \otimes \mathbb{R}^{k_m} = \mathbb{R}^{k_1 k_2 \cdots k_m}$$

by lexicographic order on the basis

$$\mathcal{B} = \{e^1_{i_1} \otimes e^2_{i_2} \otimes \cdots \otimes e^m_{i_m}\}_{\substack{1 \le i_j \le k_j \\ 1 \le j \le m}}$$

where $\{e^j_1, \dots, e^j_{k_j}\}$ is the standard basis of $\mathbb{R}^{k_j}$, $1 \le j \le m$.

The system $\{U_\alpha, \psi^i_\alpha\}_{\alpha \in \mathfrak{A}}$ gives rise to a $\mathrm{Gl}(k_i)$-cocycle $\{\gamma^i_{\alpha\beta}\}_{\alpha\beta \in \mathfrak{A}}$, for each $i = 1, \dots, k$, with the aid of which the bundle E_i can be assembled from the products $U_\alpha \times \mathbb{R}^{k_i}$, $\alpha \in \mathfrak{A}$. We try to define a cocycle for assembling the tensor product of these bundles by setting

$$\gamma_{\alpha\beta} : U_\alpha \cap U_\beta \to \mathrm{Gl}(k_1 k_2 \cdots k_m),$$
$$\gamma_{\alpha\beta}(x) = \gamma^1_{\alpha\beta}(x) \otimes \gamma^2_{\alpha\beta}(x) \otimes \cdots \otimes \gamma^m_{\alpha\beta}(x).$$

Lemma 7.4.1. *The map $\gamma_{\alpha\beta}$ is smooth and $\gamma = \{\gamma_{\alpha\beta}\}_{\alpha,\beta \in \mathfrak{A}}$ is a cocycle.*

Proof. The cocycle property is rather obvious. For smoothness, it is enough to show that the vector-valued function

$$x \mapsto \gamma_{\alpha\beta}(x)(e^1_{i_1} \otimes \cdots \otimes e^m_{i_m})$$

is smooth on $U_\alpha \cap U_\beta$. But

$$\gamma^j_{\alpha\beta}(x)(e^j_{i_j}) = \sum_{\ell=1}^{k_j} a^j_\ell(x) e^j_\ell$$

and the real-valued functions $a^j_\ell(x)$ are smooth. Expanding

$$\begin{aligned} \gamma_{\alpha\beta}(x)(e^1_{i_1} \otimes \cdots \otimes e^m_{i_m}) &= \gamma^1_{\alpha\beta}(x)(e^1_{i_1}) \otimes \cdots \otimes \gamma^m_{\alpha\beta}(x)(e^m_{i_m}) \\ &= \left(\sum_{\ell=1}^{k_1} a^1_\ell(x) e^1_\ell\right) \otimes \cdots \otimes \left(\sum_{\ell=1}^{k_m} a^m_\ell(x) e^m_\ell\right), \end{aligned}$$

we obtain a linear combination of the elements of the basis B with smooth functions of x as coefficients. ☐

Exercise 7.4.2. Let $\pi : E \to M$ be the bundle determined by the cocycle $\{U_\alpha, \gamma_{\alpha\beta}\}_{\alpha\in\mathfrak{A}}$ and produce a *canonical* isomorphism

$$E_x = E_{1\,x} \otimes E_{2\,x} \otimes \cdots \otimes E_{m\,x},$$

for each $x \in M$.

In particular, given an n-plane bundle $\pi : E \to M$, we can form the tensor powers

$$\pi : \mathcal{T}^k(E) \to M,$$

the fiber over $x \in M$ being, canonically, $\mathcal{T}^k(E_x)$. If E has an associated cocycle $\{\gamma_{\alpha\beta}\}_{\alpha,\beta\in\mathfrak{A}}$, then $\mathcal{T}^k(E)$ has $\{\mathcal{T}^k(\gamma_{\alpha\beta})\}_{\alpha,\beta\in\mathfrak{A}}$ as an associated cocycle.

Lemma 7.4.3. *The* 0*th tensor power* $\mathcal{T}^0(E)$ *is canonically isomorphic to the trivial bundle* $M \times \mathbb{R}$.

Proof. Indeed, for each $x \in M$, $\mathcal{T}^0(E_x) = \mathbb{R}$ and, if $x \in U_\alpha \cap U_\beta$, then $\mathcal{T}^0(\gamma_{\alpha\beta}(x)) = \mathrm{id}_\mathbb{R}$. □

The following is proven in the same way.

Lemma 7.4.4. *The* 1*st tensor power* $\mathcal{T}^1(E)$ *is canonically isomorphic to* E.

We denote by

$$\pi : \mathcal{T}(E) \to M$$

the collection $\{\pi : \mathcal{T}^k(E) \to M\}_{k=0}^\infty$ and interpret this system as a "bundle" of graded $\mathbb{R}$-algebras over M.

In complete analogy with these constructions, we form the exterior powers

$$\pi : \Lambda^k(E) \to M$$

of the bundle E over M and the "bundle"

$$\pi : \Lambda(E) \to M$$

of exterior $\mathbb{R}$-algebras. Again, $\Lambda^0(E) = M \times \mathbb{R}$ and $\Lambda^1(E) = E$. Finally, one forms the symmetric powers $\mathcal{S}^k(E)$ and the bundle $\mathcal{S}(E)$ of symmetric algebras, noting the identities $\mathcal{S}^0(E) = M \times \mathbb{R}$ and $\mathcal{S}^1(E) = E$.

Exercise 7.4.5. Let $\dim M = n \geq 2$. Prove that M is orientable if and only if $\Lambda^n(T^*(M))$ admits a nowhere 0 section.

By identities proven in the previous sections, we obtain the following.

Lemma 7.4.6. *If* E *is a vector bundle over* M, *then*

$$\mathcal{T}^k(E^*) = \mathcal{T}^k(E)^*,$$

$$\Lambda^k(E^*) = \Lambda^k(E)^*,$$

$$\mathcal{S}^k(E^*) = \mathcal{S}^k(E)^*,$$

canonically for each integer $k \geq 0$.

Of particular note are the tensor bundles

$$\mathcal{T}^r_s(E) = \mathcal{T}^r(E^*) \otimes \mathcal{T}^s(E),$$

of covariant degree r and contravariant degree s.

Similarly, one can define bundles of linear homomorphisms. If E and F are vector bundles over M, we construct a vector bundle $\mathrm{Hom}(E, F)$, the fiber of which over each $x \in M$ is the vector space $\mathrm{Hom}_{\mathbb{R}}(E_x, F_x)$. Since this vector space is canonically equal to $E_x^* \otimes F_x$ (Exercise 7.1.28), we define

$$\mathrm{Hom}(E, F) = E^* \otimes F.$$

Likewise, the bundle of k-linear maps $(E_x)^k \to F_x$, $\forall\, x \in M$, can be defined as

$$L^k(E, F) = \mathcal{T}^k(E^*) \otimes F,$$

and its antisymmetric and symmetric cousins are

$$AL^k(E, F) = \Lambda^k(E^*) \otimes F,$$
$$SL^k(E, F) = \mathcal{S}^k(E^*) \otimes F,$$

respectively.

Exercise 7.4.7. Recall the operation of *direct sum* $V_1 \oplus V_2$ of vector spaces. If E_1 and E_2 are vector bundles over M, show how to define a bundle $E_1 \oplus E_2$ over M with fibers $(E_1 \oplus E_2)_x = E_{1\,x} \oplus E_{2\,x}$, $\forall\, x \in M$. This is called the "Whitney sum" of the bundles.

Exercise 7.4.8. If V_1, V_2, and W are finite dimensional vector spaces, construct a natural bilinear map

$$\theta : (V_1 \oplus V_2) \times W \to (V_1 \otimes W) \oplus (V_2 \otimes W),$$

hence a natural linear map

$$\theta : (V_1 \oplus V_2) \otimes W \to (V_1 \otimes W) \oplus (V_2 \otimes W).$$

Prove that θ is a linear isomorphism. Use this to prove that there is a canonical bundle isomorphism

$$(E_1 \oplus E_2) \otimes F = (E_1 \otimes F) \oplus (E_2 \otimes F).$$

Remark. Recall the general philosophical principal mentioned when we constructed the dual bundle. If one views vector spaces as vector bundles over a point, then all "natural" linear and multilinear constructions for combining vector spaces to get new ones extend to analogous operations, fiber by fiber, on vector bundles. Here "natural" means that the constructions can be carried out without reference to choices of bases. Such constructions are also "canonical". Similarly, natural relations between vector spaces, such as the relation

$$\mathrm{Hom}(V, W) = V^* \otimes W,$$

extend to analogous relations between vector bundles.

7.5. The Module of Sections

We are going to view the set of all vector bundles over a fixed manifold M as the objects of a category $\mathcal{V}_M$. For this, we need to define the morphisms of the category. Let

$$\pi : E \to M,$$
$$\rho : F \to M$$

be vector bundles (of possibly differing fiber dimensions).

Definition 7.5.1. A homomorphism of the n-plane bundle E to the m-plane bundle F is a commutative diagram

$$\begin{array}{ccc} E & \xrightarrow{\varphi} & F \\ \pi\downarrow & & \downarrow\rho \\ M & \xrightarrow{\text{id}} & M \end{array}$$

where φ is smooth and, for each $x \in M$, $\varphi_x = \varphi|E_x : E_x \to F_x$ is linear. We denote by $\text{HOM}(E, F)$ the set of all bundle homomorphisms from E to F.

Example 7.5.2. The canonical fiberwise inclusions

$$A^k : \Lambda^k(E_x) \hookrightarrow \mathcal{T}^k(E_x),$$
$$S^k : \mathcal{S}^k(E_x) \hookrightarrow \mathcal{T}^k(E_x)$$

(Sections 7.2 and 7.3) assemble to give bundle monomorphisms

$$A^k : \Lambda^k(E) \hookrightarrow \mathcal{T}^k(E),$$
$$S^k : \mathcal{S}^k(E) \hookrightarrow \mathcal{T}^k(E),$$

the smoothness of these maps being easily checked via local trivializations.

It is clear that the composition of bundle homomorphisms, whenever defined, is a bundle homomorphism and that $\text{id}_E : E \to E$ is a bundle homomorphism, so $\mathcal{V}_M$ is a category with bundle homomorphisms as its morphisms.

To each object $E \in \mathcal{V}_M$ we associate a $C^\infty(M)$-module $\Gamma(E)$, the space of smooth sections of E. If $\varphi : E \to F$ is a bundle homomorphism, there is induced a $C^\infty(M)$-linear map $\varphi_* : \Gamma(E) \to \Gamma(F)$, defined by

$$\varphi_*(s)(x) = \varphi(s(x)), \quad \forall\, x \in M, \quad \forall\, s \in \Gamma(E).$$

It is obvious that $(\text{id}_E)_* = \text{id}_{\Gamma(E)}$ and that $(\psi \circ \varphi)_* = \psi_* \circ \varphi_*$, so Γ is a covariant functor

$$\Gamma : \mathcal{V}_M \rightsquigarrow \mathcal{M}(C^\infty(M)).$$

(The squiggly arrow "$\rightsquigarrow$" is commonly used for functors.)

Exercise 7.5.3. If $E, F \in \mathcal{V}_M$, exhibit a canonical identification

$$\text{HOM}(E, F) = \Gamma(\text{Hom}(E, F)) = \Gamma(E^* \otimes F).$$

In particular, the set $\text{HOM}(E, F)$ of homomorphisms of the bundle E to the bundle F is naturally a $C^\infty(M)$-module.

The main purpose of this section is to show that the functor Γ transforms the $\mathbb{R}$-multilinear bundle constructions of Section 7.4 into the corresponding $C^\infty(M)$-multilinear module constructions of Sections 7.1, 7.2 and 7.3. A fairly easy case in point is the following.

Proposition 7.5.4. *If E is a vector bundle over M, then there is a canonical isomorphism*

$$\Gamma(E^*) = \Gamma(E)^*$$

of $C^\infty(M)$-modules.

Indeed, Proposition 6.2.11 was a particular case of this proposition and the proof for the general case remains the same.

Consider the tensor product $\Gamma(E) \otimes \Gamma(F)$ of $C^\infty(M)$-modules. Since these are also vector spaces over $\mathbb{R}$, this notation can be ambiguous. In order to avoid such ambiguities the tensor product of R-modules A and B is often written $A \otimes_R B$. For the time being, we will denote this $C^\infty(M)$-module by $\Gamma(E) \otimes_{C^\infty(M)} \Gamma(F)$ and its decomposable elements by $s \otimes_{C^\infty(M)} \sigma$. The vector space tensor product will be $\Gamma(E) \otimes_{\mathbb{R}} \Gamma(F)$ and its decomposables will be $s \otimes_{\mathbb{R}} \sigma$. Since one seldom thinks of the set of sections as a real vector space, we will ultimately drop the subscript $C^\infty(M)$, but retain $\otimes_{\mathbb{R}}$ for the vector space case.

Given $s \in \Gamma(E)$ and $\sigma \in \Gamma(F)$, one produces $\alpha(s, \sigma) \in \Gamma(E \otimes F)$ by setting

$$\alpha(s, \sigma)(x) = s(x) \otimes \sigma(x) \in E_x \otimes F_x = (E \otimes F)_x, \ \ \forall x \in M.$$

We will write $\alpha(s, \sigma) = s \otimes \sigma$, the pointwise tensor product of sections. It is not hard to check that this defines a smooth section of $E \otimes F$ and that

$$\alpha : \Gamma(E) \times \Gamma(F) \to \Gamma(E \otimes F)$$

is $C^\infty(M)$-bilinear. Denote also by α the associated $C^\infty(M)$-linear map

$$\alpha : \Gamma(E) \otimes_{C^\infty(M)} \Gamma(F) \to \Gamma(E \otimes F).$$

We emphasize that $s \otimes_{C^\infty(M)} \sigma$ and $s \otimes \sigma = \alpha(s \otimes_{C^\infty(M)} \sigma)$ are conceptually distinct. The following theorem asserts that this conceptual distinction can safely be disregarded.

Theorem 7.5.5. *The $C^\infty(M)$-linear map α is a canonical isomorphism of $C^\infty(M)$-modules*

$$\Gamma(E) \otimes_{C^\infty(M)} \Gamma(F) = \Gamma(E \otimes F).$$

Corollary 7.5.6. *There are canonical isomorphisms*

$$\Gamma(\mathcal{T}^k(E)) = \mathcal{T}^k(\Gamma(E)),$$
$$\Gamma(\Lambda^k(E)) = \Lambda^k(\Gamma(E)),$$
$$\Gamma(\mathcal{S}^k(E)) = \mathcal{S}^k(\Gamma(E)),$$

of $C^\infty(M)$-modules.

Proof. Indeed, the first of these identities is an immediate consequence of Theorem 7.5.5. There are canonical inclusions

$$A^k : \Lambda^k(\Gamma(E)), \hookrightarrow \mathcal{T}^k(\Gamma(E)),$$
$$A^k : \Gamma(\Lambda^k(E)) \hookrightarrow \Gamma(\mathcal{T}^k(E)).$$

The first of these is by Lemma 7.2.19 and the observation that $\mathbb{Q} \subset C^\infty(M)$ as a subring of constant functions. The second comes from the bundle inclusion of Example 7.5.2. The images of these inclusions correspond perfectly under the identification $\mathcal{T}^k(\Gamma(E)) = \Gamma(\mathcal{T}^k(E))$, proving the second identity. The third has exactly the same proof as the second. □

Combining this corollary with Proposition 7.5.4 gives

Corollary 7.5.7. *There is a canonical isomorphism*

$$\mathcal{T}^r_k(\Gamma(E)) = \Gamma(\mathcal{T}^r_k(E))$$

of $C^\infty(M)$-modules.

Example 7.5.8. There are many other natural identifications now available. For example,

$$\begin{aligned}
(\mathcal{T}^k(\Gamma(E)))^* &= (\Gamma(\mathcal{T}^k(E)))^* && \text{by Corollary 7.5.6}\\
&= \Gamma(\mathcal{T}^k(E)^*) && \text{by Proposition 7.5.4}\\
&= \Gamma(\mathcal{T}^k(E^*)) && \text{by Lemma 7.4.6}\\
&= \mathcal{T}^k(\Gamma(E^*)) && \text{by Corollary 7.5.6}\\
&= \mathcal{T}^k(\Gamma(E)^*) && \text{by Proposition 7.5.4,}
\end{aligned}$$

and there are similar identities for Λ^k and $\mathcal{S}^k$.

Example 7.5.9. A Riemannian metric $\langle\cdot,\cdot\rangle$ on a manifold M is a smooth section of $\mathcal{S}^2(T^*(M))$ that is positive definite. In local coordinates, the metric has a formula

$$\sum_{1\le i\le j\le n} g_{ij}(x)\,dx_i \cdot dx_j.$$

By the above, the metric can also be thought of as an element of

$$\mathcal{S}^2(\Gamma(T^*(M))) = \mathcal{S}^2(A^1(M)).$$

We begin the proof of Theorem 7.5.5. It is surprisingly delicate.

Lemma 7.5.10. *If both F and E are trivial bundles, then α is an isomorphism of $C^\infty(M)$-modules.*

Proof. In this case, choose global sections $\{\sigma_1,\dots,\sigma_n\}$ of E and $\{\tau_1,\dots,\tau_m\}$ of F that trivialize these bundles. These are *free* bases of the respective $C^\infty(M)$-modules $\Gamma(E)$ and $\Gamma(F)$ (Example 7.1.2), so

$$\{\sigma_i \otimes_{C^\infty(M)} \tau_j\}_{i,j=1}^{n,m}$$

is a free basis of $\Gamma(E)\otimes_{C^\infty(M)}\Gamma(F)$ (Corollary 7.1.12). The set

$$\{\sigma_i \otimes \tau_j\}_{i,j=1}^{n,m}$$

of pointwise tensor products of sections trivializes the bundle $E\otimes F$, hence is a free basis of $\Gamma(E\otimes F)$. Since

$$\alpha(\sigma_i \otimes_{C^\infty(M)} \tau_j) = \sigma_i \otimes \tau_j,$$

for all relevant indices, we see that α is an isomorphism of $C^\infty(M)$-modules. □

In light of this lemma, we will reduce the general case of Theorem 7.5.5 to the case in which both bundles are trivial. The key to this is the following.

Theorem 7.5.11. *Given a vector bundle E over M, there exists a vector bundle $E^\perp$ over M such that the bundle $E\oplus E^\perp$ is trivial.*

For the moment, we accept this.

Let E_1 and E_2 be vector bundles over M and define bundle homomorphisms

$$\begin{aligned}
\iota &: E_1 \to E_1\oplus E_2 && \iota(v) = (v,0),\\
\rho &: E_1\oplus E_2 \to E_1 && \rho(v,w) = v.
\end{aligned}$$

It is clear that $\rho\circ\iota = \mathrm{id}_{E_1}$, so the functoriality of Γ implies the following.

Lemma 7.5.12. *The composition $\rho_* \circ \iota_*$ is equal to $\mathrm{id}_{\Gamma(E_1)}$. In particular,*

$$\iota_* : \Gamma(E_1) \to \Gamma(E_1 \oplus E_2)$$

is injective and

$$\rho_* : \Gamma(E_1 \oplus E_2) \to \Gamma(E_1)$$

is surjective.

Proof of theorem 7.5.5. By Exercise 7.4.8, $(E \oplus E^\perp) \otimes (F \oplus F^\perp)$ splits off a direct summand $E \otimes F$. Consider the commutative diagram

$$\begin{array}{ccc} \Gamma((E \oplus E^\perp) \otimes (F \oplus F^\perp)) & \xleftarrow{\alpha} & \Gamma(E \oplus E^\perp) \otimes_{C^\infty(M)} \Gamma(F \oplus F^\perp) \\ {\scriptstyle \iota_*}\uparrow & & \uparrow{\scriptstyle \iota_* \otimes \iota_*} \\ \Gamma(E \otimes F) & \xleftarrow[\alpha]{} & \Gamma(E) \otimes_{C^\infty(M)} \Gamma(F) \end{array}.$$

By Lemma 7.5.10, the top arrow is an isomorphism of $C^\infty(M)$-modules. The above lemma guarantees that the leftmost vertical arrow is injective. Since

$$(\rho_* \otimes \rho_*) \circ (\iota_* \otimes \iota_*) = (\rho_* \circ \iota_*) \otimes (\rho_* \circ \iota_*) = \mathrm{id}_{\Gamma(E)} \otimes \mathrm{id}_{\Gamma(F)},$$

the rightmost vertical arrow is also injective. It follows that

$$\alpha : \Gamma(E) \otimes_{C^\infty(M)} \Gamma(F) \to \Gamma(E \otimes F)$$

is injective. Similarly, the diagram

$$\begin{array}{ccc} \Gamma((E \oplus E^\perp) \otimes (F \oplus F^\perp)) & \xleftarrow{\alpha} & \Gamma(E \oplus E^\perp) \otimes_{C^\infty(M)} \Gamma(F \oplus F^\perp) \\ {\scriptstyle \rho_*}\downarrow & & \downarrow{\scriptstyle \rho_* \otimes \rho_*} \\ \Gamma(E \otimes F) & \xleftarrow[\alpha]{} & \Gamma(E) \otimes_{C^\infty(M)} \Gamma(F) \end{array}$$

commutes and the vertical arrows are surjective, implying that

$$\alpha : \Gamma(E) \otimes_{C^\infty(M)} \Gamma(F) \to \Gamma(E \otimes F)$$

is surjective. □

Everything now hinges on Theorem 7.5.11. We prove the case in which M is compact and then quote a theorem from dimension theory that extends this proof to the noncompact case.

Suppose that E is an n-plane bundle over a compact manifold M. Compactness of M will be used only to find a finite open cover $\{U_i\}_{i=1}^r$ such that $E|U_i$ is trivial, $1 \le i \le r$. Let $\{\lambda_i\}_{i=1}^r$ be a subordinate partition of unity and, for each $i = 1, 2, \dots, r$, let $s_i^1, \dots, s_i^n \in \Gamma(E|U_i)$ be everywhere linearly independent (hence a basis at each $x \in U_i$). Let $\sigma_i^j \in \Gamma(E)$ be the extension by 0 of $\lambda_i s_i^j$, $1 \le i \le r$, $1 \le j \le n$. Then, for each $x \in M$, $\{\sigma_i^j(x)\}_{i,j=1}^{r,n}$ spans E_x. View $\Gamma(E)$ as a vector space over $\mathbb{R}$ and consider the finite dimensional subspace

$$V = \mathrm{span}_{\mathbb{R}}\{\sigma_i^j\}_{i,j=1}^{r,n}.$$

Then $p_1 : M \times V \to M$ is a trivial vector bundle. We define a surjective homomorphism

$$\rho : M \times V \to E$$

of vector bundles by setting

$$\rho(x, \sigma) = \sigma(x).$$

The smoothness of ρ is elementary, as is the fact that it carries $\{x\} \times V$ linearly onto the fiber E_x, $\forall\, x \in M$. Let $E_x^\perp \subset \{x\} \times V$ be the kernel of this linear surjection and set

$$E^\perp = \bigcup_{x \in M} E_x^\perp.$$

The fact that $E^\perp$ is a vector subbundle of $M \times V$ is a case of the following result.

Lemma 7.5.13. *If $\rho : F \to E$ is a surjective homomorphism of vector bundles over M, then*

$$E^\perp = \bigcup_{x \in M} \ker(\rho_x)$$

is a subbundle of F.

Proof. It is enough to produce local trivializations. Let $x \in M$ and choose vectors $v_1, \dots, v_r \in F_x$ that are carried by ρ_x one-to-one onto a basis of E_x. Extend these to a basis $\{v_1, \dots, v_r, v_{r+1}, \dots, v_n\}$ of F_x. By the local triviality of F, there is a neighborhood U of x in M and sections σ_i of $F|U$ such that $\sigma_i(x) = v_i$, $1 \le i \le n$, and such that $\{\sigma_i(y)\}_{i=1}^n$ is a basis of F_y, $\forall\, y \in U$. Consider the sections $s_i = \rho \circ \sigma_i$ of $E|U$. Taking U smaller, if necessary, and appealing to the continuity of ρ, we arrange that $\{s_i(y)\}_{i=1}^r$ is a basis of E_y, $\forall\, y \in U$. Then there are unique expressions

$$s_{r+1}(y) = \sum_{i=1}^r f_{r+1}^i(y) s_i(y),$$

$$\vdots$$

$$s_n(y) = \sum_{i=1}^r f_n^i(y) s_i(y),$$

where the coefficient functions f_j^i are all smooth. Consider the smooth sections

$$\tau_1(y) = \sigma_{r+1}(y) - \sum_{i=1}^r f_{r+1}^i(y)\sigma_i(y),$$

$$\vdots$$

$$\tau_{n-r}(y) = \sigma_n(y) - \sum_{i=1}^r f_n^i(y)\sigma_i(y).$$

It should be clear that these give a basis of $\ker(\rho_y)$, $\forall\, y \in U$, hence define a local trivialization of $E^\perp$ over U. □

Fix a positive definite inner product on V, viewing it as a Riemannian metric on the bundle $M \times V$. For each $x \in M$, let $\widetilde{E}_x \subset \{x\} \times V$ be the subspace orthogonal to $E_x^\perp$. We claim that the set

$$\widetilde{E} = \bigcup_{x \in M} \widetilde{E}_x$$

is a subbundle of $M \times V$. More generally,

Lemma 7.5.14. *If $F \subseteq E$ is a vector subbundle and if there is given a Riemannian metric on E, then the subset $\widetilde{F} \subseteq E$, fiberwise perpendicular to F, is a subbundle.*

Proof. Again, local triviality is all that needs to be proven. There are sections $\sigma_1, \dots, \sigma_r, \sigma_{r+1}, \dots, \sigma_n$ of $E|U$, trivializing that bundle, where U is a neighborhood of an arbitrary point of M. These can be chosen so that $\sigma_1, \dots, \sigma_r$ are sections of $F|U$ that trivialize that bundle. An application of Gram–Schmidt turns these into fiberwise orthonormal sections $s_1, \dots, s_r, s_{r+1}, \dots, s_n$ with the same properties. It follows that $s_{r+1}, \dots, s_n$ are trivializing sections of $\widetilde{F}|U$, proving that $\widetilde{F}$ is a subbundle of E as desired. □

The bundle homomorphism $\rho|\widetilde{E} : \widetilde{E} \to E$ is an isomorphism, this being true fiber by fiber, so

$$M \times V = E^{\perp} \oplus \widetilde{E} \cong E^{\perp} \oplus E.$$

This completes the proof of Theorem 7.5.11 in the case that M is a compact manifold. The compactness assumption on M is removed by showing that, whether M is compact or not, there is always a finite trivializing cover $\{U_i\}_{i=1}^r$ for E. This will follow from a theorem in dimension theory.

Theorem 7.5.15. *Let M be a manifold of dimension r. Then every open cover $\mathcal{V}$ of M admits a refinement $\mathcal{W} = \{W_\alpha\}_{\alpha \in \mathfrak{A}}$ such that, whenever the indices $\alpha_i \in \mathfrak{A}$, $1 \le i \le r+2$, are all distinct, then*

$$W_{\alpha_1} \cap W_{\alpha_2} \cap \cdots \cap W_{\alpha_{r+2}} = \emptyset.$$

For the proof, see [**20**, Theorem V.8, page 67], together with Example III.4 on page 25 of that same reference.

Theorem 7.5.16. *Let M be a manifold of dimension r and let E be a vector bundle over M. Then there is an open cover $\{U_k\}_{k=1}^{r+1}$ such that $E|U_k$ is trivial, $1 \le k \le r+1$.*

Proof. Let $\mathcal{V}$ be an open cover of M trivializing E. Let $\mathcal{W} = \{W_\alpha\}_{\alpha \in \mathfrak{A}}$ be the refinement given by Theorem 7.5.15. In particular, $E|W_\alpha$ is trivial, $\forall\, \alpha \in \mathfrak{A}$. Let $\{\lambda_\alpha\}_{\alpha \in \mathfrak{A}}$ be a partition of unity subordinate to $\mathcal{W}$.

If $S \subseteq \mathfrak{A}$ is a finite subset, define

$$U_S = \{x \in M \mid \min_{\alpha \in S} \lambda_\alpha(x) > \max_{\beta \in \mathfrak{A} \smallsetminus S} \lambda_\beta(x)\}.$$

Let $|S|$ denote the cardinality of S. The following are elementary:

(*i*) $\{U_S \mid S \subseteq \mathfrak{A} \text{ is finite}\}$ is an open cover of M;
(*ii*) if $S_1 \ne S_2$ are finite subsets of $\mathfrak{A}$ with $|S_1| = |S_2|$, then $U_{S_1} \cap U_{S_2} = \emptyset$;
(*iii*) $|S| > r+1 \Rightarrow U_S = \emptyset$;
(*iv*) $U_S \subseteq W_\alpha$, $\forall\, \alpha \in S$.

For each integer $k \ge 1$, set

$$U_k = \bigcup_{|S|=k} U_S.$$

By (*i*), $\{U_k\}_{k=1}^\infty$ is an open cover of M. For each $k \ge 1$, $E|U_k$ is trivial ((*ii*) and (*iv*)) and, by (*iii*), $U_k = \emptyset$ for $k > r+1$. □

The proof of Theorem 7.5.5 is now complete.

Definition 7.5.17. The space of covariant tensors of degree k on M is

$$\mathfrak{T}^k(M) = \Gamma(\mathfrak{T}^k(T^*(M))) = \mathfrak{T}^k(\Gamma(T^*(M))) = \mathfrak{T}^k(A^1(M)).$$

The graded algebra $\mathfrak{T}^*(M) = \{\mathfrak{T}^k(M)\}_{k=0}^\infty$ is called the covariant tensor algebra of M.

Remark. If $\mathcal{T}^*(M)$ is viewed as $\Gamma(\mathcal{T}(T^*(M)))$, the multiplication is by pointwise tensor product of sections. If it is viewed as $\mathcal{T}(\Gamma(T^*(M)))$, the multiplication is just that of the tensor algebra of the $C^\infty(M)$-module $\Gamma(T^*(M))$. By the proof of Theorem 7.5.5, the two graded algebra structures agree.

Remark. The use of the asterisk to denote graded structures is standard, as is its use to denote duals. Which meaning is intended will usually be clear from the context.

Definition 7.5.18. The space of contravariant tensors of degree k on M is

$$\mathcal{T}_k(M) = \Gamma(\mathcal{T}^k(T(M))) = \mathcal{T}^k(\Gamma(T(M))) = \mathcal{T}^k(\mathfrak{X}(M)),$$

and $\mathcal{T}_*(M) = \{\mathcal{T}_k(M)\}_{k=0}^\infty$ is called the contravariant tensor algebra of M.

Definition 7.5.19. The space of (mixed) tensors of type (r, s) on M is

$$\begin{aligned}\mathcal{T}^r_s(M) &= \Gamma(\mathcal{T}^r(T^*(M)) \otimes \mathcal{T}^s(T(M)))\\ &= \mathcal{T}^r(\Gamma(T^*(M))) \otimes \mathcal{T}^s(\Gamma(T(M)))\\ &= \mathcal{T}^r(A^1(M)) \otimes \mathcal{T}^s(\mathfrak{X}(M)).\end{aligned}$$

Definition 7.5.20. The space of k-forms on M is

$$A^k(M) = \Gamma(\Lambda^k(T^*(M))) = \Lambda^k(\Gamma(T^*(M))) = \Lambda^k(A^1(M)).$$

The exterior algebra

$$A^*(M) = \Gamma(\Lambda(T^*(M))) = \Lambda(\Gamma(T^*(M))) = \Lambda(A^1(M))$$

is called the Grassmann algebra of M.

Definition 7.5.21. The space of (covariant) symmetric tensors on M is

$$\mathcal{S}^k(M) = \Gamma(\mathcal{S}^k(T^*(M))) = \mathcal{S}^k(\Gamma(T^*(M))) = \mathcal{S}^k(A^1(M)).$$

The graded algebra

$$\mathcal{S}^*(M) = \Gamma(\mathcal{S}(T^*(M))) = \mathcal{S}(\Gamma(T^*(M))) = \mathcal{S}(A^1(M))$$

is called the symmetric algebra of M.

Remark that the graded algebras $\mathcal{T}^*(M)$, $\mathcal{T}_*(M)$, $A^*(M)$, and $\mathcal{S}^*(M)$ are all connected and that $\mathcal{T}_1(M) = \mathfrak{X}(M)$ and $\mathcal{T}^1(M) = A^1(M) = \mathcal{S}^1(M)$.

Exercise 7.5.22. Let M be an oriented n–manifold, let $(U, x^1, \dots, x^n)$ and $(V, y^1, \dots, y^n)$ be coordinate neighborhoods in M respecting the orientation, and let $\omega \in A^n(M)$ be such that $\operatorname{supp}(\omega)$ is a compact subset of $U \cap V$. Let

$$\begin{aligned}\omega &= f\, dx^1 \wedge \cdots \wedge dx^n,\\ \omega &= h\, dy^1 \wedge \cdots \wedge dy^n\end{aligned}$$

be the respective formulas for ω in these coordinate systems. Finally, let $g(x^1, \dots, x^n) = (y^1, \dots, y^n)$ be the formula for the change of coordinates.

(1) Show that, on $\varphi(U \cap V)$, $f = (h \circ g)\det(Jg)$.
(2) Show how to define the integral $\int_M \omega \in \mathbb{R}$ and prove that your definition is independent of choices.
(3) Denote the oppositely oriented M by $-M$ and show that

$$\int_{-M} \omega = -\int_M \omega.$$

[illegible] is [illegible] as [illegible] tensor product of [illegible] [illegible] that of the tensor [illegible] [illegible] the proof of Theorem [illegible] of the two [illegible] algebra [illegible]

Remark [illegible] the use of the asterisk [illegible] [illegible]

Definition 7.6.[illegible] The [illegible]

[illegible]

and [illegible] is [illegible]

Definition 7.6.[illegible] The [illegible] tensor [illegible]

[illegible]

[illegible]

Definition 7.6.[illegible] The [illegible] bundle [illegible]

[illegible]

[illegible]

[illegible]

is [illegible] the Grassmann algebra [illegible] M.

Definition 7.6.[illegible] The [illegible] space [illegible]

[illegible]

the graded [illegible]

[illegible]

is called the [illegible] algebra of M.

[illegible]

Exercise 7.6.2[illegible] Let M [illegible]

[illegible]

[illegible]

[illegible]

[illegible]

CHAPTER 8

Integration of Forms and de Rham Cohomology

In Chapter 6, we studied the first de Rham cohomology $H^1(M)$ of a manifold. This measures the difference between exactness and local exactness of 1-forms on M and was shown to have interesting topological applications. Here we generalize these ideas, using the full Grassmann algebra $A^*(M)$ to produce a graded algebra $H^*(M)$, the de Rham cohomology algebra. The proper generalization of "locally exact 1-form" is "closed p-form", defined as a p-form that is annihilated by "exterior differentiation". Exact forms are closed and $H^p(M)$ measures the extent to which closed p-forms may fail to be exact. By Stokes' theorem, the geometric boundary operator and exterior differentiation of forms are mutually adjoint operations in a certain precise sense. This is a generalization of the fundamental theorem of calculus and a powerful tool for computing cohomology. The reader who would like to pursue this theory further could hardly do better than to consult [**5**].

8.1. The Exterior Derivative

Let $U \subseteq \mathbb{R}^n$ be an open subset. Since $A^0(U) = C^\infty(U)$, we have already defined the exterior derivative

$$d : A^0(U) \to A^1(U)$$

(Definition 6.2.14). For $p \geq 1$, we can define the exterior derivative

$$d : A^p(U) \to A^{p+1}(U)$$

by the following formula:

$$d\Big(\sum_{1\leq i_1<\cdots<i_p\leq n} f_{i_1\cdots i_p}\, dx^{i_1} \wedge \cdots \wedge dx^{i_p}\Big) = \sum_{1\leq i_1<\cdots<i_p\leq n} d(f_{i_1\cdots i_p}) \wedge dx^{i_1} \wedge \cdots \wedge dx^{i_p}.$$

It is clear that this operator is $\mathbb{R}$-linear. It is not clear, but will be proven shortly, that the definition is invariant under changes of coordinates.

Lemma 8.1.1. *If $V \subseteq U$ is open, then $(d\omega)|V = d(\omega|V)$.*

Lemma 8.1.2. *The composition*

$$A^p(U) \xrightarrow{d} A^{p+1}(U) \xrightarrow{d} A^{p+2}(U)$$

is trivial ($d^2 = 0$).

Proof. By the antisymmetry of exterior multiplication, the above formula for d gives the same answer whether or not the indices are in increasing order or are distinct. Thus

$$\begin{aligned} d(d(f\,dx^{i_1}\wedge\cdots\wedge dx^{i_p})) &= d\Big(\sum_{j=1}^n \frac{\partial f}{\partial x^j}\,dx^j\wedge dx^{i_1}\cdots\wedge dx^{i_p}\Big)\\ &= \sum_{j=1}^n\sum_{k=1}^n \frac{\partial^2 f}{\partial x^k\partial x^j}\,dx^k\wedge dx^j\wedge dx^{i_1}\wedge\cdots\wedge dx^{i_p}, \end{aligned}$$

and this vanishes by the equality of mixed partials and the antisymmetry of exterior multiplication. □

Remark. The equation $d^2=0$ is equivalent to the equality of mixed partials which, in turn, is equivalent to $[\partial/\partial x^k,\partial/\partial x^j]=0$, the commutativity of coordinate fields.

The proof of the following is mechanical, hence is left to the reader.

Lemma 8.1.3. *If $\omega\in A^p(U)$ and $\eta\in A^q(U)$, then*

$$d(\omega\wedge\eta)=(d\omega)\wedge\eta+(-1)^p\omega\wedge d\eta.$$

Remark. In particular, writing $f\eta$ for $f\wedge\eta$ when $f\in A^0(U)=C^\infty(U)$, we get

$$d(f\eta)=df\wedge\eta+f\,d\eta.$$

If $\eta=df_1\wedge\cdots\wedge df_p$, where $f_i\in A^0(U)$, $1\le i\le p$, then repeated use of the above two lemmas yields $d\eta=0$ and

$$d(f\eta)=df\wedge\eta.$$

Corollary 8.1.4. *If $U\subseteq\mathbb{R}^n$ and $V\subseteq\mathbb{R}^m$ are open subsets and if the map $\varphi: U\to V$ is smooth, then*

$$d\circ\varphi^*=\varphi^*\circ d: A^p(V)\to A^{p+1}(U),$$

for all $p\ge 0$.

Proof. In the following computation, the third equality is by the above remark:

$$\begin{aligned} d(\varphi^*(f\,dy^{i_1}\wedge\cdots\wedge dy^{i_p})) &= d(\varphi^*(f)\varphi^*(dy^{i_1})\wedge\cdots\wedge\varphi^*(dy^{i_p}))\\ &= d(\varphi^*(f)\,d(\varphi^*(y^{i_1}))\wedge\cdots\wedge d(\varphi^*(y^{i_p})))\\ &= d(\varphi^*(f))\wedge d(\varphi^*(y^{i_1}))\wedge\cdots\wedge d(\varphi^*(y^{i_p}))\\ &= \varphi^*(df)\wedge\varphi^*(dy^{i_1})\wedge\cdots\wedge\varphi^*(dy^{i_p})\\ &= \varphi^*(df\wedge dy^{i_1}\wedge\cdots\wedge dy^{i_p})\\ &= \varphi^*(d(f\,dy^{i_1}\wedge\cdots\wedge dy^{i_p})). \end{aligned}$$

Since every $\eta\in A^p(V)$ is a sum of forms of the type used in the above computation, the claim follows. □

We want to extend the exterior derivative to an $\mathbb{R}$-linear operator

$$d: A^p(M)\to A^{p+1}(M)$$

on all manifolds M and for all nonnegative integers p. We take an axiomatic approach, requiring that this operator satisfy the following:

(1) $f\in A^0(M)$ and $X\in\mathfrak{X}(M)\Rightarrow df(X)=X(f)$.
(2) $\omega\in A^p(M)$ and $\eta\in A^q(M)\Rightarrow d(\omega\wedge\eta)=d\omega\wedge\eta+(-1)^p\omega\wedge d\eta$.
(3) $d^2=0$.
(4) $U\subset M$ open and $\omega\in A^p(M)\Rightarrow(d\omega)|U=d(\omega|U)$.

(5) $\varphi : M \to N$ smooth $\Rightarrow \varphi^* \circ d = d \circ \varphi^*$.

Remark. If one considers only manifolds that are open subsets of Euclidean spaces, then all of the axioms hold for the exterior derivative as already defined. For $d : A^0(M) \to A^1(M)$, which has been defined on all manifolds, we have seen the truth of the first axiom (Lemma 6.2.3).

Definition 8.1.5. If an $\mathbb{R}$-linear operator $d : A^p(M) \to A^{p+1}(M)$, defined for all smooth manifolds M and all integers $p \geq 0$, satisfies the above axioms, it is called an exterior derivative.

Theorem 8.1.6. *There is a unique exterior derivative.*

Proof. First we prove *uniqueness.* By Axiom (4), it will be enough to show that, whenever $U \subseteq M$ is a coordinate neighborhood and $\omega \in A^p(M)$, the form $d(\omega|U) \in A^{p+1}(U)$ is uniquely determined.

Let $\varphi : U \to \mathbb{R}^n$ be a diffeomorphism onto an open subset V and set $\varphi^i = \varphi^*(x^i) = x^i \circ \varphi$, $1 \leq i \leq n$. Also, by functoriality, $(\varphi^{-1})^* = (\varphi^*)^{-1}$, which is to say that $\varphi^* : A^*(V) \to A^*(U)$ is an isomorphism. Thus, since $d\varphi^i = \varphi^*(dx^i)$ (Axiom (5)), the set

$$\{d\varphi^{i_1} \wedge \cdots \wedge d\varphi^{i_p}\}_{1 \leq i_1 < \cdots < i_p \leq n}$$

is a free basis of the $C^\infty(U)$-module $A^p(U)$. Thus,

$$\omega|U = \sum_{1 \leq i_1 < \cdots < i_p \leq n} f_{i_1 \cdots i_p}\, d\varphi^{i_1} \wedge \cdots \wedge d\varphi^{i_p}$$

and

$$\begin{aligned} d(\omega|U) &= d\Bigg(\sum_{1 \leq i_1 < \cdots < i_p \leq n} f_{i_1 \cdots i_p} d\varphi^{i_1} \wedge \cdots \wedge d\varphi^{i_p} \Bigg) \\ &= \sum_{1 \leq i_1 < \cdots < i_p \leq n} df_{i_1 \cdots i_p} \wedge d\varphi^{i_1} \wedge \cdots \wedge d\varphi^{i_p}, \end{aligned}$$

where the second equality uses Axioms (2) and (3). But $df_{i_1 \cdots i_p}$ is uniquely specified by Axiom (1). Thus, $d(\omega|U)$ is unique and, as remarked above, the uniqueness in general of the operator d follows. Remark that all five axioms have been used in this argument.

We turn to the proof of *existence.* Let

$$\{U_\alpha, \varphi_\alpha\}_{\alpha \in \mathfrak{A}}$$

be the maximal coordinate atlas for M. The diagrams

$$\begin{array}{ccc} A^p(\varphi_\alpha(U_\alpha \cap U_\beta)) & \xrightarrow{g^*_{\alpha\beta}} & A^p(\varphi_\beta(U_\alpha \cap U_\beta)) \\ {\scriptstyle d}\downarrow & & \downarrow{\scriptstyle d} \\ A^{p+1}(\varphi_\alpha(U_\alpha \cap U_\beta)) & \xrightarrow[g^*_{\alpha\beta}]{} & A^{p+1}(\varphi_\beta(U_\alpha \cap U_\beta)) \end{array}$$

make sense and commute, $\forall\, \alpha, \beta \in \mathfrak{A}$, since everything is written for open subsets of Euclidean space. For $\omega \in A^p(M)$, define

$$(d\omega)|U_\alpha = \varphi^*_\alpha(d(\varphi_\alpha^{-1*}(\omega|U_\alpha)))\ \ \forall\, \alpha \in \mathfrak{A}.$$

If $U_\alpha \cap U_\beta \neq \emptyset$, we check that the two definitions agree on $U_\alpha \cap U_\beta$. Indeed, by the commutative diagram,

$$\begin{aligned}\varphi_\alpha^*(d(\varphi_\alpha^{-1*}(\omega|U_\alpha \cap U_\beta))) &= \varphi_\alpha^*(g_{\alpha\beta}^{*-1} \circ d \circ g_{\alpha\beta}^*(\varphi_\alpha^{-1*}(\omega|U_\alpha \cap U_\beta))) \\ &= (g_{\alpha\beta}^{-1} \circ \varphi_\alpha)^* \circ d \circ (\varphi_\alpha^{-1} \circ g_{\alpha\beta})^*(\omega|U_\alpha \cap U_\beta) \\ &= \varphi_\beta^*(d(\varphi_\beta^{-1*}(\omega|U_\alpha \cap U_\beta))).\end{aligned}$$

Thus, the local definitions of $d\omega$ piece together to give $d\omega \in A^{p+1}(M)$. Since the axioms are true for open subsets of Euclidean space, they are true locally on M, hence globally. □

Definition 8.1.7. A form $\omega \in A^p(M)$, $p \geq 0$, such that $d\omega = 0$ is called a closed p-form on M. Closed p-forms are also called (de Rham) p-cocycles and the real vector space of all such forms is denoted by $Z^p(M)$. We set $Z^*(M) = \{Z^p(M)\}_{p=0}^\infty$.

Definition 8.1.8. A form $\omega \in A^p(M)$, $p \geq 1$, is said to be exact if there is a form $\eta \in A^{p-1}(M)$ such that $d\eta = \omega$. Exact p-forms are also called p-coboundaries and the real vector space of all such forms is denoted $B^p(M)$. For $p = 0$, we define $B^0(M) = \{0\} \subset A^0(M)$. Finally, we set $B^*(M) = \{B^p(M)\}_{p=0}^\infty$.

Remark. A form $\omega \in A^p(M)$ is *locally exact* if every point $x \in M$ has an open neighborhood U such that $\omega|U \in B^p(U)$. One version of the Poincaré lemma (Section 8.3) asserts that the set of locally exact p-forms on M is precisely $Z^p(M)$. In particular, our earlier definition of $Z^1(M)$ (Definition 6.4.1) agrees with our present one.

Exercise 8.1.9. This exercise is in anticipation of the Poincaré lemma. Define a manifold M to be contractible if it is homotopy equivalent to a point. Prove that the following three versions of the Poincaré lemma for 1-forms are equivalent, and verify (3) when $n = 2$. (Here, $Z^1(M)$ denotes the space of closed 1-forms, not the space of locally exact ones.)

(1) The form $\omega \in A^1(M)$ is locally exact if and only if it is closed.
(2) If M is contractible, then $Z^1(M) = B^1(M)$.
(3) $Z^1(\mathbb{R}^n) = B^1(\mathbb{R}^n)$.

The formula $d^2 = 0$ is equivalent to the inclusion $B^p(M) \subseteq Z^p(M)$, $p \geq 0$.

Definition 8.1.10. For each integer $p \geq 0$, the pth (de Rham) cohomology space of M is the real vector space $H^p(M) = Z^p(M)/B^p(M)$.

If $\varphi : M \to N$ is smooth, the formula $\varphi^* \circ d = d \circ \varphi^*$ implies that

$$\varphi^*(Z^p(N)) \subseteq Z^p(M),$$
$$\varphi^*(B^p(N)) \subseteq B^p(M),$$

so φ^* induces an $\mathbb{R}$-linear map

$$\varphi^* : H^p(N) \to H^p(M).$$

As usual, we have functoriality.

Lemma 8.1.11. *The pth cohomology H^p is a contravariant functor from the category of differentiable manifolds and smooth maps to the category of real vector spaces and linear maps.*

If $\omega \in Z^p(M)$ and $\eta \in Z^q(M)$, then

$$d(\omega \wedge \eta) = d\omega \wedge \eta + (-1)^p \omega \wedge d\eta = 0.$$

That is, the exterior product of a closed p-form with a closed q-form is a closed $(p+q)$-form. Thus, $Z^*(M)$ is a graded algebra over $\mathbb{R}$ under exterior multiplication. Both $A^*(M)$ and $Z^*(M)$ are anticommutative in the sense of Definition 7.2.9.

Lemma 8.1.12. *The graded $\mathbb{R}$-algebra $Z^*(M)$ is connected if and only if M is a connected manifold.*

Proof. The space $Z^0(M)$ consists of all $f \in C^\infty(M)$ with exterior derivative $df = 0$. That is, $Z^0(M)$ is the space of locally constant, real-valued functions on M. Identifying $\mathbb{R}$ with the space of constant functions in $C^\infty(M)$, we have $\mathbb{R} \subseteq Z^0(M)$. The product in $Z^*(M)$ of a constant function and a form becomes naturally identified with scalar multiplication. But locally constant functions are all constant if and only if M is connected. □

Corollary 8.1.13. *The space $H^0(M)$ is one-dimensional if and only if M is connected. In this case, $H^0(M) = \mathbb{R}$ canonically. Generally, $H^0(M)$ is a direct product of copies of $\mathbb{R}$, one for each component of M.*

Proof. Indeed, $H^0(M) = Z^0(M)/B^0(M) = Z^0(M)$, the space of locally constant functions, and all claims follow easily. □

Lemma 8.1.14. *If $\dim M = n$, then $H^p(M) = 0$, $\forall p > n$.*

Proof. Indeed, $A^p(M) = 0$ for all integers p greater than the dimension of M. □

Lemma 8.1.15. *The graded subspace $B^*(M) \subseteq Z^*(M)$ is a 2-sided ideal, hence $H^*(M) = Z^*(M)/B^*(M)$ is a graded, anticommutative algebra over the field $\mathbb{R}$.*

Proof. If $\omega \in Z^p(M)$ and $\eta \in B^q(M)$, $q \geq 1$, then $\eta = d\alpha$ for some $\alpha \in A^{p-1}(M)$, hence

$$\begin{aligned}\omega \wedge \eta &= \omega \wedge d\alpha \\ &= d\omega \wedge (-1)^p \alpha + (-1)^p \omega \wedge d((-1)^p \alpha) \\ &= d(\omega \wedge (-1)^p \alpha).\end{aligned}$$

Since $\eta \wedge \omega = (-1)^{pq} \omega \wedge \eta$, it follows that $B^*(M)$ is a 2-sided ideal in $Z^*(M)$. □

Since smooth maps $\varphi : M \to N$ preserve exterior multiplication and pass to well-defined maps in cohomology, we see that

$$\varphi^* : H^*(N) \to H^*(M)$$

is a homomorphism of graded algebras. We have completed the proof of the following.

Theorem 8.1.16. *The graded cohomology construction defines a contravariant functor H^* from the category of differentiable manifolds and smooth maps to the category of anticommutative graded algebras over $\mathbb{R}$ and graded algebra homomorphisms. The graded algebra $H^*(M)$ is connected if and only if M is connected.*

The graded algebra $H^*(M)$ is called the (de Rham) cohomology algebra of M. Whether or not it is connected, $H^*(M)$ has a unity, namely the constant function $1 \in Z^0(M) = H^0(M)$.

Definition 8.1.17. The space of compactly supported p-forms on M is denoted by $A^p_c(M)$.

It is clear that the exterior product of two compactly supported forms is compactly supported. Indeed, it is enough that one of them be compactly supported. Thus, each $A^p_c(M)$ is a module over $C^\infty(M)$ and these assemble into a graded algebra $A^*_c(M)$ over $C^\infty(M)$. It is also a graded algebra over $\mathbb{R}$, which is more to our purpose. Furthermore,

$$d(A^p_c(M)) \subseteq A^{p+1}_c(M),$$

so one can define the space

$$Z^p_c(M) = \{\omega \in A^p_c(M) \mid d\omega = 0\}$$

and the vector subspace

$$B^p_c(M) = \{\omega = d\alpha \mid \alpha \in A^{p-1}_c(M)\}.$$

If we use the common convention that $A^{-1}(M) = A^{-1}_c(M) = 0$, the above definition of $B^p_c(M)$ includes the case $p = 0$. As before, $Z^*_c(M)$ is a graded subalgebra of $A^*_c(M)$ and $B^*_c(M) \subseteq Z^*_c(M)$ is a 2-sided ideal.

Definition 8.1.18. The (de Rham) cohomology algebra with compact support is $H^*_c(M) = Z^*_c(M)/B^*_c(M)$.

Remark that $H^*(M) = H^*_c(M)$ if M is compact. At the other extreme, if M has no compact component, the space $Z^0_c(M)$ of compactly supported, locally constant functions on M is trivial, so $H^0_c(M) = 0$. In any event, each element of $Z^0_c(M)$ will vanish on all but finitely many components of M. These observations establish the following.

Lemma 8.1.19. *The vector space $H^0_c(M)$ is isomorphic to a direct sum of copies of $\mathbb{R}$, one for each compact component of M.*

Note the different conclusions in Lemma 8.1.19 and Corollary 8.1.13. Each element of a *direct sum* has terms from only finitely many summands, while elements of a *direct product* are allowed to have terms from infinitely many of the factors.

Remark. The graded algebra $H^*_c(M)$ generally does not have a unity unless M is compact.

Recall that a smooth map $\varphi : M \to N$ is said to be proper if, for each compact set $C \subseteq N$, the set $\varphi^{-1}(C)$ is also compact. For example, $\mathrm{id} : M \to M$ is always proper. If M is compact, φ is always proper. In any event, the composition of proper maps is proper, so the class of differentiable manifolds and smooth, proper maps between them is a category.

If $\varphi : M \to N$ is proper and if $\omega \in A^p_c(N)$, then $\varphi^*(\omega) \in A^p_c(M)$. As usual, $\varphi^* \circ d = d \circ \varphi^*$, so we get an induced homomorphism of graded algebras

$$\varphi^* : H^*_c(N) \to H^*_c(M).$$

Theorem 8.1.20. *Cohomology with compact supports is a contravariant functor H^*_c from the category of differentiable manifolds and smooth, proper maps to the category of anticommutative graded algebras over $\mathbb{R}$ and graded algebra homomorphisms.*

Exercise 8.1.21. Prove that $H^1_c(\mathbb{R}) \cong \mathbb{R}$. This is the one-dimensional case of the Poincaré lemma for compactly supported cohomology.

8.2. Stokes' Theorem and Singular Homology

In this section, we define integration of forms and give a detailed treatment of two versions of Stokes' theorem. As an application of the second (combinatorial) version, we define the singular homology of a manifold and relate it to de Rham cohomology, stating the celebrated de Rham theorem. A detailed sketch of the proof of this theorem will appear in Section 8.9.

Throughout what follows, we assume that $\dim M = n$ and that M is oriented. We also allow $\partial M \neq \emptyset$.

Theorem 8.2.1. *For each oriented n-manifold M, there is a unique $\mathbb{R}$-linear functional*

$$\int_M : A^n_c(M) \to \mathbb{R},$$

called the integral and having the following property: if (U, φ) is an orientation-respecting coordinate chart, if $\omega \in A^n_c(M)$ has $\operatorname{supp}(\omega) \subset U$, and if

$$\varphi^{-1*}(\omega) = g\, dx^1 \wedge \cdots \wedge dx^n \in A^n_c(\varphi(U)),$$

then

$$\int_M \omega = \int_{\varphi(U)} g \quad \text{(the Riemann integral).}$$

Proof. First we prove *uniqueness*. Let $\{(U_\alpha, \varphi_\alpha)\}_{\alpha\in\mathfrak{A}}$ be a smooth $\mathbb{H}^n$-atlas on M respecting the orientation. Let $\{\lambda_\alpha\}_{\alpha\in\mathfrak{A}}$ be a smooth partition of unity subordinate to the atlas. If $\omega \in A^n_c(M)$, then $\lambda_\alpha\omega \in A^n_c(M)$ and $\lambda_\alpha\omega \neq 0$ for only a finite number of $\alpha \in \mathfrak{A}$. This is because $\operatorname{supp}(\omega)$ is compact and the partition of unity is locally finite. Thus,

$$\omega = \sum_{\alpha\in\mathfrak{A}} \lambda_\alpha\omega$$

and this sum is actually finite. Then, if $\int_M$ exists, linearity gives

$$\int_M \omega = \sum_{\alpha\in\mathfrak{A}} \int_M \lambda_\alpha\omega$$

and $\operatorname{supp}(\lambda_\alpha\omega) = \operatorname{supp}(\lambda_\alpha) \cap \operatorname{supp}(\omega)$ is a compact subset of U_α. By the local property of $\int_M$, each $\int_M \lambda_\alpha\omega$ is *uniquely* given as

$$\int_M \lambda_\alpha\omega = \int_{\varphi_\alpha(U_\alpha)} (\lambda_\alpha \circ \varphi_\alpha^{-1}) g_\alpha,$$

where $g_\alpha\, dx^1 \wedge \cdots \wedge dx^n = \varphi_\alpha^{-1*}(\omega|U_\alpha)$.

We give one way to define $\int_M$, establishing *existence*. This will depend on a choice of orientation-respecting coordinate atlas $\{U_\alpha, \varphi_\alpha\}_{\alpha\in\mathfrak{A}}$ and of subordinate partition of unity $\{\lambda_\alpha\}_{\alpha\in\mathfrak{A}}$. (One could remove some of this arbitrariness by requiring the atlas to be the maximal one, but the choice of partition of unity still could not be made canonical.) We appeal to the *uniqueness* already proven to show independence of the choices. If $\omega \in A^n_c(M)$, only finitely many $\lambda_\alpha\omega$ are not identically 0. Define

$$\int_M \lambda_\alpha\omega = \int_{\varphi_\alpha(U_\alpha)} (\lambda_\alpha \circ \varphi_\alpha^{-1}) g_\alpha,$$

where $g_\alpha\, dx^1 \wedge \cdots \wedge dx^n = \varphi_\alpha^{-1*}(\omega|U_\alpha)$. Then *define*

$$\int_M \omega = \sum_{\alpha\in\mathfrak{A}} \int_M \lambda_\alpha \omega,$$

a finite sum. Defined in this way,

$$\int_M : A_c^n(M) \to \mathbb{R}$$

is an $\mathbb{R}$-linear map.

We must check that, if $\operatorname{supp}(\omega) \subset U$, where (U, φ) is an arbitrary orientation-respecting coordinate chart (not necessarily in our atlas), and if

$$\varphi^{-1*}(\omega) = g\, dx^1 \wedge \cdots \wedge dx^n,$$

then

$$\int_M \omega = \int_{\varphi(U)} g.$$

First remark that

$$\begin{aligned}\int_M \omega &= \sum_{\alpha\in\mathfrak{A}} \int_{\varphi_\alpha(U_\alpha)} (\lambda_\alpha \circ \varphi_\alpha^{-1}) g_\alpha \\ &= \sum_{\alpha\in\mathfrak{A}} \int_{\varphi_\alpha(U_\alpha\cap U)} (\lambda_\alpha \circ \varphi_\alpha^{-1}) g_\alpha,\end{aligned}$$

since $\operatorname{supp}(\omega) \subset U$. By Exercise 7.5.22,

$$\int_{\varphi_\alpha(U_\alpha\cap U)} (\lambda_\alpha \circ \varphi_\alpha^{-1}) g_\alpha = \int_{\varphi(U_\alpha\cap U)} (\lambda_\alpha \circ \varphi^{-1}) g,$$

for each $\alpha \in \mathfrak{A}$. The fact that the charts are compatibly oriented is essential. Thus,

$$\begin{aligned}\int_M \omega &= \sum_{\alpha\in\mathfrak{A}} \int_{\varphi(U_\alpha\cap U)} (\lambda_\alpha \circ \varphi^{-1}) g \\ &= \sum_{\alpha\in\mathfrak{A}} \int_{\varphi(U)} (\lambda_\alpha \circ \varphi^{-1}) g \\ &= \int_{\varphi(U)} \underbrace{\left\{\sum_{\alpha\in\mathfrak{A}} \lambda_\alpha \circ \varphi^{-1}\right\}}_{\equiv 1} g \\ &= \int_{\varphi(U)} g.\end{aligned}$$

□

Remark. By Exercise 7.5.22 and the above proof,

$$\int_{-M} \omega = -\int_M \omega.$$

The orientation of M induces an orientation of ∂M in the following way. Let $\{U_\alpha, x_\alpha^1, \dots, x_\alpha^n\}_{\alpha\in\mathfrak{A}}$ be an $\mathbb{H}^n$-atlas on M respecting the orientation. By the definition of $\mathbb{H}^n$, $x_\alpha^1 \le 0$, wherever defined, and $x_\alpha^1 = 0$ exactly on $U_\alpha \cap \partial M$, $\forall\, \alpha \in \mathfrak{A}$. Let $\mathfrak{A}' = \{\alpha \in \mathfrak{A} \mid U_\alpha \cap \partial M \neq \emptyset\}$ and consider the $\mathbb{R}^{n-1}$-atlas

$$\{U_\alpha \cap \partial M, x_\alpha^2, \dots, x_\alpha^n\}_{\alpha\in\mathfrak{A}'}$$

of ∂M. Let $g_{\alpha\beta}$ and $g^{\partial}_{\alpha\beta}$ denote the respective changes of coordinates for these atlases. At $x \in U_\alpha \cap U_\beta \cap \partial M$,

$$(Jg_{\alpha\beta})_x = \begin{bmatrix} \frac{\partial x^1_\alpha}{\partial x^1_\beta} & 0 \\ * & Jg^{\partial}_{\alpha\beta} \end{bmatrix}_x$$

Here, since x^1_α decreases with x^1_β, the upper left-hand entry in this matrix is strictly positive. Since $\det(Jg_{\alpha\beta})_x > 0$, it follows that $\det(Jg^{\partial}_{\alpha\beta})_x > 0$. Thus, this $\mathbb{R}^{n-1}$-atlas on ∂M defines an orientation of ∂M.

Definition 8.2.2. The orientation of ∂M, produced as above, is said to be induced by the given orientation of M.

We always assume that, when M is oriented, ∂M has the induced orientation.

The following fundamental result asserts that exterior differentiation is the "adjoint" of passing to the boundary. For this reason, d is often referred to as the "coboundary operator".

Theorem 8.2.3 (Stokes' Theorem). *Let M be an oriented n-manifold and let $i : \partial M \hookrightarrow M$ be the inclusion. Then, if $\omega \in A^{n-1}_c(M)$,*

$$\int_M d\omega = \int_{\partial M} i^*(\omega),$$

where, if $\partial M = \emptyset$, the right-hand side is interpreted as 0.

Proof. First we prove the *local case*. That is, $M = \mathbb{H}^n$ and $\partial M = \mathbb{R}^{n-1}$. Any $\omega \in A^{n-1}_c(\mathbb{H}^n)$ can be written as

$$\omega = \sum_{j=1}^{n} (-1)^{j-1} f_j \, dx^1 \wedge \cdots \wedge \widehat{dx^j} \wedge \cdots \wedge dx^n,$$

where f_j has compact support, $1 \le j \le n$, and $\widehat{dx^j}$ indicates that this term is omitted. Then

$$i^*(\omega) = (f_1 \circ i) \, dx^2 \wedge \cdots \wedge dx^n \in A^{n-1}_c(\partial \mathbb{H}^n)$$

and

$$d\omega = \left(\sum_{j=1}^{n} \frac{\partial f_j}{\partial x^j} \right) dx^1 \wedge \cdots \wedge dx^n \in A^n_c(\mathbb{H}^n).$$

By the fundamental theorem of calculus and the compactness of $\operatorname{supp}(f_j)$, for $j = 2, \dots, n$,

$$\int_{-\infty}^{\infty} \cdots \int_{-\infty}^{0} \frac{\partial f_j}{\partial x^j} \, dx^1 \cdots dx^n = \int_{-\infty}^{\infty} \cdots \int_{-\infty}^{0} \left(\int_{-\infty}^{\infty} \frac{\partial f_j}{\partial x^j} \, dx^j \right) dx^1 \cdots \widehat{dx^j} \cdots dx^n = 0.$$

Therefore,

$$\begin{aligned}
\int_{\mathbb{H}^n} d\omega &= \int_{-\infty}^{\infty} \cdots \int_{-\infty}^{\infty} \int_{-\infty}^{0} \frac{\partial f_1}{\partial x^1}\, dx^1 \cdots dx^n \\
&= \int_{-\infty}^{\infty} \cdots \int_{-\infty}^{\infty} \underbrace{f_1(0, x^2, \ldots, x^n)}_{f_1 \circ i}\, dx^2 \cdots dx^n \\
&= \int_{\partial \mathbb{H}^n} i^* \omega.
\end{aligned}$$

Now we can prove the *global case.* Let

$$i_M : \partial M \hookrightarrow M,$$

$$i_{\mathbb{H}^n} : \partial \mathbb{H}^n \hookrightarrow \mathbb{H}^n$$

denote the respective boundary inclusions and let

$$\{U_\alpha, \varphi_\alpha = (x^1_\alpha, \ldots, x^n_\alpha)\}_{\alpha \in \mathfrak{A}},$$

$$\{U_\alpha \cap \partial M, \varphi^\partial_\alpha = (x^2_\alpha, \ldots, x^n_\alpha)\}_{\alpha \in \mathfrak{A}'}$$

be the orientation-respecting atlases on M and ∂M, respectively, as chosen above. If $\{\lambda_\alpha\}_{\alpha \in \mathfrak{A}}$ is a smooth partition of unity subordinate to the atlas on M, then $\{\lambda_\alpha \circ i_M\}_{\alpha \in \mathfrak{A}'}$ is a smooth partition of unity subordinate to the atlas on ∂M. Note that $\varphi_\alpha^{-1} \circ i_{\mathbb{H}^n} = i_M \circ (\varphi^\partial_\alpha)^{-1}$, $\forall\, \alpha \in \mathfrak{A}'$.

For $\omega \in A^{n-1}_c(M)$, the fact that $\omega = \sum_\alpha \lambda_\alpha \omega$ is a finite sum gives finite sums

$$\begin{aligned}
d\omega &= \sum_{\alpha \in \mathfrak{A}} d(\lambda_\alpha \omega), \\
i_M^* \omega &= \sum_{\alpha \in \mathfrak{A}'} i_M^*(\lambda_\alpha \omega).
\end{aligned}$$

Therefore,

$$\begin{aligned}
\int_M d\omega &= \sum_{\alpha \in \mathfrak{A}} \int_{U_\alpha} d(\lambda_\alpha \omega) \\
&= \sum_{\alpha \in \mathfrak{A}} \int_{\varphi_\alpha(U_\alpha)} (\varphi_\alpha^{-1})^*(d(\lambda_\alpha \omega)) \\
&= \sum_{\alpha \in \mathfrak{A}} \int_{\varphi_\alpha(U_\alpha)} d(\varphi_\alpha^{-1*}(\lambda_\alpha \omega)) \\
&= \sum_{\alpha \in \mathfrak{A}} \int_{\mathbb{H}^n} d(\varphi_\alpha^{-1*}(\lambda_\alpha \omega)) \\
&= \sum_{\alpha \in \mathfrak{A}'} \int_{\partial \mathbb{H}^n} i^*_{\mathbb{H}^n} \varphi_\alpha^{-1*}(\lambda_\alpha \omega).
\end{aligned}$$

The last equality is by the local version proven above. If $\partial M = \emptyset$, then

$$\operatorname{supp}(\omega) \cap \partial M = \emptyset,$$

and this integral vanishes. Otherwise, we get

$$\begin{aligned}\int_M d\omega &= \sum_{\alpha\in\mathfrak{A}'}\int_{\partial\mathbb{H}^n}(\varphi_\alpha^\partial)^{-1*}i_M^*(\lambda_\alpha\omega)\\ &= \sum_{\alpha\in\mathfrak{A}'}\int_{\varphi_\alpha(U_\alpha\cap\partial M)}(\varphi_\alpha^\partial)^{-1*}i_M^*(\lambda_\alpha\omega)\\ &= \sum_{\alpha\in\mathfrak{A}'}\int_{U_\alpha\cap\partial M} i_M^*(\lambda_\alpha\omega)\\ &= \sum_{\alpha\in\mathfrak{A}'}\int_{U_\alpha\cap\partial M}(\lambda_\alpha\circ i_M)i_M^*(\omega)\\ &= \int_{\partial M} i_M^*\omega.\end{aligned}$$

□

Theorem 8.2.4. *Let M be an oriented n-manifold with $\partial M = \emptyset$. Then*

$$\int_M : H_c^n(M) \to \mathbb{R}$$

is a well-defined, $\mathbb{R}$-linear surjection.

Proof. Since $A_c^{n+1}(M) = 0$, we have $Z_c^n(M) = A_c^n(M)$. If $\omega = d\eta \in B_c^n(M)$, then Stokes' theorem and the fact that $\partial M = \emptyset$ imply that

$$\int_M \omega = \int_M d\eta = \int_{\partial M} \eta = 0.$$

Thus, the linear map

$$\int_M : Z_c^n(M) \to \mathbb{R}$$

induces a well-defined linear map

$$\int_M : H_c^n(M) \to \mathbb{R}.$$

To prove surjectivity, we only need prove that this map is nontrivial. Let $(U, x^1, \dots, x^n)$ be a compatibly oriented chart and let $\lambda \in C^\infty(M)$ have compact support contained in U, with $\lambda \geq 0$ everywhere and $\lambda > 0$ somewhere. Thus $\omega = \lambda\, dx^1 \wedge \cdots \wedge dx^n$ can be interpreted as an element of $Z_c^n(M)$ and of $A_c^n(\mathbb{R}^n)$, so

$$\int_M \omega = \int_{\mathbb{R}^n} \lambda > 0.$$

□

A deeper fact, to be proven later (Theorem 8.6.4), is that, if M is both oriented and connected, then $\int_M$ is a bijection from $H_c^n(M)$ to $\mathbb{R}$.

In order to integrate p-forms, where $p < \dim M$, it is necessary to define suitable p-dimensional domains of integration. For the case $p = 1$, we have already studied line integrals, the domain of integration being a (piecewise) smooth curve in M. In general, it is convenient to use singular p-simplices (defined below) as domains for integrating p-forms. A singular 1-simplex is simply a smooth curve.

Recall that a subset $\Delta \subset \mathbb{R}^p$ is *convex* if, for each pair of points $v, w \in \Delta$, the straight line segment joining v and w lies entirely in Δ. If $C \subseteq \mathbb{R}^p$ is an arbitrary subset, the *convex hull* of C is defined to be the smallest convex set $\widehat{C}$ containing C. Since an arbitrary intersection of convex sets is convex, and $\mathbb{R}^p$ is itself convex, $\widehat{C}$ is just the intersection of all convex sets containing C.

Definition 8.2.5. The standard p-simplex $\Delta_p \subset \mathbb{R}^p$ is the convex hull of the set $\{e_0, e_1, \dots, e_p\}$, where e_i is the ith standard basis vector, $1 \le i \le p$, and $e_0 = 0$.

Thus, $\Delta_0 = \{0\}$, a single point, and $\Delta_1 = [0, 1]$. The cases $p = 2$ and $p = 3$ are pictured in Figures 8.2.1 and 8.2.2, respectively.

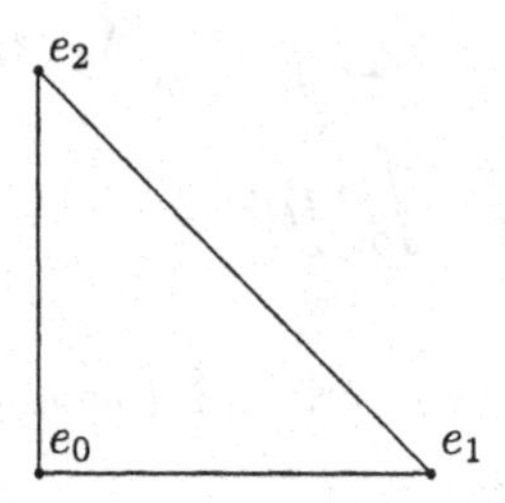

Figure 8.2.1. The standard 2-simplex

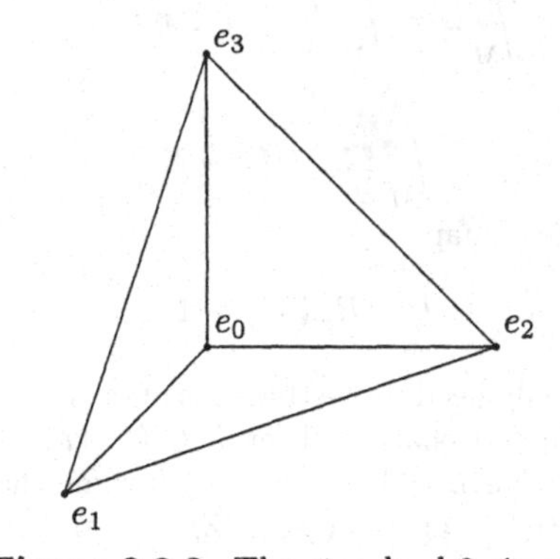

Figure 8.2.2. The standard 3-simplex

A more explicit definition of the standard p-simplex is

$$\Delta_p = \left\{ (x^1, \dots, x^p) \mid \text{each } x^i \ge 0 \text{ and } \sum_{i=1}^{p} x^i \le 1 \right\}.$$

It is sometimes convenient to set $\Delta_{-1} = \emptyset$.

Definition 8.2.6. A (smooth) singular p-simplex in a manifold M is a smooth map $s : \Delta_p \to M$.

Thus, each point of M can be thought of as a singular 0-simplex and smooth curves, up to parametrization, are singular 1-simplices. One could also define piecewise smooth singular p-simplices, but we will not do so.

Definition 8.2.7. For $0 \le i \le p$, the ith face of the standard p-simplex Δ_p is the singular $(p-1)$-simplex $F_i : \Delta_{p-1} \to \Delta_p$ defined by

$$F_i(x^1, \dots, x^{p-1}) = \begin{cases} (x^1, \dots, x^{i-1}, 0, x^i, \dots, x^{p-1}) & \text{if } i > 0, \\ (1 - x^1 - \cdots - x^{p-1}, x^1, \dots, x^{p-1}) & \text{if } i = 0. \end{cases}$$

The 0th face of Δ_0 is considered to be defined but empty. If $s : \Delta_p \to M$ is a singular p-simplex, the ith face of s is the singular $(p-1)$-simplex $\partial_i s = s \circ F_i$.

It is clear that $F_i : \Delta_{p-1} \to \Delta_p$ is a topological imbedding and that the image of F_i is exactly the subset ordinarily thought of as the "face" of Δ_p opposite the vertex e_i. The ith face $\partial_i s$ of a singular p-simplex s is essentially the restriction of s to the ith face of Δ_p, but parametrized on the standard Δ_{p-1}.

Definition 8.2.8. If $s : \Delta_p \to M$ is a singular p-simplex and $\omega \in A^p(M)$, then $s^*(\omega)$ has the form $g\, dx^1 \wedge \cdots \wedge dx^p$ and we set

$$\int_s \omega = \int_{\Delta_p} g,$$

where the right-hand side is the Riemann integral. If $s : \{0\} \to M$ is a singular 0-simplex and $\omega = f \in A^0(M)$, the integral is interpreted to mean

$$\int_s f = f(s(0)).$$

There is a combinatorial version of Stokes' theorem, according to which the integral of an exact p-form $d\eta$ over a singular p-simplex s is equal to the integral of η over the "boundary" of s.

Theorem 8.2.9 (Combinatorial Stokes' Theorem). *If $s : \Delta_p \to M$ is a singular p-simplex and $\eta \in A^{p-1}(M)$, then*

$$\int_s d\eta = \sum_{i=0}^{p} (-1)^i \int_{\partial_i s} \eta.$$

Remarks. The signs in the combinatorial Stokes' theorem are dictated by comparing the standard orientation of Δ_{p-1} with the induced orientation of $F_i(\Delta_{p-1})$ as a part of the boundary of Δ_p. We write the formal combinatorial expression

$$\partial s = \sum_{i=1}^{p} (-1)^i \partial_i s$$

and express Stokes' formula as

$$\int_s d\eta = \int_{\partial s} \eta.$$

This highlights the analogy with Theorem 8.2.3 and agrees with established usage in algebraic topology.

For $f \in A^0(M)$ and a smooth curve $s : [0,1] \to M$, Theorem 8.2.9 asserts that

$$\int_s df = f(s(1)) - f(s(0)),$$

which is just Lemma 6.3.3. That lemma was a thinly disguised version of the fundamental theorem of calculus and Theorem 8.2.9 is a somewhat less thinly disguised version of the same fundamental theorem.

Proof of theorem 8.2.9. It is clearly sufficient to prove that

$$\int_{\Delta_p} d\eta = \sum_{i=0}^{p} (-1)^i \int_{\Delta_{p-1}} F_i^*(\eta),$$

where η is a $(p-1)$-form defined on an open neighborhood of Δ_p in $\mathbb{R}^p$. We can write

$$\eta = \sum_{j=1}^{p} f_j \, dx^1 \wedge \cdots \wedge \widehat{dx^j} \wedge \cdots \wedge dx^p$$

and, by the linearity of the integral, prove the formula for each term of the sum. That is, without loss of generality, we assume that $1 \le j \le p$ and

$$\eta = f \, dx^1 \wedge \cdots \wedge \widehat{dx^j} \wedge \cdots \wedge dx^p.$$

By the local formula for exterior differentiation,

$$d\eta = (-1)^{j-1} \frac{\partial f}{\partial x^j} \, dx^1 \wedge \cdots \wedge dx^p,$$

and we are reduced to proving the formula

$$\int_{\Delta_p} (-1)^{j-1} \frac{\partial f}{\partial x^j} \, dx^1 \wedge \cdots \wedge dx^p = \sum_{i=0}^{p} (-1)^i \int_{\Delta_{p-1}} F_i^*(f \, dx^1 \wedge \cdots \wedge \widehat{dx^j} \wedge \cdots \wedge dx^p). \tag{8.1}$$

The right-hand side of equation (8.1) can be simplified. For this, it will be helpful to let x^i denote the coordinates in $\mathbb{R}^p$ and z^i the coordinates in $\mathbb{R}^{p-1}$. Remark that

$$F_0^*(dx^j) = \begin{cases} dz^{j-1} & \text{if } j > 1, \\ -\sum_{i=1}^{p-1} dz^i & \text{if } j = 1, \end{cases}$$

and, if $i > 0$,

$$F_i^*(dx^j) = \begin{cases} dz^j & \text{if } j < i, \\ 0 & \text{if } j = i, \\ dz^{j-1} & \text{if } j > i. \end{cases}$$

One obtains the formula

$$F_0^*(f \, dx^1 \wedge \cdots \wedge \widehat{dx^j} \wedge \cdots \wedge dx^p) = (-1)^{j-1} (f \circ F_0) dz^1 \wedge \cdots \wedge dz^{p-1}$$

and, if $i > 0$,

$$F_i^*(f \, dx^1 \wedge \cdots \wedge \widehat{dx^j} \wedge \cdots \wedge dx^p) = \begin{cases} 0 & \text{if } i \ne j, \\ (f \circ F_j) \, dz^1 \wedge \cdots \wedge dz^{p-1} & \text{if } i = j. \end{cases}$$

Substituting these terms in the equation (8.1) and multiplying both sides by $(-1)^{j-1}$ reduces us to proving

$$\int_{\Delta_p} \frac{\partial f}{\partial x^j} \, dx^1 \wedge \cdots \wedge dx^p = \int_{\Delta_{p-1}} (f \circ F_0) \, dz^1 \wedge \cdots \wedge dz^{p-1} - \int_{\Delta_{p-1}} (f \circ F_j) \, dz^1 \wedge \cdots \wedge dz^{p-1},$$

which, rewritten in terms of the Riemann integral, becomes

$$\int_{\Delta_p} \frac{\partial f}{\partial x^j} = \int_{\Delta_{p-1}} f(1 - z^1 - \cdots - z^{p-1}, z^1, \dots, z^{p-1}) - \int_{\Delta_{p-1}} f(z^1, \dots, z^{j-1}, 0, z^j, \dots, z^{p-1}). \tag{8.2}$$

The linear change of coordinates in $\mathbb{R}^{p-1}$, defined by

$$w^i = \begin{cases} 1 - z^1 - \cdots - z^{p-1} & \text{if } i = 1, \\ z^{i-1} & \text{if } 2 \le i \le j-1, \\ z^i & \text{if } j \le i \le p-1, \end{cases}$$

preserves volume (*i.e.*, the Jacobian determinant is ± 1) and carries Δ_{p-1} diffeomorphically onto itself, as the reader will easily check. Also,

$$z^{j-1} = 1 - w^1 - \cdots - w^{p-1}.$$

While this coordinate change reverses orientation when j is odd, the Riemann integral is insensitive to orientation, so (8.2) becomes

$$\int_{\Delta_p} \frac{\partial f}{\partial x^j} = \int_{\Delta_{p-1}} f(w^1, \dots, w^{j-1}, 1 - w^1 - \cdots - w^{p-1}, w^j \dots, w^{p-1}) - \int_{\Delta_{p-1}} f(z^1, \dots, z^{j-1}, 0, z^j, \dots, z^{p-1}).$$

To avoid notational confusion, replace the dummy variables w^i with z^i:

$$\int_{\Delta_p} \frac{\partial f}{\partial x^j} = \int_{\Delta_{p-1}} f(z^1, \dots, z^{j-1}, 1 - z^1 - \cdots - z^{p-1}, z^j \dots, z^{p-1}) - \int_{\Delta_{p-1}} f(z^1, \dots, z^{j-1}, 0, z^j, \dots, z^{p-1}). \tag{8.3}$$

Let

$$\Delta_p^j = F_j(\Delta_{p-1}) = \{(x^1, \dots, x^p) \in \Delta_p \mid x^j = 0\},$$

the face of Δ_p opposite the vertex e_j, $1 \le j \le p$. Since $j > 0$, it is evident that the linear diffeomorphism

$$F_j : \Delta_{p-1} \to \Delta_p^j,$$
$$F_j(z^1, \dots, z^{p-1}) = (z^1, \dots, z^{j-1}, 0, z^j, \dots, z^{p-1})$$

preserves $(p-1)$-dimensional volume, so the three integrals in (8.3) can be computed, respectively, by the multiple integrals

$$\int_{\Delta_p} \frac{\partial f}{\partial x^j} \, dx^1 \cdots dx^p,$$

$$\int_{\Delta_p^j} f\left(x^1, \dots, x^{j-1}, 1 - \sum_{i \ne j} x^i, x^{j+1}, \dots, x^p\right) dx^1 \cdots \widehat{dx^j} \cdots dx^p,$$

$$\int_{\Delta_p^j} f(x^1, \dots, x^{j-1}, 0, x^{j+1}, \dots, x^p) \, dx^1 \cdots \widehat{dx^j} \cdots dx^p.$$

With these substitutions, (8.3) is checked by standard manipulation of iterated integrals and an application of the fundamental theorem of calculus. The proof of Theorem 8.2.9 is complete. □

Corollary 8.2.10. *A form $\omega \in A^p(M)$ is closed if and only if $\int_{\partial s} \omega = 0$, for every singular $(p+1)$-simplex s in M.*

Proof. If ω is closed, then

$$\int_{\partial s} \omega = \int_s d\omega = \int_s 0 = 0.$$

For the converse, suppose that $d\omega = \eta \neq 0$. Choose a point $x \in M$ such that $\eta_x \neq 0$. Choose vectors $v_1, \dots, v_{p+1} \in T_x(M)$ such that $\eta_x(v_1 \wedge \cdots \wedge v_{p+1}) > 0$. These vectors must be linearly independent, so we can find a local coordinate chart $(U, x^1, \dots, x^n)$ about x in which v_i is the value of the ith coordinate field $\xi_i = \partial/\partial x^i$ at x, $1 \leq i \leq p+1$. By making this chart sufficiently small, we can guarantee that $\eta(\xi_1 \wedge \cdots \wedge \xi_{p+1}) > 0$ on all of U. Let $s : \Delta_p \to U$ be any orientation-preserving, smooth imbedding into the coordinate $(p+1)$-plane $\{(x^1, \dots, x^n) \in U \mid x^{p+2} = \cdots = x^n = 0\}$. It follows that

$$\int_{\partial s} \omega = \int_s d\omega = \int_s \eta > 0.$$

□

Singular simplices are used to detect topological features of a manifold. For instance, a piecewise smooth, closed curve $s = s_1 + \cdots + s_q$ in the punctured plane $\mathbb{R}^2 \setminus \{(0,0)\}$ can detect the missing point, provided that s has nonzero "winding number" $w(s)$ about the origin. The closed curve s is assembled from the singular 1-simplices $s_1, \dots, s_q$ which join together, end to end, to form a "1-cycle" and the winding number itself is defined by integrating the locally exact (hence, closed) 1-form η of Example 6.3.12 over s.

Piecewise smooth closed loops s in $\mathbb{R}^3 \setminus \{(0,0,0)\}$ do not detect the missing point (Example 6.4.9). However, a map $s : S^2 \to \mathbb{R}^3 \setminus \{(0,0,0)\}$ *can* snag the missing point. One effective way to use this observation is to triangulate S^2 (Section 1.3) and form singular 2-simplices $s_1, \dots, s_q$ by restricting s to these triangles. It is necessary to assume only that each s_i is smooth, so we get a piecewise smooth map

$$s : S^2 \to \mathbb{R}^3 \setminus \{(0,0,0)\}$$

and write $s = s_1 + \cdots + s_q$ by analogy with the case of loops. We call this a "singular 2-cycle". One should test whether or not the singular 2-cycle has snagged the missing point x by integrating a suitable closed 2-form ω over this cycle:

$$\int_s \omega = \sum_{i=1}^{q} \int_{s_i} \omega.$$

The possibilities for singular 2-cycles are richer than for 1-cycles. For instance, triangulations of T^2 and corresponding piecewise smooth maps s of T^2 into M define "toroidal" singular 2-cycles $s = s_1 + \cdots + s_q$ in the manifold M and such a cycle might well detect a topological feature that would be missed by a "spherical" cycle. Again, a test of what this cycle detects is made by integrating closed 2-forms over the cycle.

These remarks are extended and made precise by defining the *singular homology* of a manifold, a covariant functor H_* from the category of smooth manifolds to the category of graded vector spaces over $\mathbb{R}$. The celebrated de Rham theorem asserts that this functor is dual to de Rham cohomology. We sketch the main facts, illustrating the importance of the combinatorial Stokes' theorem for algebraic topology.

Definition 8.2.11. The set of all singular p-simplices in M, $p \geq 0$, is denoted by $\Delta_p(M)$. The space $C_p(M)$ of singular p-chains on M is the free $\mathbb{R}$-module (real vector space) generated by the set $\Delta_p(M)$. By convention, if $p < 0$, $\Delta_p(M) = \emptyset$ and $C_p(M) = 0$.

Each p-form $\omega \in A^p(M)$ can be viewed as a linear functional

$$\omega : C_p(M) \to \mathbb{R}$$

as follows. An arbitrary p-chain $c \in C_p(M)$ can be written uniquely (up to terms with coefficient 0) as a linear combination

$$c = \sum_{j=1}^{m} a_j s_j,$$

where $s_j \in \Delta_p(M)$, $1 \leq j \leq m$. One then defines the value of ω on c to be

$$\omega(c) = \int_c \omega = \sum_{j=1}^{m} a_j \int_{s_j} \omega.$$

The face operators $\partial_i s = s \circ F_i$ have already been defined, $\forall\, s \in \Delta_p(M)$, and can be viewed as set maps

$$\partial_i : \Delta_p(M) \to C_{p-1}(M).$$

Since $\Delta_p(M)$ is a basis of $C_p(M)$, these set maps extend uniquely to linear maps

$$\partial_i : C_p(M) \to C_{p-1}(M).$$

Definition 8.2.12. The boundary operator $\partial : C_p(M) \to C_{p-1}(M)$, $p \geq 0$, is the linear map

$$\partial = \sum_{i=0}^{p} (-1)^i \partial_i.$$

For $\omega \in A^{p-1}(M)$ and $c \in C_p(M)$, the combinatorial Stokes' theorem asserts that

$$\omega(\partial c) = \int_{\partial c} \omega = \int_c d\omega = d\omega(c).$$

That is, the operators d and ∂ are adjoint to one another.

The boundary operator is an algebraic analogue of the geometric notion of a boundary. The following crucial property can be viewed as the algebraic analogue of the fact that the boundary of the boundary of a manifold is empty.

Exercise 8.2.13. Prove that the composition

$$C_p(M) \xrightarrow{\partial} C_{p-1}(M) \xrightarrow{\partial} C_{p-2}(M)$$

is trivial ($\partial^2 = 0$).

Definition 8.2.14. The subspace $Z_p(M) \subseteq C_p(M)$ of all (singular) p-cycles is the kernel of the boundary operator $\partial : C_p(M) \to C_{p-1}(M)$. The subspace $B_p(M) \subseteq C_p(M)$ of all (singular) p-boundaries is the image of the boundary operator $\partial : C_{p+1}(M) \to C_p(M)$.

An immediate corollary of Lemma 8.2.13 is that $B_p(M) \subseteq Z_p(M)$.

Definition 8.2.15. The pth singular homology of M is the vector space

$$H_p(M) = Z_p(M)/B_p(M).$$

If $z \in Z_p(M)$, the homology class of z is the coset $[z] \in H_p(M)$ represented by the cycle z.

Example 8.2.16. Since $C_{-1}(M) = 0$, the boundary operator vanishes identically on $C_0(M)$. That is, $Z_0(M) = C_0(M)$ is the real vector space with basis the set of points of M. Define $\epsilon : Z_0(M) \to \mathbb{R}$ by

$$\epsilon\left(\sum_{i=1}^{m} a_i x_i\right) = \sum_{i=1}^{m} a_i,$$

where all $a_i \in \mathbb{R}$ and all $x_i \in M$. If $s \in \Delta_1(M)$, $\epsilon(\partial s) = \epsilon(s(1) - s(0)) = 0$, so $B_0(M) \subseteq \ker(\epsilon)$ and ϵ passes to a well-defined linear map $\tilde{\epsilon} : H_0(M) \to \mathbb{R}$. We assume that $M \neq \emptyset$, so there is a point $x \in M$ and $\tilde{\epsilon}([x]) = 1$, proving that $\tilde{\epsilon}$ is a surjection and $[x] \neq 0$, $\forall x \in M$. If M is *connected*, then every two points $x, y \in M$ can be joined by a piecewise smooth path $s = s_1 + \cdots + s_r$ and $\partial s = y - x$. This implies that $[x] = [y]$, hence that $H_0(M)$ has basis consisting of a single element $[x]$. We have proven that the 0th singular homology of a nonempty, connected manifold is canonically isomorphic to $\mathbb{R}$.

Example 8.2.17. Let M be contractible (*cf.* Exercise 8.1.9) with contraction $\varphi_t : M \to M$. This is a homotopy of $\varphi_0 = \mathrm{id}_M$ to a constant map φ_1. Using this contraction, we are going to define linear maps

$$L_p : C_p(M) \to C_{p+1}(M),$$

$\forall p \geq 0$, with a remarkable property.

The standard inclusion $\mathbb{R}^p \hookrightarrow \mathbb{R}^{p+1}$ restricts to an inclusion $\Delta_p \hookrightarrow \Delta_{p+1}$ which is just the face map F_{p+1}. For each point $v \in \Delta_{p+1} \smallsetminus \{e_{p+1}\}$, there is a unique point $v' \in \Delta_p$ and a unique number $t \in [0,1]$ such that

$$v = te_{p+1} + (1-t)v',$$

and every point of $\mathbb{R}^{p+1}$ of such a form is a point in Δ_{p+1}. (For $v = e_{p+1}$, v' is not unique, but $t = 1$, so this will cause no problem in what follows.) If $s : \Delta_p \to M$ is smooth, define a smooth map $L_p(s) : \Delta_{p+1} \to M$ by the formula

$$L_p(s)(te_{p+1} + (1-t)v') = \varphi_t(s(v')).$$

The fact that this is well defined when $t = 1$ is due to φ_1 being a constant map. We view $s \mapsto L_p(s)$ as a set map

$$L_p : \Delta_p(M) \to C_{p+1}(M)$$

and take the linear map L_p to be the unique linear extension of this set map to all of $C_p(M)$. In Exercise 8.2.18, you are invited to check that

$$(*) \qquad \partial \circ L_p = L_{p-1} \circ \partial + (-1)^{p+1}\, \mathrm{id}_{C_p(M)},$$

provided that $p \geq 1$. This is the remarkable property promised above. If $z \in Z_p(M)$ and $p \geq 1$, it follows that

$$\partial(L_p(z)) = L_{p-1}(\partial z) + (-1)^{p+1} z = (-1)^{p+1} z,$$

hence that $Z_p(M) \subseteq B_p(M)$. The reverse inclusion also holds, so we have the result that the singular p-cycles and the singular p-boundaries in a contractible space are exactly the same, $\forall p \geq 1$. That is, $H_p(M) = 0$ in all degrees $p > 0$.

Exercise 8.2.18. Prove the identity $(*)$ in Example 8.2.17.

The above two examples give

Theorem 8.2.19. *If M is a contractible n-manifold, then*

$$H_p(M) = \begin{cases} \mathbb{R}, & p = 0, \\ 0, & p > 0. \end{cases}$$

In particular, this is true for $M = \mathbb{R}^n$.

Proposition 8.2.20. *If $\omega \in Z^p(M)$ and $z \in Z_p(M)$, then the real number $\int_z \omega$ depends only on the cohomology class $[\omega] \in H^p(M)$ and the homology class $[z] \in H_p(M)$.*

Proof. Indeed, $[\omega]$ is the set of all closed p-forms $\omega + d\eta$, where $\eta \in A^{p-1}(M)$. We have

$$\int_z d\eta = \int_{\partial z} \eta = \int_0 \eta = 0,$$

by Stokes' theorem and the fact that z is a cycle, so

$$\int_z \omega + d\eta = \int_z \omega.$$

Similarly, $[z]$ is the set of all p-cycles of the form $z + \partial c$, where $c \in C_{p+1}(M)$. Since

$$\int_{\partial c} \omega = \int_c d\omega = \int_c 0 = 0,$$

we obtain

$$\int_{z+\partial c} \omega = \int_z \omega + \int_{\partial c} \omega = \int_z \omega.$$

□

Thus, we can define an $\mathbb{R}$-linear map $\mathrm{DR} : H^p(M) \to H_p(M)^*$, by

$$(\mathrm{DR}[\omega])([z]) = \int_{[z]} [\omega].$$

Theorem 8.2.21 (The de Rham Theorem). *The linear map* DR *is a canonical isomorphism of vector spaces.*

This is a deep result. For the case in which M is compact, the proof will be discussed in some detail in Section 8.9. In that case, the vector spaces are finite dimensional (Theorem 8.5.8), so we also get $H_p(M) = H^p(M)^*$.

The following corollary generalizes Proposition 6.4.3.

Corollary 8.2.22. *Let $\omega, \widetilde{\omega} \in Z^p(M)$. Then $[\omega] = [\widetilde{\omega}]$ if and only if $\int_z \omega = \int_z \widetilde{\omega}$ as z ranges over all singular p-cycles in M. These numbers are called the periods of ω and of the cohomology class $[\omega]$.*

In particular, ω is an exact form if and only if all of its periods are 0, which generalizes the equivalence of properties (1) and (2) in Theorem 6.3.10.

Exercise 8.2.23. Let M be an n-manifold and let $z \in Z_{n+1}(M)$. Assuming the de Rham theorem, prove that there is a chain $c \in C_{n+2}(M)$ such that $z = \partial c$.

Exercise 8.2.24. Show that singular homology is a covariant functor, proceeding as follows.

(1) If $f : M \to N$ is a smooth map between manifolds, exhibit a canonical way to induce a linear map $f_\# : C_p(M) \to C_p(N)$, $\forall\, p \geq 0$.
(2) Prove that the diagram

$$\begin{array}{ccc} C_{p+1}(M) & \xrightarrow{f_\#} & C_{p+1}(N) \\ \partial \downarrow & & \downarrow \partial \\ C_p(M) & \xrightarrow[f_\#]{} & C_p(N) \end{array}$$

commutes, $\forall\, p \geq 0$. Conclude that the linear map $f_\#$ passes to a linear map $f_* : H_*(M) \to H_*(N)$ of graded vector spaces.
(3) Verify the properties $(f \circ g)_* = f_* \circ g_*$ and $\mathrm{id}_* = \mathrm{id}$.
(4) Under the de Rham isomorphism of $H^k(M)$ with $H_k(M)^*$, show that

$$f^* : H^k(N) \to H^k(M),$$
$$f_* : H_k(M) \to H_k(N)$$

are adjoint to each other.

Exercise 8.2.25. Without appealing to the de Rham theorem, extend the argument in Example 8.2.16 to show that $H_0(M)$ is a direct sum of copies of $\mathbb{R}$, one for each connected component of M.

8.3. The Poincaré Lemma

In the following discussion, R will stand for any nondegenerate, compact interval $[a, b]$ or for $\mathbb{R}$. We consider an arbitrary n-manifold M, not necessarily orientable. If M has nonempty boundary, $M \times R$ will always denote $M \times \mathbb{R}$, thereby avoiding manifolds with corners. Homotopies, therefore, will be understood in the sense of Definition 3.8.9 whenever convenient. We will agree to denote the standard projections by

$$\pi : M \times R \to M$$

and

$$p : M \times R \to R.$$

The coordinate of R will be denoted by t and $p^*(dt) \in A^1(M \times R)$ will be denoted by dt (an abuse).

A locally finite atlas $\{(W_\alpha, x_\alpha)\}_{\alpha\in\mathfrak{A}}$ on M determines such an atlas $\{(W_\alpha \times R, (x_\alpha, t))\}_{\alpha\in\mathfrak{A}}$ on $M \times R$. Here, if $\partial M = \emptyset$ and $R = [a, b]$, we model $(n+1)$-manifolds with boundary on $\mathbb{R}^n \times [a, b]$ instead of on $\mathbb{H}^{n+1}$.

A smooth partition of unity $\{\lambda_\alpha\}_{\alpha\in\mathfrak{A}}$, subordinate to $\{W_\alpha\}_{\alpha\in\mathfrak{A}}$, determines a smooth partition of unity

$$\{\overline{\lambda}_\alpha = \pi^*(\lambda_\alpha) = \lambda_\alpha \circ \pi\}_{\alpha\in\mathfrak{A}}$$

on $M \times R$ subordinate to $\{W_\alpha \times R\}_{\alpha\in\mathfrak{A}}$. Here,

$$\overline{\lambda}_\alpha(x, t) = \lambda_\alpha(x).$$

If $\omega \in A^k(M \times R)$, let

$$\omega_\alpha = \omega|(W_\alpha \times R)$$

and write

$$\omega_\alpha = \sum_I f_I^\alpha \, dx_\alpha^I \wedge dt + \sum_J g_J^\alpha \, dx_\alpha^J,$$

where we use the conventions

$$I = i_1, i_2, \dots, i_{k-1}, \qquad 1 \le i_1 < i_2 < \dots < i_{k-1} \le n,$$
$$J = j_1, j_2, \dots, j_k, \qquad 1 \le j_1 < j_2 < \dots < j_k \le n,$$

and

$$dx^I_\alpha = dx^{i_1}_\alpha \wedge dx^{i_2}_\alpha \wedge \dots \wedge dx^{i_{k-1}}_\alpha,$$
$$dx^J_\alpha = dx^{j_1}_\alpha \wedge dx^{j_2}_\alpha \wedge \dots \wedge dx^{j_k}_\alpha.$$

Since

$$\omega = \sum_{\alpha\in\mathfrak{A}} \overline{\lambda}_\alpha \omega_\alpha,$$

we write

$$\omega = \sum_{\alpha\in\mathfrak{A}} \left(\sum_I \overline{\lambda}_\alpha f^\alpha_I \, dx^I_\alpha \wedge dt + \sum_J \overline{\lambda}_\alpha g^\alpha_J \, dx^J_\alpha \right).$$

For each $\alpha \in \mathfrak{A}$, choose

$$\theta_\alpha : M \to [0,1]$$

with $\operatorname{supp}(\theta_\alpha) \subset W_\alpha$ and $\theta_\alpha | \operatorname{supp}(\lambda_\alpha) \equiv 1$. Let $\overline{\theta}_\alpha = \pi^*(\theta_\alpha)$. Then, $\overline{\lambda}_\alpha \overline{\theta}_\alpha = \overline{\lambda}_\alpha$ and

$$\omega = \sum_{\alpha\in\mathfrak{A}} \left(\sum_I \overline{\lambda}_\alpha f^\alpha_I (\overline{\theta}_\alpha \, dx^I_\alpha) \wedge dt + \sum_J \overline{\lambda}_\alpha g^\alpha_J (\overline{\theta}_\alpha dx^J_\alpha) \right).$$

Each $x \in M$ has an open neighborhood U such that only finitely many indices $\alpha \in \mathfrak{A}$ correspond to nonzero terms in the expression for $\omega | \pi^{-1}(U)$. Also, for $q = k$ or $k-1$, $\overline{\theta}_\alpha \, dx^{i_1}_\alpha \wedge \dots \wedge dx^{i_q}_\alpha = \pi^*(\eta)$, for some $\eta \in A^q(M)$.

Lemma 8.3.1. *Each form $\omega \in A^k(M \times R)$ can be expressed as a locally finite sum of k-forms, each being one of the following two types:*

(i) $f(x,t)\, dt \wedge \pi^*(\eta)$, $\eta \in A^{k-1}(M)$,
(ii) $f(x,t) \pi^*(\eta)$, $\eta \in A^k(M)$.

We construct an important operator which "integrates out" the dt component of forms on $M \times R$. This operator is a special case of an operator in algebraic topology called "integration over the fiber".

Lemma 8.3.2. *For each $\tau \in R$ and each integer $k \ge 0$, there is a unique $\mathbb{R}$-linear map*

$$S_\tau : A^k(M \times R) \to A^{k-1}(M \times R)$$

which is additive over locally finite sums and satisfies

(a) $S_\tau(f(x,t)\, dt \wedge \pi^*(\eta)) = \left(\int_\tau^t f(x,u)\, du\right) \pi^*(\eta)$,
(b) $S_\tau(f(x,t)\pi^*(\eta)) = 0$.

(Here we understand that $A^{-1}(M \times R) = 0$, so $S_\tau : A^0(M \times R) \to 0$ is trivial, consistent with the fact that all forms in $A^0(M \times R)$ are of type (ii).)

Proof. By the existence of a decomposition of $\omega \in A^k(M \times R)$ into a locally finite sum of forms of the types (i) and (ii), the stipulated properties of S_τ force that operator to be *unique*, provided that it *exists*. But, if we fix the choice of locally finite atlas $\{W_\alpha, x_\alpha\}_{\alpha\in\mathfrak{A}}$, as well as the choice of subordinate partition of unity $\{\lambda_\alpha\}_{\alpha\in\mathfrak{A}}$ and of the functions $\{\theta_\alpha\}_{\alpha\in\mathfrak{A}}$, we then have an algorithm for producing a locally finite decomposition of $\omega \in A^k(M \times R)$ into the desired types of summands. We use (a) and (b) to define S_τ on each of these summands and remark that the result is a locally finite system of $(k-1)$-forms on $M \times R$, hence that their sum is a well defined element $S_\tau(\omega) \in A^{k-1}(M \times R)$. It is clear that S_τ, defined in this way, is $\mathbb{R}$-linear. The crucial fact that it is also additive on locally finite sums is left as Exercise 8.3.3. Uniqueness shows that the definition of S_τ is really independent of the choices. $\square$

Exercise 8.3.3. Prove that the operator S_τ in Lemma 8.3.2 is, indeed, additive on locally finite sums.

For each $\tau \in R$, let $i_\tau : M \to M \times R$ be given by $i_\tau(x) = (x, \tau)$. One version of the Poincaré lemma is that, at the cohomology level, π^* and i_τ^* are mutually inverse isomorphisms. Indeed, it is clear that $i_\tau^* \circ \pi^* = (\pi \circ i_\tau)^*$ is the identity at the level of forms, so it remains to show that $\pi^* \circ i_\tau^*$ is the identity on cohomology. The main step is the following.

Exercise 8.3.4. On $A^k(M \times R)$, prove that the operator S_τ satifies the identity

$$d \circ S_\tau + S_\tau \circ d = \mathrm{id} - \pi^* \circ i_\tau^*,$$

$\forall \tau \in R, \forall k \geq 0$.

Theorem 8.3.5 (Poincaré Lemma, Version I). *The map*

$$\pi^* : H^*(M) \to H^*(M \times R)$$

is an isomorphism and its inverse is i_τ^*, $\forall \tau \in R$. *In particular, at the cohomology level,* i_τ^* *is independent of* τ.

Proof. As remarked above, we only need to prove that, at the cohomology level, $\pi^* \circ i_\tau^* = \mathrm{id}$. If $\omega \in Z^p(M \times R)$, we apply Exercise 8.3.4 to obtain

$$\omega - \pi^*(i_\tau^*(\omega)) = d(S_\tau(\omega)) + S_\tau(d(\omega)) = d(S_\tau(\omega)).$$

That is, ω and $\pi^*(i_\tau^*(\omega))$ differ by a coboundary, and we are done. $\square$

Theorem 8.3.6 (Poincaré Lemma, Version II). *If* $f_0, f_1 : M \to N$ *are homotopic, then* $f_0^* = f_1^* : H^*(N) \to H^*(M)$.

Proof. Let $F : M \times \mathbb{R} \to N$ be the homotopy. Then $f_0 = F \circ i_0$ and $f_1 = F \circ i_1$. By functoriality,

$$f_0^* = i_0^* \circ F^*,$$
$$f_1^* = i_1^* \circ F^*.$$

But $i_0^* = i_1^*$ by Theorem 8.3.5, so $f_0^* = f_1^*$. $\square$

Here are four more versions of the Poincaré lemma. The first of these is immediate by Theorem 8.3.6, and each implies the next. All of the implications are rather obvious.

Theorem 8.3.7 (Poincaré Lemma, Version III). *If* $f : M \to N$ *is a homotopy equivalence, then* $f^* : H^*(N) \to H^*(M)$ *is an isomorphism of graded algebras.*

Theorem 8.3.8 (Poincaré Lemma, Version IV). *If* M *is a contractible manifold,*

$$H^k(M) = \begin{cases} \mathbb{R}, & k = 0, \\ 0, & k > 0. \end{cases}$$

In particular, this holds for $M = \mathbb{R}^n$.

Theorem 8.3.9 (Poincaré Lemma, Version V). *For $k > 0$, every closed k-form on a contractible manifold is exact.*

Since manifolds are locally contractible (each point has a neighborhood diffeomorphic to $\mathbb{R}^n$), the next version follows.

Theorem 8.3.10 (Poincaré Lemma, Version VI). *If $k > 0$, a k-form on a manifold M is closed if and only if it is locally exact.*

Thus, the definition of $H^1(M)$ given in Chapter 6 agrees with our current definition.

It can be shown that Version VI implies Version I, so all versions are mutually equivalent. We will not prove this.

Definition 8.3.11. If $\partial M = \emptyset$ and $f_0, f_1 : M \to N$ are proper smooth maps, they are said to be properly homotopic if there is a proper smooth map

$$F : M \times [0,1] \to N$$

such that $F(x,0) = f_0(x)$ and $F(x,1) = f_1(x)$, $\forall x \in M$. The map F is called a proper homotopy between f_0 and f_1.

For compactly supported cohomology on manifolds without boundary, define the term "homotopy" using $R = [0,1]$ and remark that both π^* and i_τ^* are proper maps. Theorem 8.3.5 continues to hold in this situation and a suitably reworded version of Theorem 8.3.6 also holds (Exercise 8.3.19). One could call this the Poincaré lemma, but what usually goes by that name for compact cohomology takes a rather different form. This is our next topic.

First note that $\omega \in A_c^k(M \times \mathbb{R})$ is a *finite* linear combination of forms of the types

(*i*) $f(x,t)\, dt \wedge \pi^*(\eta)$, where $f(x,t)$ is compactly supported, but the form $\eta \in A^{k-1}(M)$ may not be compactly supported.

(*ii*) $f(x,t)\pi^*(\eta)$, where $f(x,t)$ is compactly supported, but the form $\eta \in A^k(M)$ may not be compactly supported.

One then defines an $\mathbb{R}$-linear map

$$\pi_* : A_c^k(M \times \mathbb{R}) \to A_c^{k-1}(M),$$

called "integration along $\mathbb{R}$", by requiring that

(*a*) $\pi_*(f(x,t)\, dt \wedge \pi^*(\eta)) = \left(\int_{-\infty}^{\infty} f(x,t)\, dt\right)\eta$,

(*b*) $\pi_*(f(x,t)\pi^*(\eta)) = 0$.

As before, there is a unique $\mathbb{R}$-linear operator π_* with these properties. Remark that requiring the operator to be additive over locally finite sums is no longer necessary.

It will also be convenient to define $A_c^{-q}(M) = 0$, $\forall q > 0$, to agree that d $(= 0)$ is defined on this trivial module, hence to have $H^{-q}(M)$ defined and trivial.

Lemma 8.3.12. *With the above definitions,*

$$d \circ \pi_* = -\pi_* \circ d : A_c^k(M \times \mathbb{R}) \to A_c^k(M),$$

$\forall k \in \mathbb{Z}$. *That is, π_* anticommutes with d.*

The proof is a straightforward computation on forms of types (*i*) and (*ii*). It is analogous to Exercise 8.3.4, only easier.

Corollary 8.3.13. *The linear map* π_* *passes to a well-defined linear map*

$$\pi_* : H_c^k(M \times \mathbb{R}) \to H_c^{k-1}(M),$$

$\forall k \in \mathbb{Z}$.

We want to prove that this map is an isomorphism, so we need a candidate for its inverse. Choose a compactly supported function $b : \mathbb{R} \to \mathbb{R}$ such that $\int_{-\infty}^{\infty} b(t)\, dt = 1$. Let $\beta = b(t)\, dt \in A_c^1(\mathbb{R})$. Finally, for each $k \in \mathbb{Z}$, define

$$\beta_* : A_c^k(M) \to A_c^{k+1}(M \times \mathbb{R})$$

by

$$\beta_*(\eta) = b(t)\, dt \wedge \pi^*(\eta).$$

It is practically immediate that

$$d \circ \beta_* = -\beta_* \circ d,$$

and we draw the following conclusion.

Lemma 8.3.14. *The linear map* β_* *passes to a well-defined linear map*

$$\beta_* : H_c^k(M) \to H_c^{k+1}(M \times \mathbb{R}),$$

$\forall k \in \mathbb{Z}$.

Clearly,

$$\pi_* \beta_*(\eta) = \left(\int_{-\infty}^{\infty} b(t)\, dt \right) \eta = \eta$$

and we will show that, at the level of compact cohomology, $\beta_* \circ \pi_*$ is also the identity. Once again, we construct an operator

$$S : A_c^k(M \times \mathbb{R}) \to A_c^{k-1}(M \times \mathbb{R})$$

such that

$$d \circ S + S \circ d = \mathrm{id} - \beta_* \circ \pi_*.$$

To define S, set

$$B(t) = \int_{-\infty}^{t} b(u)\, du$$

and define S on forms of type (ii) by

$$S(f(x,t)\pi^*(\eta)) = 0$$

and, on those of type (i), by

$$S(f(x,t)\, dt \wedge \pi^*(\eta)) = \left(\int_{-\infty}^{t} f(x,u)\, du - B(t) \int_{-\infty}^{\infty} f(x,u)\, du \right) \pi^*(\eta).$$

Remark that this form is, indeed, compactly supported. This is obvious in the x variable and, for $t \downarrow -\infty$, it is also clear. But, as $t \uparrow \infty$, the function in the parentheses ultimately becomes

$$\int_{-\infty}^{\infty} f(x,u)\, du - 1 \cdot \int_{-\infty}^{\infty} f(x,u)\, du = 0,$$

so the support is bounded in all directions, hence compact.

Exercise 8.3.15. Prove that the formula

$$d \circ S + S \circ d = \mathrm{id} - \beta_* \circ \pi_*$$

holds on $A_c^k(M \times \mathbb{R})$, $\forall k \in \mathbb{Z}$.

As in the proof of Theorem 8.3.5, this is all that is needed to establish the following.

Theorem 8.3.16 (Poincaré lemma for compact supports). *The map*

$$\pi_* : H_c^k(M \times \mathbb{R}) \to H_c^{k-1}(M)$$

is a canonical isomorphism with inverse β_*, $\forall\, k \in \mathbb{Z}$.

In particular, β_* does not depend (in compact cohomology) on the choice of the compactly supported function $b(t)$ such that $\int_{-\infty}^{\infty} b(t)\, dt = 1$.

Remark. The operators S_τ and S, used to prove the Poincaré lemmas for ordinary and compact de Rham theory, are examples of *cochain homotopies* in algebraic topology. One says that $\mathrm{id}_{A^*(M\times R)}$ is cochain homotopic to $\pi^* \circ i_\tau^*$, writing

$$\pi^* \circ i_\tau^* \sim \mathrm{id}_{A^*(M\times R)}\,.$$

Similarly,

$$\beta_* \circ \pi_* \sim \mathrm{id}_{A_c^*(M\times\mathbb{R})}\,.$$

Exactly as in the proof of Theorem 8.3.5, cochain homotopic maps induce the same map in cohomology. In the present situation, since one of the maps is the identity, they both induce the identity. In Example 8.2.17, we used a *chain homotopy* between the identity and a map that is 0 in positive degrees to show that the singular homology of a contractible manifold is trivial.

Corollary 8.3.17. *For each integer* $n \geq 0$,

$$H_c^k(\mathbb{R}^n) = \begin{cases} \mathbb{R}, & k = n\,, \\ 0, & \textit{otherwise.} \end{cases}$$

Proof. This is clearly true for $n = 0$. Inductively, suppose that it is true for a given value of $n \geq 0$ and appeal to Theorem 8.3.16 to get

$$H_c^k(\mathbb{R}^{n+1}) = H_c^k(\mathbb{R}^n \times \mathbb{R}) = H_c^{k-1}(\mathbb{R}^n),$$

$\forall\, k \in \mathbb{Z}$. □

Corollary 8.3.18. *The linear map*

$$\int_{\mathbb{R}^n} : H_c^n(\mathbb{R}^n) \to \mathbb{R}$$

is an isomorphism.

Indeed, since $\mathbb{R}^n$ is orientable, we have seen that this is a surjection (Theorem 8.2.4). Since the cohomology space is one-dimensional, it is an isomorphism.

Exercise 8.3.19. Show that Theorem 8.3.5 makes sense and holds for compactly supported cohomology, provided that $\partial M = \emptyset$ and $R = [a, b]$. Using proper maps and proper homotopies, formulate and prove the analogue of Theorem 8.3.6.

8.4. Exact Sequences

A basic tool for computing cohomology will be the Mayer–Vietoris sequence (Section 8.5). In order to develop and apply this sequence, we will need some properties of exact sequences. This purely algebraic section may be a review for many readers. At any rate, the proofs are elementary and will be relegated to exercises.

We fix a commutative ring R and consider modules A, B, C, *etc.*, over R. All maps $\varphi : A \to B$ will be R-linear. We also consider graded R-modules A^*, B^*, *etc.*, over R, in which case $\varphi : A^* \to B^*$ will denote a homomorphism of graded R-modules. We will generally assume that the grading is indexed by $\mathbb{Z}$ rather than just $\mathbb{Z}^+$. No generality is lost since $A^* = \{A^k\}_{k=0}^{\infty}$ can be replaced by $\{A^k\}_{k=-\infty}^{\infty}$ by setting $A^{-p} = 0, \forall p > 0$.

Definition 8.4.1. A sequence

$$\cdots \longrightarrow A \xrightarrow{\alpha} B \xrightarrow{\beta} C \longrightarrow \cdots$$

of module homomorphisms is said to be exact at B if $\operatorname{im}(\alpha) = \ker(\beta)$. If a sequence of module homomorphisms is exact at each module (except the first and last), it is called an exact sequence. An exact sequence of the form

$$0 \longrightarrow A \xrightarrow{i} B \xrightarrow{j} C \longrightarrow 0$$

is called a short exact sequence. Similarly, the notions of exact sequence and of short exact sequence are defined for graded module homomorphisms.

Remark that, in the short exact sequence, i is injective and j is surjective.

Exercise 8.4.2. Let

$$\begin{array}{ccccccccc} A & \xrightarrow{\alpha} & B & \xrightarrow{\beta} & C & \xrightarrow{\gamma} & D & \xrightarrow{\delta} & E \\ \lambda\downarrow & & \mu\downarrow & & \nu\downarrow & & \rho\downarrow & & \pi\downarrow \\ A' & \xrightarrow[\alpha']{} & B' & \xrightarrow[\beta']{} & C' & \xrightarrow[\gamma']{} & D' & \xrightarrow[\delta']{} & E' \end{array}$$

be a commutative diagram in which the two rows are exact. If λ, μ, ρ, and π are isomorphisms, prove that ν is an isomorphism. This is called the *five lemma*.

Definition 8.4.3. A cochain complex (A^*, δ) is a graded R-module, together with a sequence

$$\cdots \xrightarrow{\delta} A^p \xrightarrow{\delta} A^{p+1} \xrightarrow{\delta} A^{p+2} \xrightarrow{\delta} \cdots$$

such that $\delta^2 = 0$. Similarly, a chain complex (C_*, ∂) is a graded R-module and a sequence

$$\cdots \xrightarrow{\partial} C_p \xrightarrow{\partial} C_{p-1} \xrightarrow{\partial} C_{p-2} \xrightarrow{\partial} \cdots$$

such that $\partial^2 = 0$.

As usual, one defines $Z^p = \ker(\delta) \cap A^p$ (respectively, $Z_p = \ker(\partial) \cap C_p$) and $B^p = \operatorname{im}(\delta) \cap A^p$ (respectively, $B_p = \operatorname{im}(\partial) \cap C_p$). In what follows, we explicitly consider cochain complexes, but everything goes through, with the obvious modifications, for chain complexes. In this book, we are mainly interested in the de Rham cochain complexes $(A^*(M), d)$ and $(A^*_c(M), d)$ and in the singular chain complex $(C_*(M), \partial)$, although others will be mentioned on occasion.

The condition that $\delta^2 = 0$ implies that $B^* \subseteq Z^*$ and the cohomology of the cochain complex is defined to be

$$H^*(A^*, \delta) = Z^*/B^*,$$

a graded R-module. This can be viewed as a measure of the extent to which the sequence in Definition 8.4.3 fails to be exact. The corresponding construction for a chain complex (C_*, ∂) is called the homology of the complex and denoted by $H_*(C_*, \partial)$.

Definition 8.4.4. A homomorphism $\varphi : (A^*, \delta) \to (C^*, \delta)$ of (co)chain complexes is a homomorphism of the graded R-modules such that $\varphi \circ \delta = \delta \circ \varphi$.

Evidently, a homomorphism $\varphi : (A^*, \delta) \to (C^*, \delta)$ of cochain complexes induces a homomorphism $\varphi^* : H^*(A^*, \delta) \to H^*(C^*, \delta)$ of graded R-modules.

Definition 8.4.5. A homomorphism $\lambda : H^*(A^*, \delta) \to H^*(C^*, \delta)$ of degree $p \in \mathbb{Z}$ is a sequence of R-linear maps

$$\lambda : H^k(A^*, \delta) \to H^{k+p}(C^*, \delta),$$

$-\infty < k < \infty$. This is sometimes written $\lambda : H^*(A^*, \delta) \to H^{*+p}(C^*, \delta)$.

For instance, in the Poincaré lemma for compactly supported cohomology, we defined a homomorphism

$$\pi_* : H_c^*(M \times \mathbb{R}) \to H_c^{*-1}(M)$$

of degree -1.

Lemma 8.4.6. *Let*

$$0 \longrightarrow (C^*, \delta) \xrightarrow{i} (D^*, \delta) \xrightarrow{j} (E^*, \delta) \longrightarrow 0$$

be a short exact sequence of homomorphisms of cochain complexes. Then, there is canonically induced a homomorphism

$$\delta^* : H^*(E^*, \delta) \to H^{*+1}(C^*, \delta)$$

of degree +1, called the connecting homomorphism. This homomorphism is "natural" in the following sense: if

$$\begin{array}{ccccccccc}
0 & \longrightarrow & (C^*, \delta) & \xrightarrow{i} & (D^*, \delta) & \xrightarrow{j} & (E^*, \delta) & \longrightarrow & 0 \\
 & & \downarrow{\scriptstyle\varphi} & & \downarrow{\scriptstyle\theta} & & \downarrow{\scriptstyle\pi} & & \\
0 & \longrightarrow & (J^*, \delta) & \xrightarrow[i]{} & (K^*, \delta) & \xrightarrow[j]{} & (L^*, \delta) & \longrightarrow & 0
\end{array}$$

is a commutative diagram with both rows exact, then

$$\begin{array}{ccc}
H^*(E^*, \delta) & \xrightarrow{\delta^*} & H^{*+1}(C^*, \delta) \\
\downarrow{\scriptstyle\pi^*} & & \downarrow{\scriptstyle\varphi^*} \\
H^*(L^*, \delta) & \xrightarrow[\delta^*]{} & H^{*+1}(J^*, \delta)
\end{array}$$

also commutes.

In the case of a short exact sequence of chain complexes, the connecting homomorphism has degree -1.

We show how to find $\delta^*[e] \in H^{k+1}(C^*, \delta)$, where $[e] \in H^k(E^*, \delta)$. Consider the commutative diagram

$$\begin{array}{ccccccccc} 0 & \longrightarrow & C^k & \xrightarrow{i} & D^k & \xrightarrow{j} & E^k & \longrightarrow & 0 \\ & & \Big\downarrow{\scriptstyle \delta} & & \Big\downarrow{\scriptstyle \delta} & & \Big\downarrow{\scriptstyle \delta} & & \\ 0 & \longrightarrow & C^{k+1} & \xrightarrow{i} & D^{k+1} & \xrightarrow{j} & E^{k+1} & \longrightarrow & 0 \\ & & \Big\downarrow{\scriptstyle \delta} & & \Big\downarrow{\scriptstyle \delta} & & \Big\downarrow{\scriptstyle \delta} & & \\ 0 & \longrightarrow & C^{k+2} & \xrightarrow{i} & D^{k+2} & \xrightarrow{j} & E^{k+2} & \longrightarrow & 0 \end{array}$$

and choose $e \in E^k$ representing $[e]$. In particular, $\delta(e) = 0$. Since j is surjective, choose $e' \in D^k$ such that $j(e') = e$. Then $j(\delta(e')) = \delta(j(e')) = \delta(e) = 0$ and exactness of the middle row implies that there is a unique $c \in C^{k+1}$ such that $i(c) = \delta(e')$. Then $i(\delta(c)) = \delta(i(c)) = \delta(\delta(e')) = 0$. Since i is one-to-one, it follows that $\delta(c) = 0$, so we define $\delta^*[e] = [c] \in H^{k+1}(C^*, \delta)$. More diagram chasing proves that $[c]$ is independent of the choices of $e \in [e]$ and of $e' \in D^k$ such that $j(e') = e$.

Exercise 8.4.7. Prove that the connecting homomorphism δ^* is natural as defined in the statement of Lemma 8.4.6.

Exercise 8.4.8. Prove that a short exact sequence

$$0 \longrightarrow (C^*, \delta) \xrightarrow{i} (D^*, \delta) \xrightarrow{j} (E^*, \delta) \longrightarrow 0$$

of cochain complexes induces a long exact sequence

$$\cdots \xrightarrow{\delta^*} H^k(C^*, \delta) \xrightarrow{i^*} H^k(D^*, \delta) \xrightarrow{j^*} H^k(E^*, \delta) \xrightarrow{\delta^*} H^{k+1}(C^*, \delta) \xrightarrow{i^*} \cdots$$

in cohomology.

One sometimes writes the long exact sequence more compactly as an exact triangle:

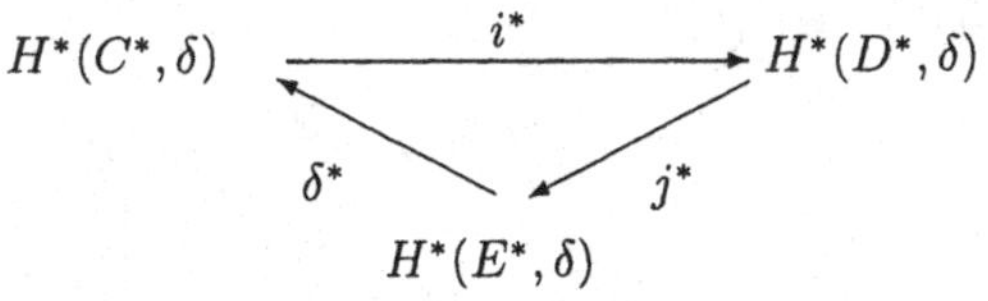

$$\begin{array}{ccc} H^*(C^*, \delta) & \xrightarrow{\quad i^* \quad} & H^*(D^*, \delta) \\ & {\scriptstyle \delta^*} \nwarrow \quad \swarrow {\scriptstyle j^*} & \\ & H^*(E^*, \delta) & \end{array}$$

Exercise 8.4.9. Let R be a field. If

$$A \xrightarrow{i} B \xrightarrow{j} C$$

is an exact sequence of vector spaces over R, prove that the dual sequence

$$A^* \xleftarrow{i^*} B^* \xleftarrow{j^*} C^*,$$

where i^* and j^* are the respective adjoints, is also exact. Find an example showing that this may fail for modules over a commutative ring.

8.5. Mayer–Vietoris Sequences

Let U_1 and U_2 be open subsets of the n-manifold M and consider the inclusions

$$\begin{aligned} j_1 &: U_1 \cap U_2 \hookrightarrow U_1, \\ j_2 &: U_1 \cap U_2 \hookrightarrow U_2, \end{aligned}$$

and

$$\begin{aligned} i_1 &: U_1, \hookrightarrow U_1 \cup U_2 \\ i_2 &: U_2 \hookrightarrow U_1 \cup U_2. \end{aligned}$$

Lemma 8.5.1. *The above inclusions give rise to a short exact sequence*

$$0 \to (A^*(U_1 \cup U_2), d) \xrightarrow{i} (A^*(U_1) \oplus A^*(U_2), d \oplus d) \xrightarrow{j} (A^*(U_1 \cap U_2), d) \to 0$$

of cochain complexes, where

$$i(\omega) = (i_1^*(\omega), i_2^*(\omega)), \quad \forall\, \omega \in A^*(U_1 \cup U_2),$$

and

$$j(\omega_1, \omega_2) = j_1^*(\omega_1) - j_2^*(\omega_2), \quad \forall\, \omega_\ell \in A^*(U_\ell), \ \ell = 1, 2.$$

Proof. Indeed, a nontrivial form on $U_1 \cup U_2$ must be nontrivial on either U_1 or U_2, so i is one-to-one. Since $j_1^* \circ i_1^* = j_2^* \circ i_2^*$, it is clear that $\operatorname{im}(i) \subseteq \ker(j)$. For the reverse inclusion, let $(\omega_1, \omega_2) \in \ker(j)$. Then $\omega_1|(U_1 \cap U_2) = \omega_2|(U_1 \cap U_2)$, so these forms fit together smoothly to define a form ω on $U_1 \cup U_2$ and $(\omega_1, \omega_2) = i(\omega)$. Finally we must prove that j is surjective. Let ω be a form on $U_1 \cap U_2$. Let $\{\lambda_1, \lambda_2\}$ be a partition of unity on $U_1 \cup U_2$ subordinate to $\{U_1, U_2\}$ and set $\omega_1 = \lambda_2 \omega$, $\omega_2 = \lambda_1 \omega$. (Note that, since λ_2 is supported in U_2, $\lambda_2 \omega$ extends smoothly by 0 to all of U_1. Similarly, $\lambda_1 \omega$ is a form on U_2.) Then, $j(\omega_1, -\omega_2) = \omega_1 + \omega_2 = \omega$. □

Theorem 8.5.2. *There is a long exact sequence*

$$\cdots \xrightarrow{d^*} H^q(U_1 \cup U_2) \xrightarrow{i^*} H^q(U_1) \oplus H^q(U_2)$$
$$\xrightarrow{j^*} H^q(U_1 \cap U_2) \xrightarrow{d^*} H^{q+1}(U_1 \cup U_2) \xrightarrow{i^*} \cdots$$

called the Mayer–Vietoris sequence.

Indeed, the cohomology of the cochain complex $(A^*(U_1) \oplus A^*(U_2), d \oplus d)$ is clearly $H^*(U_1) \oplus H^*(U_2)$, so we apply Proposition 8.4.8.

We turn to the Mayer–Vietoris sequence for compactly supported cohomology. Again, U_1 and U_2 are open subsets of some n-manifold M. Clearly, there are inclusions $\alpha_\ell : A_c^*(U_1 \cap U_2) \hookrightarrow A_c^*(U_\ell)$, $\ell = 1, 2$, and $\beta_\ell : A_c^*(U_\ell) \hookrightarrow A_c^*(U_1 \cup U_2)$, $\ell = 1, 2$. It is evident that these inclusions commute with exterior differentiation, hence induce linear maps $\alpha_\ell^*, \beta_\ell^*$ in compact cohomology, $\ell = 1, 2$.

Lemma 8.5.3. *The above inclusions induce a short exact sequence*

$$0 \to (A_c^*(U_1 \cap U_2), d) \xrightarrow{\alpha} (A_c^*(U_1) \oplus A_c^*(U_2), d \oplus d) \xrightarrow{\beta} (A_c^*(U_1 \cup U_2), d) \to 0$$

of cochain complexes, where

$$\alpha(\omega) = (\alpha_1(\omega), -\alpha_2(\omega)), \quad \forall\, \omega \in A_c^*(U_1 \cap U_2),$$

and

$$\beta(\omega_1, \omega_2) = \beta_1(\omega_1) + \beta_2(\omega_2), \quad \forall\, \omega_\ell \in A_c^*(U_\ell), \ \ell = 1, 2.$$

Proof. Everything is clear except, perhaps, the fact that β is a surjection. If ω is a compactly supported form on $U_1 \cup U_2$ and $\{\lambda_1, \lambda_2\}$ is a partition of unity on $U_1 \cup U_2$ subordinate to $\{U_1, U_2\}$, then $\lambda_1\omega$ has compact support in U_1 and $\lambda_2\omega$ has compact support in U_2 (note the switch from the proof of Lemma 8.5.1). Then, $\beta(\lambda_1\omega, \lambda_2\omega) = \omega$ and β is surjective. □

Remark. We have chosen the signs differently than in Lemma 8.5.1. This is not necessary for our present needs, but will be useful in our treatment of Poincaré duality.

Theorem 8.5.4. *There is a long exact sequence*

$$\cdots \xrightarrow{d^*} H_c^k(U_1 \cap U_2) \xrightarrow{\alpha^*} H_c^k(U_1) \oplus H_c^k(U_2)$$
$$\xrightarrow{\beta^*} H_c^k(U_1 \cup U_2) \xrightarrow{d^*} H_c^{k+1}(U_1 \cap U_2) \xrightarrow{\alpha^*} \cdots$$

called the Mayer–Vietoris sequence for compactly supported cohomology.

Definition 8.5.5. Let M be an n-manifold without boundary. An open cover $\{U_\alpha\}_{\alpha \in \mathfrak{A}}$ of M is said to be simple if it is locally finite and every nonempty, finite intersection

$$U = U_{\alpha_0} \cap U_{\alpha_1} \cap \cdots \cap U_{\alpha_r}$$

is contractible and has $H_c^*(U) = H_c^*(\mathbb{R}^n)$.

By Theorem 8.2.19 and Theorem 8.3.8, simple covers have the following property.

Lemma 8.5.6. *If $\mathfrak{U}$ is a simple cover of M and U is any nonempty, finite intersection of elements of $\mathfrak{U}$, then*

$$H^*(U) = H^*(\mathbb{R}^n),$$
$$H_*(U) = H_*(\mathbb{R}^n).$$

Theorem 8.5.7. *If M is a manifold with $\partial M = \emptyset$, then every open cover of M admits a simple refinement.*

We will postpone the proof of this theorem to Section 10.5, since it requires methods from Riemannian geometry. The idea is to produce a locally finite refinement by geodesically convex open sets and to prove that a geodesically convex open set U has the property in Definition 8.5.5. Since finite, nonempty intersections of geodesically convex sets are geodesically convex, we obtain a simple refinement.

Using Theorem 8.5.7 and Mayer–Vietoris sequences, we obtain the following interesting result.

Theorem 8.5.8. *If M admits a finite simple cover, then $H^*(M)$ and $H_c^*(M)$ are finite dimensional. In particular, if M is a compact manifold without boundary, then $H^*(M)$ is finite dimensional.*

Proof. Select a finite simple cover $\{U_i\}_{i=1}^r$ of M. We proceed by induction on r. If $r = 1$, then $M = U_1$ has the ordinary and compact cohomology of $\mathbb{R}^n$. Thus, $H^*(U_1) = H^*(\text{singleton})$, hence is finite dimensional. For $H_c^*(U_1)$, the assertion is given by Corollary 8.3.17. Suppose, then, that it has been shown that $H_c^*(N)$ and $H^*(N)$ are finite dimensional whenever N has a simple cover by $r-1$ elements, some $r \geq 2$. Let M have the simple cover $\{U_i\}_{i=1}^r$, let $U = \bigcup_{i=1}^{r-1} U_i$ and remark that $\{U_1 \cap U_r, U_2 \cap U_r, \dots, U_{r-1} \cap U_r\}$ is a simple cover of $U \cap U_r$. We consider the compactly supported case. By the inductive hypothesis, $H_c^*(U)$ and $H_c^*(U \cap U_r)$ are both finite dimensional as, of course, is $H_c^*(U_r)$. Since $M = U \cup U_r$, the Mayer–Vietoris sequence gives an exact sequence

$$H_c^*(U) \oplus H_c^*(U_r) \xrightarrow{\beta^*} H_c^*(M) \xrightarrow{d^*} H_c^{*+1}(U \cap U_r).$$

By standard linear algebra,

$$H_c^*(M) \cong \ker(d^*) \oplus \operatorname{im}(d^*) = \operatorname{im}(\beta^*) \oplus \operatorname{im}(d^*).$$

Since β^* has finite dimensional domain and d^* has finite dimensional range, the assertion for $H_c^*(M)$ follows. The proof for ordinary cohomology uses the appropriate Mayer–Vietoris sequence in the same way. □

Remark. Even if the compact manifold M has boundary, it is true that $H^*(M)$ is finite dimensional. One way to prove this is to show that $\operatorname{int}(M)$ has a finite simple cover and that M and $\operatorname{int}(M)$ are homotopically equivalent.

Exercise 8.5.9. If the manifold M is connected, but not necessarily compact, prove that the real vector spaces $H_c^*(M)$ and $H_*(M)$ have dimension at most countably infinite. You may use the de Rham theorem.

Exercise 8.5.10. Prove that

$$H^k(S^n) = \begin{cases} \mathbb{R}, & k = 0, n, \\ 0, & \text{otherwise}, \end{cases}$$

for all $n \geq 1$.

There is also a Mayer–Vietoris sequence for singular homology. The proof is similar to those for cohomology except for one technical point, the proof of which is very tedious and would take us too far afield. Since we will need this sequence for the proof of the de Rham theorem, we derive it here, referring the reader to standard references in algebraic topology for the bothersome technicality.

The inclusions

$$j_1 : U_1 \cap U_2 \hookrightarrow U_1,$$
$$j_2 : U_1 \cap U_2 \hookrightarrow U_2$$

and

$$i_1 : U_1 \hookrightarrow U_1 \cup U_2,$$
$$i_2 : U_2 \hookrightarrow U_1 \cup U_2$$

induce an exact sequence

$$(8.4) \quad 0 \to (C_*(U_1 \cap U_2), \partial) \xrightarrow{j} (C_*(U_1) \oplus (C_*(U_2), \partial \oplus \partial) \xrightarrow{i} i(C_*(U_1 \cup U_2), \partial),$$

where $j(c) = (j_{1\#}(c), -j_{2\#}(c))$ and $i(c_1, c_2) = i_{1\#}(c_1) + i_{2\#}(c_2)$ (and the induced homomorphisms $i_{1\#}$, $j_{1\#}$, *etc.*, are as in Exercise 8.2.24. The exactness is immediate. If $(*)$ were a short exact sequence, the Mayer–Vietoris sequence for singular homology would follow immediately, but it is generally false that i is a surjection. This brings us to the technical point.

Definition 8.5.11. Let $\mathfrak{U} = \{U_\alpha\}_{\alpha \in \mathfrak{A}}$ be an open cover of the manifold M. A singular p-simplex $s : \Delta_p \to M$ is said to be $\mathfrak{U}$-small if, for some $\alpha \in \mathfrak{A}$, $s(\Delta_p) \subseteq U_\alpha$. The set of $\mathfrak{U}$-small singular p-simplices is denoted by $\Delta_p^{\mathfrak{U}}(M)$. The vector subspace of $C_p(M)$ spanned by $\Delta_p^{\mathfrak{U}}(M)$ is denoted by $C_p^{\mathfrak{U}}(M)$.

By the definition of the singular boundary operator, it is immediate that

$$\partial : C_p^{\mathfrak{U}}(M) \to C_{p-1}^{\mathfrak{U}}(M).$$

Definition 8.5.12. The chain complex $(C_*^{\mathfrak{U}}(M), \partial)$ is called the complex of $\mathfrak{U}$-small chains. The homology of this complex is $H_*^{\mathfrak{U}}(M)$, the $\mathfrak{U}$-small homology of M.

It is clear that the natural inclusion of the space of $\mathfrak{U}$-small chains into the space of all chains is a homomorphism of chain complexes

$$\imath^{\mathfrak{U}} : (C_*^{\mathfrak{U}}(M), \partial) \hookrightarrow (C_*(M), \partial),$$

so there is induced a canonical homomorphism in homology

$$\imath_*^{\mathfrak{U}} : H_*^{\mathfrak{U}}(M) \to H_*(M).$$

We arrive at the technical result.

Proposition 8.5.13. *The homomorphism $\imath_*^{\mathfrak{U}}$ is a canonical isomorphism*

$$H_*^{\mathfrak{U}}(M) = H_*(M).$$

Proofs of Proposition 8.5.13 will be found in the standard references in algebraic topology, such as [**13**, pp. 85–88] and [**46**, pp. 207–208]. The idea is to *subdivide* the singular simplices in each cycle z until all simplices in the subdivision are $\mathfrak{U}$-small. With appropriate choices of signs, there results a $\mathfrak{U}$-small cycle z' with $[z'] = [z]$.

Thus, homology can be computed using $\mathfrak{U}$-small chains, for any open cover $\mathfrak{U}$ of M. In our situation, $\mathfrak{U} = \{U_1, U_2\}$ is an open cover of the manifold $U_1 \cup U_2$ and we replace the sequence $(*)$ with the short exact sequence

$$0 \to (C_*(U_1 \cap U_2), \partial) \xrightarrow{j} (C_*(U_1) \oplus (C_*(U_2), \partial \oplus \partial) \xrightarrow{i} (C_*^{\mathfrak{U}}(U_1 \cup U_2), \partial) \to 0.$$

Theorem 8.5.14. *There is a long exact sequence*

$$\cdots \xrightarrow{\partial_*} H_p(U_1 \cap U_2) \xrightarrow{j_*} H_p(U_1) \oplus H_p(U_2) \xrightarrow{i_*} H_p(U_1 \cup U_2) \xrightarrow{\partial_*} H_{p-1}(U_1 \cap U_2) \xrightarrow{j_*} \cdots$$

called the Mayer–Vietoris homology sequence.

Using the result of Exercise 8.4.9, we obtain dual Mayer–Vietoris sequences.

Theorem 8.5.15. *The Mayer–Vietoris sequences dualize to exact sequences*

$$\cdots \xrightarrow{i'} H^q(U_1)^* \oplus H^q(U_2)^* \xrightarrow{j'} H^q(U_1 \cap U_2)^* \xrightarrow{d'} H^{q+1}(U_1 \cup U_2)^* \xrightarrow{i'} \cdots$$

$$\cdots \xrightarrow{\alpha'} H_c^q(U_1)^* \oplus H_c^q(U_2)^* \xrightarrow{\beta'} H_c^q(U_1 \cup U_2)^* \xrightarrow{d'} H_c^{q+1}(U_1 \cap U_2)^* \xrightarrow{i'} \cdots$$

$$\cdots \xrightarrow{j'} H_p(U_1)^* \oplus H_p(U_2)^* \xrightarrow{i'} H_p(U_1 \cup U_2)^* \xrightarrow{\partial'} H_{p-1}(U_1 \cap U_2)^* \xrightarrow{j'} \cdots$$

where i' is the adjoint of i^* (respectively, of i_*), etc.

8.6. Computations of Cohomology

In this section, we compute the top dimensional cohomology of connected manifolds.

Lemma 8.6.1. *If the open subsets U_1, U_2, and $U_1 \cap U_2$ of an n-manifold M all have the same compact cohomology as $\mathbb{R}^n$ and are coherently oriented, then*

$$\int_{U_1 \cup U_2} : H_c^n(U_1 \cup U_2) \to \mathbb{R}$$

is an isomorphism.

Proof. Consider the diagram

$$\begin{array}{ccccccc}
H_c^n(U_1 \cap U_2) & \xrightarrow{\alpha^*} & H_c^n(U_1) \oplus H_c^n(U_2) & \xrightarrow{\beta^*} & H_c^n(U_1 \cup U_2) & \xrightarrow{d^*} & 0 \\
{\scriptstyle \int_{U_1 \cap U_2}} \downarrow & & {\scriptstyle \int_{U_1} \oplus \int_{U_2}} \downarrow & & {\scriptstyle \int_{U_1 \cup U_2}} \downarrow & & \downarrow {\scriptstyle \mathrm{id}} \\
\mathbb{R} & \xrightarrow[\Delta]{} & \mathbb{R} \oplus \mathbb{R} & \xrightarrow[\delta]{} & \mathbb{R} & \longrightarrow & 0
\end{array}$$

where $\Delta(t) = (t, -t)$ and $\delta(s,t) = s + t$, $\forall\, s, t \in \mathbb{R}$. Commutativity of the diagram is obvious (coherency of orientations is essential), exactness of the bottom row is obvious, and the top row is exact by Theorem 8.5.4. The map $\int_{U_1 \cap U_2}$ is an isomorphism by the hypothesis that $H_c^n(U_1 \cap U_2) = \mathbb{R}$ (as in Corollary 8.3.18). Similarly, $\int_{U_1} \oplus \int_{U_2}$ is an isomorphism. Since the diagram can be extended harmlessly by a commutative square of 0s on the right, it follows from the Five Lemma that $\int_{U_1 \cup U_2}$ is an isomorphism. □

Lemma 8.6.2. *Let $\{U_\alpha\}_{\alpha \in \mathfrak{A}}$ be a simple open cover of a connected, oriented n-manifold M without boundary, each U_α being oriented coherently with the orientation of M. Let $\omega_\alpha, \omega_\beta \in A_c^n(M)$ have respective supports in U_α and U_β, $\alpha, \beta \in \mathfrak{A}$. Then $[\omega_\alpha] = [\omega_\beta]$ if and only if $\int_M \omega_\alpha = \int_M \omega_\beta$.*

Proof. If $[\omega_\alpha] = [\omega_\beta]$, then we know that $\int_M \omega_\alpha = \int_M \omega_\beta$. For the converse, assume equality of the integrals. Since M is connected, we can find a sequence of indices $\alpha = \alpha_0, \alpha_1, \dots, \alpha_r = \beta$ in $\mathfrak{A}$ such that $U_{\alpha_{i-1}} \cap U_{\alpha_i} \neq \emptyset$, $1 \le i \le r$. Choose $\omega_{\alpha_i} \in A_c^n(M)$ such that $\operatorname{supp}(\omega_{\alpha_i}) \subset U_{\alpha_i}$, $1 \le i \le r$, and such that $\omega_{\alpha_0} = \omega_\alpha$, $\omega_{\alpha_r} = \omega_\beta$, and

$$\int_{U_{\alpha_0}} \omega_{\alpha_0} = \int_{U_{\alpha_1}} \omega_{\alpha_1} = \cdots = \int_{U_{\alpha_r}} \omega_{\alpha_r}.$$

This is clearly possible. By Lemma 8.6.1,

$$\int_{U_{\alpha_{i-1}} \cup U_{\alpha_i}} : H_c^n(U_{\alpha_{i-1}} \cup U_{\alpha_i}) \to \mathbb{R}$$

is an isomorphism, $1 \le i \le r$, so

$$[\omega_{\alpha_{i-1}}] = [\omega_{\alpha_i}] \in H_c^n(U_{\alpha_{i-1}} \cup U_{\alpha_i}).$$

That is, there is a form $\eta \in A_c^{n-1}(U_{\alpha_{i-1}} \cup U_{\alpha_i})$ such that $\omega_{\alpha_i} = \omega_{\alpha_{i-1}} + d\eta$. These forms all live in $A_c^*(M)$, so $[\omega_{\alpha_i}] = [\omega_{\alpha_{i-1}}] \in H_c^n(M)$, $1 \le i \le r$. In particular, $[\omega_\alpha] = [\omega_\beta]$ as desired. □

Lemma 8.6.3. *If $\{U_\alpha\}_{\alpha\in\mathfrak{A}}$ is a simple cover of the connected, oriented n-manifold M with $\partial M = \emptyset$, then, for each $\alpha_0 \in \mathfrak{A}$, the natural inclusion*

$$e : A^n_c(U_{\alpha_0}) \hookrightarrow A^n_c(M)$$

induces an isomorphism

$$e_* : H^n_c(U_{\alpha_0}) \to H^n_c(M).$$

Proof. Indeed, since $\int_{U_{\alpha_0}} : H^n_c(U_{\alpha_0}) \to \mathbb{R}$ is an isomorphism, $[\omega] \in H^n_c(U_{\alpha_0})$ is nontrivial if and only if $\int_M \omega = \int_{U_{\alpha_0}} \omega \neq 0$. Since $\int_M$ vanishes on $B^n_c(M)$, it follows that $e_*[\omega] \neq 0$, so e_* is injective.

We prove surjectivity. Let $\omega \in A^n_c(M)$ and use a partition of unity $\{\lambda_\alpha\}_{\alpha\in\mathfrak{A}}$, subordinate to the simple cover, to write

$$\omega = \sum_{\alpha\in\mathfrak{A}} \lambda_\alpha \omega = \sum_{i=1}^{r} \lambda_{\alpha_i}\omega,$$

where $\lambda_{\alpha_i}\omega \in A^n_c(U_{\alpha_i})$, $1 \le i \le r$. Choose $\omega_i \in A^n_c(U_{\alpha_0})$ so that

$$\int_{U_{\alpha_0}} \omega_i = \int_{U_{\alpha_i}} \lambda_{\alpha_i}\omega.$$

Let

$$\widetilde{\omega} = \sum_{i=1}^{r} \omega_i \in A^n_c(U_{\alpha_0})$$

and remark that, by the above lemma, $[\omega_i] = [\lambda_{\alpha_i}\omega] \in H^n_c(M)$, $1 \le i \le r$. Thus, as classes in $H^n_c(M)$,

$$\begin{aligned} [\widetilde{\omega}] &= \sum_{i=1}^{r} [\omega_i] \\ &= \sum_{i=1}^{r} [\lambda_{\alpha_i}\omega] \\ &= \left[\sum_{i=1}^{r} \lambda_{\alpha_i}\omega\right] \\ &= [\omega]. \end{aligned}$$

That is, viewing $[\widetilde{\omega}] \in H^n_c(U_{\alpha_0})$ and $[\omega] \in H^n_c(M)$, we have proven that $e_*[\widetilde{\omega}] = [\omega]$, so e_* is surjective. □

Theorem 8.6.4. *If M is a connected, oriented n-manifold, $\partial M = \emptyset$, then the linear map*

$$\int_M : H^n_c(M) \to \mathbb{R}$$

is an isomorphism.

Proof. Fix a simple cover and let U be an element of that cover. Consider the commutative diagram

$$\begin{array}{ccc} H^n_c(U) & \xrightarrow{\int_U} & \mathbb{R} \\ e_* \downarrow & & \downarrow \mathrm{id} \\ H^n_c(M) & \xrightarrow[\int_M]{} & \mathbb{R} \end{array}$$

Since e_* and $\int_U$ are isomorphisms, so is $\int_M$. □

Corollary 8.6.5. *If M is a compact, connected, oriented n-manifold without boundary, then*

$$\int_M : H^n(M) \to \mathbb{R}$$

is an isomorphism.

Theorem 8.6.6. *If M is a connected, nonorientable n-manifold with empty boundary, then $H^n_c(M) = 0$. In particular, if M is also compact, $H^n(M) = 0$.*

Proof. Let $\omega \in A^n_c(M)$. We must show that $\omega = d\theta$ for suitable $\theta \in A^{n-1}_c(M)$. Choose a simple cover $\{U_\alpha\}_{\alpha\in\mathfrak{A}}$. By a partition of unity argument, write ω as a finite sum of forms, each compactly supported in one or another element of the cover. If each of these is the exterior derivative of a compactly supported form, we are done. Thus, without loss of generality, we assume that $\text{supp}(\omega) \subset U_{\alpha_0}$. By nonorientability of M, there is a sequence $U_{\alpha_0}, U_{\alpha_1}, \dots, U_{\alpha_r}$ of elements of the simple cover and orientations μ_i of U_{α_i}, $0 \le i \le r$, with the following properties:

1. $U_{\alpha_{i-1}} \cap U_{\alpha_i} \neq \emptyset$, $1 \le i \le r$;
2. μ_{i-1} and μ_i restrict to the same orientation of $U_{\alpha_{i-1}} \cap U_{\alpha_i}$, $1 \le i \le r$;
3. $U_{\alpha_r} = U_{\alpha_0}$ and $\mu_r = -\mu_0$.

Choose forms $\omega_i \in A^n_c(U_{\alpha_i})$, $0 \le i \le r$, such that $\omega = \omega_0$ and

$$\int_{(U_{\alpha_i},\mu_i)} \omega_i = \int_{(U_{\alpha_{i-1}},\mu_{i-1})} \omega_{i-1},$$

$1 \le i \le r$. Thus,

$$\omega_i = \omega_{i-1} + d\eta_{i-1}, \quad \eta_{i-1} \in A^{n-1}_c(U_{\alpha_{i-1}} \cup U_{\alpha_i}), \quad 1 \le i \le r.$$

That is,

$$\omega_r = \omega_0 + d\eta, \quad \eta \in A^{n-1}_c(M).$$

On the other hand,

$$\int_{(U_{\alpha_0},\mu_0)} \omega_0 = \int_{(U_{\alpha_r},\mu_r)} \omega_r = -\int_{(U_{\alpha_0},\mu_0)} \omega_r,$$

implying that

$$\omega_r = -\omega_0 + d\gamma, \quad \gamma \in A^{n-1}_c(U_{\alpha_0}).$$

Combining these equations, we conclude that $\omega = \omega_0 = d\theta$ for suitable $\theta \in A^{n-1}_c(M)$. □

If M is compact, any simple cover, being locally finite, is finite. If M is noncompact, it may or may not admit finite simple covers, but it always admits infinite ones. An easy variation on the proof of Theorem 8.5.7 gives

Lemma 8.6.7. *If M is noncompact and connected, $\partial M = \emptyset$, then there is a countably infinite simple cover $\{U_i\}_{i=1}^\infty$ by relatively compact sets such that $U_i \cap U_{i+1} \neq \emptyset$, $1 \le i < \infty$.*

Exercise 8.6.8. Use Lemma 8.6.7 to prove that, if M is a noncompact, connected, boundaryless n-manifold (orientable or not), then $H^n(M) = 0$.

Exercise 8.6.9. If M is an n-manifold with ∂M compact and nonempty, prove that M and $\text{int}(M)$ are homotopically equivalent. If M is connected, conclude that $H^n(M) = 0$. (Hint: There is a compactly supported vector field on M, pointing inward and nowhere 0 along ∂M. This generates a "half flow", parametrized on $[0, \infty)$ and stationary outside of a neighborhood of ∂M.)

8.7. Degree Theory*

Let M and N be connected, oriented n-manifolds without boundary and $f : M \to N$ a proper map. Let $y \in N$ be a regular value of f. Then $f^{-1}(y)$ is compact and discrete, hence finite. Set $f^{-1}(y) = \{y_1, \dots, y_q\}$ and let

$$\epsilon_i = \begin{cases} +1 & \text{if } f_{*y_i} \text{ preserves orientation,} \\ -1 & \text{if } f_{*y_i} \text{ reverses orientation,} \end{cases}$$

$1 \le i \le q$.

Definition 8.7.1. With the above conventions, $\deg_y(f) = \sum_{i=1}^q \epsilon_i$. This is called the local degree of f at the regular value y.

The canonical isomorphisms $\int_M$ and $\int_N$ allow us to identify $H^n_c(M)$ and $H^n_c(N)$ with $\mathbb{R}$ unambiguously. This identification is understood in the following discussion.

Proposition 8.7.2. *If $f : M \to N$ is proper, where M and N are connected, oriented n-manifolds without boundary, and if $y \in N$ is a regular value, then $f^* : H^n_c(N) \to H^n_c(M)$ is multiplication by* $\deg_y(f)$.

Proof. By Theorem 3.9.1, there is an open, connected neighborhood $U \subset N$ of y such that

$$f^{-1}(U) = U_1 \cup \cdots \cup U_q,$$

a union of disjoint open sets such that $y_i \in U_i$ and f carries U_i diffeomorphically onto U, $1 \le i \le q$. If $[\omega] \in H^n_c(N)$, we can choose a representative n-form ω so that $\text{supp}(\omega)$ is a compact subset of U. Then, $\omega_i = f^*(\omega)|U_i$ is compactly supported in U_i, $1 \le i \le q$, and $f : U_i \to U$ preserves or reverses orientation according as $\epsilon_i = 1$ or -1. Thus,

$$\begin{aligned} \int_M f^*(\omega) &= \sum_{i=1}^q \int_{U_i} \omega_i \\ &= \sum_{i=1}^q \epsilon_i \int_U \omega \\ &= \left(\sum_{i=1}^q \epsilon_i\right) \int_N \omega \\ &= \deg_y(f) \int_N \omega. \end{aligned}$$

It follows that $f^* : H^n_c(N) \to H^n_c(M)$ is multiplication by $\deg_y(f)$. □

Remark. This has several obvious consequences:

1. $f^* : H^n_c(N) \to H^n_c(M)$ is multiplication by an integer.
2. $\deg_y(f) = \deg(f)$ (the degree of f) is independent of y.
3. $\deg(f)$ is a proper homotopy invariant of f.

4. If M is compact, $\deg(f)$ is a homotopy invariant of f.
5. $\deg(f \circ g) = \deg(f)\deg(g)$.

Theorem 8.7.3. *Let W be an oriented $(n+1)$-manifold with nonempty, connected boundary. Let N be a connected, oriented n-manifold without boundary and let $f : \partial W \to N$ be proper. If f extends to a proper map $F : W \to N$, then* $\deg(f) = 0$.

Proof. Suppose that f extends to a proper map $F : W \to N$. If $\omega \in A^n_c(N)$, then $F^*(\omega) \in A^n_c(W)$ and $d(F^*(\omega)) = F^*(d\omega) = 0$, since $d\omega \in A^{n+1}_c(N) = 0$. Thus, by Stokes theorem,

$$\int_{\partial W} f^*(\omega) = \int_W d(F^*(\omega)) = \int_W 0 = 0.$$

By Proposition 8.7.2, it follows that $\deg(f) = 0$. □

This theorem partially generalizes Theorem 6.5.6.

Let $S^n \subset \mathbb{R}^{n+1}$ be the unit sphere and let $\alpha_n : S^n \to S^n$ denote the antipodal interchange map

$$\alpha_n(x^1, \dots, x^{n+1}) = (-x^1, \dots, -x^{n+1}).$$

Proposition 8.7.4. $\deg(\alpha_n) = (-1)^{n+1}$.

Proof. Since α_n is a diffeomorphism, every $x \in S^n$ is a regular value and has pre-image a singleton. Thus, the question reduces to whether α_n preserves or reverses orientation. The linear extension $A : \mathbb{R}^{n+1} \to \mathbb{R}^{n+1}$ of α_n is represented by the matrix $-I_{n+1}$ with determinant $(-1)^{n+1}$. This transformation, therefore, is orientation-preserving if and only if n is odd. The restriction of the transformation to the unit ball D^{n+1} is orientation-preserving if and only if n is odd. But $S^n = \partial D^{n+1}$ has orientation induced by the orientation of D^n, hence $A|S^n = \alpha_n : S^n \to S^n$ is orientation-preserving if and only if n is odd. □

Theorem 8.7.5. *The sphere S^n has a nowhere zero tangent vector field if and only if n is odd.*

Proof. For n odd, you constructed a nowhere vanishing vector field in Exercise 4.3.12. Suppose, therefore, that $s : S^n \to \mathbb{R}^{n+1} \setminus \{0\}$ is smooth with $s(v) \perp v$, $\forall v \in S^n$. Equivalently, s is a nowhere zero section of $T(S^n)$. Note that

$$v \cos\theta + s(v) \sin\theta \neq 0,$$

$\forall \theta \in \mathbb{R}$, so we can define a smooth map

$$F : S^n \times [0,1] \to S^n$$

by

$$F(v,t) = \frac{v \cos t\pi + s(v) \sin t\pi}{\|v \cos t\pi + s(v) \sin t\pi\|}.$$

Then,

$$\left.\begin{aligned} F(v,0) &= v \\ F(v,1) &= -v \end{aligned}\right\} \ \forall v \in S^n,$$

so $\alpha_n \sim \mathrm{id}$ and $1 = \deg(\alpha_n) = (-1)^{n+1}$, implying that n is odd. □

Corollary 8.7.6. *Every smooth flow on S^{2n} has at least one stationary point.*

For the sphere S^2, Theorem 8.7.5 and its corollary are sometimes stated facetiously as the previously quoted "you can't comb the hair on a billiard ball". (People whose sensibilities are offended by hairy billiard balls substitute "coconut".)

Theorem 8.7.7. *If $f : S^n \to S^n$ is smooth and* $\deg(f) \neq (-1)^{n+1}$, *then f has a fixed point.*

Proof. In fact, we will prove that, in the case that f has no fixed point, $f \sim \alpha_n$, hence $\deg(f) = (-1)^{n+1}$.

We claim that $t(f(v) + v) \neq v$, $\forall\, v \in S^n$, $0 \leq t \leq 1$. Otherwise, $t \neq 0$ and

$$f(v) = (1 - t)v/t.$$

Since $\|f(v)\| = 1 = \|v\|$, it follows that $t = 1/2$ and $f(v) = v$, contrary to assumption. Therefore, we can define $F : S^n \times [0, 1] \to S^n$ by

$$F(v, t) = \frac{t(f(v) + v) - v}{\|t(f(v) + v) - v\|}.$$

This is a homotopy between f $(t = 1)$ and α_n $(t = 0)$. □

Exercise 8.7.8. If $f : S^n \to S^n$ has $|\deg(f)| \neq 1$, prove that f has a fixed point *and* that there is a point that f carries to its antipode.

Exercise 8.7.9. If $f : P^{2n} \to P^{2n}$ is smooth, prove that f has a fixed point. (Hint. Lift to the universal cover.)

Exercise 8.7.10. If M is a compact, connected, orientable, boundaryless manifold of dimension n and $f : M \to \mathbb{R}^{n+1}$ is a smooth imbedding, use degree theory to prove that $\mathbb{R}^{n+1} \smallsetminus f(M)$ has exactly two connected components and $f(M)$ is the set-theoretic boundary of each. This is the Jordan–Brouwer separation theorem. (Proceed in analogy with Exercise 3.9.23. In fact, the mod 2 degree theory is adequate for this.)

Exercise 8.7.11. Let $\sigma_1, \sigma_2 : S^1 \to \mathbb{R}^3$ be smooth maps with disjoint images. Define the *linking number* $\mathrm{Lk}(\sigma_1, \sigma_2)$ to be the degree of the map $f : S^1 \times S^1 \to S^2$ defined by

$$f(x, y) = \frac{\sigma_1(x) - \sigma_2(y)}{\|\sigma_1(x) - \sigma_2(y)\|}.$$

Intuitively, it seems reasonable to define σ_1 and σ_2 to be *topologically unlinked* if there is a compact, orientable 2-manifold N with $\partial N = S^1$ and σ_1 extends to a smooth map

$$\overline{\sigma}_1 : N \to \mathbb{R}^3 \smallsetminus \sigma_2(S^1)$$

(or if the parallel condition holds, in which the roles of σ_1 and σ_2 are interchanged). Prove that, if σ_1 and σ_2 are topologically unlinked, then $\mathrm{Lk}(\sigma_1, \sigma_2) = 0$. Give an example showing, at least intuitively, that the requirement that N be orientable is necessary. (Hint: Consider the Möbius strip.)

8.8. Poincaré Duality*

Assume that M is a connected, oriented n-manifold with empty boundary. We study the pairing

$$\int_M : A^k_c(M) \otimes_{\mathbb{R}} A^{n-k}(M) \to \mathbb{R}$$

defined by

$$\omega \otimes_{\mathbb{R}} \eta \mapsto \int_M \omega \wedge \eta.$$

The fact that one of the forms is compactly supported guarantees that the integral is defined. One can view this pairing as an $\mathbb{R}$-linear map

$$\mathrm{PD} : A_c^k(M) \to A^{n-k}(M)^*,$$

where $A^{n-k}(M)^*$ is the vector space dual of $A^{n-k}(M)$ and

$$\mathrm{PD}(\omega) = \int_M \omega \wedge \{\cdot\}.$$

If $\omega = d\gamma$, some $\gamma \in A_c^{k-1}(M)$ and $d\eta = 0$, it is clear that

$$\int_M \omega \wedge \eta = \int_M d(\gamma \wedge \eta) = 0$$

by Stokes' theorem. Similarly, if ω is closed and η is exact, the integral is 0. Thus, our pairing passes to

$$\int_M : H_c^k(M) \otimes H^{n-k}(M) \to \mathbb{R}.$$

Again, this can be interpreted as a linear map

$$\mathrm{PD} : H_c^k(M) \to H^{n-k}(M)^*,$$

called the *Poincaré duality operator.*

Theorem 8.8.1 (Poincaré Duality Theorem). *Suppose that M is a connected, oriented n-manifold, $\partial M = \emptyset$, and that M admits a finite simple cover. Then the Poincaré duality operator*

$$\mathrm{PD} : H_c^k(M) \to H^{n-k}(M)^*$$

is an isomorphism, $\forall\, k$. In particular, this holds for M compact and defines a canonical isomorphism $H^k(M) = H^{n-k}(M)^* = H_{n-k}(M)$.

Of course, the last equality depends on the de Rham theorem. Remark that, in the compact case, we can also say that $H^k(M) \cong H^{n-k}(M)$, but the isomorphism is not canonical. Theorem 8.8.1 will be proven by a series of lemmas.

Lemma 8.8.2. *If M has the same ordinary and compact cohomology as $\mathbb{R}^n$, the Poincaré duality operator*

$$\mathrm{PD} : H_c^k(M) \to H^{n-k}(M)^*$$

is an isomorphism.

Proof. Indeed, if $k \neq n$, both $H_c^k(M)$ and $H^{n-k}(M)^*$ are 0, so the assertion is trivially true. If $k = n$, then $[\omega] \in H_c^n(M) = \mathbb{R}$ is uniquely determined by $\int_M \omega$, while $c \in H^0(M) = \mathbb{R}$ is just the constant function c. Then,

$$\mathrm{PD}[\omega](c) = \int_M c\omega = c \int_M \omega.$$

That is, $\mathrm{PD}[\omega] : \mathbb{R} \to \mathbb{R}$ is just multiplication by $\int_M \omega$, proving that PD is also an isomorphism in this case. □

The proof of Theorem 8.8.1 will use the Mayer–Vietoris sequences. Let $U_1, U_2 \subset M$ be open subsets. Consider the diagram

$$\begin{array}{ccc} H^k_c(U_1 \cap U_2) & \xrightarrow{\alpha^*} & H^k_c(U_1) \oplus H^k_c(U_2) \\ {\scriptstyle \mathrm{PD}}\big\downarrow & & \big\downarrow{\scriptstyle \mathrm{PD}\oplus\mathrm{PD}} \\ H^{n-k}(U_1 \cap U_2)^* & \xrightarrow[j']{} & H^{n-k}(U_1)^* \oplus H^{n-k}(U_2)^* \end{array}$$

where α^* is as in the definition of the Mayer–Vietoris sequence for compact supports and j' is as in the dual of the ordinary Mayer–Vietoris sequence.

Lemma 8.8.3. *The above diagram is commutative.*

Proof. Let $[\omega] \in H^k_c(U_1 \cap U_2)$ and $([\eta_1], [\eta_2]) \in H^{n-k}(U_1) \oplus H^{n-k}(U_2)$. Remark that

$$\int_{U_1 \cap U_2} \omega \wedge j_\ell^*(\eta_\ell) = \int_{U_\ell} \alpha_\ell(\omega) \wedge \eta_\ell,$$

$\ell = 1, 2$. Thus,

$$\begin{aligned} j' \,\mathrm{PD}[\omega]([\eta_1], [\eta_2]) &= \mathrm{PD}[\omega](j^*([\eta_1], [\eta_2])) \\ &= \mathrm{PD}[\omega]([j_1^*(\eta_1)] - [j_2^*(\eta_2)]) \\ &= \int_{U_1 \cap U_2} \omega \wedge (j_1^*(\eta_1) - j_2^*(\eta_2)) \\ &= \int_{U_1 \cap U_2} \omega \wedge j_1^*(\eta_1) - \int_{U_1 \cap U_2} \omega \wedge j_2^*(\eta_2) \\ &= \int_{U_1} \alpha_1(\omega) \wedge \eta_1 - \int_{U_2} \alpha_2(\omega) \wedge \eta_2. \end{aligned}$$

But

$$\begin{aligned} (\mathrm{PD} \oplus \mathrm{PD})\alpha^*[\omega]([\eta_1], [\eta_2]) &= (\mathrm{PD} \oplus \mathrm{PD})(\alpha_1^*[\omega], -\alpha_2^*[\omega])([\eta_1], [\eta_2]) \\ &= \mathrm{PD}(\alpha_1^*[\omega])[\eta_1] - \mathrm{PD}(\alpha_2^*[\omega])[\eta_2] \\ &= \int_{U_1} \alpha_1(\omega) \wedge \eta_1 - \int_{U_2} \alpha_2(\omega) \wedge \eta_2, \end{aligned}$$

giving the asserted commutativity. □

Exercise 8.8.4. Prove that the diagram

$$\begin{array}{ccc} H^k_c(U_1) \oplus H^k_c(U_2) & \xrightarrow{\beta^*} & H^k_c(U_1 \cup U_2) \\ {\scriptstyle \mathrm{PD}\oplus\mathrm{PD}}\big\downarrow & & \big\downarrow{\scriptstyle \mathrm{PD}} \\ H^{n-k}(U_1)^* \oplus H^{n-k}(U_2)^* & \xrightarrow[i']{} & H^{n-k}(U_1 \cup U_2)^* \end{array}$$

is commutative.

Remark that the different sign conventions for the Mayer–Vietoris sequences in ordinary and compactly supported cohomology are needed for this exercise and the previous lemma.

Lemma 8.8.5. *The diagram*

$$\begin{array}{ccc} H_c^k(U_1 \cup U_2) & \xrightarrow{d^*} & H_c^{k+1}(U_1 \cap U_2) \\ \text{PD}\downarrow & & \downarrow\text{PD} \\ H^{n-k}(U_1 \cup U_2)^* & \xrightarrow[d']{} & H^{n-k-1}(U_1 \cap U_2)^* \end{array}$$

commutes up to sign. More precisely, $d' \circ \text{PD} = (-1)^{k+1}\,\text{PD} \circ d^*$.

Proof. We fix $[\omega] \in H_c^k(U_1 \cup U_2)$ and $[\eta] \in H^{n-k-1}(U_1 \cap U_2)$ and verify that

$$\text{PD}(d^*[\omega])([\eta]) = \int_{U_1 \cap U_2} d\lambda_1 \wedge \omega \wedge \eta, \tag{8.5}$$

$$d'(\text{PD}[\omega])([\eta]) = (-1)^{k+1} \int_{U_1 \cap U_2} d\lambda_1 \wedge \omega \wedge \eta, \tag{8.6}$$

where $\{\lambda_1, \lambda_2\}$ is a partition of unity on $U_1 \cup U_2$ subordinate to $\{U_1, U_2\}$.

In the diagram

$$\begin{array}{ccc} & A_c^k(U_1) \oplus A_c^k(U_2) & \xrightarrow{\beta} A_c^k(U_1 \cup U_2) \\ & d\oplus d \downarrow & \\ A_c^{k+1}(U_1 \cap U_2) \xrightarrow[\alpha]{} & A_c^{k+1}(U_1) \oplus A_c^{k+1}(U_2) & \end{array}$$

we see that

$$\begin{aligned} \beta(\lambda_1\omega, \lambda_2\omega) &= \omega, \\ d \oplus d(\lambda_1\omega, \lambda_2\omega) &= (d\lambda_1 \wedge \omega, -d\lambda_1 \wedge \omega), \\ \alpha(d\lambda_1 \wedge \omega) &= (d\lambda_1 \wedge \omega, -d\lambda_1 \wedge \omega), \end{aligned}$$

where the second equation uses the fact that $d\lambda_1 + d\lambda_2 = d(\lambda_1 + \lambda_2) = 0$. Therefore, $d^*[\omega] = [d\lambda_1 \wedge \omega]$ and equation (8.5) follows.

In order to compute $d'(\text{PD}[\omega])([\eta]) = \text{PD}([\omega])(d^*[\eta])$, we first compute $d^*[\eta]$. Here we set $r = n - k$ and consider

$$\begin{array}{ccc} & A^{r-1}(U_1) \oplus A^{r-1}(U_2) & \xrightarrow{j} A^{r-1}(U_1 \cap U_2) \\ & d\oplus d \downarrow & \\ A^r(U_1 \cup U_2) \xrightarrow[i]{} & A^r(U_1) \oplus A^r(U_2) & \end{array}$$

and note that

$$\begin{aligned} j(\lambda_2\eta, -\lambda_1\eta) &= \eta, \\ d \oplus d(\lambda_2\eta, -\lambda_1\eta) &= (-d\lambda_1 \wedge \eta, -d\lambda_1 \wedge \eta), \\ i(-d\lambda_1 \wedge \eta) &= (-d\lambda_1 \wedge \eta, -d\lambda_1 \wedge \eta). \end{aligned}$$

It follows that $d^*[\eta] = [-d\lambda_1 \wedge \eta]$. Then

$$\begin{aligned} d'(\mathrm{PD}[\omega])([\eta]) &= \mathrm{PD}([\omega])(d^*[\eta]) \\ &= \mathrm{PD}([\omega])([-d\lambda_1 \wedge \eta]) \\ &= \int_{U_1 \cap U_2} \omega \wedge (-d\lambda_1 \wedge \eta) \\ &= (-1)^{k+1} \int_{U_1 \cap U_2} d\lambda_1 \wedge \omega \wedge \eta, \end{aligned}$$

which is equation (8.6). Here, the sign is due to permuting the 1-form $d\lambda_1$ past the k-form ω. □

Proof of theorem 8.8.1. Let $\{U_i\}_{i=1}^r$ be a simple cover of M and proceed by induction on r. By Lemma 8.5.6 and the definition of simple covers, the case $r = 1$ is given by Lemma 8.8.2. If, for a given $r \geq 2$, the assertion has been proven whenever a manifold has a simple cover with $r - 1$ elements, then it holds for the manifold $U = U_1 \cup \cdots \cup U_{r-1}$, for U_r, and for $U \cap U_r$ (which has the simple cover $\{U_1 \cap U_r, \dots, U_{r-1} \cap U_r\}$). We must prove it for $M = U \cup U_r$. By the Mayer–Vietoris compact cohomology sequence of Theorem 8.5.4 and the dual cohomology sequence of Theorem 8.5.15, together with Lemmas 8.8.3, 8.8.5, Exercise 8.8.4 and the Five Lemma (for which commutativity up to sign is fine),

$$\mathrm{PD} : H^k_c(M) \to H^{n-k}(M)^*$$

is an isomorphism. □

Exercise 8.8.6. Let M be connected, oriented and n-dimensional with $\partial M = \emptyset$. Let $N \subset M$ be a compact, oriented, k-dimensional submanifold with $\partial N = \emptyset$, $0 \leq k < n$, and denote the inclusion map by $i : N \hookrightarrow M$. If $[\omega] \in H^k(M)$, we will write $\int_N [\omega]$ for $\int_N i^*[\omega]$.

(1) Show that there is a unique compactly supported cohomology class $[\eta_N] \in H^{n-k}_c(M)$ such that
$$\int_N [\omega] = \int_M \eta_N \wedge \omega,$$
$\forall\, [\omega] \in H^k(M)$. For fairly obvious reasons, $[\eta_N]$ is called the *Poincaré dual* of N.

(2) If $U \subset M$ is any open neighborhood of N, prove that the representative, compactly supported form $\eta_N \in [\eta_N]$ can be chosen so that $\mathrm{supp}(\eta_N) \subset U$. This is the *localization principle* for the Poincaré dual of N.

(3) If $i_0, i_1 : N \to M$ are two smooth imbeddings, we say that they are isotopic if there is a homotopy $i_t : N \to M$ between i_0 and i_1 such that i_t is a smooth imbedding, $0 \leq t \leq 1$. In this case we also say that $N_\ell = i_\ell(N)$ are isotopic submanifolds of M, $\ell = 0, 1$. If N_0 and N_1 are isotopic submanifolds of M, prove that $[\eta_{N_0}] = [\eta_{N_1}]$.

(4) Suppose that P_1 and P_2 are compact, oriented, boundaryless submanifolds of M of respective dimensions k_1 and k_2 such that $n = k_1 + k_2$. Show that
$$\int_{P_1} [\eta_{P_2}] = (-1)^{k_1 k_2} \int_{P_2} [\eta_{P_1}].$$
This is called the *algebraic intersection number* $\iota(P_1, P_2)$ of P_1 with P_2 and is an integer (but you are probably not prepared to prove that).

(5) If P_1 and P_2 as above are isotopic to submanifolds P_1' and P_2', respectively, such that $P_1' \cap P_2' = \emptyset$, prove that $\iota(P_1, P_2) = 0$.

(6) If P is a compact, orientable n-manifold without boundary, let $\Delta_P \subset P \times P$ be the diagonal, $\Delta_P = \{(x, x) \mid x \in P\}$. If P has a nowhere vanishing vector field, prove that $\iota(\Delta_P, \Delta_P) = 0$.

Exercise 8.8.7. In part (6) of Exercise 8.8.6, you proved half of the Poincaré–Hopf theorem: *There is a nowhere vanishing vector field on P if and only if $\iota(\Delta_P, \Delta_P) = 0$.* Assuming this theorem, prove the following.

(1) The diagonal $\Delta_P \subset P \times P$ can be isotoped completely off of itself if and only if its algebraic self intersection number $\iota(\Delta_P, \Delta_P)$ vanishes. (In particular, by Theorem 8.7.5, in $S^{2k} \times S^{2k}$ the diagonal cannot be isotoped completely off of itself.)

(2) Every compact, orientable, odd dimensional manifold with empty boundary has a nowhere vanishing vector field.

8.9. The de Rham Theorem*

We will prove the following case of Theorem 8.2.21.

Theorem 8.9.1 (de Rham Theorem). *If M is a manifold without boundary which has a finite simple cover, then the de Rham map*

$$\mathrm{DR} : H^*(M) \to (H_*(M))^*$$

is a canonical isomorphism of graded vector spaces.

The proof follows exactly the pattern of proof of Theorem 8.8.1.

Lemma 8.9.2. *If the n-manifold M has the same singular homology and cohomology as $\mathbb{R}^n$, then the de Rham homomorphism*

$$\mathrm{DR} : H^k(M) \to H_k(M)^*$$

is an isomorphism.

Proof. If $n = 0$, the result is immediate, so we assume $n > 0$. If $k \neq 0$, both $H^k(M)$ and $H_k(M)^*$ are 0, so the assertion is trivially true. Finally, $H^0(M) = Z^0(M) = \mathbb{R}$ is the space of constant functions on M and $H_0(M) = \mathbb{R}$ has canonical basis the singleton $\{[x]\}$, where $x \in M$ is fixed but arbitrary. If $c \in H^0(M)$, then $\mathrm{DR}(c)([x]) = c(x) = c$, so DR is an isomorphism as claimed. □

We will use the appropriate Mayer–Vietoris sequences. Let $U_1, U_2 \subset M$ be open subsets. Consider the diagram

$$\begin{array}{ccc} H^k(U_1 \cap U_2) & \xrightarrow{d^*} & H^{k+1}(U_1 \cup U_2) \\ \mathrm{DR}\downarrow & & \downarrow\mathrm{DR} \\ H_k(U_1 \cap U_2)^* & \xrightarrow[\partial']{} & H_{k+1}(U_1 \cup U_2)^* \end{array}$$

with d^* as in Theorem 8.5.2 and ∂' as in Theorem 8.5.15.

Lemma 8.9.3. *The above diagram is commutative.*

Proof. Let $[\omega] \in H^k(U_1 \cap U_2)$ and $[z] \in H_{k+1}(U_1 \cup U_2)$. Let $\mathcal{U}$ denote the open cover $\{U_1, U_2\}$ of $U_1 \cup U_2$. We can choose the representative cycle $z \in [z]$ to be $\mathcal{U}$-small. That is, $z = z_1 + z_2$, where $z_i \in C_{k+1}(U_i)$, $i = 1,2$. Note that z_1 and z_2 may not, individually, be cycles. All that is required is that $\partial(z_1 + z_2) = 0$, so

$$\partial(z_1) = -\partial(z_2) \in C_k(U_1 \cap U_2).$$

As a singular chain, z_2 is a linear combination of $s_1, \dots, s_q$, where

$$s_i : \Delta_{k+1} \to U_2, \quad 1 \le i \le q.$$

We define the *support* of this chain to be

$$|z_2| = \bigcup_{i=1}^{q} s_i(\Delta_{k+1}),$$

a compact subset of U_2. It is easy to choose a smooth partition of unity $\{\lambda_1, \lambda_2\}$ on $U_1 \cup U_2$, subordinate to $\mathcal{U}$ and having the property that $\lambda_2 \equiv 1$ on a $|z_2|$, hence $\lambda_1 \equiv 0$ on a $|z_2|$. Remark that, for $1 \le i \le q$,

$$s_i^*(d\lambda_2) = d(s_i^*(\lambda_2)) = d(1) \equiv 0,$$

with a similar remark for $d\lambda_1$.

Given these choices, we will show that

$$\mathrm{DR}(d^*[\omega])([z]) = \int_{z_1} d\lambda_2 \wedge \omega, \tag{8.7}$$

$$\partial'(\mathrm{DR}[\omega])([z]) = \int_{z_1} d\lambda_2 \wedge \omega. \tag{8.8}$$

Since $[\omega]$ and $[z]$ were arbitrary elements of the respective vector spaces, commutativity of the diagram will follow.

In the diagram

$$\begin{array}{ccccc} & & A^k(U_1) \oplus A^k(U_2) & \xrightarrow{\ j\ } & A^k(U_1 \cap U_2) \\ & & \big\downarrow{\scriptstyle d \oplus d} & & \\ A^{k+1}(U_1 \cup U_2) & \xrightarrow[i]{} & A^{k+1}(U_1) \oplus A^{k+1}(U_2) & & \end{array}$$

we see that

$$\begin{aligned} j(\lambda_2\omega, -\lambda_1\omega) &= \omega, \\ d \oplus d(\lambda_2\omega, -\lambda_1\omega) &= (d\lambda_2 \wedge \omega, d\lambda_2 \wedge \omega), \\ i(d\lambda_2 \wedge \omega) &= (d\lambda_2 \wedge \omega, d\lambda_2 \wedge \omega), \end{aligned}$$

where the second equation uses the fact that $d\lambda_1 = -d\lambda_2$. Therefore, $d^*[\omega] = [d\lambda_2 \wedge \omega]$ and

$$\mathrm{DR}(d^*[\omega])([z]) = \int_z d\lambda_2 \wedge \omega = \int_{z_1} d\lambda_2 \wedge \omega$$

since $\lambda_2 \equiv 1$ on $|z_2|$. This is equation (8.7).

In the diagram

$$\begin{array}{ccc} & C_{k+1}(U_1) \oplus C_{k+1}(U_2) & \xrightarrow{i} C^{\mathfrak{U}}_{k+1}(U_1 \cup U_2) \\ & \Big\downarrow{\scriptstyle \partial\oplus\partial} & \\ C_k(U_1 \cap U_2) \xrightarrow[j]{} & C_k(U_1) \oplus C_k(U_2) & \end{array}$$

we see that

$$\begin{aligned} i(z_1, z_2) &= z, \\ \partial \oplus \partial(z_1, z_2) &= (\partial z_1, -\partial z_1), \\ j(\partial z_1) &= (\partial z_1, -\partial z_1), \end{aligned}$$

where the second equation uses the fact that $\partial z_1 = -\partial z_2$. It follows that $\partial_*[z] = [\partial z_1]$. Then,

$$\begin{aligned} \partial'(\mathrm{DR}[\omega])([z]) &= \int_{\partial_*[z]} [\omega] \\ &= \int_{\partial z_1} \lambda_1 \omega + \int_{\partial z_1} \lambda_2 \omega \\ &= \int_{-\partial z_2} \lambda_1 \omega + \int_{\partial z_1} \lambda_2 \omega \\ &= -\int_{z_2} d\lambda_1 \wedge \omega + \int_{z_1} d\lambda_2 \wedge \omega \\ &= \int_{z_1} d\lambda_2 \wedge \omega, \end{aligned}$$

since $\partial z_1 = -\partial z_2$ and λ_1 vanishes identically on $|z_2|$. This is equation (8.8). □

Commutativity of the remaining squares is easier since the connecting homomorphisms are not involved.

Exercise 8.9.4. The diagram

$$\begin{array}{ccc} H^k(U_1) \oplus H^k(U_2) & \xrightarrow{j^*} & H^k(U_1 \cap U_2) \\ \Big\downarrow{\scriptstyle \mathrm{DR}\oplus\mathrm{DR}} & & \Big\downarrow{\scriptstyle \mathrm{DR}} \\ H_k(U_1)^* \oplus H_k(U_2)^* & \xrightarrow[j']{} & H_k(U_1 \cap U_2)^* \end{array}$$

is commutative.

Exercise 8.9.5. The diagram

$$\begin{array}{ccc} H^k(U_1 \cup U_2) & \xrightarrow{i^*} & H^k(U_1) \oplus H^k(U_2) \\ \Big\downarrow{\scriptstyle \mathrm{DR}} & & \Big\downarrow{\scriptstyle \mathrm{DR}\oplus\mathrm{DR}} \\ H_k(U_1 \cup U_2)^* & \xrightarrow[i']{} & H_k(U_1)^* \oplus H_k(U_2)^* \end{array}$$

is commutative.

Proof of theorem 8.9.1. Let $\{U_i\}_{i=1}^r$ be a simple cover of M and proceed by induction on r. By Lemma 8.5.6, the case $r = 1$ is given by Lemma 8.9.2. If, for a given $r \geq 2$, the assertion has been proven whenever a manifold has a simple cover with $r - 1$ elements, then it holds for the manifold $U = U_1 \cup \cdots \cup U_{r-1}$, for U_r, and for $U \cap U_r$ (which has the simple cover $\{U_1 \cap U_r, \ldots, U_{r-1} \cap U_r\}$). We must prove it for $M = U \cup U_r$. By the Mayer–Vietoris cohomology sequence of Theorem 8.5.2 and the dual homology sequence of Theorem 8.5.15, together with Lemma 8.9.3, Exercises 8.9.4 and 8.9.5 and the Five Lemma, $\mathrm{DR} : H^k(M) \to H_k(M)^*$ is an isomorphism. □

There are many versions of the de Rham theorem. The version we have proven identifies de Rham theory *as a graded vector space* with the dual of singular homology. Actually, this latter can be defined directly from a singular cochain complex (the dual of the singular chain complex) and is called singular cohomology. There is a natural graded algebra structure in singular cohomology (the multiplication is called "cup product" for some obscure reason) and a stronger version of the de Rham theorem asserts that DR is an isomorphism of graded algebras. Also, the requirement that $\partial M = \emptyset$ was convenient for our approach, but is quite inessential.

In Appendix D, we will prove a version of the de Rham theorem for the Čech cohomology algebra $\check{H}^*(M)$, a cohomology theory fashioned out of the family of open subsets of M. By a parallel argument, we will also show that the Čech cohomology algebra is isomorphic to the singular cohomology algebra defined using all of the continuous singular simplices instead of only the smooth ones. A very interesting consequence is the following.

Theorem 8.9.6. *The de Rham cohomology algebra $H^*(M)$ depends only on the underlying topological manifold M, not on the choice of differentiable structure.*

We discuss here the equality $H^p(M) = \check{H}^p(M)$ for $p = 0, 1$. This will also motivate the use of the term "cocycle" in our earlier discussion of differentiable structures (Definition 3.1.10) and in vector bundle theory (Definition 3.4.2), as well as the cohomology notation $H^1(M; \mathrm{Gl}(n))$ in Exercise 3.4.6.

Let $\mathfrak{U} = \{U_\alpha\}_{\alpha \in \mathfrak{A}}$ be an open cover of the manifold M. A Čech 0-cochain on $\mathfrak{U}$ is a function θ which, to each $U_{\alpha_0} \in \mathfrak{U}$, assigns a real number $\theta_{\alpha_0} = \theta(U_{\alpha_0})$. A Čech 1-cochain on $\mathfrak{U}$ is a function γ which assigns a real number

$$\gamma_{\alpha_0 \alpha_1} = \gamma(U_{\alpha_0}, U_{\alpha_1})$$

to every ordered pair $(U_{\alpha_0}, U_{\alpha_1})$ of elements $U_{\alpha_0}, U_{\alpha_1} \in \mathfrak{U}$ such that $U_{\alpha_0} \cap U_{\alpha_1} \neq \emptyset$. Similarly, a Čech 2-cochain ζ assigns a real number $\zeta_{\alpha_0\alpha_1\alpha_2}$ to each ordered triple $(U_{\alpha_0}, U_{\alpha_1}, U_{\alpha_2})$ of elements of $\mathfrak{U}$ with $U_{\alpha_0} \cap U_{\alpha_1} \cap U_{\alpha_2} \neq \emptyset$. The general pattern is clear, but we will stick with p-cochains for $p = 0, 1, 2$. The set of p-cochains is denoted by $\check{C}^p(\mathfrak{U})$.

As real-valued functions on a set, p-cochains can be added and they can be multiplied by real scalars. This makes $\check{C}^p(\mathfrak{U})$ into a vector space over $\mathbb{R}$.

Define Čech coboundary operators

$$0 \xrightarrow{\delta} \check{C}^0(\mathfrak{U}) \xrightarrow{\delta} \check{C}^1(\mathfrak{U}) \xrightarrow{\delta} \check{C}^2(\mathfrak{U})$$

by

$$(\delta\theta)_{\alpha_0\alpha_1} = \theta_{\alpha_1} - \theta_{\alpha_0} \qquad \text{if } \theta \in \check{C}^0(\mathfrak{U}),$$

$$(\delta\gamma)_{\alpha_0\alpha_1\alpha_2} = \gamma_{\alpha_1\alpha_2} - \gamma_{\alpha_0\alpha_2} + \gamma_{\alpha_0\alpha_1} \qquad \text{if } \gamma \in \check{C}^1(\mathfrak{U}).$$

It is a moment's work to check that $\delta^2 = 0$, so we obtain the space $\check{Z}^p(\mathfrak{U})$ of Čech p-cocycles, the space $\check{B}^p(\mathfrak{U})$ of Čech p-coboundaries, and the pth Čech cohomology space $\check{H}^p(\mathfrak{U})$, for $p = 0, 1$.

Remark. A cochain $\gamma \in \check{C}^1(\mathfrak{U})$ is a cocycle precisely if it satisfies the *cocycle condition*

$$\gamma_{\alpha\eta} = \gamma_{\alpha\beta} + \gamma_{\beta\eta},$$

whenever $U_\alpha \cap U_\beta \cap U_\eta \neq \emptyset$. The Gl($n$)-cocycles in bundle theory had a completely analogous definition, except for the multiplicative notation forced by the multiplicative structure of Gl(n). Indeed, if $\tilde{\gamma}$ is a $\mathrm{Gl}_+(1)$-cocycle on $\mathfrak{U}$, then $\log \circ \tilde{\gamma}$ is a Čech 1-cocycle on $\mathfrak{U}$. The set of Gl(1)-cocycles forms an abelian group under operations inherited from Gl(1), but, for $n > 1$, the Gl(n)-cocycles do not form a group of any kind because of the noncommutativity of Gl(n).

Lemma 8.9.7. *If each $U_\alpha \in \mathfrak{U}$ is connected, the space $\check{H}^0(\mathfrak{U})$ is canonically isomorphic to the space of locally constant, real valued functions on M. In particular, $\check{H}^0(\mathfrak{U}) = H^0(M)$.*

Proof. Remark that $\check{H}^0(\mathfrak{U}) = \check{Z}^0(\mathfrak{U})$ and that this is the space of 0-cochains θ such that $\theta_\alpha = \theta_\beta$ whenever $U_\alpha \cap U_\beta \neq \emptyset$. Thinking of θ_α as a constant function on U_α, $\forall\, \alpha \in \mathfrak{A}$, we see that these constant functions agree on overlaps of their domains, hence unite to form a coherent locally constant function θ on M. Conversely, if $\theta : M \to \mathbb{R}$ is a locally constant function, its restriction $\theta_\alpha = \theta | U_\alpha$ is constant by the connectivity of U_α, $\forall\, \alpha \in \mathfrak{A}$. □

Remark that simple covers $\mathfrak{U}$ satisfy Lemma 8.9.7.

Lemma 8.9.8. *If $\mathfrak{U}$ is a simple cover, there is a canonical linear isomorphism $\check{H}^1(\mathfrak{U}) = H^1(M)$.*

Remark. It is an immediate consequence of Lemmas 8.9.7 and 8.9.8 that $\check{H}^p(\mathfrak{U})$ does not depend on the choice of simple cover, hence this vector space can be denoted by $\check{H}^p(M)$, $p = 0, 1$. For the case $p = 0$, the purely topological condition on $\mathfrak{U}$ in Lemma 8.9.7 proves that $H^0(M)$ and $\check{H}^0(M)$ are topological invariants. However, the definition of a simple cover requires a differentiable structure, so we cannot conclude from Lemma 8.9.8 that $H^1(M)$ and $\check{H}^1(M)$ are topological invariants of M. The proper definition of Čech cohomology involves passing to an algebraic limit over the directed set of all open covers of M, thus obtaining a true topological invariant. In Section 10.5, we will show that every open cover has a simple refinement and, in Appendix D, use this fact to prove Theorem 8.9.6.

We sketch the construction of the isomorphism in Lemma 8.9.8 and leave verification of several details to the exercises. Fix the choice of simple cover $\mathfrak{U} = \{U_\alpha\}_{\alpha \in \mathfrak{A}}$.

We define a linear map

$$\varphi : H^1(M) \to \check{H}^1(\mathfrak{U}).$$

Given $[\omega] \in H^1(M)$, select a representative $\omega \in [\omega]$. By simplicity of the cover, $H^1(U_\alpha) = 0$, so the restriction $\omega_\alpha = \omega | U_\alpha$ of the closed 1-form ω is exact, $\forall\, \alpha \in \mathfrak{A}$. Thus, we can choose $f_\alpha \in A^0(U_\alpha)$ such that $\omega_\alpha = df_\alpha$, $\forall\, \alpha \in \mathfrak{A}$. On $U_{\alpha_0} \cap U_{\alpha_1} \neq \emptyset$, $d(f_{\alpha_0} - f_{\alpha_1}) = \omega - \omega \equiv 0$, so $f_{\alpha_0} - f_{\alpha_1}$ is locally constant on $U_{\alpha_0} \cap U_{\alpha_1}$. The cover being simple, this set is connected, so $f_{\alpha_0} - f_{\alpha_1} = c_{\alpha_0\alpha_1} \in \mathbb{R}$ is a constant. This defines a Čech 1-cochain $c \in \check{C}^1(\mathfrak{U})$. But

$$(\delta c)_{\alpha_0\alpha_1\alpha_2} = c_{\alpha_1\alpha_2} - c_{\alpha_0\alpha_2} + c_{\alpha_0\alpha_1} = (f_{\alpha_1} - f_{\alpha_2}) - (f_{\alpha_0} - f_{\alpha_2}) + (f_{\alpha_0} - f_{\alpha_1}) = 0$$

on $U_{\alpha_0} \cap U_{\alpha_1} \cap U_{\alpha_2}$, so $c \in \check{Z}^1(\mathfrak{U})$. If $[c] \in \check{H}^1(\mathfrak{U})$ depends only on $[\omega] \in H^1(M)$, we can set

$$\varphi([\omega]) = [c].$$

Exercise 8.9.9. Prove that the class $[c]$ defined above is independent of the choice of representative $\omega \in [\omega]$ and of the choices of $f_\alpha \in A^0(U_\alpha)$ such that $df_\alpha = \omega_\alpha$. Consequently, φ is a well-defined linear map.

We define a linear map

$$\psi : \check{H}^1(\mathfrak{U}) \to H^1(M).$$

For this, we will need to fix the choice of a smooth partition of unity $\{\lambda_\alpha\}_{\alpha\in\mathfrak{A}}$ subordinate to $\mathfrak{U}$. Given $[c] \in \check{H}^1(\mathfrak{U})$, choose a representative cocycle $c \in [c]$. For each $\alpha_0 \in \mathfrak{A}$, define $f_{\alpha_0} \in A^0(U_{\alpha_0})$ by

$$f_{\alpha_0} = \sum_{\alpha\in\mathfrak{A}} c_{\alpha_0\alpha}\lambda_\alpha.$$

Then, on $U_{\alpha_0} \cap U_{\alpha_1} \neq \emptyset$,

$$f_{\alpha_0} - f_{\alpha_1} = \sum_{\alpha\in\mathfrak{A}} (c_{\alpha_0\alpha} - c_{\alpha_1\alpha})\lambda_\alpha = \sum_{\alpha\in\mathfrak{A}} c_{\alpha_0\alpha_1}\lambda_\alpha = c_{\alpha_0\alpha_1}.$$

It follows that $df_{\alpha_1} = df_{\alpha_0}$ on $U_{\alpha_0} \cap U_{\alpha_1}$, so these exact forms assemble to give a well-defined locally exact 1-form $\omega \in Z^1(M)$. If $[\omega] \in H^1(M)$ depends only on $[c] \in \check{H}^1(\mathfrak{U})$, we can set $\psi([c]) = [\omega]$.

Exercise 8.9.10. Prove that the class $[\omega]$ defined above is independent of the choice of representative $c \in [c]$. Consequently, ψ is a well-defined linear map.

Lemma 8.9.11. *The homomorphisms φ and ψ are mutually inverse.*

Proof. Given $[c] \in \check{H}^1(\mathfrak{U})$, the definition of $\psi([c]) = [\omega]$ produces functions $f_\alpha \in A^0(U_\alpha)$ such that $\omega|U_\alpha = df_\alpha$, $\forall \alpha \in \mathfrak{A}$. Using this choice in the definition of $\varphi([\omega])$ gives back the representative cocycle c. That is, $\varphi \circ \psi = \mathrm{id}$. For the reverse composition, the definition of $\varphi([\omega]) = [c]$ selected the functions $f_\alpha \in A^1(U_\alpha)$ such that all $df_\alpha = \omega|U_\alpha$ and $f_{\alpha_0} - f_{\alpha_1} = c_{\alpha_0\alpha_1}$. The definition of ψ produces different functions

$$\tilde{f}_{\alpha_0} = \sum_{\alpha\in\mathfrak{A}} c_{\alpha_0\alpha}\lambda_\alpha = \sum_{\alpha\in\mathfrak{A}} (f_{\alpha_0} - f_\alpha)\lambda_\alpha = f_{\alpha_0} + h,$$

where

$$h = -\sum_{\alpha\in\mathfrak{A}} f_\alpha\lambda_\alpha \in A^0(M).$$

The closed form $\widetilde{\omega}$ obtained by piecing together the exact forms $d\tilde{f}_\alpha$ is related to ω by

$$\widetilde{\omega} - \omega = dh,$$

so $[\widetilde{\omega}] = [\omega]$, proving that $\psi \circ \varphi = \mathrm{id}$. □

In particular, although ψ was defined relative to a choice of partition of unity, it inverts φ which did not depend on that choice, so ψ is, in fact, independent of the choice also.

We close this section with some remarks about triangulations and cohomology. Let $S = \{e_0, e_1, \ldots, e_n\}$ be the set of vertices of the standard n-simplex Δ_n. The convex hull of any subset $\Sigma \subseteq S$ of cardinality $p+1$ is a p-simplex. It lies in the boundary of Δ_n and will be called a p-face of Δ_n. The natural ordering of the indices of the points $e_{i_\ell} \in \Sigma$ defines a canonical identification of this p-face with the standard p-simplex Δ_p. More precisely, this natural ordering defines a canonical linear imbedding $\Delta_p \hookrightarrow \Delta_n$ with image the given p-face. If $s : \Delta_n \to M$ is a singular n-simplex, its p-faces are the singular p-simplices obtained by restricting s to the p-faces of Δ_n.

Recall from Section 1.3 the fact that compact surfaces can be triangulated. A corresponding theorem for compact, differentiable n-manifolds also holds. That is, the manifold can be divided up into a union of smoothly imbedded n-simplices $\Delta_n^1, \ldots \Delta_n^r$, any two of which either do not meet at all or meet along exactly one common lower dimensional face. This theorem is intuitively plausible, but rather difficult to prove.

If Δ stands for a choice of triangulation of M, we obtain a chain subcomplex $(C_*^\Delta(M), \partial) \subset (C_*(M), \partial)$ by using only those singular simplices that are the inclusion maps of simplices of the triangulation. (By the simplices of the triangulation, we mean all of the p-faces of the n-simplices of Δ, $0 \le p \le n$.) This is called the *simplicial chain complex* associated to the triangulation Δ. Remark that $C_p^\Delta(M)$ is a finite dimensional vector subspace of $C_p(M)$, $0 \le p \le n$, and vanishes if $p > n$. Let

$$i^\Delta : (C_*^\Delta(M), \partial) \hookrightarrow (C_*(M), \partial)$$

be the inclusion map, a homomorphism of chain complexes, and let $H_*^\Delta(M)$ be the homology of the simplicial chain complex.

The following theorem is standard in algebraic topology (*cf.* [**39**, p. 191], where it is proven more generally for simplicial complexes.)

Theorem 8.9.12. *The inclusion homomorphism i^Δ induces a canonical isomorphism $H_*^\Delta(M) = H_*(M)$.*

The beauty of this result is that the problem of finding the homology of compact manifolds is reduced to a finite set of computations. Note that the theorem assures independence of the choice of triangulation, so one normally chooses Δ to have the fewest possible simplices. The triangulation of S^2 depicted in Figure 1.3.10 has the fewest simplices of any triangulation of S^2. Triangulations can be used to give a proof, without appeal to Riemannian geometry, of the existence of a simple cover of a compact manifold. We do not pursue this, but remark that it leads to a very simple proof that the Čech cohomology of this simple cover and the simplicial cohomology $(H_*^\Delta(M))^*$ are canonically isomorphic.

Exercise 8.9.13. Using the minimal triangulation of S^2 depicted in Figure 1.3.10, give a direct computation of the homology of S^2.

Exercise 8.9.14. Fix a triangulation Δ of the compact n-manifold M and let c_p denote the number of p-simplices of Δ, $0 \le p \le n$. Note that $c_p = \dim C_p^\Delta(M)$. Let $h_p = \dim H_p^\Delta(M) = \dim H_p(M)$ (called the pth Betti number of M). Define the *Euler characteristics* of Δ and M by

$$\chi(\Delta) = \sum_{p=0}^{n} (-1)^p c_p,$$

$$\chi(M) = \sum_{p=0}^{n} (-1)^p h_p,$$

respectively.

(1) Prove that $\chi(\Delta) = \chi(M)$. Thus, this important topological invariant can be computed from a triangulation, but does not depend on the choice of triangulation.
(2) Compute $\chi(S^2)$ and give an intuitive proof, not using part (1), that this number is independent of the choice of triangulation.
(3) Prove that $\chi(M) = 0$ for compact, odd-dimensional manifolds M.
(4) In fact, it can be proven that, if M is orientable,

$$\chi(M) = \iota(\Delta_M, \Delta_M),$$

the algebraic self intersection number of Exercise 8.8.6, part (4). The Poincaré–Hopf theorem (Exercise 8.8.7) then asserts: *There is a nowhere vanishing vector field on M if and only if the Euler characteristic of M vanishes.* (In fact, the orientability condition, required in the earlier statement, can now be dropped.) Assuming this theorem, give a new proof that S^n admits a nowhere vanishing vector field if and only if n is odd.

CHAPTER 9

Forms and Foliations

In Section 4.5, we proved the vector field version of the Frobenius integrability theorem: *a k-plane field E on a manifold M is integrable if and only if* $\Gamma(E) \subseteq \mathfrak{X}(M)$ *is a Lie subalgebra.* In this chapter, we develop an equivalent version of this theorem, stated in terms of the Grassmann algebra $A^*(M)$ of differential forms. Useful consequences of this point of view will be treated.

9.1. The Frobenius Theorem Revisited

Let M be an n-manifold without boundary and let $E \subseteq T(M)$ be a smooth k-plane distribution on M.

Definition 9.1.1. For each integer $p \geq 0$, the degree p annihilator of E is

$$I^p(E) = \{\omega \in A^p(M) \mid \omega(\zeta) = 0, \forall \zeta \in \Lambda^p(\Gamma(E))\},$$

where we understand that, for $p = 0$, $I^0(E) = 0$. The annihilator of E is

$$I(E) = \{I^p(E)\}_{p \geq 0} \subseteq A^*(M).$$

It is clear that $I(E)$ is a graded $C^\infty(M)$-submodule of $A^*(M)$, but more is true.

Lemma 9.1.2. *The annihilator $I(E)$ is a 2-sided graded ideal in $A^*(M)$.*

Indeed, this follows by applying the following lemma fiber-by-fiber.

Lemma 9.1.3. *Let V be a finite dimensional vector space and let $E \subseteq V$ be a subspace. Then, the annihilator*

$$I(E) = \{\omega \in \Lambda^p(V^*) \mid \omega(\zeta) = 0, \forall \zeta \in \Lambda^p(E)\}_{p \geq 0}$$

is a 2-sided graded ideal in $\Lambda(V^*)$.

Proof. Let $\dim V = n$, $\dim E = k$, and let $\{e_1, \dots, e_k, f_1, \dots, f_{n-k}\}$ be a basis of V with $e_i \in E$, $1 \leq i \leq k$. Let $\{e_1^*, \dots, e_k^*, f_1^*, \dots, f_{n-k}^*\}$ be the dual basis of V^*. Then, in particular, $\{f_1^*, \cdots, f_{n-k}^*\}$ is a basis of $I^1(E)$. Consider

$$\eta = e_{i_1}^* \wedge \cdots \wedge e_{i_p}^* \wedge f_{j_1}^* \wedge \cdots \wedge f_{j_{r-p}}^* \in \Lambda^r(V^*),$$

where $1 \leq i_1 < \cdots < i_p \leq k$, $1 \leq j_1 < \cdots < j_{r-p} \leq n-k$, and it is allowed that either $p = 0$ or $p = r$. If $p \neq r$, it is clear that η vanishes on anything of the form $e_{m_1} \wedge \cdots \wedge e_{m_r}$, hence it vanishes on all $\zeta \in \Lambda^r(E)$. If $p = r$, η ranges over a basis of $\Lambda^r(E^*)$. It follows that the set of all the forms η, with $p \neq r$, is a basis of $I^r(E)$, $r \geq 1$, clearly implying that $I(E)$ is a 2-sided graded ideal. □

Definition 9.1.4. A graded ideal $\mathfrak{I} \subseteq A^*(M)$ is a differential graded ideal if $d(\mathfrak{I}) \subseteq \mathfrak{I}$.

Theorem 9.1.5 (The Frobenius theorem). *The following are equivalent for a k-plane distribution $E \subseteq T(M)$:*

(1) *E is integrable;*
(2) *$I(E)$ is a differential graded ideal;*
(3) *$d(I^1(E)) \subseteq I^2(E)$.*

For the proof, we need the following.

Lemma 9.1.6. *Let $\omega \in A^1(M)$ and let $X, Y \in \mathfrak{X}(M)$. Then*

$$d\omega(X \wedge Y) = X(\omega(Y)) - Y(\omega(X)) - \omega([X,Y]).$$

Proof. Define

$$\varpi : \mathfrak{X}(M) \times \mathfrak{X}(M) \to C^\infty(M)$$

by the formula

$$\varpi(X,Y) = X(\omega(Y)) - Y(\omega(X)) - \omega([X,Y]).$$

This is clearly $\mathbb{R}$-bilinear and antisymmetric. We claim, in fact, that ϖ is $C^\infty(M)$-bilinear. Indeed, let $f \in C^\infty(M)$ and compute

$$\begin{aligned}
\varpi(fX,Y) &= fX(\omega(Y)) - Y(f\omega(X)) - \omega([fX,Y]) \\
&= fX(\omega(Y)) - fY(\omega(X)) - Y(f)\omega(X) - \omega(f[X,Y] - Y(f)X) \\
&= f(X(\omega(Y)) - Y(\omega(X)) - \omega([X,Y])) \\
&= f\varpi(X,Y).
\end{aligned}$$

By antisymmetry, we also have

$$\varpi(X, fY) = -\varpi(fY, X) = -f\varpi(Y,X) = f\varpi(X,Y).$$

Thus, $\varpi \in \mathfrak{T}^2(M)$. By antisymmetry, $\varpi \in A^2(M)$.

In order to prove the equality of ϖ and $d\omega$, it will be enough to show that these forms agree on any coordinate chart $(U, x^1, \dots, x^n)$. (The previous paragraph was needed so that $\varpi|U$ would make sense.) Write

$$\omega|U = \sum_{i=1}^n g_i\, dx^i,$$
$$d\omega|U = \sum_{i=1}^n d(g_i\, dx^i),$$

and remark that no generality is lost in assuming that ω is of the form $g\,dx^i$. By permuting the coordinates, assume that $\omega = g\,dx^1$, so

$$d\omega = -\sum_{i=2}^n \frac{\partial g}{\partial x^i}\, dx^1 \wedge dx^i.$$

Thus, if $k < j$,

$$d\omega\left(\frac{\partial}{\partial x^k} \wedge \frac{\partial}{\partial x^j}\right) = \begin{cases} 0 & \text{if } k \neq 1, \\ -\partial g/\partial x^j & \text{if } k = 1. \end{cases}$$

Since $1 \leq k < j$, we have $dx^1(\partial/\partial x^j) = 0$, so

$$\begin{aligned}
\overline{\omega}\left(\frac{\partial}{\partial x^k} \wedge \frac{\partial}{\partial x^j}\right) &= \frac{\partial}{\partial x^k}\left(g\,dx^1\left(\frac{\partial}{\partial x^j}\right)\right) - \frac{\partial}{\partial x^j}\left(g\,dx^1\left(\frac{\partial}{\partial x^k}\right)\right) \\
&\quad - g\,dx^1\left(\left[\frac{\partial}{\partial x^k}, \frac{\partial}{\partial x^j}\right]\right) \\
&= -\frac{\partial}{\partial x^j}\left(g\,dx^1\left(\frac{\partial}{\partial x^k}\right)\right) \\
&= \begin{cases} 0 & \text{if } k \neq 1 \\ -\partial g/\partial x^j & \text{if } k = 1 \end{cases} \\
&= d\omega\left(\frac{\partial}{\partial x^k} \wedge \frac{\partial}{\partial x^j}\right).
\end{aligned}$$

Since $d\omega|U = \overline{\omega}|U$ for an arbitrary coordinate neighborhood, $d\omega = \overline{\omega}$. □

Proof of theorem 9.1.5. We prove that (1) ⇒ (2). Thus, it is assumed that E is integrable and we must prove that, if $\omega \in I^q(E)$, then $d\omega \in I^{q+1}(E)$. By the integrability condition, it will be enough to prove this in a coordinate chart $(U, x^1, \cdots, x^n)$ such that $\{\partial/\partial x^1, \dots, \partial/\partial x^k\}$ spans $\Gamma(E|U)$. Then $\{dx^{k+1}, \dots, dx^n\}$ spans $I^1(E|U)$. In these coordinates, we write

$$\omega = \sum_{1 \leq i_1 < \cdots < i_q \leq n} f_{i_1 \cdots i_q}\, dx^{i_1} \wedge \cdots \wedge dx^{i_q}.$$

Then, if $1 \leq i_1 < \cdots < i_q \leq k$,

$$f_{i_1 \cdots i_q} = \omega\left(\frac{\partial}{\partial x^{i_1}} \wedge \cdots \wedge \frac{\partial}{\partial x^{i_q}}\right) = 0,$$

so every nonzero term in the expression for $\omega|U$ contains at least one $dx^{i_j} \in I^1(E|U)$. The same will then hold for $d\omega|U$ and, $I(E|U)$ being an ideal, we see that $d\omega|U \in I^{q+1}(E|U)$. Covering M with such charts, we conclude that $d\omega \in I^{q+1}(E)$.

The implication (2) ⇒ (3) is trivial. We prove that (3) ⇒ (1). Thus, we are given that $d(I^1(E)) \subseteq I^2(E)$. Let $X, Y \in \Gamma(E)$ be given and choose an arbitrary element $\omega \in I^1(E)$. Then, since $d\omega \in I^2(E)$,

$$\begin{aligned}
0 &= d\omega(X \wedge Y) \\
&= X(\underbrace{\omega(Y)}_{\equiv 0}) - Y(\underbrace{\omega(X)}_{\equiv 0}) - \omega([X, Y]) \\
&= -\omega([X, Y]).
\end{aligned}$$

Since $\omega \in I^1(E)$ is arbitrary, it follows that $[X, Y] \in \Gamma(E)$. Since $X, Y \in \Gamma(E)$ are arbitrary, it follows that $\Gamma(E) \subseteq \mathfrak{X}(M)$ is a Lie subalgebra. By the vector field version of the Frobenius theorem, E is integrable. □

Exercise 9.1.7. Let $\mathcal{F}$ be a foliation of codimension q and integral to the distribution E. Prove that the exterior product of any $q + 1$ elements of $I(E)$ vanishes identically.

We note that Lemma 9.1.6, which played a key role in the proof of Theorem 9.1.5, is a special case of the following exercise.

Exercise 9.1.8. If $\omega \in A^q(M)$ and $X_1, \cdots, X_{q+1} \in \mathfrak{X}(M)$, prove that

$$d\omega(X_1 \wedge \cdots \wedge X_{q+1}) = \sum_{i=1}^{q+1}(-1)^{i+1}X_i(\omega(X_1 \wedge \cdots \wedge \widehat{X}_i \wedge \cdots \wedge X_{q+1}))$$
$$+ \sum_{i<j}(-1)^{i+j}\omega([X_i, X_j] \wedge X_1 \wedge \cdots \wedge \widehat{X}_i \wedge \cdots \wedge \widehat{X}_j \wedge \cdots \wedge X_{q+1}).$$

Remark. If $q = 0$, the formula in Exercise 9.1.8 is understood to reduce to

$$df(X_1) = (-1)^{1+1}X_1(f),$$

$\forall f \in A^0(M)$. For $q \geq 1$, the formula is noteworthy in that it gives a completely coordinate-free definition of the exterior derivative. It is useful in other ways, one of which is given in the following example.

Example 9.1.9. The formula in Exercise 9.1.8 is closely related to one that occurs in a purely algebraic context. Let $\mathfrak{L}$ be a finite dimensional Lie algebra over $\mathbb{R}$ and let $\Lambda(\mathfrak{L}^*)$ be the exterior algebra of the dual vector space $\mathfrak{L}^*$. Thus, $\Lambda^q(\mathfrak{L}^*)$ is the dual space of $\Lambda^q(\mathfrak{L})$ or, equivalently, the space of antisymmetric q-linear functionals on $\mathfrak{L}$. One defines a coboundary operator

$$\delta : \Lambda^q(\mathfrak{L}^*) \to \Lambda^{q+1}(\mathfrak{L}^*)$$

by the formula

$$\delta\omega(v_1, \ldots, v_{q+1})$$
$$= \begin{cases} \sum_{i<j}(-1)^{i+j}\omega([v_i, v_j], v_1, \ldots, \widehat{v}_i, \ldots, \widehat{v}_j, \ldots, v_{q+1}), & q \geq 1, \\ 0, & q = 0, \end{cases}$$

where $\omega \in \Lambda^q(\mathfrak{L}^*)$ is viewed as a multilinear functional and $v_i \in \mathfrak{L}$ is arbitrary, $1 \leq i \leq q+1$. The antisymmetry of $\delta\omega$ is part of Exercise 9.1.10, so $\delta\omega \in \Lambda^{q+1}(\mathfrak{L}^*)$. In Exercise 9.1.10, you are also asked to prove that $\delta^2 = 0$, so $(\Lambda(\mathfrak{L}^*), \delta)$ is a cochain complex. The cohomology of this complex is denoted by $H^*(\mathfrak{L})$ and is called the cohomology of the Lie algebra $\mathfrak{L}$. This is of considerable interest in algebra, but we want to remark on its use in studying the de Rham cohomology of Lie groups.

If G is a Lie group and we apply the above construction to its Lie algebra $\mathfrak{L} = L(G)$, we obtain a finite dimensional cochain complex $(\Lambda(L(G)^*), \delta)$. Remark that the subspace of left-invariant q-forms in $A^q(G)$ is canonically isomorphic to $\Lambda^q(T_e(G)^*) = \Lambda^q(L(G)^*)$. This can also be identified as the subspace of all $\omega \in A^q(G)$ such that $\omega(X_1, \ldots, X_q)$ is a constant function, for each choice of left-invariant vector fields $X_1, \ldots, X_q \in \mathfrak{X}(G)$. The formula for $d\omega$ in Exercise 9.1.8, when evaluated on $L(G) \subset \mathfrak{X}(G)$, reduces to $\delta\omega$, $\forall \omega \in \Lambda^q(L(G)^*)$. It follows that there is an injective homomorphism

$$(\Lambda(L(G)^*), \delta) \hookrightarrow (A^*(G), d)$$

of cochain complexes. There is induced a canonical homomorphism

$$\iota : H^*(L(G)) \to H^*(G)$$

of graded algebras and a surprising theorem, which we will not prove, asserts that, if G is both compact and connected, then ι is an isomorphism (*cf.* [**47**, Chapter IV]). Since, as a vector space, $L(G)$ is completely determined by $T_e(G)$ and, as a Lie algebra, by the Lie derivatives at e of left-invariant vector fields, it follows that the cohomology of the manifold G is entirely determined by infinitesimal data at e. This is a remarkable case of recovering global data from linear approximations at a single point.

Exercise 9.1.10. For $\omega \in \Lambda^q(\mathfrak{L}^*)$, prove that the coboundary operator δ, defined in Example 9.1.9, does produce an element $\delta\omega \in \Lambda^{q+1}(\mathfrak{L}^*)$. Prove also that $\delta^2 = 0$.

Exercise 9.1.11. Using the theorem cited in the remark above, compute the de Rham cohomology algebra of the torus T^n.

Exercise 9.1.12. Prove that, if G is a Lie group, the set of connected Lie subgroups $K \subseteq G$ corresponds one-to-one to the set of graded ideals $\mathfrak{I} \subseteq \Lambda(L(G)^*)$ such that $\delta(\mathfrak{I}) \subseteq \mathfrak{I}$.

9.2. The Normal Bundle and Transversality

Let $I^1 \subseteq A^1(M)$ be a $C^\infty(M)$-submodule that is closed under locally finite sums. For example, if $E \subseteq T(M)$ is a k-plane distribution on M, we might take $I^1 = I^1(E)$. For each $x \in M$, set

$$Q_x = \{\omega_x \in T_x^*(M) \mid \omega \in I^1\}.$$

Lemma 9.2.1. *For each $x \in M$, Q_x is a vector subspace of $T_x^*(M)$.*

This is rather obvious. Indeed, if we view I^1 as a vector space over $\mathbb{R}$, we see that $\omega \mapsto \omega_x$ defines an $\mathbb{R}$-linear map $I^1 \to T_x^*(M)$ with image Q_x. Let $q_x = \dim Q_x$.

Definition 9.2.2. The $C^\infty(M)$-submodule $I^1 \subseteq A^1(M)$ has constant rank q if $q_x = q$, $\forall x \in M$.

Exercise 9.2.3. Let M be an n-manifold, $I^1 \subseteq A^1(M)$ a $C^\infty(M)$-submodule that is closed under locally finite sums. Prove that the following are equivalent:

(1) I^1 has constant rank q;
(2) $I^1 = \Gamma(Q)$, for some q-plane subbundle $Q \subseteq T^*(M)$;
(3) $I^1 = I^1(E)$, for some $(n-q)$-plane subbundle $E \subseteq T(M)$.

Definition 9.2.4. If E is a p-plane distribution on M, set $q = n - p$, the codimension of E. Then the q-plane subbundle $Q \subseteq T^*(M)$ such that $I^1(E) = \Gamma(Q)$ is called the normal bundle of E. If E is integrable and $\mathfrak{F}$ is the corresponding foliation, then Q is called the normal bundle of $\mathfrak{F}$ and q is called the codimension of $\mathfrak{F}$ ($\operatorname{codim} \mathfrak{F}$).

This terminology comes from the following observation.

Lemma 9.2.5. *Let E be a p-plane distribution on M and fix a choice of Riemannian metric on M. Let*

$$E_x^\perp = \{v \in T_x(M) \mid v \perp E_x\},$$

$\forall x \in M$. Then

$$E^\perp = \bigcup_{x \in M} E_x^\perp \subseteq T(M)$$

is a vector subbundle isomorphic to the normal bundle Q.

Proof. Denote the Riemannian metric on $T_x(M)$ by $\langle \cdot, \cdot \rangle_x$ and define an isomorphism $\varphi : T(M) \to T^*(M)$ of bundles by

$$\varphi_x(v) = \langle v, \cdot \rangle_x, \quad \forall x \in M, \ \forall v \in T_x(M).$$

It is clear that φ carries $E^\perp$ onto Q. □

Our way of defining the normal bundle as a subbundle of the cotangent bundle is intrinsic. Defining it in $T(M)$ via a Riemannian metric gives a subbundle that depends on the choice of the metric.

Definition 9.2.6. Let M be a manifold with a foliation $\mathcal{F}$ that is integral to the distribution E. A smooth map $f : N \to M$ is transverse to $\mathcal{F}$ if, $\forall x \in N$, $f_{*x}(T_x(N)) \cup E_{f(x)}$ spans $T_{f(x)}(M)$. In this case we write $f \pitchfork \mathcal{F}$.

Remark that no assumption is made about the relative dimensions of N and M other than what is implicit in the definition: $\dim N$ must be large enough that its sum with the dimension of the leaves of $\mathcal{F}$ is at least as large as $\dim M$. Remark also that a submersion $f : N \to M$ is automatically transverse to every foliation of M.

Lemma 9.2.7. *Let $\mathcal{F}$ be a foliation of M with normal bundle Q. A smooth map $f : N \to M$ is transverse to $\mathcal{F}$ if and only if $f_x^* : T_{f(x)}^*(M) \to T_x^*(N)$ is one-to-one on $Q_{f(x)}$, $\forall x \in N$.*

Proof. Assume that $f \pitchfork \mathcal{F}$. If $\alpha \in Q_{f(x)}$ and $f_x^*(\alpha) = 0$, then, for every $v \in T_x(N)$,

$$\alpha(f_{*x}(v)) = f_x^*(\alpha)(v) = 0.$$

But α also vanishes on $E_{f(x)}$, hence, by transversality, it vanishes on all of $T_{f(x)}(M)$. That is, $\alpha = 0$.

For the converse, suppose that $f_x^* : Q_{f(x)} \to T_x^*(N)$ is one-to-one. Let $v \in T_{f(x)}(M)$. We are to prove that v is the sum of an element of $E_{f(x)}$ and an element of $f_{*x}(T_x(N))$. Choose a basis $\{\alpha_1, \dots, \alpha_q\} \subset Q_{f(x)}$ and let $a_i = \alpha_i(v)$, $1 \le i \le q$. Since $\{f_x^*(\alpha_1), \dots, f_x^*(\alpha_q)\}$ is linearly independent, there is a vector $w \in T_x(N)$ such that $a_i = f_x^*(\alpha_i)(w)$, $1 \le i \le q$. That is,

$$\alpha_i(f_{*x}(w) - v) = a_i - a_i = 0, \ \ 1 \le i \le q,$$

from which it follows that $f_{*x}(w) - v \in E_{f(x)}$. □

Theorem 9.2.8. *Let $\mathcal{F}$ be a foliation of M, $\operatorname{codim} \mathcal{F} = q$, and suppose that $f : N \to M$ is smooth and transverse to $\mathcal{F}$. Then there is a canonically defined foliation $f^{-1}(\mathcal{F})$ (called the pullback of $\mathcal{F}$ by f) of N of codimension q such that f carries each leaf of $f^{-1}(\mathcal{F})$ into a leaf of $\mathcal{F}$.*

Proof. Let $\mathcal{F}$ be integral to the distribution E and have normal bundle Q. Then $f^*(I^1(E)) \subseteq A^1(N)$ is a vector subspace and we let $I^1 \subseteq A^1(N)$ be the $C^\infty(N)$-submodule, closed under locally finite sums, that is generated by this subspace. An arbitrary element $\eta \in I^1$ can be written locally as

$$\eta = \sum_{i=1}^{\ell} f_i \eta_i,$$

where $\eta_i = f^*(\omega_i)$ and $\omega_i \in I^1(E) = \Gamma(Q)$, $1 \le i \le \ell$. By Lemma 9.2.7, I^1 will have constant rank q, so we can write

$$I^1 = I^1(\widehat{E}) = \Gamma(\widehat{Q}),$$

as in Exercise 9.2.3. It is clear that $f^*(I(E)) \subseteq I(\widehat{E})$. Furthermore, writing $\eta \in I^1(\widehat{E})$ locally as above, we see that, locally,

$$d\eta = \sum_{i=1}^{\ell} (df_i \wedge f^*(\omega_i) + f_i f^*(d\omega_i)) \in I^2(\widehat{E}),$$

since $f^*(d\omega_i) \in f^*(I(E))$ by the integrability of E (Theorem 9.1.5). But this implies that $\widehat{E}$ is integrable, again by Theorem 9.1.5. Define $f^{-1}(\mathcal{F})$ to be the foliation integral to $\widehat{E}$. The normal bundle is $\widehat{Q}$, so $\operatorname{codim} f^{-1}(\mathcal{F}) = q$.

It remains to be shown that each leaf of $f^{-1}(\mathcal{F})$ is carried by f into a leaf of $\mathcal{F}$. Since the leaves of $\mathcal{F}$ (respectively, of $f^{-1}(\mathcal{F})$) are maximal connected integral manifolds to E (respectively, to $\widehat{E}$), it will be enough to show that $f_{*x}(\widehat{E}_x) \subseteq E_{f(x)}$, $\forall x \in N$. But, if $\alpha \in Q_{f(x)}$ and $v \in \widehat{E}_x$, then

$$\alpha(f_{*x}(v)) = f_x^*(\alpha)(v) = 0,$$

since $f_x^*(\alpha) \in \widehat{Q}_x$. Since $\alpha \in Q_{f(x)}$ is arbitrary, it follows that

$$f_{*x}(v) \in E_{f(x)}, \quad \forall v \in \widehat{E}_x.$$

□

Example 9.2.9. Let $f : N \to M$ be a submersion, $\dim M = q$, $\dim N = n$. Then, as y ranges over M, the connected components of the submanifolds $f^{-1}(y)$ range over the leaves of a foliation of N of codimension q. Indeed, the unique 0-plane distribution on M is trivially integrable, the leaves of the corresponding foliation $\mathcal{F}$ being the points of M. This foliation is of codimension q, so there is a pullback foliation $f^{-1}(\mathcal{F})$ on N of codimension q. Each leaf L of $f^{-1}(\mathcal{F})$ is a connected submanifold of N of dimension $n - q$ and is carried by f into a point y of M. By the constant rank theorem, $f^{-1}(y)$ is also a submanifold of dimension $n - q$, so L must be a connected component of $f^{-1}(y)$.

Example 9.2.10. Let $E_1, E_2 \subseteq T(M)$ be integrable subbundles with corresponding foliations $\mathcal{F}_i$ of codimension q_i, $i = 1, 2$. Let E_i have fiber dimension $p_i = n - q_i$, $i = 1, 2$, where $n = \dim M$. If, for each leaf L of $\mathcal{F}_1$, the inclusion map $\iota : L \hookrightarrow M$ (a one-to-one immersion) is transverse to $\mathcal{F}_2$, we will say that $\mathcal{F}_1$ is transverse to $\mathcal{F}_2$ and write $\mathcal{F}_1 \pitchfork \mathcal{F}_2$. This simply means that, for each $x \in M$, $E_{1x} \cup E_{2x}$ spans $T_x(M)$, so the relation is symmetric ($\mathcal{F}_1 \pitchfork \mathcal{F}_2 \Rightarrow \mathcal{F}_2 \pitchfork \mathcal{F}_1$). In this case, if $x \in M$ and $\iota_x : L_x \hookrightarrow M$ is the inclusion of the leaf through x of $\mathcal{F}_1$, then $\iota_x^{-1}(\mathcal{F}_2)$ is a foliation of L_x of codimension q_2. Thus, the leaves of $\iota_x^{-1}(\mathcal{F}_2)$ have tangent spaces contained in the restriction of E_2 to $\iota_x(L_x)$. These tangent spaces also lie in E_1 and their dimension is $p_1 - q_2 = p_1 + p_2 - n$, the fiber dimension of the bundle $E_1 \cap E_2$. Since each point of M lies in a leaf of E_1, it follows that $E_1 \cap E_2$ is integrable and that the leaf through $x \in M$ of the corresponding foliation is the the leaf through x of $\iota_x^{-1}(\mathcal{F}_2)$. These leaves are just the connected components of the intersections (when nonempty) of leaves of $\mathcal{F}_1$ with leaves of $\mathcal{F}_2$. We can denote this foliation of M by $\mathcal{F}_1 \cap \mathcal{F}_2$. It is of codimension $n - (p_1 + p_2 - n) = q_1 + q_2$.

Definition 9.2.11. A p-plane distribution E on M is transversely orientable if its normal bundle Q is orientable. A foliation $\mathcal{F}$ of M of dimension p is transversely orientable if it is integral to a transversely orientable p-plane distribution.

Remark. If $\mathcal{F}$ is transversely orientable and M is orientable, then each leaf of $\mathcal{F}$ is an orientable manifold. Indeed, $T(M) \cong E \oplus Q$ and it follows easily that E is an orientable vector bundle. Thus, the tangent bundle to a leaf L, being the restriction of E to L, is orientable.

Proposition 9.2.12. *If $\mathcal{F}$ is a transversely orientable foliation of M and $f : N \to M$ is transverse to $\mathcal{F}$, then $f^{-1}(\mathcal{F})$ is transversely orientable.*

Proof. Let Q be the normal bundle of $\mathcal{F}$ and $\widehat{Q}$ the normal bundle of $f^{-1}(\mathcal{F})$. These are q-plane subbundles of the cotangent bundles of M and N, respectively. Since Q is orientable, there is a nowhere vanishing section ω of $\Lambda^q(Q)$. By Lemma 9.2.7, $f^*(\omega)$ is a nowhere vanishing section of $\Lambda^q(\widehat{Q})$, proving that $\widehat{Q}$ is orientable. □

Exercise 9.2.13. Let $E \subseteq T(M)$ be a p-plane distribution on the n-manifold M, and let $Q \subseteq T^*(M)$ be its normal bundle, a q-plane bundle where $q = n - p$. Assume that E is transversely orientable, hence that there is a nowhere zero q-form $\omega \in \Gamma(\Lambda^q(Q))$.

(1) For each $x \in M$, prove that E_x is the set of all vectors $v \in T_x(M)$ such that $\omega_x(v \wedge v_1 \wedge \cdots \wedge v_{q-1}) = 0$, for all choices of $v_1, \dots, v_{q-1} \in T_x(M)$. We call E_x the nullspace of ω_x and we also say that E is defined by the partial differential equation (P.D.E.) $\omega = 0$. If E is integrable, the leaves of the foliation $\mathcal{F}$ integral to E are said to be the maximal solutions to the P.D.E. $\omega = 0$. In this case, ω is said to be integrable.
(2) Prove that ω is integrable if and only if there is a form $\eta \in A^1(M)$ such that $d\omega = \eta \wedge \omega$.
(3) Let the foliations $\mathcal{F}_1$ and $\mathcal{F}_2$ in Example 9.2.10 be transversely orientable and let the P.D.E. $\omega_i = 0$ define the bundle E_i, $i = 1, 2$. Show that the P.D.E. $\omega_1 \wedge \omega_2 = 0$ defines $E_1 \cap E_2$ and verify the integrability condition
$$d(\omega_1 \wedge \omega_2) = \eta \wedge \omega_1 \wedge \omega_2$$
as a direct consequence of the integrability of ω_1 and ω_2.

Exercise 9.2.14. Let $\mathcal{F}$ be a transversely orientable foliation of codimension q with normal bundle Q and tangent distribution E. Let $\omega \in \Gamma(\Lambda^q(Q))$ be nowhere zero. Let $\eta \in A^1(M)$ be such that $d\omega = \eta \wedge \omega$.

(1) Prove that $d\eta \in I^2(E)$.
(2) Prove that $\eta \wedge (d\eta)^q \in A^{2q+1}(M)$ is a closed form.
(3) Show that $[\eta \wedge (d\eta)^q] \in H^{2q+1}(M)$ does not depend on the allowable choices of ω and of η. (Hint: First hold ω fixed and prove independence of the choice of η. Then prove independence of the choice of ω.) This class is denoted by $\mathrm{gv}(\mathcal{F})$ and called the *Godbillon–Vey class of* $\mathcal{F}$.
(4) If $f : N \to M$ is transverse to $\mathcal{F}$, prove that
$$f^*(\mathrm{gv}(\mathcal{F})) = \mathrm{gv}(f^{-1}(\mathcal{F})).$$
This is called the *naturality* of the Godbillon–Vey class.

The Godbillon–Vey class was discovered in the early 1970s, leading to a formidable body of research into the algebraic topology of foliations.

9.3. Closed, Nonsingular 1-forms*

The topology of foliations is a fascinating and subtle topic. In this book we can only scratch the surface, but, with the tools developed so far, there are some interesting questions about foliations of codimension one that are accessible. One of these starts with the question: *which compact, connected, boundaryless n-manifolds M admit closed, nowhere zero 1-forms ω?* A nowhere zero form is also said to be *nonsingular*.

Throughout this section, we fix the hypothesis that the n-manifold M is compact and connected with $\partial M = \emptyset$.

Lemma 9.3.1. *In order that M, as above, admit a closed, nonsingular form $\omega \in A^1(M)$, it is necessary that $H^1(M) \neq 0$. Indeed, such a form determines a nontrivial element $[\omega] \in H^1(M)$.*

Proof. If $\omega = df$, some $f \in C^\infty(M)$, the compactness of M implies the existence of a critical point $x \in M$ of f. For instance, a point where the maximum is attained will do. But $df_x = 0$ at any critical point x, contradicting the assumption that $\omega = df$ is nonsingular. □

It is known, however, that this condition is not sufficient. The following observation relates the question to foliations.

Lemma 9.3.2. *Let $\omega \in A^1(M)$ be closed and nowhere 0. At each $x \in M$, define*

$$E_x = \{v \in T_x(M) \mid \omega_x(v) = 0\}.$$

Then $E = \bigcup_{x \in M} E_x$ is an integrable distribution on M.

Proof. Indeed, let $I^1 \subseteq A^1(M)$ be the $C^\infty(M)$-submodule generated by ω. That is, I^1 is the set of all $f\omega$, $f \in C^\infty(M)$, hence is of constant rank 1. Then E is clearly the $(n-1)$-plane distribution such that $I^1 = I^1(E)$. Since ω is closed, we have $d(f\omega) = df \wedge \omega \in I^2(E)$, so E is integrable by Theorem 9.1.5. □

We let $\mathfrak{F}_\omega$ denote the foliation of M corresponding to the closed, nonsingular 1-form ω. The codimension of $\mathfrak{F}_\omega$ is 1.

Example 9.3.3. Since S^1 is parallelizable, it admits a nowhere 0 form $\theta \in A^1(S^1)$. By default, $d\theta = 0$. The corresponding foliation is the zero-dimensional foliation having each point of S^1 as a leaf. Admittedly, this is not an interesting example, but it leads to a class of interesting examples, namely the manifolds that *fiber* over S^1. That is, we consider a compact, connected, boundaryless n-manifold M, together with a smooth map

$$\pi : M \to S^1$$

that is locally trivial in a sense quite similar to the local triviality of vector bundles and of principal bundles: there is a compact (possibly not connected) $(n-1)$-manifold F without boundary such that each point $z \in S^1$ has a neighborhood U and a commutative diagram

$$\begin{array}{ccc} \pi^{-1}(U) & \xrightarrow{\varphi} & U \times F \\ {\scriptstyle\pi}\downarrow & & \downarrow{\scriptstyle p_1} \\ U & \xrightarrow[\text{id}]{} & U \end{array}$$

such that φ is a diffeomorphism. We say that $\pi : M \to S^1$ is a fibration (or a fiber bundle with fiber F). In particular, $\pi : M \to S^1$ is a submersion and the $(n-1)$-manifolds $\pi^{-1}(z)$ are all diffeomorphic to F. The connected components of these fibers are the leaves of a (codimension one) foliation, as was observed in Example 9.2.9. Also, $\omega = \pi^*(\theta)$ is a closed, nonsingular 1-form on M and $\mathfrak{F}_\omega$ is exactly the foliation by the components of the fibers.

Example 9.3.4. Let $T^3 = S^1 \times S^1 \times S^1$. There are three obvious fibrations $\pi_i : T^3 \to S^1$ given by $\pi_i(z_1, z_2, z_3) = z_i$, $1 \leq i \leq 3$. These are trivial cases of Example 9.3.3, but there are more interesting examples of closed, nonsingular forms $\omega \in A^1(T^3)$. Indeed, let $\omega_i = \pi_i^*(\theta)$, $1 \leq i \leq 3$. These are pointwise linearly independent over $\mathbb{R}$, so every nontrivial linear combination $\omega = \sum_{i=1}^{3} a_i\omega_i$ is closed and nonsingular. If, when we view $\mathbb{R}$ as a vector space over the rationals $\mathbb{Q}$, the set $\{a_1, a_2, a_3\}$ is linearly independent, then the corresponding foliation $\mathcal{F}_\omega$ of T^3 has each leaf diffeomorphic to $\mathbb{R}^2$ and dense in T^3. Similarly, if we require that two of these numbers, say $\{a_1, a_2\}$, be linearly independent over $\mathbb{Q}$, but not all three, each leaf of $\mathcal{F}_\omega$ is diffeomorphic to $\mathbb{R} \times S^1$ and is dense in T^3. Finally, if $\{a_1\}$ is linearly independent over $\mathbb{Q}$, but $\{a_1, a_2\}$ and $\{a_1, a_3\}$ are not, each leaf of $\mathcal{F}_\omega$ is diffeomorphic to T^2 and these leaves are the fibers of a fibration of T^3 over S^1. Remark that any triple $\{a_1, a_2, a_3\}$ can be uniformly well approximated by triples of this last type. Thus, there is a sense in which the linear foliations of T^3 by dense planes or by dense cylinders can be uniformly well approximated by fibrations over S^1. These assertions are left as an exercise.

Exercise 9.3.5. Fill in the details in Example 9.3.4

Definition 9.3.6. Let $\mathcal{F}$ be a foliation of M of codimension one. Let the flow $\Phi : \mathbb{R} \times M \to M$ be smooth, nonsingular (*i.e.*, no stationary points), with flow lines everywhere transverse to $\mathcal{F}$. If, for each leaf L of $\mathcal{F}$ and each $t \in \mathbb{R}$, $\Phi_t(L)$ is also a leaf of $\mathcal{F}$, we say that Φ is a transverse, invariant flow for $\mathcal{F}$.

Lemma 9.3.7. *The flow $\Phi : \mathbb{R} \times M \to M$ (not necessarily nonsingular and not necessarily transverse to $\mathcal{F}$) carries leaves of $\mathcal{F}$ to leaves of $\mathcal{F}$ if and only if the infinitesimal generator $X \in \mathfrak{X}(M)$ of Φ satisfies*

$$[X, \Gamma(E)] \subseteq \Gamma(E),$$

where $E \subset T(M)$ is the integrable distribution of tangent spaces to $\mathcal{F}$.

Proof. If Φ carries leaves to leaves, then

$$\Phi_{-t*}(E_x) \subseteq E_{\Phi_{-t}(x)}, \quad \forall x \in M, \quad \forall t \in \mathbb{R}.$$

Thus, if $Y \in \Gamma(E)$,

$$[X, Y] = \lim_{t \to 0} \frac{\Phi_{-t*}(Y) - Y}{t} \in \Gamma(E).$$

For the converse, suppose that $[X, \Gamma(E)] \subseteq \Gamma(E)$ and remark that it will be sufficient to show that, in any Frobenius chart $(U, x^1, \dots, x^n)$ for $\mathcal{F}$, Φ_t carries plaques to plaques for small enough values of t. Here, we assume that the plaques are the level sets $x^n = \text{const}$. In these coordinates, we write

$$X = \sum_{i=1}^{n} f_i \frac{\partial}{\partial x^i}.$$

The condition that

$$\left[\frac{\partial}{\partial x^j}, X\right] \in \Gamma(E|U), \ 1 \le j \le n-1,$$

implies that $f_n = f_n(x^n)$ is independent of $x^1, \dots, x^{n-1}$. Thus, the local system of O.D.E. for Φ is

$$\begin{aligned} \frac{dx^i}{dt} &= f_i(x^1, \dots, x^n), \ 1 \le i \le n-1, \\ \frac{dx^n}{dt} &= f_n(x^n). \end{aligned}$$

Consequently, given the initial condition $a = (a^1, \dots, a^n) \in U$, the nth coordinate $x^n(a,t)$ of the flow line $\Phi_t(a)$ depends only on a^n and t. That is, the plaque $x^n = a^n$ is carried into the plaque $x^n = x^n(a^n, t)$. □

Theorem 9.3.8. *The foliation $\mathfrak{F}$ admits a transverse, invariant flow if and only if $\mathfrak{F} = \mathfrak{F}_\omega$, for some closed, nonsingular 1-form $\omega \in A^1(M)$.*

Proof. Assume that Φ is a transverse, invariant flow for $\mathfrak{F}$ and, using the infinitesimal generator $X \in \mathfrak{X}(M)$ of the flow, define a nonsingular 1-form ω by

$$\begin{aligned} \omega(X) &\equiv 1, \\ \omega|E &\equiv 0, \end{aligned}$$

where E is the distribution of tangent spaces to $\mathfrak{F}$. Clearly, ω_x spans the normal fiber Q_x, $\forall x \in M$. We must prove that $d\omega = 0$. For this, let $x \in M$ and choose a basis $\{v_1, \dots, v_{n-1}\} \subset E_x$. We can extend v_i to a field $Y_i \in \Gamma(E)$, $1 \le i \le n-1$. Since $\Gamma(E) \subset \mathfrak{X}(M)$ is a Lie subalgebra,

$$d\omega(Y_i, Y_j) = Y_i(\omega(Y_j)) - Y_j(\omega(Y_i)) - \omega[Y_i, Y_j] = 0,$$

$1 \le i, j \le n-1$. Also, since $\omega(X) \equiv 1$,

$$d\omega(X, Y_i) = X(\omega(Y_i)) - Y_i(\omega(X)) - \omega[X, Y_i] = -\omega[X, Y_i],$$

$1 \le i \le n-1$. By the invariance property and Lemma 9.3.7, $[X, Y_i] \in \Gamma(E)$, and it follows that

$$d\omega(X, Y_i) = -\omega[X, Y_i] = 0.$$

In particular, $(d\omega)_x$ vanishes on all pairs from the basis $\{v_1, \dots, v_{n-1}, X_x\}$ of $T_x(M)$, hence $(d\omega)_x = 0$. Since $x \in M$ is arbitrary, $d\omega = 0$.

Conversely, suppose that $\mathfrak{F} = \mathfrak{F}_\omega$, where $d\omega = 0$. We must produce the transverse, invariant flow. Let$\{U_\alpha, x^1_\alpha, \dots, x^n_\alpha\}_{\alpha=1}^m$ be a cover of M by Frobenius charts, the $\mathfrak{F}$-plaques in U_α being the level sets $x^n_\alpha = \text{const}$. Then,

$$\omega_\alpha = \omega|U_\alpha = f_\alpha \, dx^n_\alpha,$$

where $f_\alpha \ne 0$ on U_α. Set

$$X_\alpha = \frac{1}{f_\alpha} \frac{\partial}{\partial x^n_\alpha},$$

a vector field transverse to the plaques and satisfying

$$\omega_\alpha(X_\alpha) \equiv 1.$$

Let $\{\lambda_\alpha\}_{\alpha=1}^m$ be a partition of unity subordinate to the Frobenius atlas and set

$$X = \sum_{\alpha=1}^m \lambda_\alpha X_\alpha \in \mathfrak{X}(M).$$

This vector field satisfies $\omega(X) \equiv 1$ and, in particular, the flow Φ that it generates is everywhere transverse to $\mathfrak{F}$. Furthermore, if $Y \in \Gamma(E)$,

$$0 = d\omega(X, Y) = X(\omega(Y)) - Y(\omega(X)) - \omega[X, Y] = -\omega[X, Y],$$

implying that $[X, Y] \in \Gamma(E)$. Since $Y \in \Gamma(E)$ is arbitrary, Lemma 9.3.7 completes the proof that Φ is a transverse, invariant flow for $\mathfrak{F}_\omega$. □

Fix the hypothesis that ω, $\mathfrak{F}_\omega$, and $\Phi : \mathbb{R} \times M \to M$ are all as above. Let $\mathfrak{L}$ denote the 1-dimensional foliation of M by the flow lines of Φ.

Lemma 9.3.9. *For arbitrary leaves L and L' of $\mathcal{F}_\omega$, there are values $t \in \mathbb{R}$ such that $\Phi_t(L) \cap L' \neq \emptyset$, in which case Φ_t carries L diffeomorphically onto L'.*

Proof. It is clear that $\Phi_t(L) \cap L' \neq \emptyset$ if and only if Φ_t carries L diffeomorphically onto L'. Thus, we obtain an equivalence relation on the set of leaves by setting $L \sim L'$ if and only if there is such a value of t. Since the flow is leaf-preserving and transverse to $\mathcal{F}$, an easy application of the inverse function theorem proves that $\Phi : \mathbb{R} \times L \to M$ is a local diffeomorphism, hence it has as image an open subset of M. This image is the union of the leaves equivalent to L, hence, by the connectivity of M, all leaves are equivalent. □

If L is a leaf of $\mathcal{F}_\omega$, denote by $P(L, \omega)$ the set $\{t \in \mathbb{R} \mid \Phi_t(L) = L\}$.

Lemma 9.3.10. *If L and L' are leaves, then $P(L,\omega) = P(L',\omega)$ and this set, call it $P(\omega)$, is an additive subgroup of $\mathbb{R}$.*

Proof. Let $t \in P(L,\omega)$. Let $\tau \in \mathbb{R}$ be such that $\Phi_\tau(L) = L'$. Then

$$\Phi_t(L') = \Phi_{t+\tau}(L) = \Phi_\tau(\Phi_t(L)) = \Phi_\tau(L) = L'.$$

Thus, $P(L,\omega) \subseteq P(L',\omega)$ and the reverse inclusion is proven in the same way. This set $P(\omega)$ carries *every* leaf to itself. By the properties of a flow, it is clear that $P(\omega)$ is closed under addition and multiplication by -1, hence is an additive subgroup of $\mathbb{R}$. □

Exercise 9.3.11. Let $\{U_\alpha, x^1_\alpha, \dots, x^n_\alpha\}_{\alpha=1}^m$ be a Frobenius atlas for $\mathcal{F}_\omega$. Let $\sigma : [a,b] \to M$ be a piecewise smooth loop. Prove that σ is homotopic to a piecewise smooth loop τ with the following property: there is a partition

$$a = t_0 < t_1 < \cdots < t_p = b$$

such that, for $1 \le i \le p$, $\tau|[t_{i-1}, t_i] \subset U_{\alpha_i}$ and this segment either lies in a plaque of $\mathcal{F}$ or in a plaque of $\mathcal{L}$.

Corollary 9.3.12. *The group $P(\omega)$ is exactly the set of periods of the closed 1-form ω.*

Proof. If $a \in P(\omega)$, the segment $s_1(t) = \Phi_t(x_0)$, where t ranges from 0 to a, has both endpoints in the leaf L of $\mathcal{F}$ through x_0. Let s_2 be a piecewise smooth path in L from the endpoint of s_1 to the initial point of s_1. Then $s = s_1 + s_2$ is a piecewise smooth loop in M and it is clear that $\int_s \omega = a$. That is, a is a period of ω.

For the reverse inclusion, choose a piecewise smooth loop σ and deform it to τ as in Exercise 9.3.11. Let $\tau_k = \tau|[\mu_k, \nu_k]$, $0 \le k \le r$, (taken in increasing order) be the segments of τ that lie in $\mathcal{L}$-plaques and let $\tau|[\nu_k, \mu_{k+1}]$, $0 \le k \le r$, lie in $\mathcal{F}$-plaques, with $\tau(\mu_{r+1}) = \tau(\mu_0)$. Then

$$\int_\sigma \omega = \int_\tau \omega = \sum_{k=1}^{r} \int_{\tau_k} \omega = \sum_{k=1}^{r} a_k = a,$$

and a is a period of ω. All periods can be obtained in this way. Let L_k denote the leaf through $\tau(\mu_k)$, $0 \le k \le r+1$ (hence $L_{r+1} = L_0$). Then $\Phi_{a_k}(L_k) = L_{k+1}$, $0 \le k \le r$, and $\Phi_a(L_0) = L_0$, proving that $a \in P(\omega)$. □

Lemma 9.3.13. *The subgroup $P(\omega) \subset \mathbb{R}$ is either infinite cyclic or everywhere dense in $\mathbb{R}$.*

Proof. Indeed, if $P(\omega) \neq 0$, this follows from Corollary 4.4.10. But, by Theorem 6.3.10, $P(\omega) = 0$ implies that $[\omega] = 0$, contradicting Lemma 9.3.1. $\square$

Theorem 9.3.14. *If the period group $P(\omega)$ of a closed, nonsingular form $\omega \in A^1(M)$ is infinite cyclic, then the leaves of $\mathcal{F}_\omega$ are the fibers of a suitable fibration $\rho : M \to S^1$. If $P(\omega)$ is not infinite cyclic, each leaf of $\mathcal{F}_\omega$ is dense in M.*

Proof. If $P(\omega)$ is not infinite cyclic, then Lemma 9.3.13 implies that a dense set of real numbers t has the property that $\Phi_t(L) = L$, for an arbitrary leaf L of $\mathcal{F}_\omega$. Since $\Phi : \mathbb{R} \times L \to M$ is onto, it follows easily that L is dense in M.

Suppose that $P(\omega)$ is infinite cyclic and let $a \in (0, \infty)$ be the smallest positive period. Then, replacing ω by ω/a, we lose no generality in assuming that $P(\omega) = \mathbb{Z}$. Fix $x_0 \in M$ and consider all piecewise smooth paths $\sigma : [a, b] \to M$ with $\sigma(a) = x_0$. If $\sigma(b) = x$, define

$$\rho(x) = e^{2\pi i \int_\sigma \omega}.$$

In order to see that this is well defined, let τ be another such path from x_0 to x. Let $-\sigma$ denote the path from x to x_0 obtained by reversing the parametrization of σ and consider the loop $\zeta = \tau + (-\sigma)$. Since the period $\int_\zeta \omega = k$ is an integer, it follows that

$$e^{2\pi i \int_\tau \omega} = e^{2\pi i (k + \int_\sigma \omega)} = e^{2\pi i \int_\sigma (\omega)}.$$

Furthermore, since M is assumed to be connected, $\rho(x)$ is defined for every $x \in M$. The reader can verify that $\rho : M \to S^1$ is smooth.

We claim that $\rho(x)$ is constant as x ranges over a given leaf L of $\mathcal{F}_\omega$. Indeed, given arbitrary points $x, y \in L$, let σ be a path from x_0 to y and let τ be a path in L from y to x. But $\int_\tau \omega = 0$, since ω vanishes on vectors in E, so

$$\rho(x) = e^{2\pi i \int_{\sigma+\tau} \omega} = e^{2\pi i \int_\sigma \omega} = \rho(y).$$

This shows that $\rho|L$ is constant. By the definition of $P(\omega)$, it is clear that $\rho : M \to S^1$ sets up a one-to-one correspondence between the leaves of $\mathcal{F}_\omega$ and the points of S^1.

Finally, we prove local triviality. Indeed, given $z \in S^1$, let $U \subset S^1$ be the open arc $\{ze^{2\pi i t} \mid -1/2 < t < 1/2\}$. Let $L_z = \rho^{-1}(z)$. The map

$$\psi : U \times L_z \to \rho^{-1}(U),$$

defined by $\psi(t, w) = \Phi_t(w)$, is a diffeomorphism and the inverse diffeomorphism $\varphi = \psi^{-1}$ makes the diagram

$$\begin{array}{ccc} \rho^{-1}(U) & \xrightarrow{\ \varphi\ } & U \times L_z \\ {\scriptstyle\rho}\downarrow & & \downarrow{\scriptstyle p_1} \\ S^1 & \xrightarrow[\text{id}]{} & S^1 \end{array}$$

commute. $\square$

Thus, the possibilities illustrated in Example 9.3.4 are the typical ones. The last statement in that example, that the linear foliations of T^3 with dense leaves can be arbitrarily well approximated by fibrations over S^1, is also typical.

Theorem 9.3.15 (D. Tischler). *If ω is a closed, nonsingular 1-form on M, then there is a sequence $\{\omega_i\}_{i=1}^\infty$ of closed, nonsingular 1-forms with $P(\omega_i) \cong \mathbb{Z}$, $\forall\, i \geq 1$ such that $\lim_{i\to\infty} \omega_i = \omega$ uniformly on M.*

Corollary 9.3.16. *There is a closed, nonsingular form $\omega \in A^1(M)$ if and only if M fibers over S^1.*

This corollary is an answer to the opening question of this section. The proof of Theorem 9.3.15, which will be found in [**45**] and in [**6**, Section 9.4], is not difficult, but it uses some algebraic topology not developed in this book. Among other things, one needs the fact, mentioned (but not proven) after Exercise 6.4.19, that the loops in that exercise define an isomorphism of $H^1(M;\mathbb{Z})$ onto $\mathbb{Z}^k$ if M is compact.

Exercise 9.3.17. Let $\mathcal{F}$ be a foliation of M of codimension 1 and let

$$\mathfrak{U} = \{U_\alpha, x_\alpha^1, \dots, x_\alpha^n\}_{\alpha=1}^m$$

be an atlas of Frobenius charts. Thus, on $U_\alpha \cap U_\beta$, the change of coordinates has the form

$$\begin{aligned} x_\alpha^i &= x_\alpha^i(x_\beta^1, \dots, x_\beta^n), \qquad 1 \le i \le n, \\ x_\alpha^n &= x_\alpha^n(x_\beta^n). \end{aligned}$$

(1) If there is a closed, nonsingular 1-form ω such that $\mathcal{F} = \mathcal{F}_\omega$, prove that the Frobenius atlas can be chosen in such a way that the second equation above always takes the form
$$x_\alpha^n = x_\beta^n + c_{\alpha\beta},$$
where $c_{\alpha\beta} \in \mathbb{R}$ is a constant, defined whenever $U_\alpha \cap U_\beta \neq \emptyset$.
(2) Conversely, if the Frobenius atlas can be chosen as in part (a), prove that there is a closed, nowhere vanishing 1-form ω such that $\mathcal{F} = \mathcal{F}_\omega$.
(3) In this situation, show that $c_{\alpha\beta} = c(U_\alpha, U_\beta)$ defines a Čech cocycle on the open cover $\mathfrak{U}$.
(4) It is a fact that, in the above situation, the Frobenius cover $\mathfrak{U}$ can be chosen to be simple. Assuming this, prove that the de Rham isomorphism $H^1(M) = \check{H}^1(\mathfrak{U})$, as defined in Section 8.9, identifies $[c] \in \check{H}^1(\mathfrak{U})$ with $[\omega] \in H^1(M)$.

CHAPTER 10

Riemannian Geometry

Properly speaking, geometry is the study of manifolds that are equipped with some additional structure that permits *measurements.* For example, nowhere in the definition of a piecewise smooth curve is there anything that would enable us to measure the *length* of the curve. Likewise, on a compact, oriented n-manifold, we can integrate n-forms, but which of these integrals should be interpreted as the *volume* of the manifold? And given intersecting curves, how could we measure the *angle* they make at an intersection point? The additional structure that is needed is a *metric* tensor, Riemannian metrics and, to a lesser extent, pseudo–Riemannian metrics, being the main examples.

Such a tensor makes it easy to define the quantities mentioned above and provides much more. For instance, the metric tensor gives rise to the "Levi-Civita connection", which can be thought of as a way of parallel transporting vectors along curves. One is led to study special curves that are "straight" in the sense that the velocity field is parallel along them. These are "geodesics", the analogues in Riemannian geometry of straight lines in Euclidean geometry. These geodesics are locally length minimizing, but this may fail in the global sense. For instance, if two points on a sphere are not antipodal, then, in the standard metric on the sphere, the great circle through these points is a geodesic. It falls into two imbedded arcs joining the points, one of which is the shortest curve joining them, but the other clearly is not.

Parallel transport along curves holds some surprises for Euclidean "flatlanders". For instance, consider the geodesic triangle σ on S^2 in Figure 10.0.1. Imagine that you are a two-dimensional native on S^2. Starting at point A, you walk down the first leg of the triangle, holding the initial tangent vector straight ahead so as to keep it (in your world) always parallel to its original position. Upon arriving at point B, you start moving sideways along the equatorial geodesic, determinedly keeping the vector pointing in a direction always parallel to its earlier positions. Finally, at C, you start up the last leg of your journey, walking backwards and again holding the vector in front of you in a constant parallel direction. Upon arriving at A, you find that, despite your best efforts not to change the direction of the vector, it ends up pointing in a different direction at A than it started with! Although, at the beginning of this experiment, you may have been convinced that your world was a Euclidean plane, you now have evidence of intrinsic "curvature", something you (as a two-dimensional creature) probably cannot *imagine*, but can nevertheless *conceive.* If you are Gauss, you may even be able to figure out how to compute the curvature of your world via experiments such as the above.

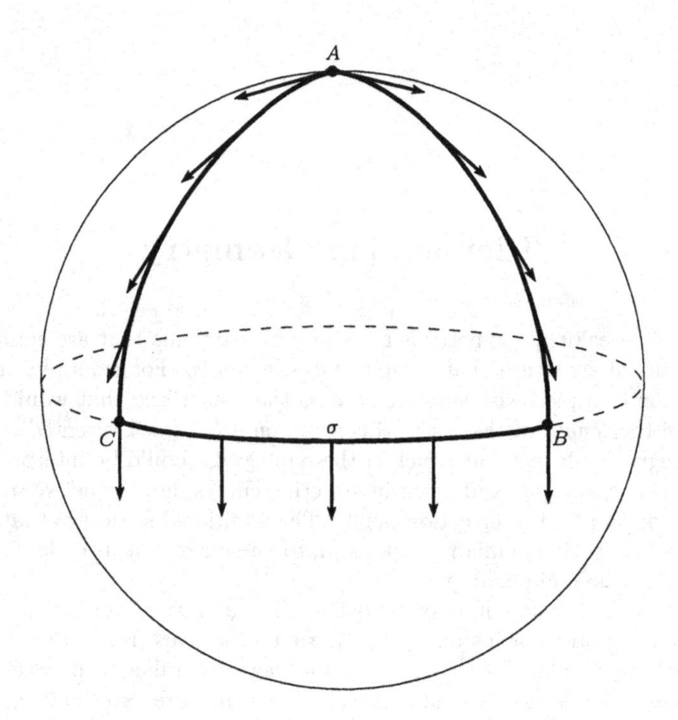

Figure 10.0.1. A parallel field along σ in S^2

10.1. Connections

Let $U \subseteq \mathbb{R}^n$ be open. Given $X, Y \in \mathfrak{X}(U)$, define $D_X(Y) \in \mathfrak{X}(U)$ as follows. Write

$$X = \sum_{i=1}^{n} f_i \frac{\partial}{\partial x^i},$$

$$Y = \sum_{j=1}^{n} g_j \frac{\partial}{\partial x^j},$$

and define

$$\begin{aligned} D_X Y &= \sum_{j=1}^{n} X(g_j) \frac{\partial}{\partial x^j} \\ &= \sum_{i,j=1}^{n} f_i \frac{\partial g_j}{\partial x^i} \frac{\partial}{\partial x^j}. \end{aligned}$$

We can view D as an $\mathbb{R}$-bilinear map

$$D : \mathfrak{X}(U) \times \mathfrak{X}(U) \to \mathfrak{X}(U).$$

It has the following properties:

(1) $D_{fX}(Y) = fD_X(Y), \forall f \in C^\infty(U), \forall X, Y \in \mathfrak{X}(U)$.
(2) $D_X(fY) = X(f)Y + fD_X(Y), \forall f \in C^\infty(U), \forall X, Y \in \mathfrak{X}(U)$.

This is an example of a *connection.*

Definition 10.1.1. Let M be a smooth manifold. A connection on M is an $\mathbb{R}$-bilinear map

$$\nabla : \mathfrak{X}(M) \times \mathfrak{X}(M) \to \mathfrak{X}(M),$$

written $\nabla(X,Y) = \nabla_X Y$ or $\nabla_X(Y)$, with the following properties:

(1) $\nabla_{fX}Y = f\nabla_X Y, \forall f \in C^\infty(M), \forall X, Y \in \mathfrak{X}(M)$;
(2) $\nabla_X(fY) = X(f)Y + f\nabla_X Y, \forall f \in C^\infty(M), \forall X, Y \in \mathfrak{X}(M)$.

The connection D, defined above on open subsets $U \subseteq \mathbb{R}^n$, is called the Euclidean connection.

Remark. By property (1), ∇ is a tensor in the first argument. Thus, if $v \in T_x(M)$ and $Y \in \mathfrak{X}(M)$, $\nabla_v Y \in T_x(M)$ is defined. By property (2), however, ∇ is not a tensor in the second argument. Property (2), together with the C^∞ Urysohn lemma, allows us to prove in standard fashion that $\nabla_v Y \in T_x(M)$ depends only on the values of Y in an arbitrarily small neighborhood of x. Thus, connections can be restricted to open subsets of M.

We describe an important class of examples, the Levi-Civita connections for submanifolds of Euclidean space. Let $M \subset \mathbb{R}^m$ be a smoothly imbedded n-manifold. If $x \in M$, then $T_x(\mathbb{R}^m) = \mathbb{R}^m$ canonically. Also,

$$T_x(\mathbb{R}^m) = T_x(M) \oplus \nu_x(M),$$

where

$$\nu_x(M) = \{v \in T_x(\mathbb{R}^m) \mid v \perp T_x(M)\},$$

perpendicularity being defined by the Euclidean metric in $\mathbb{R}^m$. Then

$$\nu(M) = \bigcup_{x \in M} \nu_x(M) \subset T(\mathbb{R}^m)|M$$

is an $(m-n)$-plane bundle over M and

$$T(\mathbb{R}^m)|M = T(M) \oplus \nu(M)$$

is a canonical direct sum bundle decomposition. The summand $\nu(M)$ is the normal bundle of M in $\mathbb{R}^m$ introduced in Subsection 3.7.C. The canonical projection

$$p : T(\mathbb{R}^m)|M = T(M) \oplus \nu(M) \to T(M)$$

is a surjective homomorphism of bundles.

Let $X, Y \in \mathfrak{X}(M)$. Given $x \in M$, there is a neighborhood U of x in $\mathbb{R}^m$ and extensions $\widetilde{X}, \widetilde{Y}$ of $X|(U \cap M)$ and $Y|(U \cap M)$, respectively, to fields on U. Then, $(D_{\widetilde{X}}\widetilde{Y})_x$ depends only on $\widetilde{X}_x = X_x$ and on $\widetilde{Y}$. Represent $X_x = \langle s \rangle_x$ as an infinitesimal curve, where $s : (-\epsilon, \epsilon) \to M$ is smooth and $s(0) = x$. Then,

$$D_{X_x}\widetilde{Y} = \frac{d}{dt}(\widetilde{Y}_{s(t)})\bigg|_{t=0} = \frac{d}{dt}(Y_{s(t)})\bigg|_{t=0}$$

and this depends only on $Y|(U \cap M)$, not on the choice of extension $\widetilde{Y}$. Therefore, $D_X Y$ is a well-defined element of $\Gamma(T(\mathbb{R}^m)|M)$.

Definition 10.1.2. If $X, Y \in \mathfrak{X}(M)$, then the operator

$$\nabla : \mathfrak{X}(M) \times \mathfrak{X}(M) \to \mathfrak{X}(M),$$

defined by

$$\nabla_X Y = p(D_X Y)$$

is called the Levi-Civita connection on $M \subset \mathbb{R}^m$.

The following is totally elementary, as the reader can check.

Lemma 10.1.3. *The Levi-Civita connection is a connection on* M.

If ∇ is a connection on M and $(U, x^1, \dots, x^n)$ is a coordinate chart, set $\xi_i = \partial/\partial x^i$ and write

$$\nabla_{\xi_i} \xi_j = \sum_{k=1}^{n} \Gamma_{ij}^k \xi_k.$$

Definition 10.1.4. The functions $\Gamma_{ij}^k \in C^\infty(U)$ are called the *Christoffel symbols* of ∇ in the given local coordinates.

Definition 10.1.5. Let ∇ be any connection on a manifold M. The torsion of ∇ is the $\mathbb{R}$-bilinear map

$$T : \mathfrak{X}(M) \times \mathfrak{X}(M) \to \mathfrak{X}(M)$$

defined by the formula

$$T(X, Y) = \nabla_X Y - \nabla_Y X - [X, Y].$$

If $T \equiv 0$, then ∇ is said to be *torsion free* or *symmetric.*

Exercise 10.1.6. Prove that the torsion T of a connection ∇ on M is $C^\infty(M)$-bilinear. That is, $T \in \mathfrak{T}_1^2(M)$ and $T(v, w) \in T_x(M)$ is defined, $\forall v, w \in T_x(M)$, $\forall x \in M$. Torsion is a *tensor.*

Exercise 10.1.7. Prove that the connection ∇ is torsion free if and only if, in every local coordinate chart $(U, x^1, \dots, x^n)$, the Christoffel symbols have the symmetry

$$\Gamma_{ij}^k = \Gamma_{ji}^k, \quad 1 \le i, j, k \le n.$$

This is the reason that torsion free connections are also said to be symmetric.

Exercise 10.1.8. If $M \subset \mathbb{R}^m$ is a smoothly imbedded submanifold, prove that the Levi-Civita connection is symmetric.

We use connections to define a way of differentiating vector fields along a curve. Indeed, if $X \in \mathfrak{X}(M)$ and $s : [a, b] \to M$ is a smooth curve, then, at each point $s(t)$, one can compute

$$X'_{s(t)} = \nabla_{\dot{s}(t)} X \in T_{s(t)}(M).$$

But we will also be interested in differentiating vector fields along s that are only defined along s. In fact, it is often natural to consider fields $X_{s(t)}$ along s that are also parametrized by the parameter t, allowing $X_{s(t_1)} \neq X_{s(t_2)}$ even if $s(t_1) = s(t_2)$, $t_1 \neq t_2$. For instance, $X_{s(t)} = \dot{s}(t)$, the velocity field, may exhibit such behavior. In these cases, it is not immediately clear how to use a connection to produce the desired derivative.

Definition 10.1.9. Let $s : [a, b] \to M$ be smooth. A vector field along s is a smooth map $v : [a, b] \to T(M)$ such that the diagram

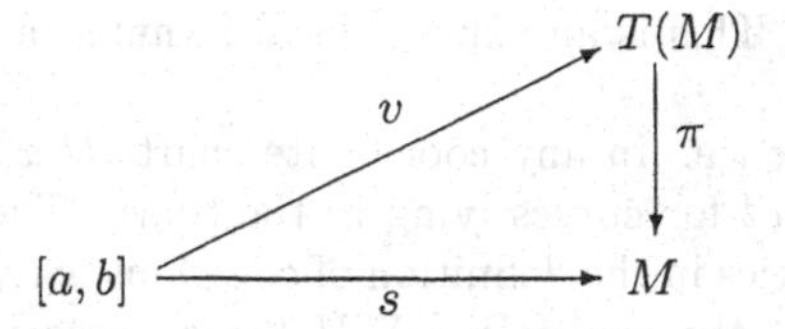

commutes. The set of vector fields along s is denoted by $\mathfrak{X}(s)$.

We have already seen two examples, namely, the restriction $(Y|s)(t) = Y_{s(t)}$ to s of a vector field $Y \in \mathfrak{X}(M)$ and the velocity field $\dot{s}(t)$.

Via pointwise operations, it is evident that $\mathfrak{X}(s)$ is a real vector space and, indeed, a $C^\infty[a,b]$–module.

Definition 10.1.10. Let ∇ be a connection on M. An associated covariant derivative is an operator

$$\frac{\nabla}{dt} : \mathfrak{X}(s) \to \mathfrak{X}(s),$$

defined for every smooth curve s on M, and having the following properties:

1. ∇/dt is $\mathbb{R}$–linear;
2. $(\nabla/dt)(fv) = (df/dt)v + f\nabla v/dt$, $\forall f \in C^\infty[a,b]$, $\forall v \in \mathfrak{X}(s)$;
3. if $Y \in \mathfrak{X}(M)$, then

$$\frac{\nabla}{dt}(Y|s)(t) = \nabla_{\dot{s}(t)}(Y) \in T_{s(t)}(M), \quad a \le t \le b.$$

Remark. By property (3), ∇/dt is associated *only* to the one connection ∇.

Theorem 10.1.11. *To each connection ∇ on M, there is associated a unique covariant derivative ∇/dt.*

Proof. We prove uniqueness first. For this, it is enough to work in an arbitrary coordinate chart $(U, x^1, \dots, x^n)$. Let Γ^k_{ij} be the Christoffel symbols for ∇, set $\xi_i = \partial/\partial x^i$, consider a smooth curve $s : [a,b] \to U$, and let $v \in \mathfrak{X}(s)$. Write

$$v(t) = \sum_{i=1}^n v^i(t)\xi_{i\,s(t)},$$

$$\dot{s}(t) = \sum_{j=1}^n u^j(t)\xi_{j\,s(t)}.$$

Then any associated covariant derivative must satisfy

$$\begin{aligned}
\frac{\nabla v}{dt} &= \sum_{i=1}^n \left(\frac{dv^i}{dt}\xi_i + v^i \frac{\nabla \xi_i}{dt} \right) \\
&= \sum_{k=1}^n \frac{dv^k}{dt}\xi_k + \sum_{i=1}^n v^i \nabla_{\dot{s}}\xi_i \\
&= \sum_{k=1}^n \frac{dv^k}{dt}\xi_k + \sum_{i,j,k=1}^n v^i u^j \Gamma^k_{ji}\xi_k \\
&= \sum_{k=1}^n \left(\frac{dv^k}{dt} + \sum_{i,j=1}^n v^i u^j \Gamma^k_{ji} \right)\xi_k,
\end{aligned}$$

evaluated along $s(t)$. This is an explicit local formula in terms of the connection, proving uniqueness.

We turn to existence. In any coordinate chart $(U, x^1, \dots, x^n)$, use the above formula to define ∇/dt for curves lying in the chart. The reader can easily check that the three properties in the definition of covariant derivative are satisfied, where $M = U$. Thus, in U, the connection $\nabla|U$ has an associated covariant derivative and, by the preceding paragraph, this covariant derivative is unique. Consequently, on overlaps $U \cap V$ of charts, the two sets of Christoffel symbols must define the *same* covariant derivative for $\nabla|U \cap V$. (Classical geometers and physicists defined connections and covariant derivatives by Christoffel symbols and they checked this invariance via explicit change of coordinate formulas.) Thus, along any smooth curve $s : [a, b] \to M$, these local definitions of ∇/dt can be pieced together to give a global definition. □

In particular, for the Euclidean connection D in $\mathbb{R}^m$, all $\Gamma^k_{ij} \equiv 0$ and we get the expected formula:

$$v(t) = \sum_{i=1}^{m} v^i(t)\xi_{i\,s(t)},$$

$$\frac{Dv}{dt} = \sum_{i=1}^{m} \frac{dv^i}{dt}\xi_{i\,s(t)}.$$

For $M \subset \mathbb{R}^m$, the covariant derivative associated to the Levi-Civita connection ∇ on M is

$$\frac{\nabla v}{dt} = p\left(\frac{Dv}{dt}\right),$$

the orthogonal projection into $T(M)$ of the usual Euclidean covariant derivative.

Convention. From now on, we adopt the "summation convention" of Einstein. According to this convention, the summation symbol is omitted and it is understood that any expression is summed over all repeating indices. For example,

$$\left(\frac{dv^k}{dt} + v^i u^j \Gamma^k_{ji}\right)\xi_k = \sum_{k=1}^{n}\left(\frac{dv^k}{dt} + \sum_{i,j=1}^{n} v^i u^j \Gamma^k_{ji}\right)\xi_k.$$

It is necessary that the indices repeat for terms in a product, not just in a sum, and it is customary that the repeated index occur once as a superscript and once as a subscript (a custom we will usually honor). Thus,

$$\frac{dv^k}{dt} + v^i u^j \Gamma^k_{ji} = \frac{dv^k}{dt} + \sum_{i,j=1}^{n} v^i u^j \Gamma^k_{ji},$$

the index k being repeated only in terms separated by $+$.

Definition 10.1.12. Let M be a manifold with a connection ∇. Let $v \in \mathfrak{X}(s)$ for a smooth path $s : [a, b] \to M$. If $\nabla v/dt \equiv 0$ on s, then v is said to be parallel along s (relative to the given connection).

Theorem 10.1.13. *Let ∇ be a connection on M, $s : [a, b] \to M$ a smooth path, $c \in [a, b]$, and $v_0 \in T_{s(c)}(M)$. Then there is a unique parallel field $v \in \mathfrak{X}(s)$ such that $v(c) = v_0$. This field is called the parallel transport of v_0 along s.*

Proof. In local coordinates, write

$$\begin{aligned}\dot{s}(t) &= u^j(t)\xi_{j\,s(t)},\\ v(t) &= v^i(t)\xi_{i\,s(t)},\\ v_0 &= a^i\xi_{i\,s(c)}.\end{aligned}$$

Here, as promised, we are using the summation convention. The condition that v be parallel along s becomes the equation

$$0 = \left(\frac{dv^k}{dt}(t) + v^i(t)u^j(t)\Gamma^k_{ji}(s(t))\right)\xi_{k\,s(t)},$$

or, equivalently, the linear O.D.E. system

$$\frac{dv^k}{dt} = -v^i u^j \Gamma^k_{ji}, \quad 1 \le k \le n, \tag{10.1}$$

with initial conditions

$$v^k(c) = a^k, \quad 1 \le k \le n.$$

By the existence and uniqueness of solutions of O.D.E., there is $\epsilon > 0$ such that the solutions $v^k(t)$ exist and are unique for $c-\epsilon < t < c+\epsilon$. In fact, these equations being *linear* in the v^ks, it is standard in O.D.E. theory (Appendix C, Theorem C.4.1) that there is no restriction on ϵ, so the unique solutions $v^k(t)$ are defined on all of $[a, b]$, $1 \le k \le n$. □

Example 10.1.14. If $s : [a, b] \to M$ is only piecewise smooth and $v_0 \in T_{s(a)}(M)$, we can parallel transport v_0 along the first smooth segment $s|[a, t_1]$, then take the value $v(t_1)$ as an initial condition, parallel transporting that vector along the second segment, *etc.* The result is a *piecewise smooth* vector field along s, called the parallel transport of v_0. An example was given in Figure 10.0.1, where we intuited the way a flatlander on S^2 would try to parallel translate a vector along a piecewise smooth loop. It should be clear that this parallel transport is exactly the one defined by the Levi-Civita connection ∇ on $S^2 \subset \mathbb{R}^3$. Indeed, on the first leg of the triangle, v is the unit tangent field to the great circle and Dv/dt is perpendicular to S^2. Thus, $\nabla v/dt = p(Dv/dt) = 0$. On the second leg of the journey, v is parallel, even from the Euclidean point of view, and $\nabla v/dt = p(Dv/dt) = p(0) = 0$. On the third leg, v is again the unit tangent field to the great circle. Note that the final vector in this field is rotated from its initial direction by exactly the angle of the geodesic triangle at A. By varying this piecewise smooth loop, we can produce any rotation we want. This is an example of *holonomy,* a concept we treat next.

Definition 10.1.15. Consider a piecewise smooth loop

$$s : [a, b] \to M$$

based at $x_0 = s(a) = s(b)$. Then the holonomy of ∇ around s is the map

$$h_s : T_{x_0}(M) \to T_{x_0}(M),$$

defined by setting

$$h_s(v_0) = v(b),$$

where $v \in \mathfrak{X}(s)$ is the parallel transport of $v_0 \in T_{x_0}(M)$. Here, $\mathfrak{X}(s)$ denotes the space of continuous, piecewise smooth fields along s.

Since the parallel transport v of v_0 along s is the solution of the linear system (10.1), it follows that, if w is also the parallel transport along s of a vector $w_0 \in T_{x_0}(M)$, then $v + w \in \mathfrak{X}(s)$ is the parallel transport of $v_0 + w_0$. Similarly, if $a \in \mathbb{R}$, $av \in \mathfrak{X}(s)$ is the parallel transport of av_0. This proves the following.

Lemma 10.1.16. *The holonomy*

$$h_s : T_{x_0}(M) \to T_{x_0}(M)$$

of ∇ around the piecewise smooth loop s is a linear transformation.

Definition 10.1.17. If $s : [a, b] \to M$ is piecewise smooth, a weak reparametrization of s is a curve $s \circ r$, where r is a piecewise smooth map $r : [c, d] \to [a, b]$ carrying $\{c, d\}$ onto $\{a, b\}$. If $r(c) = a$ and $r(d) = b$, the reparametrization is said to be orientation-preserving. If $r(c) = b$ and $r(d) = a$, it is said to be orientation-reversing.

Lemma 10.1.18. *Let $s : [a, b] \to M$ be a piecewise smooth loop at x_0 and let $\tilde{s} = s \circ r : [c, d] \to M$ be a weak reparametrization. If the reparametrization is orientation-preserving, then $h_{\tilde{s}} = h_s$ and, if it is orientation-reversing, $h_{\tilde{s}} = h_s^{-1}$.*

Proof. Without loss of generality, assume that s and r are smooth. Set

$$\begin{aligned} \tilde{s}(\tau) &= s(r(\tau)), \\ \tilde{v}(\tau) &= v(r(\tau)), \end{aligned}$$

hence

$$\begin{aligned} \tilde{u}^j(\tau) &= \frac{dr}{d\tau}(\tau) u^j(r(\tau)), \\ \frac{d\tilde{v}^k}{d\tau}(\tau) &= \frac{dr}{d\tau}(\tau) \frac{dv^k}{dt}(r(\tau)), \end{aligned}$$

and the linear system

$$\frac{d\tilde{v}^k}{d\tau} = -\tilde{v}^i \tilde{u}^j \Gamma^k_{ji} \tag{10.2}$$

is obtained from the system (10.1) by multiplying through by $dr/d\tau$. Since the system (10.1) is assumed to be satisfied, so is the system (10.2). Thus, if $r(c) = a$ and $r(d) = b$,

$$h_{\tilde{s}}(v_0) = \tilde{v}(d) = v(b) = h_s(v_0).$$

If $r(c) = b$ and $r(d) = b$, then we take the initial condition to be $\tilde{v}(c) = v(b) = h_s(v_0)$ and

$$h_{\tilde{s}}(h_s(v_0)) = h_{\tilde{s}}(v(b)) = \tilde{v}(d) = v(a) = v_0.$$

□

Let $\Omega(M, x_0)$ denote the set of all piecewise smooth loops in M based at x_0, loops being identified if one is a weak reparametrization of the other. More precisely, on the set of piecewise smooth loops, we quotient out the smallest equivalence relation that identifies each loop with all of its weak reparametrizations. If $s_1, s_2 \in \Omega(M, x_0)$, then $s_1 + s_2 \in \Omega(M, x_0)$ and

$$h_{s_1 + s_2} = h_{s_2} \circ h_{s_1}.$$

Also, $h_{-s} = h_s^{-1}$, so h_s is a nonsingular linear transformation of $T_{x_0}(M)$. These considerations give the following.

Lemma 10.1.19. *The set* $\{h_s\}_{s\in\Omega(M,x_0)}$ *is a subgroup*

$$\mathcal{H}_{x_0}(M) \subseteq \mathrm{Gl}(T_{x_0}(M)),$$

called the holonomy group (at x_0*) of the connection* ∇.

Remark. If M is connected and $x_0, x_1 \in M$, then the groups $\mathcal{H}_{x_0}(M)$ and $\mathcal{H}_{x_1}(M)$ are isomorphic, but generally not *canonically* isomorphic. Indeed, fix a piecewise smooth path $\sigma : [0,1] \to M$ with $\sigma(0) = x_0$ and $\sigma(1) = x_1$. Then, given any $s \in \Omega(M, x_1)$, note that

$$\sigma + s + (-\sigma) \in \Omega(M, x_0)$$

and that the correspondence $h_s \mapsto h_{\sigma+s+(-\sigma)}$ defines a group homomorphism from $\mathcal{H}_{x_1}(M)$ to $\mathcal{H}_{x_0}(M)$. By replacing σ with $-\sigma$, we get the inverse group homomorphism $\mathcal{H}_{x_0}(M) \to \mathcal{H}_{x_1}(M)$, proving that these are group isomorphisms. Generally, the isomorphism depends on the choice of σ, so it is not canonical.

Exercise 10.1.20. The standard imbedding of S^1 as the unit circle in $\mathbb{R}^2$ defines a standard imbedding of $T^n = S^1 \times \cdots \times S^1$ in $\mathbb{R}^{2n}$. Prove that the associated Levi-Civita connection on T^n has holonomy $\mathcal{H}_x(T^n) = \{\mathrm{id}\}$, $\forall x \in T^n$. (Hint: Find a coordinate atlas relative to which the Christoffel symbols vanish.)

Exercise 10.1.21. We say that ∇ is a *globally flat* connection if (as in Exercise 10.1.20) its holonomy group $\mathcal{H}_x(M)$ is trivial, $\forall x \in M$. (The reason for this terminology will become apparent when we study the relationship between curvature and holonomy.) Prove that a manifold M has a globally flat connection if and only if M is parallelizable.

Exercise 10.1.22. By Exercise 10.1.21, the spheres S^3 and S^7 support globally flat connections. Prove that, for $n \geq 2$, the Levi-Civita connection relative to the standard imbedding $S^n \subset \mathbb{R}^{n+1}$ is not globally flat.

Finally, we should point out that connections are ubiquitous.

Theorem 10.1.23. *Every manifold* M *has a connection.*

Proof. Let $\{U_\alpha, x^1_\alpha, \dots, x^n_\alpha\}_{\alpha\in\mathfrak{A}}$ be an atlas on M and let D^α be the Euclidean connection on U_α relative to the coordinates $x^1_\alpha, \dots, x^n_\alpha$. Let $\{\lambda_\alpha\}_{\alpha\in\mathfrak{A}}$ be a smooth partition of unity subordinate to the atlas. Given $X, Y \in \mathfrak{X}(M)$, write $X_\alpha = X|U_\alpha$ and $Y_\alpha = Y|U_\alpha$. Then, define

$$\nabla = \sum_{\alpha\in\mathfrak{A}} \lambda_\alpha D^\alpha : \mathfrak{X}(M) \times \mathfrak{X}(M) \to \mathfrak{X}(M),$$

where

$$\nabla_X Y = \sum_{\alpha\in\mathfrak{A}} \lambda_\alpha D^\alpha_{X_\alpha} Y_\alpha \in \mathfrak{X}(M).$$

It is entirely straightforward to verify that ∇ is a connection. □

10.2. Riemannian Manifolds

A Riemannian manifold is a pair $(M, \langle\cdot,\cdot\rangle)$ consisting of the manifold M and a choice of Riemannian metric on $T(M)$. From now on, such a choice of metric is fixed and we will speak of "the Riemannian manifold M".

Definition 10.2.1. If $v \in T_x(M)$, some $x \in M$, then the length of v is the non-negative number

$$\|v\| = \sqrt{\langle v, v\rangle_x}.$$

When no ambiguity is likely, we will often dispense with the subscript x on $\langle v, w\rangle_x$, where $v, w \in T_x(M)$.

Definition 10.2.2. If $v, w \in T_x(M)$, some $x \in M$, then the angle between v and w is the unique $\theta \in [0, \pi]$ such that

$$\cos\theta = \frac{\langle v, w\rangle}{\|v\|\|w\|}.$$

Definition 10.2.3. If $s : [a, b] \to M$ is smooth, then the length of s is

$$|s| = \int_a^b \|\dot{s}(t)\|\, dt.$$

If $s = s_1 + \cdots + s_q$ is piecewise smooth, each s_i being smooth, then the length is

$$|s| = |s_1| + \cdots + |s_q|.$$

Lemma 10.2.4. *If $u : [c, d] \to [a, b]$ is a weak reparametrization, then $|s| = |s \circ u|$.*

The proof should be familiar from advanced calculus. Notice that it does not matter whether u preserves orientation or reverses it.

If the Riemannian manifold M is oriented, we also get a canonical *volume form* $\Omega \in A^n(M)$ (where $n = \dim M$). Consider a local trivialization of $T(M)$. That is, we are given an open set $U \subseteq M$ and a smooth frame $(X_1, \dots, X_n)$ of vector fields defined on U that determines the correct orientation at each point of U. Relative to this trivialization, we express the Riemannian metric by

$$\langle a^i X_i, b^j X_j\rangle = a^i b^j \langle X_i, X_j\rangle = a^i b^j h_{ij}.$$

That is, relative to the given basis, this bilinear form is represented by the smooth matrix-valued function $[h_{ij}]$, where $h_{ij} = \langle X_i, X_j\rangle$. This is a symmetric matrix. Since the metric is positive definite,

$$h = \det[h_{ij}] > 0.$$

Let $\{\omega^1, \dots, \omega^n\} \subset A^1(U)$ be the dual basis: $\omega^i(X_j) = \delta^i_j$.

Definition 10.2.5. The Riemann (or Riemannian) volume element on U, relative to the given local trivialization of $T(M)$, is

$$\sqrt{h}\,\omega^1 \wedge \cdots \wedge \omega^n \in A^n(U).$$

In particular, if $(X_1, \dots, X_n)$ is an *orthonormal* frame, the volume element becomes $\omega^1 \wedge \cdots \wedge \omega^n$. This agrees with intuition. The following theorem shows that the volume element is independent of the choice of local trivializations.

Theorem 10.2.6. *If M is an oriented, Riemannian n-manifold, there is a globally defined form $\Omega \in A^n(M)$ that, relative to any orientation-respecting local trivialization of $T(M)$, coincides with the Riemann volume element.*

Proof. Indeed, let $(U, X_1, \dots, X_n)$ and $(V, Z_1, \dots, Z_n)$ be two such trivializations with $U \cap V \neq \emptyset$. Let the respective dual bases of 1-forms be $\{\omega^1, \dots, \omega^n\}$ and $\{\eta^1, \dots, \eta^n\}$. Let $\gamma(x) = [\gamma_{ij}(x)]$ be the $\mathrm{Gl}(n)$-valued function on $U \cap V$ such that

$$(Z_1, \dots, Z_n)\gamma = (X_1 \dots, X_n)$$

on $U \cap V$. Since the frames are coherently oriented, $\det \gamma > 0$. Let

$$h_{ij} = \langle X_i, X_j \rangle,$$
$$f_{ij} = \langle Z_i, Z_j \rangle.$$

Then, $[h_{ij}] = [\gamma_{ki}]^{\mathrm{T}} [f_{k\ell}][\gamma_{\ell j}]$ and it follows that

$$\sqrt{h} = (\det \gamma)\sqrt{f}.$$

Also,

$$(\det \gamma')\eta^1 \wedge \cdots \wedge \eta^n = \omega^1 \wedge \cdots \wedge \omega^n,$$

where $\gamma' = (\gamma^{\mathrm{T}})^{-1}$. Putting this information together, we obtain

$$\sqrt{h}\,\omega^1 \wedge \cdots \wedge \omega^n = (\det \gamma)\sqrt{f}(\det \gamma')\,\eta^1 \wedge \cdots \wedge \eta^n = \sqrt{f}\,\eta^1 \wedge \cdots \wedge \eta^n.$$

Thus, the locally defined volume forms fit together coherently to define Ω as desired. □

Definition 10.2.7. Let $(U, x^1, \dots, x^n)$ be a coordinate chart with coordinate fields $\xi_i = \partial/\partial x^i$. Then the functions

$$g_{ij} = \langle \xi_i, \xi_j \rangle, \ 1 \leq i, j \leq n,$$

are called the metric coefficients.

Thus, in correctly oriented local coordinate charts $(U, x^1, \dots, x^n)$, the volume element is given in terms of the metric coefficients by

$$\Omega | U = \sqrt{g}\, dx^1 \wedge \cdots \wedge dx^n,$$

where $g = \det[g_{ij}]$.

Definition 10.2.8. Let M be an oriented Riemannian n-manifold and let $U \subseteq M$ be a relatively compact, open subset. Then the volume of U is

$$|U| = \int_U \Omega.$$

Remark. On the σ-algebra of Borel sets of M, this Riemannian volume generates a measure, finite on the relatively compact Borel sets. Even if M is not orientable, Ω is defined locally up to sign and, for small, connected, open sets U,

$$|U| = \left| \int_U \Omega \right|.$$

This also leads to a Borel measure, finite on relatively compact Borel sets.

Definition 10.2.9. A connection ∇ on the Riemannian manifold M is a Riemannian connection if, for all $X, Y, Z \in \mathfrak{X}(M)$,

$$X \langle Y, Z \rangle = \langle \nabla_X Y, Z \rangle + \langle Y, \nabla_X Z \rangle.$$

Definition 10.2.10. A connection ∇ on the Riemannian manifold M is a Levi-Civita connection if it is symmetric and Riemannian.

Example 10.2.11. Let $M \subseteq \mathbb{R}^m$ be a smoothly imbedded n-manifold. The standard inner product $\langle \cdot, \cdot \rangle$ on $\mathbb{R}^m$, viewed as a Riemannian metric on the tangent bundle $T(\mathbb{R}^m)$, restricts to a Riemannian metric $\langle \cdot, \cdot \rangle_M$ on $T(M) \subseteq T(\mathbb{R}^m)|M$. By Exercise 10.1.8, the connection ∇ on M, constructed in the previous section and called there the Levi-Civita connection, is symmetric. The Euclidean connection D on $\mathbb{R}^m$ clearly satisfies

$$X \langle Y, Z \rangle = \langle D_X Y, Z \rangle + \langle Y, D_X Z \rangle, \quad \forall X, Y, Z \in \mathfrak{X}(\mathbb{R}^m).$$

If $X, Y, Z \in \mathfrak{X}(M)$, then $D_X Y = \nabla_X Y + W$, where $W \in \Gamma(\nu(M))$, hence $W \perp Z$ everywhere on M. Consequently,

$$\langle D_X Y, Z \rangle = \langle \nabla_X Y, Z \rangle_M$$

and, similarly,

$$\langle Y, D_X Z \rangle = \langle Y, \nabla_X Z \rangle_M .$$

Thus, ∇ is Riemannian, hence is a Levi-Civita connection in our new sense as well.

Exercise 10.2.12. Prove that a Riemannian manifold M has a unique Levi-Civita connection, proceeding as follows.

(1) *Uniqueness.* Show that a Levi-Civita connection ∇ must satisfy the identity

$$\begin{aligned} 2 \langle \nabla_X Y, Z \rangle = {} & X \langle Y, Z \rangle + Y \langle X, Z \rangle - Z \langle X, Y \rangle \\ & + \langle [X, Y], Z \rangle + \langle [Z, X], Y \rangle + \langle [Z, Y], X \rangle , \end{aligned}$$

$\forall X, Y, Z \in \mathfrak{X}(M)$.

(2) *Existence.* Use the identity in (1) to *define*

$$\nabla : \mathfrak{X}(M) \times \mathfrak{X}(M) \to \mathfrak{X}(M)$$

and prove that ∇ is a Levi-Civita connection.

In any local coordinate chart, the matrix $[g_{ij}]$ of metric coefficients is nonsingular, so we can define

$$[g^{k\ell}] = [g_{ij}]^{-1}.$$

The coefficients $g^{k\ell}$ are rational functions of the metric coefficients g_{ij}. By definition, they satisfy

$$g_{ik} g^{k\ell} = \delta_i^\ell .$$

Exercise 10.2.13. Let Γ_{ij}^k denote the Christoffel symbols of the Levi-Civita connection and find a formula for Γ_{ij}^k that involves only the $g^{k\ell}$'s and first derivatives of the g_{ij}'s.

Definition 10.2.14. A property of the Riemannian manifold M is *intrinsic* if it depends only on the metric. Otherwise, the property is *extrinsic.* It is *geometric* if it does not depend on choices of local coordinates.

For example, the functions g_{ij} and $g^{k\ell}$ are intrinsic, but not geometric. By Exercise 10.2.13, the Christoffel symbols Γ_{ij}^k for the Levi-Civita connection are also intrinsic, but not geometric. In particular, the Levi-Civita connection for $M \subseteq \mathbb{R}^m$ is intrinsic, even though our initial definition of it used the normal bundle $\nu(M)$, a structure that can be proven to be extrinsic.

Our definition of "intrinsic" may seem a bit too informal by current standards. For a more formal definition, one needs the notion of an *isometry.*

Definition 10.2.15. An isometry $\varphi : M_1 \to M_2$ between two Riemannian manifolds, with respective metrics $\langle \cdot, \cdot \rangle_i$, $i = 1, 2$, is a diffeomorphism such that

$$\langle \varphi_{*x}(v), \varphi_{*x}(w) \rangle_2 = \langle v, w \rangle_1 ,$$

for arbitrary $x \in M_1$ and $v, w \in T_x(M_1)$.

Now we see that a property of Riemannian manifolds should be called intrinsic if and only if it is preserved by all isometries.

Exercise 10.2.16. If $\varphi : M \to N$ is a diffeomorphism and ∇ is a connection on N, show how to define the pullback connection $\varphi^*\nabla$ on M. If φ is an isometry between Riemannian manifolds and ∇ is the Levi-Civita connection on N, prove that $\varphi^*\nabla$ is the Levi-Civita connection on M. This is the formal proof that the Levi-Civita connection is an intrinsic property of a Riemannian manifold.

Exercise 10.2.17. Let $v, w \in \mathfrak{X}(s)$. Show that the covariant derivative defined by the Levi-Civita connection (indeed, by any Riemannian connection) satisfies

$$\frac{d}{dt} \langle v, w \rangle = \left\langle \frac{\nabla v}{dt}, w \right\rangle + \left\langle v, \frac{\nabla w}{dt} \right\rangle.$$

(It follows from this exercise that fields parallel along a curve s, relative to a Levi-Civita connection, make a constant angle with each other and have constant lengths along s.)

10.3. Gauss Curvature

Throughout this section, unless otherwise stated, we assume that $\partial M = \emptyset$ and $\dim M = 2$, and that we are given a fixed imbedding $M \hookrightarrow \mathbb{R}^3$ with normal bundle $\nu(M)$. As usual, we use the metric induced on M by the Euclidean metric of $\mathbb{R}^3$ and the associated Levi-Civita connection ∇. The Euclidean connection will be denoted by D.

Given $x \in M$, we find a connected neighborhood $U \subseteq M$ of x and a smooth section $\vec{n} \in \Gamma(\nu(M)|U)$ such that $\|\vec{n}\| \equiv 1$. Remark that $\vec{n}$ is determined up to sign. Remark also that one can interpret $\vec{n}$ as a smooth map

$$\vec{n} : U \to S^2.$$

Definition 10.3.1. The map $\vec{n} : U \to S^2$ is called the Gauss map.

Intuitively, the area of $\vec{n}(U) \subseteq S^2$ seems to have something to do with the *curvature* of M in U. Thus, if U is an open subset of a 2-plane in $\mathbb{R}^3$, $\vec{n}(U)$ degenerates to a single point, hence has area 0 (Figure 10.3.1). We say that the plane has (Gauss) curvature 0. Similarly, if U lies in a right circular cylinder, $\vec{n}(U)$ will lie in a great circle in S^2 (Figure 10.3.2). Again the area of $\vec{n}(U)$ is 0 and we say that the cylinder has (Gauss) curvature 0. The point here is that the cylinder can be "unrolled" to a portion of a plane without any metric distortions. On the other hand, if $U \subset S^2$, then $\vec{n} = \pm\,\mathrm{id}$ and the area of $\vec{n}(U)$ is the same as the area of U, this being a positive number. We say that the sphere has positive (Gauss) curvature. It is not possible to flatten out any portion of S^2 to be planar without distorting the metric properties. For similar reasons, every convex surface has nonnegative curvature everywhere (Figure 10.3.3). By introducing a notion of "signed area", one obtains cases of negative curvature, a saddle shaped surface

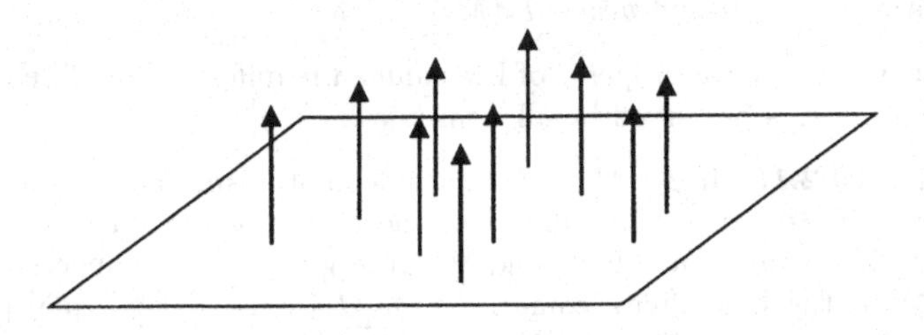

Figure 10.3.1. The flat plane

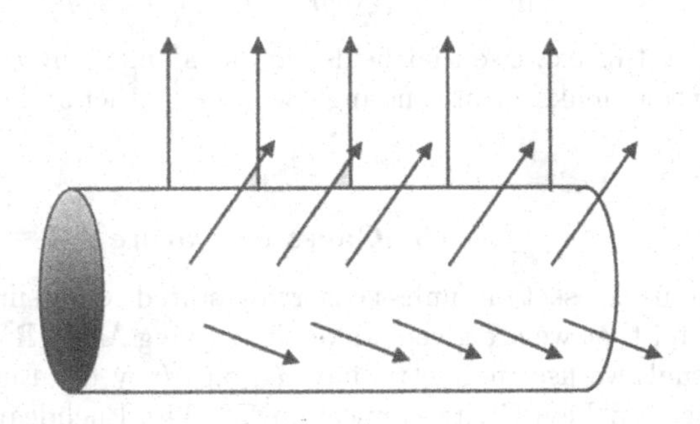

Figure 10.3.2. The flat cylinder

being the typical example (Figure 10.3.4 and Exercise 10.3.8). In order to put these ideas into precise form, we introduce the *Weingarten map.*

If $v \in T_y(M)$, some $y \in U$, then $D_v\vec{n} \in \mathbb{R}^3$ makes sense. The equation

$$0 = v\langle \vec{n}, \vec{n}\rangle = \langle D_v\vec{n}, \vec{n}\rangle + \langle \vec{n}, D_v\vec{n}\rangle = 2\langle D_v\vec{n}, \vec{n}\rangle$$

proves that $D_v\vec{n} \perp \vec{n}$, hence $D_v\vec{n} \in T_y(M)$.

Definition 10.3.2. The linear map $L : T_y(M) \to T_y(M)$, defined by

$$L(v) = D_v\vec{n},$$

is called the Weingarten map.

This map is well defined up to sign. As soon as the sign of the Gauss map $\vec{n}$ has been fixed, the sign of the Weingarten map is determined also. Remark that $T_y(M) = T_{\vec{n}(y)}(S^2)$ in $\mathbb{R}^3$, since these 2-planes have the common normal vector $\vec{n}(y)$. Thus, we are allowed to view the Weingarten map as

$$L : T_y(M) \to T_{\vec{n}(y)}(S^2).$$

Lemma 10.3.3. *The Weingarten map* $L : T_y(M) \to T_{\vec{n}(y)}(S^2)$ *is the differential at* y *of the Gauss map.*

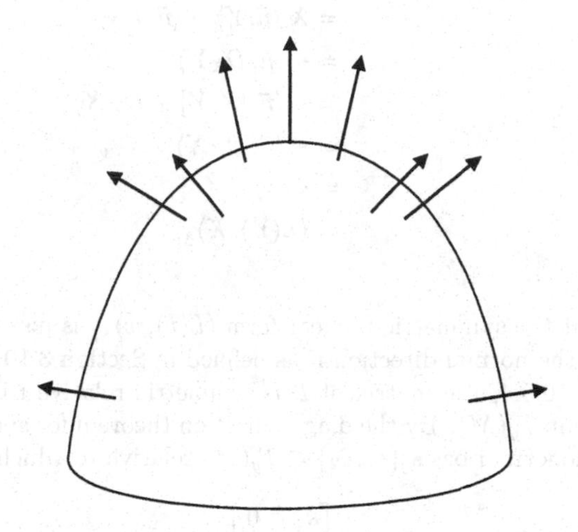

Figure 10.3.3. Nonnegative curvature

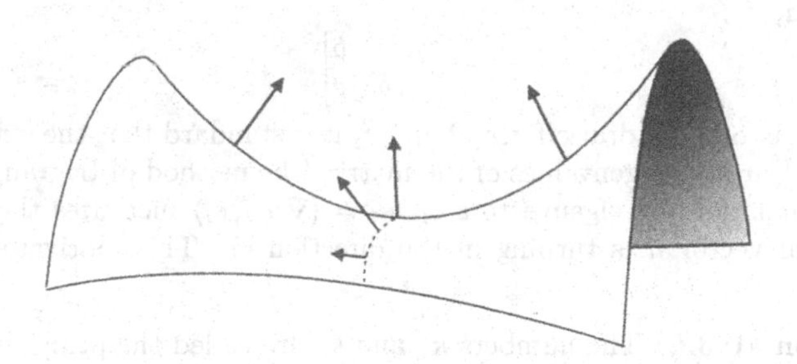

Figure 10.3.4. Negative curvature

Proof. Represent an arbitrary vector in $T_y(M)$ as $\dot{s}(0)$ for a suitable arc $s : (-\epsilon, \epsilon) \to M$. Then

$$\vec{n}_{*y}(\dot{s}(0)) = \frac{d}{dt}\vec{n}(s(t))\bigg|_{t=0} = D_{\dot{s}(0)}\vec{n} = L(\dot{s}(0)).$$

□

Lemma 10.3.4. *The Weingarten map is self adjoint. That is,*

$$\langle L(v), w \rangle_M = \langle v, L(w) \rangle_M ,$$

$\forall v, w \in T_y(M)$.

Proof. Let $v, w \in T_y(M)$ and, by making the open set $U \subseteq M$ smaller, if necessary, extend these vectors to fields $X, Y \in \mathfrak{X}(U)$. Then

$$\begin{aligned}\langle L(X), Y\rangle_M &= \langle D_X\vec{n}, Y\rangle \\ &= X\,\langle \vec{n}, Y\rangle - \langle \vec{n}, D_X Y\rangle \\ &= -\,\langle \vec{n}, D_X Y\rangle \\ &= -\,\langle \vec{n}, [X,Y] + D_Y X\rangle \\ &= -\,\langle \vec{n}, D_Y X\rangle \\ &= \cdots \\ &= \langle L(Y), X\rangle_M .\end{aligned}$$

□

Remark that the symmetric bilinear form $\langle L(v), w\rangle_M$ is just the second fundamental form in the normal direction $\vec{n}$ as defined in Section 3.10.

By Lemma 10.3.4, the matrix of L is symmetric relative to any choice of orthonormal basis in $T_y(M)$. By the diagonalization theorem for symmetric matrices, there is an orthonormal basis $\{e_1, e_2\} \subset T_y(M)$ relative to which the matrix for L is

$$\begin{bmatrix}\kappa_1 & 0 \\ 0 & \kappa_2\end{bmatrix}.$$

That is, e_i is an eigenvector for L with eigenvalue κ_i, $i = 1, 2$. We agree to number these so that $\kappa_1 \le \kappa_2$.

Lemma 10.3.5. *As v ranges over the unit circle in $T_y(M)$, the quadratic form $\langle L(v), v\rangle_M$ takes minimum value κ_1 and maximum value κ_2.*

Indeed, if

$$\begin{bmatrix}a & b \\ b & c\end{bmatrix}$$

is the matrix of a quadratic form Q on $\mathbb{R}^2$, it is standard that the extreme values of $Q(v)$ on S^1 are the eigenvalues of the matrix (the method of Lagrange multipliers). Remark that, for the eigenvectors e_i, $\kappa_i = \langle \nabla_{e_i}\vec{n}, e_i\rangle$ measures the rate at which the normal vector $\vec{n}$ is turning in the direction e_i. This motivates the following definition.

Definition 10.3.6. The numbers κ_1 and κ_2 are called the principal curvatures of M at y. The product $\kappa_1\kappa_2 = \det L$ is called the Gauss curvature of M at y and is denoted by κ (or $\kappa(y)$).

Remark. There is another important kind of curvature, the *mean curvature* h of M, which we will not treat in any detail. It is defined by

$$h = \operatorname{tr} L = \kappa_1 + \kappa_2.$$

Unlike the Gauss curvature, this quantity depends, up to sign, on the choice of the unit normal field ν. It can be shown that surfaces of mean curvature $h \equiv 0$ are exactly the ones that, in a certain precise sense, locally minimize surface area. Such surfaces are called *minimal surfaces.* They arise, for instance, when one considers the possible shapes of soap films spanning wire loops of various configurations. Such a soap film will be modeled by a 2-manifold S with boundary and $M = S \setminus \partial S$ will be a minimal surface. The work of D. Hoffman and W. H. Meeks (**[19]**, **[17]**,

[18], *et al.*), inspired and illuminated by some spectacular computer graphics, has revealed an astounding array of complete, unbounded minimal surfaces.

Let Ω' denote the Riemann volume form on S^2 and let Ω denote the volume form on M. Define the "signed area" of the Gauss map on U to be

$$A(\vec{n}(U)) = \int_U \vec{n}^*(\Omega')$$

and, as usual, let the area of U be

$$A(U) = \int_U \Omega.$$

As U shrinks down on $\{y\}$, one can try to form a kind of "derivative" of the Gauss map $\vec{n}$ with respect to area:

$$\frac{d\vec{n}(y)}{d\Omega} = \lim_{U \to \{y\}} \frac{A(\vec{n}(U))}{A(U)}.$$

In order to define this precisely, define

$$\delta(U, y) = \sup_{x \in U} \|x - y\|,$$

using the ordinary Euclidean norm in $\mathbb{R}^3$, and characterize this derivative, if it exists, as the unique number such that, for each $\epsilon > 0$, there is $\delta > 0$ for which

$$U \ni y \text{ and } \delta(U, y) < \delta \Rightarrow \left| \frac{d\vec{n}(y)}{d\Omega} - \frac{A(\vec{n}(U))}{A(U)} \right| < \epsilon.$$

Exercise 10.3.7. For each $y \in M$, prove that this derivative exists and that

$$\frac{d\vec{n}(y)}{d\Omega} = \kappa(y).$$

Exercise 10.3.8. Let $M \subset \mathbb{R}^3$ be the graph of the equation $z = x^2 - y^2$. Prove that $\kappa(x, y, z) < 0$, $\forall\, (x, y, z) \in M$.

In the above discussion, the normal field $\vec{n}$ played a central role. The curvature of M was seen as a measure of how much this field "spreads" infinitesimally at a point. Thus, curvature appears to be an extrinsic property of the surface. But Gauss proved a remarkable theorem (he called it his "Theorema Egregium") that showed the Gauss curvature to be intrinsic. A two-dimensional inhabitant of the surface can take measurements leading to the computation of curvature. We turn to this theorem.

Let $X, Y, Z \in \mathfrak{X}(U)$ and extend these to fields $\widetilde{X}, \widetilde{Y}, \widetilde{Z} \in \mathfrak{X}(\widetilde{U})$, where $\widetilde{U} \subseteq \mathbb{R}^3$ is an open set such that $\widetilde{U} \cap M = U$.

Lemma 10.3.9. *For fields chosen as above,*

$$D_{\widetilde{X}} \widetilde{Y} = \nabla_X Y - \langle L(X), Y \rangle\, \vec{n}$$

along U.

Proof. Along U, $D_{\widetilde{X}} \widetilde{Y}$ depends only on X and Y. As in the proof of Lemma 10.3.4,

$$\langle D_X Y, \vec{n} \rangle = -\langle L(X), Y \rangle,$$

so the component of $D_X Y$ perpendicular to M is $-\langle L(X), Y \rangle\, \vec{n}$. By the definition of the Levi-Civita connection, $\nabla_X Y$ is the component of $D_X Y$ tangent to M. □

The Euclidean connection D satisfies a simple commutator relation:

$$[D_{\widetilde{X}}, D_{\widetilde{Y}}] = D_{[\widetilde{X},\widetilde{Y}]}.$$

This is because $D_{\widetilde{X}}$ and $D_{\widetilde{Y}}$ operate on a vector field $\widetilde{Z}$ by applying $\widetilde{X}$ and $\widetilde{Y}$ respectively to the individual components of $\widetilde{Z}$. It turns out that, on M, *curvature* is an obstruction to this commutator relation for ∇.

Definition 10.3.10. The curvature operator

$$R(X,Y) : \mathfrak{X}(M) \to \mathfrak{X}(M)$$

is defined, for arbitrary $X, Y \in \mathfrak{X}(M)$, by

$$R(X,Y)Z = \nabla_X \nabla_Y Z - \nabla_Y \nabla_X Z - \nabla_{[X,Y]} Z,$$

$\forall Z \in \mathfrak{X}(M)$.

The fact that this operator is related to curvature is far from obvious. It is the content of the Theorema Egregium.

By Lemma 10.3.9, at every point of U we have

$$\begin{aligned} D_{\widetilde{X}}(D_{\widetilde{Y}}\widetilde{Z}) &= D_{\widetilde{X}}(\nabla_Y Z - \langle L(Y), Z\rangle \vec{n}) \\ &= D_X(\nabla_Y Z) - \langle L(Y), Z\rangle L(X) - X\,\langle L(Y), Z\rangle \vec{n} \\ &= \nabla_X(\nabla_Y Z) - \langle L(Y), Z\rangle L(X) - X\,\langle L(Y), Z\rangle \vec{n} \\ &\quad - \langle L(X), \nabla_Y Z\rangle \vec{n}. \end{aligned}$$

Similarly, at every point of U,

$$\begin{aligned} - D_{\widetilde{Y}}(D_{\widetilde{X}}\widetilde{Z}) = -\nabla_Y(\nabla_X Z) &+ \langle L(X), Z\rangle L(Y) \\ &+ Y\,\langle L(X), Z\rangle \vec{n} + \langle L(Y), \nabla_X Z\rangle \vec{n}. \end{aligned}$$

Finally, at every point of U,

$$-D_{[\widetilde{X},\widetilde{Y}]}\widetilde{Z} = -\nabla_{[X,Y]} Z + \langle L([X,Y]), Z\rangle \vec{n}.$$

The sum of left sides of these equations being 0, the tangential parts and normal parts of the right sides separately sum to 0. The first of these zero sums gives the *Gauss equation*:

$$R(X,Y)Z = \langle L(Y), Z\rangle L(X) - \langle L(X), Z\rangle L(Y). \tag{10.3}$$

Summing the coefficients of $\vec{n}$ also gives 0. Applying the fact that ∇ is a Riemannian connection and using the fact that Z varies freely over all tangent fields, the reader can obtain the *Codazzi–Mainardi equation*

$$L([X,Y]) = \nabla_Y L(X) - \nabla_X L(Y). \tag{10.4}$$

One immediate consequence of equation (10.3) is the following.

Lemma 10.3.11. *The expression $R(X,Y)Z$ has value at $y \in U$ depending only on the vectors X_y, Y_y, Z_y. Thus, this expression is a tensor, called the Riemann curvature tensor R.*

Indeed, the right-hand side of equation (10.3) is clearly a tensor in all three vector fields. That is, $R \in \mathcal{T}_1^3(M)$. Note also that, since ∇ is an intrinsic and geometric property of the surface, so is the Riemann curvature tensor.

Theorem 10.3.12 (Theorema Egregium). *Let $y \in U$ and let $\{e_1, e_2\}$ be an orthonormal basis of $T_y(M)$. Then*

$$\langle R(e_1, e_2)e_2, e_1\rangle = \kappa(y).$$

In particular, the Gauss curvature of a surface in $\mathbb{R}^3$ is intrinsic and geometric.

Proof. By equation (10.3),

$$\langle R(e_1, e_2)e_2, e_1\rangle = \langle L(e_2), e_2\rangle \langle L(e_1), e_1\rangle - \langle L(e_1), e_2\rangle \langle L(e_2), e_1\rangle .$$

Since the basis is orthonormal, it is true (and easily checked) that the corresponding matrix representation of L is the 2×2 matrix with (i, j)th entry $\langle L(e_j), e_i\rangle$. Thus,

$$\langle R(e_1, e_2)e_2, e_1\rangle = \det L = \kappa(y).$$

□

By Exercise 10.2.16, the Levi-Civita connection is preserved by isometries, hence we obtain the more formal version of the statement that κ is an intrinsic property.

Corollary 10.3.13. *If $f : M \to N$ is an isometry between two surfaces in $\mathbb{R}^3$ and if κ_M and κ_N are the respective Gauss curvatures of these surfaces, then*

$$\kappa_N(f(x)) = \kappa_M(x), \quad \forall x \in M.$$

Theorem 10.3.12 suggests a definition of curvature for a general connection on an n-manifold M.

Definition 10.3.14. Let M be an n-manifold and let ∇ be a connection on M. The curvature operator R of ∇ is given, for each choice of the fields $X, Y, Z \in \mathfrak{X}(M)$, by

$$R(X,Y)Z = \nabla_X \nabla_Y Z - \nabla_Y \nabla_X Z - \nabla_{[X,Y]} Z.$$

Exercise 10.3.15. Prove that the curvature operator $R(X, Y)Z$ of a connection ∇ is $C^\infty(M)$-trilinear in the fields X, Y, Z. Consequently, for each $x \in M$ and $u, v, w \in T_x(M)$, $R(u, v)w \in T_x(M)$ is well defined.

If M is a Riemannian manifold and ∇ is the Levi-Civita connection, then R is called the Riemann tensor. We will return to the study of this tensor later. It turns out that, in Riemannian geometry, the Riemann tensor is exactly the obstruction to the geometry being locally Euclidean. In the non-Riemannian geometry of spacetime, there is an analogue of the Levi-Civita connection and Einstein represents gravity by the curvature tensor of this connection. Special relativity ("flat" spacetime) is the case in which this curvature tensor vanishes identically.

Exercise 10.3.16. Let $M \subset \mathbb{R}^3$ be a compact 2-manifold. You are to prove that it is not possible that $\kappa \leq 0$ on all of M. Proceed as follows.

(1) By compactness, choose a point $v_0 \in M$ at which the function

$$\lambda : M \to \mathbb{R},$$
$$\lambda(v) = \|v\|^2$$

assumes its maximum. Prove that $0 \neq v_0 \perp T_{v_0}(M)$.

(2) If $s : (-\epsilon, \epsilon) \to M$ is smooth with $s(0) = v_0$ and $\dot{s}(0) \neq 0$, prove that $\langle \ddot{s}(0), \vec{n}(v_0)\rangle$ is strictly negative.

(3) Using the above, prove that the principal curvatures κ_1 and κ_2 are both nonzero and have the same sign at v_0, hence $\kappa(v_0) > 0$.

Exercise 10.3.17. One calls a point $v \in M$ at which $\kappa_1 = \kappa_2$ an *umbilic* point. Let $U \subseteq M$ be the set of points that are not umbilic.

(1) Prove that U is open in M and that κ_1 and κ_2 are smooth functions on U.
(2) Prove that each $v \in U$ has a neighborhood $V \subseteq U$ on which there is a smooth, orthonormal frame field (X_1, X_2) such that $L(X_i) = \kappa_i X_i$, $i = 1, 2$.
(3) For $V \subseteq U$ and $X_1, X_2 \in \mathfrak{X}(V)$ as in part (2), define $f_1, f_2 \in C^\infty(V)$ by
$$f_1 = \frac{X_2(\kappa_1)}{\kappa_2 - \kappa_1},$$
$$f_2 = \frac{-X_1(\kappa_2)}{\kappa_2 - \kappa_1}.$$
Prove that
$$\nabla_{X_1} X_1 = -f_1 X_2, \qquad \nabla_{X_1} X_2 = f_1 X_1,$$
$$\nabla_{X_2} X_1 = f_2 X_2, \qquad \nabla_{X_2} X_2 = -f_2 X_1,$$
and that
$$[X_1, X_2] = f_1 X_1 - f_2 X_2.$$
(4) Using the formulas in part (3), show that, if $v \in U$ is a critical point for both κ_1 and κ_2, then, at v,
$$\kappa = \frac{X_1^2(\kappa_2) - X_2^2(\kappa_1)}{\kappa_2 - \kappa_1}.$$

Exercise 10.3.18. Let $M \subset \mathbb{R}^3$ be a compact, connected 2-manifold with constant curvature $\kappa \equiv a$. By Exercise 10.3.16, $a > 0$. You are to prove that M is a 2-sphere, centered at some point $w_0 \in \mathbb{R}^3$ and of radius $1/\sqrt{a}$. Proceed as follows.

(1) Prove that $v \in M$ is a point at which κ_2 is maximum if and only if it is a point at which κ_1 is minimum.
(2) Use part (4) of Exercise 10.3.17 to show that κ_2 can be maximum only at an umbilic point.
(3) Prove that every point of M is an umbilic and that $\kappa_1 \equiv \sqrt{a} \equiv \kappa_2$. (Hint: This is the maximum value of κ_2.)
(4) Deduce the form of the Gauss map, drawing the desired conclusion. Be sure to make clear how you use the hypothesis that M is connected.

10.4. Complete Riemannian Manifolds

This section presents the Hopf–Rinow theorems and related matters. The author first learned this material from J. Milnor's beautiful exposition [**28**, pp. 55–64], and its influence will be evident in what follows. The goal here is to use the Riemannian metric to obtain a topological metric on M and to relate the topological notion of "completeness" to the problem of extending geodesics indefinitely. In the process, one also discovers, without the use of variational calculus, that geodesics locally minimize arc length.

In Euclidean geometry, straight lines play a central role. They can be characterized as the unique smooth curves whose tangent fields are parallel. Here, parallelism under the Euclidean connection is clearly identical with the absolute parallelism in Euclidean space. Taking our cue from this observation, we define the notion of a *geodesic* for a general connection.

Definition 10.4.1. Let M be an n-manifold with a connection ∇. A smooth curve $s : [a, b] \to M$ is a geodesic for ∇ if $\dot{s}(t)$ is parallel along $s(t)$, $a \le t \le b$.

We are interested in the case in which ∇ is the Levi-Civita connection of a Riemannian manifold, so we make that assumption from here on. We emphasize that we are considering general Riemannian manifolds, not just surfaces in $\mathbb{R}^3$.

Definition 10.4.2. A smooth curve $\sigma : [a, b] \to M$ is said to be evenly parametrized if $\|\dot{\sigma}(t)\| \equiv c$, $a \le t \le b$, for some constant $c \ge 0$.

By Exercise 10.2.17, the following is immediate.

Lemma 10.4.3. *A geodesic s on a Riemannian manifold M is necessarily evenly parametrized.*

In local coordinates, the definition of a geodesic translates into a system of nonlinear, second order, ordinary differential equations. Indeed, write

$$\begin{aligned} s(t) &= (x^1(t), \dots, x^n(t)), \\ \dot{s}(t) &= (\dot{x}^1(t), \dots, \dot{x}^n(t)), \\ &= \dot{x}^i(t)\xi_{i\, s(t)}, \end{aligned}$$

and write down the parallelism condition for $\dot{s}(t)$:

$$\begin{aligned} 0 &= \frac{\nabla(\dot{s}(t))}{dt} \\ &= \ddot{x}^i(t)\xi_{i\, s(t)} + \dot{x}^i(t)\frac{\nabla}{dt}\left(\xi_{i\, s(t)}\right) \\ &= \ddot{x}^i(t)\xi_{i\, s(t)} + \dot{x}^i(t)\nabla_{\dot{s}(t)}\left(\xi_i\right) \\ &= \ddot{x}^k(t)\xi_{k\, s(t)} + \dot{x}^i(t)\dot{x}^j(t)\Gamma^k_{ij}(s(t))\xi_{k\, s(t)} \\ &= (\ddot{x}^k + \dot{x}^i\dot{x}^j\Gamma^k_{ij})\xi_k. \end{aligned}$$

Equivalently, this is the second order system

$$\ddot{x}^k + \dot{x}^i\dot{x}^j\Gamma^k_{ij} = 0, \quad 1 \le k \le n. \tag{10.5}$$

By setting $u^i = \dot{x}^i$, $1 \le i \le n$, we get an equivalent, nonlinear, first order system

$$\begin{aligned} \dot{u}^k + u^i u^j \Gamma^k_{ij} &= 0, \quad 1 \le k \le n, \\ \dot{x}^\ell &= u^\ell, \quad 1 \le \ell \le n. \end{aligned}$$

Given initial conditions

$$\begin{aligned} 2x^\ell(0) &= a^\ell, \quad 1 \le \ell \le n, \\ u^k(0) &= b^k, \quad 1 \le k \le n, \end{aligned}$$

there is a unique solution

$$s(t) = (x^1(t), \dots, x^n(t)),$$

defined on some interval $-\epsilon < t < \epsilon$. The initial condition can be written

$$\begin{aligned} s(0) &= (a^1, \dots, a^n), \\ \dot{s}(0) &= (b^1, \dots, b^n). \end{aligned}$$

The existence and uniqueness theorem for solutions of O.D.E., together with the smooth dependence of the solutions on initial conditions, gives the following.

Theorem 10.4.4. *Let $x_0 \in M$. Then, there is an open neighborhood U of x_0 in M and numbers $\epsilon_1, \epsilon_2 > 0$ such that, for every $x \in U$ and every $v_x \in T_x(M)$ with $\|v_x\| < \epsilon_1$, there is a unique geodesic*

$$\sigma_{v_x} : (-\epsilon_2, \epsilon_2) \to M$$

with

$$\sigma_{v_x}(0) = x,$$
$$\dot{\sigma}_{v_x}(0) = v_x.$$

Furthermore, in local coordinates, $x = (x^1, \dots, x^n)$, $v_x = v^i \xi_{i\,x}$, and

$$\sigma_{v_x}(t) = \varphi(x^1, \dots, x^n, v^1, \dots, v^n, t)$$

defines a smooth function of $2n + 1$ variables.

The domain of the function φ is $W \times (-\epsilon_2, \epsilon_2)$, where

$$W = \{v \in T(U) \mid \|v\| < \epsilon_1\},$$

an open neighborhood of 0_{x_0}.

Exercise 10.4.5. Let $M \subset \mathbb{R}^3$ be a smoothly imbedded surface with the relativized metric and Levi-Civita connection. Let $\sigma : [a, b] \to M$ be evenly parametrized and suppose that $\operatorname{im}(\sigma) \subseteq P \cap M$, where P is a 2-plane in $\mathbb{R}^3$ such that $\nu_{\sigma(t)}(M) \subset P$, $a \le t \le b$. Prove that σ is a geodesic.

Exercise 10.4.6. If M is a surface of revolution (as defined in freshman calculus), identify a natural, infinite family of geodesics. Discuss whether or when the "circles of latitude" are geodesics.

Exercise 10.4.7. Show that every geodesic on the unit sphere $S^2 \subset \mathbb{R}^3$ must lie along a great circle.

Remark. It will be convenient to reparametrize geodesics so that their domain of definition always contains the closed interval $[-1, 1]$. The following trick will be found in [**28**, p. 57]. The system (10.5) is homogeneous of degree 2. This implies that, if $\sigma(t)$ is a geodesic, so is $\sigma^c(t) = \sigma(ct)$, for a fixed constant c, and $\dot{\sigma}^c(t) = c\dot{\sigma}(ct)$. Let $0 < \epsilon < \epsilon_1\epsilon_2/2$. If $\|v\| < \epsilon$ and $|t| < 2$, then

$$\|2v/\epsilon_2\| < \epsilon_1,$$
$$|\epsilon_2 t/2| < \epsilon_2,$$

so the curve

$$\gamma_v(t) = \sigma_{2v/\epsilon_2}(\epsilon_2 t/2), \quad -2 < t < 2,$$

defines a geodesic

$$\gamma_v : (-2, 2) \to M$$

such that

$$\gamma_v(0) = x,$$
$$\dot{\gamma}_v(0) = v.$$

If

$$B_x(\epsilon) = \{v_x \in T_x(M) \mid \|v_x\| < \epsilon\},$$

then $\gamma_v(t)$ varies smoothly with

$$(v,t) \in \bigcup_{x \in U} B_x(\epsilon) \times (-2,2),$$

an open neighborhood of $(0_{x_0}, 0)$ in $T(M) \times \mathbb{R}$.

Definition 10.4.8. For $x \in U$, $v \in T_x(M)$, $\|v\| < \epsilon$, all as above,

$$\exp_x(v) = \gamma_v(1).$$

By the homogeneity of the system (10.5), we obtain $\gamma_v(\tau) = \gamma_{\tau v}(1)$. The following is an immediate consequence.

Lemma 10.4.9. *For $x \in U$, $v \in T_x(M)$, $\|v\| < \epsilon$, all as above,*

$$\gamma_v(t) = \exp_x(tv), \quad -1 \le t \le 1.$$

Conventions. In what follows, we routinely view $M \subset T(M)$ by the imbedding of M as the 0-section of $T(M)$. Thus, $x \in M$ is identified with $0_x \in T_x(M)$. Since the tangent space $T_v(V)$ at any point v of a finite dimensional vector space V is identified canonically with V itself, we will also identify $T_{0_x}(T_x(M))$ with $T_x(M)$.

Theorem 10.4.10. *If $\epsilon_x > 0$ is sufficiently small, then*

$$\exp_x : B_x(\epsilon_x) \to M,$$

called the exponential map at x, is a diffeomorphism onto an open neighborhood of x in M.

Proof. Write $\eta = \exp_x : B_x(\epsilon) \to M$. Clearly, $\eta(0_x) = x$. We compute

$$\eta_{*0_x} : T_{0_x}(B_x(\epsilon)) = T_x(M) \to T_x(M).$$

For $0 \neq v \in T_{0_x}(B_x(\epsilon)) = T_x(M)$, set $s(t) = tv$, $-\epsilon/\|v\| < t < \epsilon/\|v\|$, a smooth curve on $B_x(\epsilon)$ with $s(0) = 0_x$, $\dot{s}(0) = v$. Then $\eta \circ s$ defines a curve on M, $\eta(s(0)) = \eta(0_x) = x$ and $\eta(s(t)) = \exp_x(tv) = \gamma_v(t)$. Thus,

$$\eta_{*0_x}(\dot{s}(0)) = \eta_{*0_x}(v) = \dot{\gamma}_v(0) = v.$$

That is, η_{*0_x} is the identity under the identification $T_{0_x}(B_x(\epsilon)) = T_x(M)$. By the inverse function theorem, η will be a diffeomorphism for $\epsilon = \epsilon_x > 0$ sufficiently small. □

Remark that, if the manifold is compact, the number $\epsilon = \epsilon_x > 0$ can be chosen to be independent of x.

Definition 10.4.11. The Riemannian manifold M is geodesically complete if $\exp_x(v)$ is defined for all $x \in M$ and for all $v \in T_x(M)$. Equivalently, every geodesic segment extends (uniquely) to a geodesic $\gamma(t)$, $-\infty < t < \infty$.

Definition 10.4.12. A path $\sigma : [a,b] \to M$ is regular if it is a smooth immersion. If there is a partition $a = t_0 < t_1 < \cdots < t_r = b$ such that $\sigma|[t_{i-1}, t_i]$ is regular, $1 \le t \le r$, then σ is piecewise regular.

Nondegenerate geodesics σ are regular. Indeed, if $\dot{\sigma}(t_0) = 0$, for some t_0, then the fact that σ is evenly parametrized implies that $\dot{\sigma} \equiv 0$ and σ degenerates to a constant path.

Definition 10.4.13. Let M be a connected Riemannian manifold. Then the Riemann (or Riemannian) distance function $\rho : M \times M \to [0, \infty)$ is defined by

$$\rho(x, y) = \inf_{\sigma} |\sigma|,$$

where σ ranges over all piecewise regular paths from x to y in M.

Fixing the hypothesis that M is a connected, Riemannian n-manifold with $\partial M = \emptyset$, we are going to prove the following results.

Proposition 10.4.14. *The Riemann distance function is a topological metric on the Riemannian manifold M and the metric space topology is the same as the manifold topology.*

Theorem 10.4.15 (Hopf–Rinow I). *If the Riemannian manifold M is geodesically complete, $x, y \in M$, then there is a geodesic γ from x to y such that $|\gamma| = \rho(x, y)$. In particular, $\exp_x : T_x(M) \to M$ is surjective.*

Theorem 10.4.16 (Hopf–Rinow II). *The Riemannian manifold M is a complete metric space in the metric ρ if and only it is geodesically complete.*

Corollary 10.4.17. *If M is compact, then M is geodesically complete in any Riemann metric.*

Because of these results, it is standard to use the term "complete Riemannian manifold" when either geodesic completeness or metric completeness is intended.

A number of preliminary considerations are necessary for the proofs of Proposition 10.4.14 and the Hopf–Rinow theorems. To begin with, remark that the distance function ρ has the following properties:

1. $\rho(x, x) = 0$, $\forall\, x \in M$;
2. $\rho(x, y) = \rho(y, x)$, $\forall\, x, y \in M$;
3. $\rho(x, y) \le \rho(x, z) + \rho(z, y)$, $\forall\, x, y, z \in M$.

Thus, to prove that ρ is a topological metric, it is only necessary to prove that

$$\rho(x, y) = 0 \Rightarrow x = y.$$

Let $x_0 \in M \subset T(M)$, choose an open neighborhood $U \subseteq M$ of x_0, and let $\epsilon > 0$. Then

$$W = \{v_x \in T_x(M) \mid x \in U \text{ and } \|v_x\| < \epsilon\}$$

is an open neighborhood of $x_0 = 0_{x_0}$ in $T(M)$. If U and ϵ are chosen small enough,

$$G(v) = (\pi(v), \exp_{\pi(v)}(v))$$

is defined and smooth as a function $G : W \to M \times M$. We can take U to be a coordinate neighborhood, with coordinates $x^1, \dots, x^n$. We can also assume that there is a smooth, orthonormal frame field $Z_1, \dots, Z_n$ defined on U. We obtain, thereby, coordinates

$$(x^1, \dots, x^n, y^1, \dots, y^n) \leftrightarrow y^i Z_{i\,(x^1, \dots, x^n)}$$

on $\pi^{-1}(U)$ such that

$$\begin{aligned} W &= \{(x^1, \dots, x^n, y^1, \dots, y^n) \mid (x^1, \dots, x^n) \in U,\ \sum_{i=1}^{n} (y^i)^2 < \epsilon^2\} \\ &= U \times B(\epsilon). \end{aligned}$$

Lemma 10.4.18. *Given $x_0 \in M$, the neighborhood W of x_0 in $T(M)$ can be chosen, as above, so small that $G : W \to M \times M$ carries W diffeomorphically onto an open neighborhood of (x_0, x_0) in $M \times M$.*

Proof. Indeed, let ξ_i represent the basic coordinate fields for x^i and ζ_j those for y^j, $1 \le i, j \le n$, and observe that

$$G_{*0_{x_0}}(\xi_{i\,0_{x_0}}) = (\xi_{i\,x_0}, \xi_{i\,x_0}),$$
$$G_{*0_{x_0}}(\zeta_{j\,0_{x_0}}) = (0_{x_0}, Z_{j\,x_0}).$$

Thus, $G_{*0_{x_0}}$ is bijective and the assertion follows by the inverse function theorem. □

We fix the choice of W as in this lemma. Let V be an open neighborhood of x_0 in M such that

$$V \times V \subseteq G(W) \text{ and } V \subseteq U.$$

Given $(x, y) \in V \times V$, let $v_x \in W$ be the *unique* element such that $(x, y) = G(v_x) = (x, \exp_x(v_x))$. That is, $\exp_x(tv_x)$, $0 \le t \le 1$, is the *unique* geodesic of length $< \epsilon$ going from x to y and parametrized on $[0, 1]$. The point $\exp_x(tv_x)$ depends smoothly on $(v_x, t) \in W \times [0, 1]$ and G is a diffeomorphism on W, so $\exp_x(tv_x)$ also depends smoothly on (x, y, t). Less formally, we say that the unique geodesic of length $< \epsilon$ joining $x, y \in V$ depends smoothly on its endpoints.

Remark that, even though $x, y \in V$, there is no reason why the geodesic $\exp_x(tv_x)$ should stay in V for all values of $t \in [0, 1]$. It is a fact, as we will see later, that V can be chosen so that the geodesics of length $< \epsilon$ joining any points $x, y \in V$ do remain in V. This is J. H. C. Whitehead's theorem on the existence of geodesically convex neighborhoods which will give us the existence theorem for simple covers that we used in the discussion of de Rham cohomology.

Finally, note that this discussion also shows that, for each $x \in V \subseteq U$, $\exp_x$ maps the open ϵ-ball $B_x(\epsilon) \subset T_x(M)$ diffeomorphically onto an open neighborhood of x in M that contains V.

Let $U \subseteq \mathbb{R}^2$ be open and consider smooth maps $s : U \to M$. In complete analogy with the case of curves, we obtain the space $\mathfrak{X}(s)$ of all vector fields along s, these being all smooth maps $v : U \to T(M)$ such that the diagram

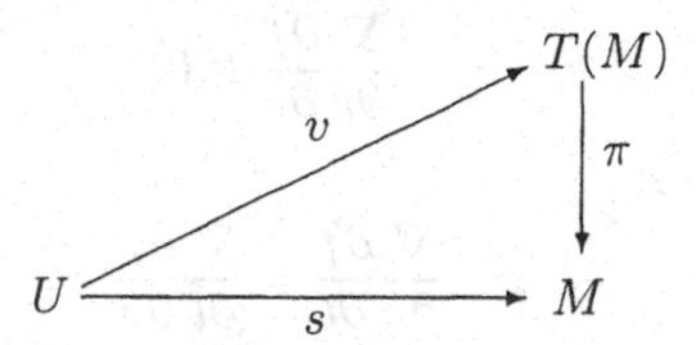

commutes.

Example 10.4.19. Let (r, t) be the coordinates of U. Then we define $\partial s/\partial r$ and $\partial s/\partial t \in \mathfrak{X}(s)$ by

$$\frac{\partial s}{\partial r}(r, t) = s_{*(r,t)}\left(\left.\frac{\partial}{\partial r}\right|_{(r,t)}\right),$$
$$\frac{\partial s}{\partial t}(r, t) = s_{*(r,t)}\left(\left.\frac{\partial}{\partial t}\right|_{(r,t)}\right).$$

Just as for curves, given $v \in \mathfrak{X}(s)$, we have partial covariant derivatives

$$\frac{\nabla v}{\partial r}, \frac{\nabla v}{\partial t} \in \mathfrak{X}(s),$$

obtained by taking covariant derivatives along the respective curve families

$$\begin{aligned} t &\equiv \text{ constant}, \\ r &\equiv \text{ constant}. \end{aligned}$$

Exercise 10.4.20. For $s : U \to M$ as above, prove that

$$\frac{\nabla}{\partial r}\left(\frac{\partial s}{\partial t}\right) = \frac{\nabla}{\partial t}\left(\frac{\partial s}{\partial r}\right).$$

Let $x \in V$ and $\epsilon > 0$ be as before. For $0 < a < \epsilon$, define the spherical shell around x of radius a to be

$$S_a = \{\exp_x(v) \mid v \in T_x(M) \text{ and } \|v\| = a\},$$

clearly a smooth hypersurface (submanifold of codimension one).

Lemma 10.4.21. *The geodesics out of x meet the spherical shell S_a orthogonally, $0 < a < \epsilon$.*

Proof. Let $v(t)$ be a smooth curve in $T_x(M)$, parametrized on $\mathbb{R}$, with $\|v(t)\| \equiv 1$. For $|r| < \epsilon$, define a smooth function

$$f(r,t) = \exp_x(rv(t)).$$

For fixed $t_0 \in \mathbb{R}$, $f(r,t_0)$ describes the geodesic out of x with initial velocity $v(t_0)$. For fixed $r_0 \in (-\epsilon, \epsilon)$, $f(r_0,t)$ describes a smooth curve on the shell S_{r_0}. Since $v(t)$ was arbitrary, it will be enough to prove that the curves $f(r_0,t)$ and $f(r,t_0)$ meet orthogonally at the point $f(r_0,t_0)$. Since (r_0,t_0) is to be arbitrary, what we really need to show is that

$$\left\langle \frac{\partial f}{\partial r}, \frac{\partial f}{\partial t} \right\rangle \equiv 0.$$

Since the r-curves ($t \equiv$ constant) are geodesics, we have

$$\frac{\nabla}{\partial r}\frac{\partial f}{\partial r} \equiv 0.$$

Also, by Exercise 10.4.20,

$$\frac{\nabla}{\partial r}\frac{\partial f}{\partial t} = \frac{\nabla}{\partial t}\frac{\partial f}{\partial r},$$

so

$$\begin{aligned} \frac{\partial}{\partial r}\left\langle \frac{\partial f}{\partial r}, \frac{\partial f}{\partial t} \right\rangle &= \left\langle \frac{\nabla}{\partial r}\frac{\partial f}{\partial r}, \frac{\partial f}{\partial t} \right\rangle + \left\langle \frac{\partial f}{\partial r}, \frac{\nabla}{\partial r}\frac{\partial f}{\partial t} \right\rangle \\ &= \left\langle \frac{\partial f}{\partial r}, \frac{\nabla}{\partial t}\frac{\partial f}{\partial r} \right\rangle \\ &= \frac{1}{2}\frac{\partial}{\partial t}\left\langle \frac{\partial f}{\partial r}, \frac{\partial f}{\partial r} \right\rangle \\ &= 0. \end{aligned}$$

The last equality is due to the fact that $\|\partial f/\partial r\|$ is the length of the velocity field along the geodesic $r \mapsto \exp_x(rv(t))$ and this length is the constant $\|v(t)\| \equiv 1$. It follows that $\langle \partial f/\partial r, \partial f/\partial t\rangle$ is constant in r. But, at $r = 0$, $\partial f/\partial t \equiv 0$, so

$$\left\langle \frac{\partial f}{\partial r}, \frac{\partial f}{\partial t} \right\rangle \equiv 0.$$

□

Let $x \in V$ and set $V_x = \exp_x(B_x(\epsilon))$. Then $V \subseteq V_x$. Let

$$\sigma : [a,b] \to V_x \smallsetminus \{x\}$$

be a piecewise regular curve. Since $\exp_x : B_x(\epsilon) \to V_x$ is a diffeomorphism, there is a unique piecewise regular curve $\tilde{\sigma}$ in $B_x(\epsilon) \smallsetminus \{0\}$, parametrized on $[a, b]$ and such that $\sigma = \exp_x \circ \tilde{\sigma}$. We write

$$\tilde{\sigma}(t) = \|\tilde{\sigma}(t)\| \frac{\tilde{\sigma}(t)}{\|\tilde{\sigma}(t)\|} = r(t)v(t),$$

where

$$0 < r(t) = \|\tilde{\sigma}(t)\| < \epsilon,$$
$$\|v(t)\| \equiv 1.$$

Thus, $\sigma(t) = \exp_x(r(t)v(t))$ with $r(t)$ and $v(t)$ subject to these conditions.

Lemma 10.4.22. *For σ as above,*

$$|\sigma| = \int_a^b \|\dot{\sigma}(t)\|\, dt \geq |r(b) - r(a)|.$$

If equality holds, then $r(t)$ is strictly monotone and piecewise regular and $v(t)$ is constant. Consequently, any shortest path joining two concentric shells around x is a piecewise regular reparametrization of a radial geodesic segment.

Proof. Let $f(r,t) = \exp_x(rv(t))$. Thus, $\sigma(t) = f(r(t),t)$, so

$$\dot{\sigma}(t) = \frac{dr}{dt}\frac{\partial f}{\partial r} + \frac{\partial f}{\partial t}.$$

Clearly, $\|\partial f/\partial r\| \equiv 1$ and, by Lemma 10.4.21, $\partial f/\partial r \perp \partial f/\partial t$. Thus,

$$\|\dot{\sigma}(t)\|^2 = \left|\frac{dr}{dt}\right|^2 + \left\|\frac{\partial f}{\partial t}\right\|^2 \geq \left|\frac{dr}{dt}\right|^2.$$

If equality holds, then $\partial f/\partial t \equiv 0$, so $\dot{v}(t) \equiv 0$ and $v(t)$ is constant. Therefore,

$$\begin{aligned} |\sigma| &= \int_a^b \|\dot{\sigma}(t)\|\, dt \\ &\geq \int_a^b \left|\frac{dr}{dt}\right| dt \\ &\geq \left|\int_a^b \frac{dr}{dt}\, dt\right| \\ &= |r(b) - r(a)|. \end{aligned}$$

If equality holds, not only is $v(t) = v$ constant, but dr/dt cannot change sign and, since $r(t)v = \tilde{\sigma}(t)$ is piecewise regular, dr/dt can never be 0. That is, $r(t)$ is strictly monotonic and piecewise regular. □

Theorem 10.4.23. *For V and $\epsilon > 0$ as above, let $\gamma : [0,1] \to M$ be the unique geodesic of length $< \epsilon$ joining two points $x, y \in V$. Let $\sigma : [0,1] \to M$ be an arbitrary piecewise regular path joining the same two points. Then $|\gamma| \leq |\sigma|$, where equality holds if and only if σ is obtained from γ by a piecewise regular reparametrization.*

Proof. Set $y = \exp_x(rv)$, where $0 < r < \epsilon$ and $\|v\| = 1$. Then, if $0 < \delta < r < \epsilon$, σ contains a segment joining the spherical shell S_δ to S_r and lying between these shells, hence lying in V_x. By Lemma 10.4.22, this segment has length at least $r - \delta$, so $|\sigma| \geq r - \delta$. Letting $\delta \downarrow 0$, we conclude that $|\sigma| \geq r = |\gamma|$. If $|\sigma| = r$, then the segment from each S_δ to S_r must be a reparametrization of (the same) radial geodesic. Since σ is piecewise regular, the reparametrization $r(t)$ is continuous and $\dot{r}(t)$ has only jump discontinuities, occurring only finitely often as $\delta \downarrow 0$, so σ itself is a piecewise regular reparametrization of γ. Conversely, if σ is a piecewise regular reparametrization of γ, then $|\sigma| = |\gamma|$. □

Corollary 10.4.24. *Let $\sigma : [a,b] \to M$ be piecewise regular and have minimal length for any piecewise regular path from $\sigma(a)$ to $\sigma(b)$. Then σ is obtained from a geodesic by piecewise regular reparametrization. If σ is regular, it is a regular reparametrization of a geodesic. If $\|\dot{\sigma}\|$ is constant, σ is a geodesic.*

Proof. Consider any segment of σ lying in an open set V as above and having length $< \epsilon$. By the above, this segment must be a piecewise regular reparametrization of a geodesic. Since every interior point of σ lies in the interior of such a segment and σ is made up of finitely many such segments, σ must be a piecewise regular reparametrization $\gamma(r(t))$ of a geodesic γ. If σ is regular, then

$$0 \neq \dot{\sigma}(t) = \frac{dr}{dt}\dot{\gamma}(r(t))$$

and this implies that $r(t)$ is a regular change of parameter. Since $\|\dot{\gamma}(r(t))\|$ is constant, dr/dt will be constant if $\|\dot{\sigma}(t)\|$ is. In this case, $r(t) = ct + e$ for suitable constants $c \neq 0$ and e, so $\sigma(t) = \gamma(ct + e)$ is a geodesic. □

The next two results complete the proof of Proposition 10.4.14.

Proposition 10.4.25. *If $x, y \in M$ and $\rho(x,y) = 0$, then $x = y$.*

Proof. Suppose $x \neq y$. Choose V and $\epsilon > 0$ as usual, but such that $x \in V$, $y \notin V$. For a suitable choice of $\eta \in (0, \epsilon)$, $\exp_x(B_x(\eta)) \subset V$. Then, every piecewise regular path σ from x to y must meet the spherical η-shell centered at x, so $|\sigma| \geq \eta$ and $\rho(x,y) \geq \eta > 0$. □

Thus, the Riemann distance function is a topological metric on M.

Proposition 10.4.26. *The topology induced on M by the metric ρ coincides with the manifold topology.*

Proof. Let $x \in M$ and choose $\epsilon > 0$ so small that $\exp_x : B_x(\epsilon) \to M$ is a diffeomorphism onto an open neighborhood $U_x(\epsilon)$ of x in the manifold topology. The set of all such $U_x(\epsilon)$ is a base for the manifold topology of M. But

$$U_x(\epsilon) = \{y \in M \mid \rho(x,y) < \epsilon\}. \tag{10.6}$$

Indeed, if $y \in U_x(\epsilon)$, then $\rho(x,y) < \epsilon$. Furthermore, if $z \in M \setminus U_x(\epsilon)$, every piecewise regular σ from x to z must meet every η-shell, $0 < \eta < \epsilon$, centered at x, so $|\sigma| \geq \epsilon$ and, consequently, $\rho(x,z) \geq \epsilon$. This proves the assertion (10.6), showing

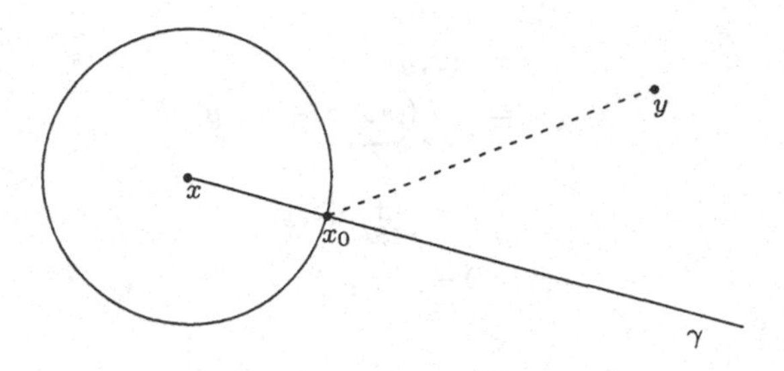

Figure 10.4.1. The Hopf–Rinow setup

that $\{U_x(\epsilon) \mid x \in M,\ \epsilon > 0\}$ is also a base for the topology induced by the metric ρ. □

We note also the following useful fact.

Proposition 10.4.27. *If $C \subseteq M$ is compact, there exists $\delta > 0$ such that any two points $x, y \in C$ with $\rho(x,y) < \delta$ are joined by a unique geodesic (parametrized on $[0,1]$) of length $< \delta$. This geodesic is the shortest piecewise regular path from x to y and depends smoothly on its endpoints. In particular, if M is compact, δ can be chosen uniformly for all of M.*

Proof. Cover C by open sets V_α with corresponding ϵ_α as in the above discussion. Select a finite subcover $\{V_{\alpha_i}\}_{i=1}^r$. Let $\delta > 0$ be a Lebesgue number for this cover (*i.e.*, if $\rho(x,y) < \delta$ and $x, y \in C$, then x and y lie in a common V_{α_i}). We can also demand that $\delta \le \min_{1\le i\le r} \epsilon_{\alpha_i}$. All assertions now follow from the previous discussion. □

We turn to the Hopf–Rinow theorems.

Proof of theorem 10.4.15. We assume that M is geodesically complete and choose arbitrary $x, y \in M$, $r = \rho(x,y)$. We must prove that there is a geodesic γ from x to y such that $|\gamma| = r$. This is trivial if $x = y$, so we exclude that possibility.

Let V and $\epsilon > 0$ be chosen as usual, with $x \in V$, and let $S_\delta \subset V$ be a spherical shell around x, $\delta < r$ sufficiently small. Since S_δ is compact, there is $x_0 \in S_\delta$ such that

$$\rho(x_0, y) = \min_{z \in S_\delta} \rho(z, y).$$

Write $x_0 = \exp_x(\delta v_0)$, $\|v_0\| = 1$. The geodesic $\gamma(t) = \exp_x(tv_0)$ is defined for *all* real values of t because M is geodesically complete (see Figure 10.4.1).

We will prove that, contrary to the possibility allowed in Figure 10.4.1,

$$y = \gamma(r) = \exp_x(rv_0),$$

thereby proving Theorem 10.4.15. Our procedure will be to prove, for each $t \in [\delta, r]$, the proposition

$$\mathbf{F}_t: \ \rho(\gamma(t), y) = r - t.$$

In particular, $\mathbf{F}_r$ asserts that $\rho(\gamma(r), y) = 0$, giving $y = \gamma(r)$.

(*a*) We prove $\mathbf{F}_\delta$. Every piecewise regular curve from x to y must cross S_δ, hence

$$\begin{aligned} r &= \rho(x,y) \\ &= \inf_{z\in S_\delta} (\underbrace{\rho(x,z)}_{\delta} + \rho(z,y)) \\ &= \delta + \inf_{z\in S_\delta} \rho(z,y) \\ &= \delta + \rho(x_0,y). \end{aligned}$$

Therefore,

$$\rho(\gamma(\delta),y) = \rho(x_0,y) = r - \delta.$$

(*b*) Assuming the truth of $\mathbf{F}_{t_0}$, some $t_0 \in [\delta, r)$, we will prove the existence of a maximal half-open interval $[t_0,\eta)$ such that $\eta \le r$ and $\mathbf{F}_t$ is true for all $t \in [t_0,\eta)$.

Indeed, for $\delta' > 0$ sufficiently small (as usual), let $S_{\delta'}$ be the spherical shell of radius δ' around $\gamma(t_0)$. Let $x_0' \in S_{\delta'}$ be a point with $\rho(x_0',y)$ minimum. Then, as before,

$$\begin{aligned} r - t_0 &= \rho(\gamma(t_0),y) \\ &= \inf_{z\in S_{\delta'}} (\underbrace{\rho(\gamma(t_0),z)}_{\delta'} + \rho(z,y)) \\ &= \delta' + \rho(x_0',y), \end{aligned}$$

so

$$\rho(x_0',y) = r - t_0 - \delta' = r - (t_0 + \delta').$$

We will show that $x_0' = \gamma(t_0 + \delta')$. Indeed,

$$\begin{aligned} \rho(x,x_0') &\ge \rho(x,y) - \rho(y,x_0') \\ &= r - (r - t_0 - \delta') \\ &= t_0 + \delta'. \end{aligned}$$

But the path consisting of the segment of γ from x to $\gamma(t_0)$, followed by a minimal geodesic from $\gamma(t_0)$ to x_0', has length $t_0 + \delta'$, hence is a piecewise geodesic, parametrized by arc length, joining x to x_0' and of minimal length. By Corollary 10.4.24, this path is a geodesic. It coincides with γ on $[0,t_0]$ and $t_0 > 0$, hence it coincides with γ on $[0, t_0 + \delta']$. That is, $x_0' = \gamma(t_0 + \delta')$, as claimed.

We have proven that

$$\rho(\gamma(t_0 + \delta'),y) = \rho(x_0',y) = r - (t_0 + \delta'),$$

which is the assertion $\mathbf{F}_{t_0+\delta'}$. Since $\delta' > 0$ was arbitrarily small, there is a half-open interval $[t_0,\eta')$, $\eta' < r$, on which $\mathbf{F}_t$ holds. The union of all such intervals produces the maximal one $[t_0,\eta)$.

(*c*) Since $\mathbf{F}_\delta$ holds, let $[\delta,\eta)$ be the maximal half-open interval on which $\mathbf{F}_t$ holds. But the truth of $\mathbf{F}_t$ on $[\delta,\eta)$ implies $\mathbf{F}_\eta$, by continuity. Thus, if $\eta < r$, we could apply (*b*) to obtain a contradiction to the maximality of $[\delta,\eta)$. Consequently, $\eta = r$ and $\mathbf{F}_r$ holds. □

Proof of theorem 10.4.16. We first assume that M is geodesically complete and we prove that M, as a metric space under ρ, is complete. For this, it will be enough to prove that, whenever $B \subseteq M$ is a ρ-bounded subset, the closure $\overline{B}$ is compact. Choose any $x \in B$ and consider the continuous map

$$\exp_x : T_x(M) \to M,$$

defined because M is geodesically complete. Since B is bounded, there is a number $r > \sup_{y \in B} \rho(x, y)$. If $D \subset T_x(M)$ is the closed ball of radius r, then $B \subseteq \exp_x(D)$ (Theorem 10.4.15) and $\exp_x(D)$ is compact. Thus, $\overline{B} \subseteq \exp_x(D)$ is compact.

Next, assuming that M is complete in the metric ρ, we prove that M is geodesically complete. Let $x \in M$ and let $v \in T_x(M)$ have unit norm. Let (a, b) denote the maximal open interval about 0 in $\mathbb{R}$ such that $\exp_x(tv)$ is defined, $\forall t \in (a, b)$. We must show that $a = -\infty$ and $b = \infty$.

If $b < \infty$, choose $\{t_k\}_{k=1}^{\infty} \subset (a, b)$ such that $t_k \uparrow b$ strictly. This is a Cauchy sequence. Set $x_k = \exp_x(t_k v)$ and remark that $\rho(x_\ell, x_k) \leq t_\ell - t_k$, whenever $k < \ell$. Indeed, the segment of $\exp_x(t)$, $t_k \leq t \leq t_\ell$, is a geodesic of length $t_\ell - t_k$ joining these two points. Therefore, $\{x_k\}_{k=1}^{\infty}$ is Cauchy in the metric ρ and, by the completeness of this metric, $x_k \to y \in M$.

Define $\gamma : [0, b] \to M$ by

$$\gamma(t) = \begin{cases} \exp_x(tv), & 0 \leq t < b, \\ y, & t = b. \end{cases}$$

By the previous paragraph, γ is continuous on $[0, b]$. It is smooth on $[0, b)$. If γ is also smooth at b, it is a geodesic and can be extended as a geodesic to $[0, b + \eta)$, some $\eta > 0$. This would contradict the maximality of (a, b), proving that $b = \infty$.

In order to prove that γ is smooth at b, choose a neighborhood $V \subset M$ of y and a number $\epsilon > 0$ such that $\exp_z(w)$ is defined, $\forall z \in V$, $\forall w \in T_x(M)$ with $\|w\| < \epsilon$. For a Cauchy sequence $\{x_k = \exp_x(t_k v)\}_{k=1}^{\infty}$, chosen as above, $x_k \in V$ and $b - t_k < \epsilon$, for all sufficiently large values of k. Let k be large enough and set $v_k = \dot{\gamma}(t_k) \in T_{x_k}(M)$. Since $\|v_k\| = 1$, $\exp_{x_k}(tv_k)$ is defined for $0 \leq t \leq b - t_k < \epsilon$. But, for $0 \leq t < b - t_k$, this curve coincides with $\gamma(t_k + t)$. By continuity, $\gamma(b) = \exp_{x_k}((b - t_k)v_k)$, completing the proof that γ is smooth at b.

A completely parallel argument shows that $a = -\infty$. □

Exercise 10.4.28. Let M be a complete Riemannian manifold, $\mathfrak{F}$ a foliation of M, and let L be a leaf of $\mathfrak{F}$. The Riemannian metric $\langle \cdot, \cdot \rangle$ on M induces a Riemannian metric $\langle \cdot, \cdot \rangle_L$ on L via the one-to-one immersion $i : L \hookrightarrow M$. Let ρ_L denote the corresponding topological metric on L. Generally, this is *not* the restriction of the metric ρ of M. Let $\{x_k\}_{k=1}^{\infty} \subset L$ be ρ_L-Cauchy.

(1) Prove that $\{x_k\}_{k=1}^{\infty}$ is ρ-Cauchy, hence that $x_k \to x \in M$.
(2) Let $(U, y^1, \ldots, y^n)$ be a Frobenius neighborhood of x. Prove that all but finitely many of the points x_k lie on the same $\mathfrak{F}$-plaque in U as x.
(3) Using the above, conclude that, as a Riemannian manifold in the induced metric $\langle \cdot, \cdot \rangle_L$, L is complete.

In particular, this exercise implies that a leaf L in a compact, foliated manifold $(M, \mathfrak{F})$ is complete in any metric $\langle \cdot, \cdot \rangle_L$ that arises, as above, by restricting to L an arbitrary Riemannian metric $\langle \cdot, \cdot \rangle$ on M. The leaf L need not, itself, be compact. The geodesics on the Riemannian manifold $(L, \langle \cdot, \cdot \rangle_L)$ are not, generally, geodesics in $(M, \langle \cdot, \cdot \rangle)$.

10.5. Geodesic Convexity

We will prove a theorem of J. H. C. Whitehead that, in particular, will guarantee the existence of simple refinements of open covers. This result was anticipated and used in our treatment of de Rham cohomology (Chapter 8).

Throughout this section, M is a Riemannian n-manifold with empty boundary.

Definition 10.5.1. A subset $X \subseteq M$ is star shaped with respect to a point $x_0 \in X$ if each $x \in X$ can be joined to x_0 by a unique shortest geodesic in M and if this geodesic always lies in X.

Definition 10.5.2. A subset $X \subseteq M$ is geodesically convex if it is star shaped with respect to each of its points.

Equivalently, X is geodesically convex if any two of its points are joined by a unique shortest geodesic in M and this geodesic lies in X. The following is immediate.

Lemma 10.5.3. *An arbitrary intersection of geodesically convex sets is geodesically convex.*

Theorem 10.5.4 (Whitehead). *Let $W \subseteq M$ be open, $x \in W$. Then there is a geodesically convex, open neighborhood $U \subset W$ of x.*

Before proving this result, we show how it implies the existence of simple refinements.

Lemma 10.5.5. *A set $X \subseteq M$, star shaped with respect to $x_0 \in X$, is contractible.*

Proof. Indeed, each $x \in X$ determines uniquely $v_x \in T_{x_0}(M)$ and $t_x > 0$ such that $\|v_x\| = 1$ and $x = \exp_{x_0}(t_x v_x)$. Then, $F : X \times [0,1] \to X$, defined by

$$F(x, \tau) = \exp_{x_0}(\tau t_x v_x),$$

is the desired contraction. □

It seems that open, star shaped sets $U \subseteq M$ are always diffeomorphic to $\mathbb{R}^n$, but this is extremely difficult to prove. The problem is that the set theoretic boundary ∂U may be very badly behaved. For instance, the "radius function"

$$r : S^{n-1}_{x_0} \to [0, \infty],$$

even if it takes only finite values, may not be continuous, let alone smooth. This function is defined on the sphere of unit vectors in $T_{x_0}(M)$ and assigns to $v \in S^{n-1}_{x_0}$ the supremum $r(v)$ of the numbers $\tau > 0$ such that

$$\exp_{x_0}(tv) \in U, \quad 0 < t \leq \tau.$$

Keep this possible bad behavior of r in mind while attempting the following exercise.

Exercise 10.5.6. Let $U \subseteq M$ be open and star shaped with respect to $x_0 \in U$ and let $C \subset U$ be compact. Prove that there is an open set $V \subset U$, also star shaped with respect to x_0, such that $\overline{V}$ is a compact subset of M and $C \subset V \subset \overline{V} \subset U$.

Proposition 10.5.7. *If $U \subseteq M$ is an open set, star shaped with respect to x_0, then $H^*_c(U) = H^*_c(\mathbb{R}^n)$.*

Proof. For a suitable value $\rho_0 > 0$, the open ball $B_{x_0}(\rho_0) \subset T_{x_0}(M)$, centered at 0 with radius ρ_0, is carried by $\exp_{x_0}$ diffeomorphically onto an open set $U_{\rho_0} \subset U$ with compact closure in U. Since $B_{x_0}(\rho_0)$ is diffeomorphic to $\mathbb{R}^n$, the same is true for U_{ρ_0}. In particular, $H_c^*(U_{\rho_0}) = H_c^*(\mathbb{R}^n)$. The inclusion $i : U_{\rho_0} \hookrightarrow U$ induces homomorphisms

$$i_* : A_c^*(U_{\rho_0}) \to A_c^*(U),$$
$$i_* : H_c^*(U_{\rho_0}) \to H_c^*(U),$$

so it will be enough to prove that the second of these is an isomorphism.

Let $C \subset U$ be compact. By Exercise 10.5.6, find open sets $V \subset W \subset U$, star shaped with respect to x_0 and such that $\overline{V} \subset W \subset \overline{W} \subset U$, where $\overline{V}$ and $\overline{W}$ are compact, and $\overline{U}_{\rho_0} \cup C \subset V$. Also, fix $0 < a < b < \rho_0$ and the corresponding open balls $U_a \subset U_b \subset U_{\rho_0}$. By the smooth Urysohn lemma, find $f : U \to [0,1]$ such that

$$f|(U \smallsetminus W) \equiv 0,$$
$$f|(\overline{V} \smallsetminus U_b) > 0,$$
$$f|\overline{U}_a \equiv 0.$$

Let $Z \in \mathfrak{X}(U \smallsetminus \{x_0\})$ be nowhere 0, tangent to the radial geodesics out of x_0, and everywhere pointing toward x_0. Let $F_t : U \to U$ denote the flow, defined for all time t, generated by the compactly supported vector field $fZ \in \mathfrak{X}(U)$. Then, since $C \smallsetminus U_{\rho_0}$ is contained in the interior of the support of fZ, as is $\partial\overline{U}_{\rho_0}$, there is a value $\tau > 0$ such that $F_\tau(C) \subset U_{\rho_0}$. Set $\psi = F_{-\tau}$. Since $F_{-\tau}$ is a compactly supported diffeomorphism of U onto itself, isotopic through such diffeomorphisms F_t to $F_0 = \mathrm{id}_U$, it follows that $\psi^* : H_c^*(U) \to H_c^*(U)$ is the identity.

Let $\omega \in Z_c^p(U)$ and let $C = \mathrm{supp}(\omega)$. By the previous paragraph, we obtain $\psi : U \to U$ such that $\psi^*(\omega) \in Z_c^p(U_{\rho_0})$ and $\psi^*[\omega] = [\omega] \in H_c^p(U)$. It follows that $[\omega] \in \mathrm{im}(i_*)$, hence that i_* carries $H_c^*(U_{\rho_0})$ *onto* $H_c^*(U)$.

Suppose that $\omega \in Z_c^p(U_{\rho_0})$ and that $i_*[\omega] = 0$. Choose $a > 0$ as above such that $\mathrm{supp}(\omega) \subset U_a$. Viewing $\omega = i_*(\omega)$ in $Z_c^p(U)$, we find $\eta \in A_c^{p-1}(U)$ such that $\omega = d\eta$. Let $C = \mathrm{supp}(\eta)$ and obtain $\psi : U \to U$, as above, so that

$$\psi^*(\omega) = \omega,$$
$$\psi^*(\eta) = \eta_0 \in A_c^{p-1}(U_{\rho_0}),$$

but

$$d\eta_0 = d\psi^*(\eta) = \psi^*(d\eta) = \psi^*(\omega) = \omega.$$

That is, $[\omega] = 0$ in $H_c^*(U_{\rho_0})$, proving that i_* is one-to-one. □

Corollary 10.5.8. *Every open cover of M admits a simple refinement.*

Proof. By Theorem 10.5.4, each open cover admits a refinement by open, geodesically convex sets. With a little care, one chooses this refinement to be locally finite (Exercise 10.5.9). By Lemma 10.5.3, any finite intersection of elements of this refinement is also an open, geodesically convex set, hence star shaped. By Lemma 10.5.5 and Proposition 10.5.7, this refinement is simple. □

Exercise 10.5.9. Check the assertion in the proof of Corollary 10.5.8 that the refinement by open, geodesically convex sets can be chosen to be locally finite.

Exercise 10.5.10. Let $x \in U \subseteq M$ where U is open in M and star shaped with respect to x. Let $r : S_x^{n-1} \to [0,\infty]$ be the radius function for U. That is, $S_x^{n-1} \subset T_x(M)$ is the unit sphere and

$$U = \{\exp_x(tv) \mid v \in S_x^{n-1} \text{ and } 0 \le t < r(v)\}.$$

If r is finite-valued of class C^∞, prove that U is diffeomorphic to $\mathbb{R}^n$.

Exercise 10.5.11. Let $x \in U \subseteq M$ and $r : S_x^{n-1} \to [0,\infty]$ be as in the preceding exercise, but do not assume that r is smooth or even continuous. Prove that r is lower semicontinuous. That is, $r^{-1}(a,\infty]$ is open in S_x^{n-1}, $\forall\, a \in \mathbb{R}$.

Exercise 10.5.12. Let $x \in U \subseteq M$ and $r : S_x^{n-1} \to [0,\infty]$ be as in the preceding exercises. Construct an example in which r is finite-valued everywhere and discontinuous on a dense subset of S_x^{n-1}.

We turn to the proof of Theorem 10.5.4. Let $x \in W$, as in the statement of the theorem. As in Section 10.4, choose a neighborhood V of x in W and a number $\epsilon > 0$ such that any two points $y, z \in V$ can be joined by a unique geodesic $\sigma_{y,z}$ in M of length $< \epsilon$. As usual, $\sigma_{y,z}$ is parametrized on $[0,1]$ and depends smoothly on $(y,z) \in V \times V$. Choose $\delta > 0$ such that the open ball $B_x(\delta) \subset T_x(M)$ of radius δ is carried diffeomorphically by $\exp_x$ onto a neighborhood $U_x \subseteq V$ of x.

Let $(v_1,\dots,v_n)$ be an orthonormal frame for $T_x(M)$ and coordinatize this vector space by

$$(x^1,\dots,x^n) \leftrightarrow x^i v_i.$$

Under the diffeomorphism $\exp_x^{-1} : U_x \to B_x(\delta)$, these become coordinates on U_x (called a *normal* coordinate system on U_x). The corresponding coordinate fields are $\xi_i \in \mathfrak{X}(U_x)$, $1 \le i \le n$. If $y \in U_x$ has coordinates $(b_1,\dots,b_n)$, then

$$\rho(x,y)^2 = \sum_{i=1}^n b_i^2.$$

If $0 < \delta_* < \delta$, if $S_{\delta_*} \subset U_x$ is the spherical shell of radius δ_*, centered at x, if $y = (b_1,\dots,b_n) \in S_{\delta*}$, and if $v = a^j \xi_j \in T_y(S_{\delta_*})$, then

$$b_i a^i = 0.$$

The key lemma for the proof of Theorem 10.5.4 is the following.

Lemma 10.5.13. *If $\delta_* \in (0,\delta)$ is small enough, then every geodesic*

$$\gamma : (-\eta,\eta) \to M,$$

such that $\gamma(0) = y \in S_{\delta_*}$ and $\dot\gamma(0) \in T_y(S_{\delta_*})$, has the property that $\rho(x,\gamma(t)) > \delta_*$, for all sufficiently small values of $|t| \ne 0$.

Proof. As above, denote the normal coordinates of y by $(b_1,\dots,b_n)$. Let δ_* be so small that, for $\sum_{i=1}^n b_i^2 \le \delta_*$, the symmetric matrix

$$Q = 2[\delta_{k\ell} - b_i\Gamma^i_{k\ell}(b_1,\dots,b_n)]$$

is so close to $[2\delta_{k\ell}]$ as to be positive definite. Let

$$\gamma(t) = (x^1(t),\dots,x^n(t)), \quad -\eta < t < \eta,$$

be a geodesic in M, tangent to S_{δ_*} at $\gamma(0) = y = (b_1,\dots,b_n)$, and let $\dot\gamma(0) = a^i\xi_i$. Define

$$F(t) = \rho(x,\gamma(t))^2 - \delta_*^2 = \sum_{i=1}^n x^i(t)^2 - \delta_*^2.$$

For small values of $|t|$, this is a smooth function and

$$\begin{aligned} F(0) &= 0, \\ F'(0) &= 2x^i(0)\dot{x}^i(0) \\ &= 2b_i a^i \\ &= 0, \\ F''(t) &= 2(\dot{x}^i(t)\dot{x}^i(t) + x^i(t)\ddot{x}^i(t)). \end{aligned}$$

Since $\gamma(t)$ is a geodesic, it satisfies

$$\ddot{x}^i = -\dot{x}^k\dot{x}^\ell\Gamma^i_{k\ell}, \quad 1 \le i \le n,$$

giving

$$\begin{aligned} F''(0) &= 2(\dot{x}^i(0)^2 - x^i(0)\dot{x}^k(0)\dot{x}^\ell(0)\Gamma^i_{k\ell}(x^1(0),\dots,x^n(0))) \\ &= 2((a^i)^2 - a^k a^\ell(b_i\Gamma^i_{k\ell}(b_1,\dots,b_n))) \\ &= [a^1,\dots,a^n]Q\begin{bmatrix} a^1 \\ \vdots \\ a^n \end{bmatrix}, \end{aligned}$$

the value of a positive definite quadratic form on the vector $\dot{\gamma}(0) \neq 0$. That is, $F''(0) > 0$. Plugging this data into the 2nd order Taylor expansion of $F(t)$ about $t = 0$, we see that

$$F(t) = \frac{t^2}{2}F''(0) + O(t^3) > 0,$$

for small enough values of $|t| \neq 0$. □

Proof of theorem 10.5.4. Choose $N_x = \exp_x(B_x(\delta_*))$, where δ_* is chosen by the above lemma. Let $R \subseteq N_x \times N_x$ be the subset of all (y,z) such that $\sigma_{y,z}$ lies entirely in N_x. By the smooth dependence of this geodesic on its endpoints, R is an open subset. It is also clear that $R \neq \emptyset$. If we prove that R is also a closed subset, then, by the connectivity of $N_x \times N_x$, $R = N_x \times N_x$ and N_x is geodesically convex. Let $\{(y_k, z_k)\}_{k=1}^\infty \subset R$ with $\lim_{k\to\infty}(y_k, z_k) = (y_0, z_0)$ in $N_x \times N_x$. If $(y_0, z_0) \notin R$, then σ_{y_0,z_0} meets $\partial N_x = S_{\delta_*}$. If σ_{y_0,z_0} is tangent to the spherical shell at some point of intersection, an application of the lemma shows that σ_{y_0,z_0} contains points in $U_x \smallsetminus \overline{N}_x$. But smooth dependence on (y_0, z_0) implies that this remains true for all values of (y,z) sufficiently near (y_0, z_0), hence for (y_k, z_k), k sufficiently large. This contradicts the fact that $(y_k, z_k) \in R$. But if an intersection point of σ_{y_0,z_0} with the shell is not a point of tangency, it is clear that σ_{x_0,y_0} contains points in $U_x \smallsetminus \overline{N}_x$, leading to the same contradiction. Thus, $(x_0, y_0) \in R$, proving that R is closed in N_x. □

10.6. The Cartan Structure Equations

We return to the torsion and curvature tensors that were introduced earlier for a connection ∇. The key to understanding the geometric significance of these tensors is a pair of equations, written in terms of differential forms, called the equations of structure. In this section, we derive these structure equations in local coordinate charts. (In the next chapter, where we treat *principal bundles*, we will be able to obtain global, coordinate-free versions of these equations by lifting them to the frame bundle.) As an application, we will prove that the Riemann tensor is exactly the obstruction to the integrability of the Riemannian structure. That is, the vanishing of curvature is equivalent to the existence of a coordinate atlas $\{U_\alpha, x_\alpha^1, \dots, x_\alpha^n\}_{\alpha\in\mathfrak{A}}$ such that the coordinate fields $\partial/\partial x_\alpha^i$, $1 \le i \le n$, form an orthonormal frame field on U_α, for each $\alpha \in \mathfrak{A}$. Equivalently, the Riemannian manifold is locally isometric to Euclidean space.

In what follows, ∇ is a general connection on the n-manifold M, $n \ge 2$. To begin with, we will work in an open, trivializing neighborhood U for $T(M)$ and fix the trivialization by a choice of a smooth frame $(X_1, \dots, X_n)$ on U. Define $\theta^i \in A^1(U)$ by $\theta^i(X_j) = \delta^i_j$, $1 \le i, j \le n$. Then, each $X \in \mathfrak{X}(U)$ can be written

$$X = \theta^i(X) X_i.$$

Remark. Elie Cartan called the frame field a "moving frame". The discussion in this section concerns his "method of moving frames". In the next chapter, we will use principal bundles to globalize this method and give some geometric applications.

Define forms $\omega^i_j \in A^1(U)$, $1 \le i, j \le n$, by

$$\nabla_X X_j = \omega^i_j(X) X_i, \quad \forall X \in \mathfrak{X}(U).$$

The fact that these are forms, *i.e.*, that $\omega^i_j(fX) = f\omega^i_j(X)$, $\forall f \in C^\infty(U)$, follows from the fact that $\nabla_X X_j$ is a tensor in X.

In order to express the torsion and curvature tensors of ∇ in terms of the frame field, we introduce forms $\tau^i, \Omega^i_j \in A^2(U)$ by the formulas

$$T(X,Y) = \tau^i(X,Y) X_i,$$
$$R(X,Y) X_j = \Omega^i_j(X,Y) X_i.$$

The fact that these are antisymmetric tensors (*i.e.*, 2-forms) follows from the same properties of T and R.

Theorem 10.6.1 (Cartan structure equations). *The above forms satisfy the identities*

$$d\theta^i = -\omega^i_j \wedge \theta^j + \tau^i, \tag{10.7}$$
$$d\omega^i_j = -\omega^i_k \wedge \omega^k_j + \Omega^i_j. \tag{10.8}$$

Proof. The proof is a computation. We carry this out for equation (10.7). The computation of equation (10.8) is more of the same and will be left as an exercise.

For arbitrary $X, Y \in \mathfrak{X}(U)$,

$$\begin{aligned}
\tau^i(X,Y)X_i &= \nabla_X Y - \nabla_Y X - [X,Y] \\
&= \nabla_X(\theta^j(Y)X_j) - \nabla_Y(\theta^j(X)X_j) - \theta^j([X,Y])X_j \\
&= (X(\theta^j(Y)) - Y(\theta^j(X)) - \theta^j([X,Y]))X_j \\
&\quad + \theta^j(Y)\nabla_X X_j - \theta^j(X)\nabla_Y X_j \\
&= d\theta^j(X,Y)X_j + \theta^j(Y)\omega^i_j(X)X_i - \theta^j(X)\omega^i_j(Y)X_i \\
&= (d\theta^i(X,Y) + \theta^j(Y)\omega^i_j(X) - \theta^j(X)\omega^i_j(Y))X_i.
\end{aligned}$$

But we claim that

$$\omega^i_j(X)\theta^j(Y) - \omega^i_j(Y)\theta^j(X) = \omega^i_j \wedge \theta^j(X,Y).$$

Indeed, the standard inclusion $A^2(U) \hookrightarrow \mathcal{T}_0^2(U)$ takes

$$\omega_j^i \wedge \theta^j \mapsto \omega_j^i \otimes \theta^j - \theta^j \otimes \omega_j^i$$

(Lemma 7.2.18). Thus, the coefficients of X_i on each side of the above being equal, $1 \le i \le n$, we obtain

$$\tau^i(X,Y) = (d\theta^i + \omega_j^i \wedge \theta^j)(X,Y).$$

Since X and Y are arbitrary, equation (10.7) follows. □

Exercise 10.6.2. Verify equation (10.8).

Using matrix notation, we can write the equations of structure more compactly. Set

$$\theta = \begin{bmatrix} \theta^1 \\ \vdots \\ \theta^n \end{bmatrix},$$

$$\tau = \begin{bmatrix} \tau^1 \\ \vdots \\ \tau^n \end{bmatrix},$$

$$\omega = \left[\omega_j^i\right],$$

$$\Omega = \left[\Omega_j^i\right].$$

The n-tuples θ and τ can be thought of as $\mathbb{R}^n$-valued forms. The matrices ω and Ω can be thought of as $L(\mathrm{Gl}(n))$-valued forms. In Chapter 11, we will lift these local forms to the frame bundle of $T(M)$, where they will fit together coherently to define global forms.

Definition 10.6.3. The $\mathbb{R}^n$-valued forms θ and τ are called the trivializing coframe field and the torsion form, respectively. The $L(\mathrm{Gl}(n))$-valued forms ω and Ω are called the connection form and the curvature form, respectively.

The structure equations are

$$d\theta = -\omega \wedge \theta + \tau, \tag{10.9}$$

$$d\omega = -\omega \wedge \omega + \Omega, \tag{10.10}$$

where we multiply matrices of forms by the usual rules of matrix multiplication, but use exterior multiplication for products of entries.

Lemma 10.6.4 (Key Lemma). *Let $U \subseteq M$ be an open, connected subset, together with a connection ∇ and frame field $(X_1, \dots, X_n)$ on U with associated curvature form $\Omega \equiv 0$. Then, given $q \in U$ and $B \in \mathcal{M}(n)$, there is a connected neighborhood V of q and a unique smooth map*

$$A : V \to \mathcal{M}(n)$$

such that

$$A(q) = B \text{ and } dA = A \wedge \omega.$$

Proof. By hypothesis, the second structure equation becomes

$$d\omega = -\omega \wedge \omega. \tag{$*$}$$

Let $(V, x^1, \ldots, x^n)$ be a coordinate chart about q in U, let $P = V \times \mathcal{M}(n)$ and let $p : P \to V$ be projection onto the first factor. Coordinatize P by the coordinates x^i on the V factor and the standard coordinates z^i_j on the $\mathcal{M}(n)$ factor. Then $Z = [z^i_j]$ can be interpreted as a matrix-valued function on P, constant in the coordinates x^i, and the matrix ω of connection forms can be interpreted as a matrix of 1-forms on P, constant in the coordinates z^i_j. Define

$$\Lambda = dZ - Z \wedge \omega,$$

a matrix of 1-forms on P. Then

$$d\Lambda = -dZ \wedge \omega - Z \wedge d\omega = -(dZ - Z \wedge \omega) \wedge \omega = -\Lambda \wedge \omega,$$

where we have used $(*)$. This equation looks very much like a Frobenius integrability condition. Indeed, the n^2 entries of the matrix Λ are of the form

$$\lambda^i_j = dz^i_j - z^i_k \wedge \omega^k_j,$$

hence are linearly independent pointwise on P. Then

$$E_u = \bigcap_{1 \le i,j \le n} \ker \lambda^i_{j\,u}, \quad \forall u \in P,$$

defines an n-plane distribution on P. The 1-forms λ^i_j generate the annihilator ideal $I(E) \subset A^*(P)$ and the equations

$$d\lambda^i_j = -\lambda^i_k \wedge \omega^k_j$$

show that $dI^1(E) \subset I^2(E)$, so E is integrable by Theorem 9.1.5. Let $\mathcal{F}$ denote the associated foliation. Notice that the restriction of λ^i_j to any factor $\{x\} \times \mathcal{M}(n)$ is just dz^i_j, so a vector tangent to this factor and annihilated by every λ^i_j must be 0. It follows that $p_{*u} : E_u \to T_{p(u)}(V)$ is an isomorphism, for each $u \in P$, hence that p restricts to a local diffeomorphism on each leaf of $\mathcal{F}$. In particular, making V smaller, if necessary, we can assure that the leaf through (q, B) is the graph Γ_A of a smooth function $A : V \to \mathcal{M}(n)$. That is, the leaf is the image of the section

$$\alpha : V \to P,$$
$$\alpha(x) = (x, A(x)).$$

The necessary and sufficient condition that Γ_A be a leaf is that $\alpha^*(\Lambda) \equiv 0$. Since $\alpha^*(Z) = Z \circ \alpha = A$ and $\alpha^*(\omega) = \omega$, we see that Γ_A is a leaf if and only if

$$0 \equiv \alpha^*(dZ - Z \wedge \omega) = dA - A \wedge \omega.$$

The condition that $A(q) = B$ is equivalent to the condition that the leaf Γ_A pass through the point (q, B). This uniquely determines the leaf, hence the function A. □

The Frobenius theorem played a key role in the above proof, showing that curvature is the obstruction to integrability of a certain n-plane distribution. In the case of the Levi-Civita connection for a Riemannian metric, we are about to show that this integrability is equivalent to the integrability of the infinitesimal O(n) structure defined by the metric.

Let ∇ be the Levi-Civita connection of a Riemannian metric. In this case, the frame field $(X_1, \ldots, X_n)$ can and will be chosen to be *orthonormal.* Since this connection is torsion free, the torsion form τ vanishes. The following exercises are straightforward computations.

Exercise 10.6.5. Let R be the Riemann tensor and let $(X_1, \dots, X_n)$ be an orthonormal frame field. For arbitrary $X, Y \in \mathfrak{X}(U)$, prove that

$$\langle R(X,Y)X_j, X_k\rangle = -\langle X_j, R(X,Y)X_k\rangle.$$

Exercise 10.6.6. Prove that the connection form ω and curvature form Ω of ∇, relative to an orthonormal frame field, take values in the Lie algebra $L(\mathrm{O}(n))$ of antisymmetric matrices. For the curvature form, appeal to the previous exercise.

Theorem 10.6.7. *The Riemannian manifold M is locally isometric to Euclidean space (also said to be flat) if and only if the Riemann tensor $R \equiv 0$.*

Since the Riemann tensor for the Euclidean metric does vanish identically, the "only if" part of this result is evident. Thus, we assume that $R \equiv 0$ and prove that every point $q \in M$ has a coordinate neighborhood $(V, y^1, \dots, y^n)$ such that the coordinate frame field $(\xi_1, \dots, \xi_n)$ is orthonormal on V.

In the Key Lemma 10.6.4, choose the matrix B to be an element of $\mathrm{O}(n)$. Under this and our other current hypotheses, we have the following.

Lemma 10.6.8. *The function $A : V \to \mathcal{M}(n)$ takes its image in the group $\mathrm{O}(n)$.*

Proof. Since $A(q) = B$ and $BB^{\mathrm{T}} = I$, it will be enough to show that AA^{T} is constant on V. But

$$\begin{aligned} d(AA^{\mathrm{T}}) &= dA \wedge A^{\mathrm{T}} + A \wedge (dA)^{\mathrm{T}} \\ &= A \wedge \omega \wedge A^{\mathrm{T}} + A \wedge (A \wedge \omega)^{\mathrm{T}} \\ &= A \wedge \omega \wedge A^{\mathrm{T}} - A \wedge \omega \wedge A^{\mathrm{T}} \\ &\equiv 0, \end{aligned}$$

since, by Exercise 10.6.6, $\omega^{\mathrm{T}} = -\omega$. By the connectivity of V, it follows that AA^{T} is constant. □

Proof of Theorem 10.6.7. Write $A = [a^i_k]$, an orthogonal matrix of functions by Lemma 10.6.8. Define

$$\varphi = A \wedge \theta,$$

a column vector of 1-forms $\varphi^i = a^i_k \theta^k$. Since A is nonsingular and θ is the coframe to $X_1, \dots, X_n$, we see that the 1-forms φ^i are linearly independent pointwise on V. Furthermore,

$$\begin{aligned} d\varphi &= dA \wedge \theta + A \wedge d\theta \\ &= A \wedge \omega \wedge \theta + A \wedge (-\omega \wedge \theta) \\ &\equiv 0, \end{aligned}$$

where we use the fact that, the Levi-Civita connection being torsion-free, the first structure equation becomes

$$d\theta = -\omega \wedge \theta.$$

We lose no generality in taking V to be contractible, so the closed forms φ^i will be exact: $\varphi^i = dy^i$ for suitable $y^i \in C^\infty(V)$, $1 \le i \le n$. Since these forms are linearly independent at q, we can choose V smaller, if necessary, to guarantee that

$$(y^1, \dots, y^n) : V \to \mathbb{R}^n$$

is a diffeomorphism of V onto an open subset of $\mathbb{R}^n$. That is, $(V, y^1, \dots, y^n)$ is a coordinate chart and we will show that the associated coordinate frame field $(\xi_1, \dots, \xi_n)$ is everywhere orthonormal. Remark that

$$\varphi^i(X_j) = a^i_k \theta^k(X_j) = a^i_j,$$

implying that $X_j = a^i_j \xi_i$. That is,

$$(\xi_1, \dots, \xi_n)A = (X_1, \dots, X_n).$$

Since A is $\mathrm{O}(n)$-valued and $(X_1, \dots, X_n)$ is an orthonormal frame, it follows that $(\xi_1, \dots, \xi_n)$ is also orthonormal. □

Remark. This theorem is a special case of the fact that, in a certain precise sense, curvature determines the Riemannian geometry up to local isometry. We will not prove this more general result, but the reader will find an extensive treatment of these matters in [**40**, Chapter 7].

10.7. Riemannian Homogeneous Spaces*

This will be a very quick, introductory look at a large topic. We assume that $(M, \langle \cdot, \cdot \rangle)$ is a connected, Riemannian manifold. The group of all isometries $\varphi : M \to M$ will be denoted by $I(M)$. The action of $I(M)$ on M preserves all intrinsic properties. In particular, it preserves the Riemann distance function, hence is a group of metric space isometries. We note, without proof, the following classical result [**32**].

Theorem 10.7.1 (Myers and Steenrod). *If M is a Riemannian manifold, the group $I(M)$, with the compact-open topology, is isomorphic, as a topological group, to a Lie group such that the natural action,*

$$I(M) \times M \to M,$$

defined by $(\varphi, x) \mapsto \varphi(x)$, is smooth.

Definition 10.7.2. A Riemannian manifold M is homogeneous if the action $I(M) \times M \to M$ is transitive.

The fact that $I(M)$ is a Lie group implies that a homogeneous Riemannian manifold is of the form $M = I(M)/K$, where K is a closed subgroup, hence a properly imbedded Lie subgroup of $I(M)$. We will not use this fact.

Proposition 10.7.3. *If M is a homogeneous Riemannian manifold, then it is a complete Riemannian manifold.*

Proof. At any $x \in M$, let $B_x(\epsilon)$ be the diffeomorphic image, under $\exp_x$, of the open ϵ-ball in $T_x(M)$. For every $y \in M$, choose $\varphi^y_x \in I(M)$ such that $\varphi^y_x(x) = y$. Then $B_y(\epsilon) = \varphi^y_x(B_x(\epsilon))$ is the diffeomorphic image, under $\exp_y$, of the open ϵ-ball in $T_y(M)$. The point is that this ϵ is uniform for all points of M.

If $s : (a, b) \to M$ is a geodesic with maximal parameter interval, we must show that $b = \infty$. The same proof will give $a = -\infty$. We may assume that s is parametrized by arclength. If $b < \infty$, find $c \in (a, b)$ such that $b - c < \epsilon$. Then $B_{s(c)}(\epsilon)$ is as above, so there is a geodesic segment σ of length ϵ out of $s(c)$, having initial velocity $\dot{s}(c)$. Then σ must agree with s on $[c, b)$ and $|\sigma| = \epsilon > b - c$, so s extends to the interval $(a, c + \epsilon)$ where $c + \epsilon > b$, contradicting maximality. □

Definition 10.7.4. A (Riemannian) symmetric space M is a Riemannian manifold with the property that, for each $x \in M$, there is $\psi_x \in I(M)$ such $\psi_x(x) = x$ and $(\psi_x)_{*x} = -\operatorname{id}_{T_x(M)}$.

Remark that ψ_x reverses every geodesic s through x. That is,

$$s(0) = x \Rightarrow \psi_x(s(t) = s(-t).$$

As obvious examples, $\mathbb{R}^n$ with the Euclidean metric and S^n with its usual metric are both symmetric spaces. There are many other examples and the theory of symmetric spaces is highly developed (*cf.* [**14**]).

Proposition 10.7.5. *If M is a symmetric space, then M is a complete Riemannian manifold.*

Proof. Let $s : (a, b) \to M$ be a maximal geodesic. If $b < \infty$, choose $c \in (a, b)$ closer to b than to a. By reparametrizing, if necessary, we can assume that $c = 0$, hence that $b < -a$. Then, for $-b < t < -a$,

$$\sigma(t) = \psi_{s(0)}(s(-t))$$

is a geodesic that coincides with s on $(-b, b)$. These fit together to form a geodesic parametrized on $(a, -a)$, contradicting the maximality of b. Thus, $b = \infty$ and, similarly, $a = -\infty$. □

Corollary 10.7.6. *If M is a connected symmetric space, then M is a homogeneous Riemannian manifold.*

Proof. Indeed, if $x, y \in M$, completeness and connectedness allow us to find a (shortest) geodesic s joining them (Theorem 10.4.15). Parametrize this geodesic on $[-1, 1]$, $s(-1) = x$ and $s(1) = y$. Then $\psi_{s(0)} \in I(M)$ reverses this geodesic, hence carries x to y. □

Definition 10.7.7. Let $\langle \cdot, \cdot \rangle$ be a Riemannian metric on a Lie group G. This metric is left-invariant if every left translation in G is an isometry. It is right-invariant if all right translations are isometries. If the metric is both right- and left-invariant it is said to be bi-invariant.

Lemma 10.7.8. *Relative to a left-invariant (respectively, right-invariant) Riemannian metric, a Lie group is a homogeneous Riemannian manifold.*

This is clear, as is the existence of such metrics.

Lemma 10.7.9. *If $\langle \cdot, \cdot \rangle$ is a right-invariant metric on the Lie group G, if $Y, Z \in \mathfrak{X}(G)$, and if $X \in L(G)$, then*

$$\langle [X, Y], Z \rangle = -\langle Y, [X, Z] \rangle.$$

Proof. Since X is left-invariant, it generates the flow

$$\Phi_t(x) = x \cdot \exp(tX),$$

where $\exp(tX)$ is the one-parameter group of X. Thus,

$$\begin{aligned}\langle [X,Y], Z\rangle &= \lim_{t\to 0}\left\langle \frac{Y\cdot \exp(-tX) - Y}{t}, Z\right\rangle \\ &= \lim_{t\to 0}\frac{1}{t}\left(\langle Y\cdot\exp(-tX), Z\rangle - \langle Y, Z\rangle\right) \\ &= \lim_{t\to 0}\frac{1}{t}\left(\langle Y, Z\cdot \exp(tX)\rangle - \langle Y, Z\rangle\right) \\ &= -\langle Y, [X,Z]\rangle .\end{aligned}$$

□

Corollary 10.7.10. *If the Lie group G admits a bi-invariant metric and ∇ is the corresponding Levi-Civita connection, then*

$$\nabla_X X \equiv 0, \quad \forall X \in L(G).$$

Proof. Indeed, by Exercise 10.2.12, if $X, Y \in L(G)$, we get

$$\langle \nabla_X X, Y\rangle = \langle [Y,X], X\rangle = -\langle [X,Y], X\rangle = \langle Y, [X,X]\rangle \equiv 0.$$

□

Corollary 10.7.11. *The geodesics on G, relative to a bi-invariant metric, are the cosets (left or right) of the one-parameter subgroups.*

Proof. Indeed, due to the bi-invariance, we only need to show that the geodesics out of e are exactly the one-parameter subgroups. But $s(t) = \exp(tX)$ is integral to the left-invariant field X and $\nabla_X X \equiv 0$, so the one-parameter groups are geodesics. But these groups are in one-to-one correspondence with their initial velocity vectors X_e, as are the geodesics, so every geodesic out of e must be of this form. □

Corollary 10.7.12. *If the connected Lie group G has a bi-invariant metric, then every element of G can be reached from e by a one-parameter subgroup.*

This follows from the geodesic completeness of G by Theorem 10.4.15.

Example 10.7.13. This last corollary gives a necessary condition for a Lie group to admit a bi-invariant metric. Consider the connected Lie group $\mathrm{Sl}(2)$. If

$$A = \begin{bmatrix} a & b \\ c & d\end{bmatrix} \in \mathrm{Sl}(2),$$

then $\det(A) = 1$ and

$$\begin{aligned} ab + bd &= b\,\mathrm{tr}(A), \\ ac + dc &= c\,\mathrm{tr}(A), \\ a^2 + bc &= a^2 + ad - 1 \\ &= a\,\mathrm{tr}(A) - 1 \\ d^2 + bc &= d^2 + ad - 1, \\ &= d\,\mathrm{tr}(A) - 1. \end{aligned}$$

These give

$$A^2 = \mathrm{tr}(A)A - I,$$

hence

$$\mathrm{tr}(A^2) = \mathrm{tr}(A)^2 - 2, \quad \forall A \in \mathrm{Sl}(2).$$

Let

$$A = \begin{bmatrix} 1 & -1 \\ 5 & -4 \end{bmatrix} \in \mathrm{Sl}(2),$$

so $\operatorname{tr}(A) = -3$. If there is $E \in L(\mathrm{Sl}(2))$ such that $A = \exp(E) = e^E$, set $B = e^{E/2}$ and get $A = B^2$. This gives

$$-3 = \operatorname{tr}(B^2) = \operatorname{tr}(B)^2 - 2 \geq -2,$$

a clear contradiction. Thus, this element $A \in \mathrm{Sl}(2)$ cannot be reached from I by a one–parameter Lie group. The group Sl(2) cannot have a bi-invariant metric.

Proposition 10.7.14. *If G has a bi-invariant metric, it is a symmetric space relative to that metric.*

Proof. By homogeneity, it will only be necessary to produce ψ_x for $x = e$. For this, we take $\psi_e(y) = y^{-1}, \forall\, y \in G$. To see that this is an isometry, let $X, Y \in L(G)$, remark that $\psi_{e*}(X)$ and $\psi_{e*}(Y)$ are right-invariant fields, hence the constant $\langle X, Y\rangle$ and the constant $\langle \psi_{e*}(X), \psi_{e*}(Y)\rangle$ are both equal to $\langle X_e, Y_e\rangle = \langle -X_e, -Y_e\rangle$. □

Theorem 10.7.15. *Every compact Lie group has a bi-invariant metric.*

Proof. Choose a left invariant Riemannian metric $\langle \cdot, \cdot\rangle$ on G. The corresponding Riemann volume form Ω is left invariant also, so the Borel measure μ that it defines is left invariant. The corresponding integral satisfies

$$\int_G f\, d\mu = \int_G f \circ L_a\, d\mu, \quad \forall\, a \in G.$$

By multiplying the metric by a suitable positive constant, we normalize this integral:

$$\int_G 1\, d\mu = 1.$$

We define an inner product on $L(G)$ by

$$\langle X, Y\rangle' = \int_G \langle R_{x*}(X), R_{x*}(Y)\rangle\, d\mu(x).$$

It is trivial that this is a positive definite bilinear form. As an inner product on $L(G)$, it determines a left invariant Riemannian metric on G. But we claim that this is also a right-invariant metric. Indeed, for all $a \in G$ and $X, Y \in L(G)$,

$$\begin{aligned}
\langle R_{a*}(X), R_{a*}(Y)\rangle' &= \int_G \langle R_{x*}R_{a*}(X), R_{x*}R_{y*}(Y)\rangle\, d\mu(x) \\
&= \int_G \langle R_{ax*}(X), R_{ax*}(Y)\rangle\, d\mu(x) \\
&= \int_G \langle R_{x*}(X), R_{x*}(Y)\rangle\, d\mu(x) \quad \text{(left-invariance of } \mu) \\
&= \langle X, Y\rangle'.
\end{aligned}$$

□

Thus, the compact Lie groups provide an interesting set of examples of symmetric spaces.

Exercise 10.7.16. Let G be an n-dimensional, connected Lie group with a given bi-invariant Riemannian metric. Prove that G is flat if and only if G is abelian. (By Exercise 5.1.30, $G = T^k \times \mathbb{R}^{n-k}$.)

Exercise 10.7.17. Let G be a compact, connected Lie group, $\sigma : G \to G$ a Lie group automorphism such that $\sigma^2 = \text{id}$, and let

$$K = \{x \in G \mid \sigma(x) = x\}.$$

(1) Prove that K is a compact Lie subgroup of G.
(2) Find a bi-invariant Riemannian metric g on G relative to which σ is an isometry.
(3) Show that g passes to a Riemannian metric on the homogeneous space G/K, making that manifold a symmetric space.
(4) Show that the map $\varphi : G \to G$ defined by $\varphi(x) = x\sigma(x^{-1})$ passes to a smooth imbedding $\overline{\varphi} : G/K \hookrightarrow G$.
(5) Write $L(G) = L(K) \oplus \mathfrak{M}$, where $\mathfrak{M} \perp L(K)$ relative to the metric g. Prove that $\exp(\mathfrak{M}) = \varphi(G/K)$.
(6) Prove that φ carries each geodesic in G/K onto a geodesic in G, exactly doubling arclength and preserving angles.
(7) Conclude that the properly imbedded submanifold $M = \exp(\mathfrak{M})$ of G, under the metric $g|T(M)$, is itself a symmetric space and a *totally geodesic* submanifold of G (*i.e.*, the geodesics of M are also geodesics in G).

CHAPTER 11

Principal Bundles*

A good command of the theory of principal bundles is essential for mastery of modern differential geometry (*cf.* [**22**], [**23**]). In recent years, principal bundles have also become central to key advances in mathematical physics (Yang–Mills theory) which have in turn generated exciting new mathematics (*e.g.*, S. K. Donaldson's work on differentiable 4-dimensional manifolds [**7**], [**11**]). This chapter will be a brief introduction to principal bundles.

11.1. The Frame Bundle

Before defining the term "principal bundle", we discuss the central motivating example.

Let V be a real vector space of dimension n. An *n-frame* in V is an ordered basis $\mathfrak{v} = (v_1, \dots, v_n)$. If $V = \mathbb{R}^n$, each v_i is a column vector and the frame $\mathfrak{v}$ is a nonsingular matrix. That is, the set of all n-frames in $\mathbb{R}^n$ is naturally identified with the manifold $\mathrm{Gl}(n)$. Generally, we denote the set of all n-frames in V by $F(V)$ and topologize this set as a subset of $V^n = V \times \cdots \times V$. Let $\varphi : V \to \mathbb{R}^n$ be an isomorphism of vector spaces and define a diffeomorphism $\overline{\varphi} : V^n \to \mathfrak{M}(n)$ by

$$\overline{\varphi}(v_1, \dots, v_n) = (\varphi(v_1), \dots, \varphi(v_n)).$$

Clearly $\overline{\varphi}(F(V)) = \mathrm{Gl}(n)$, hence $F(V) \subset V^n$ is an open subset, diffeomorphic via $\overline{\varphi}$ to $\mathrm{Gl}(n)$.

Definition 11.1.1. The smooth manifold $F(V)$ is called the frame manifold of V.

One might try to make $F(V)$ into a Lie group via $\overline{\varphi}$, but this is a bad idea since, if φ_1 and φ_2 are two isomorphisms of V to $\mathbb{R}^n$, it is not generally true that the diffeomorphism $\overline{\varphi}_1 \circ (\overline{\varphi}_2)^{-1} : \mathrm{Gl}(n) \to \mathrm{Gl}(n)$ is a group isomorphism. That is, there is no *canonical* way to make $F(V)$ into a Lie group.

There is, however, a natural right action

$$F(V) \times \mathrm{Gl}(n) \to F(V)$$

defined by formal matrix multiplication

$$(v_1, \dots, v_n) \cdot \begin{bmatrix} a_{11} & \cdots & a_{1n} \\ \vdots & & \vdots \\ a_{n1} & \cdots & a_{nn} \end{bmatrix} = \left(\sum_{i=1}^n a_{i1} v_i, \dots, \sum_{i=1}^n a_{in} v_i \right).$$

This is a smooth, transitive action with trivial isotropy group (such an action is said to be *simply transitive*). Remark that the right action of $\mathrm{Gl}(n)$ on itself, defined by the group multiplication, is also smooth and simply transitive and that $\overline{\varphi} : F(V) \to \mathrm{Gl}(n)$ respects these actions in the following sense.

Lemma 11.1.2. *If* $\varphi : V \to \mathbb{R}^n$ *is a linear isomorphism and if* $\mathfrak{v} \in F(V)$, $A \in \mathrm{Gl}(n)$, *then*

$$\overline{\varphi}(\mathfrak{v} \cdot A) = \overline{\varphi}(\mathfrak{v})A.$$

One says that the map $\overline{\varphi}$ is $\mathrm{Gl}(n)$-*equivariant* or, more simply, *equivariant*.

There is a natural map $F(V) \times \mathbb{R}^n \to V$ defined by formal matrix multiplication

$$(\mathfrak{v}, \vec{a}) \mapsto \mathfrak{v} \cdot \vec{a} = (v_1, \dots, v_n) \cdot \begin{bmatrix} a^1 \\ \vdots \\ a^n \end{bmatrix} = \sum_{i=1}^{n} a^i v_i. \tag{11.1}$$

For each fixed $\mathfrak{v} \in F(V)$, this defines a linear isomorphism $\mathfrak{v} : \mathbb{R}^n \to V$, and every linear isomorphism is of this form. We summarize.

Lemma 11.1.3. *The map defined by* (11.1) *sets up a canonical identification of* $F(V)$ *with the set of all linear isomorphisms* $\mathfrak{v} : \mathbb{R}^n \to V$.

One can canonically reconstruct the vector space V from the frame manifold $F(V)$ as follows. Define the left action of $\mathrm{Gl}(n)$ on the space $F(V) \times \mathbb{R}^n$ to be the "diagonal action"

$$A \cdot (\mathfrak{v}, \vec{a}) = (\mathfrak{v} \cdot A^{-1}, A\vec{a}), \tag{11.2}$$

where A ranges over $\mathrm{Gl}(n)$ and $(\mathfrak{v}, \vec{a})$ ranges over $F(V) \times \mathbb{R}^n$. It is elementary that this is a smooth, left action of the group $\mathrm{Gl}(n)$, so there is an associated equivalence relation on $F(V) \times \mathbb{R}^n$ with the $\mathrm{Gl}(n)$-orbits as equivalence classes. We denote the quotient set by $F(V) \times_{\mathrm{Gl}(n)} \mathbb{R}^n$. The equivalence class of $(\mathfrak{v}, \vec{a})$ will be denoted by $[\mathfrak{v}, \vec{a}]$. Remark that

$$[\mathfrak{v} \cdot A, \vec{a}] = [\mathfrak{v}, A\vec{a}], \quad \forall A \in \mathrm{Gl}(n). \tag{11.3}$$

We put a vector space structure on the quotient set $F(V) \times_{\mathrm{Gl}(n)} \mathbb{R}^n$ by fixing a frame $\mathfrak{v} \in F(V)$ and defining

$$\begin{aligned} [\mathfrak{v}, \vec{a}_1] + [\mathfrak{v}, \vec{a}_2] &= [\mathfrak{v}, \vec{a}_1 + \vec{a}_2], \\ c[\mathfrak{v}, \vec{a}] &= [\mathfrak{v}, c\vec{a}]. \end{aligned}$$

To see that these definitions do not depend on the choice of frame, let $\mathfrak{u}$ be another choice, let $A \in \mathrm{Gl}(n)$ be the unique matrix such that $\mathfrak{u} = \mathfrak{v} \cdot A$, and use the relation (11.3) to obtain

$$\begin{aligned} [\mathfrak{u}, \vec{a}_1] + [\mathfrak{u}, \vec{a}_2] &= [\mathfrak{v}, A\vec{a}_1] + [\mathfrak{v}, A\vec{a}_2] \\ &= [\mathfrak{v}, A(\vec{a}_1 + \vec{a}_2)] \\ &= [\mathfrak{u}, \vec{a}_1 + \vec{a}_2], \\ c[\mathfrak{u}, \vec{a}] &= c[\mathfrak{v}, A\vec{a}] \\ &= [\mathfrak{v}, Ac\vec{a}] \\ &= [\mathfrak{u}, c\vec{a}]. \end{aligned}$$

The map defined by (11.1) passes to a well-defined linear map

$$\psi : F(V) \times_{\mathrm{Gl}(n)} \mathbb{R}^n \to V.$$

Fixing a frame $\mathfrak{v} \in F(V)$, we define

$$j : V \to F(V) \times_{\mathrm{Gl}(n)} \mathbb{R}^n$$

by

$$j(\mathfrak{v}\cdot\vec{a}) = [\mathfrak{v},\vec{a}].$$

It is clear that ψ and j are mutual inverses, hence that j does not really depend on the choice of frame $\mathfrak{v}$. We summarize these remarks in the following.

Lemma 11.1.4. *The vector space operations on $F(V) \times_{\mathrm{Gl}(n)} \mathbb{R}^n$ are well defined, independently of the choice of the frame $\mathfrak{v}$, and the map ψ is a canonical isomorphism of vector spaces, the inverse j being well defined independently of the choice of frame.*

These remarks generalize to vector bundles fiberwise. Let $\pi : E \to M$ be an n-plane bundle over an m-manifold. Each fiber $E_x = \pi^{-1}(x)$ is an n-dimensional real vector space, so we form the associated frame manifold $F(E_x)$ and the disjoint union

$$F(E) = \bigsqcup_{x\in M} F(E_x).$$

We also define $p : F(E) \to M$ by $p(F(E_x)) = x$, $\forall\, x \in M$. If

$$\begin{array}{ccc} \pi^{-1}(U) & \xrightarrow{\varphi} & U \times \mathbb{R}^n \\ {\scriptstyle\pi}\downarrow & & \downarrow{\scriptstyle p_1} \\ U & \xrightarrow[\mathrm{id}]{} & U \end{array}$$

is a local trivialization of E, then $\varphi_x : E_x \to \mathbb{R}^n$ is a linear isomorphism, $\forall\, x \in U$, inducing a diffeomorphism

$$\overline{\varphi}_x : F(E_x) \to \mathrm{Gl}(n).$$

This determines a commutative diagram

$$\begin{array}{ccc} p^{-1}(U) & \xrightarrow{\overline{\varphi}} & U \times \mathrm{Gl}(n) \\ {\scriptstyle p}\downarrow & & \downarrow{\scriptstyle p_1} \\ U & \xrightarrow[\mathrm{id}]{} & U \end{array}$$

where $\overline{\varphi}$ is bijective, hence defines a topology and smooth structure on $p^{-1}(U)$. It is straightforward to check that, on overlaps $p^{-1}(U) \cap p^{-1}(V)$, corresponding to two local trivializations of E, the two smooth structures and underlying topologies coincide and that this makes $F(E)$ into a smooth manifold of dimension $m + n^2$. Furthermore, $p : F(E) \to M$ is a smooth submersion and the inclusion map, $i_x : F(E_x) \hookrightarrow F(E)$ smoothly imbeds the fiber $p^{-1}(x)$ as a proper submanifold, $\forall\, x \in M$.

We have defined a new kind of bundle over M, called the *frame bundle* of E. The fibers $F(E_x)$ are not vector spaces. Instead, they are manifolds diffeomorphic to $\mathrm{Gl}(n)$ that admit a canonical right action of $\mathrm{Gl}(n)$. This defines a right action

$$F(E) \times \mathrm{Gl}(n) \to F(E)$$

that is simply transitive on each fiber. By Lemma 11.1.2, the local trivializations turn this action into the action

$$(U \times \mathrm{Gl}(n)) \times \mathrm{Gl}(n) \to U \times \mathrm{Gl}(n)$$

defined by

$$(x, B) \cdot A = (x, BA).$$

This is clearly a smooth action, so

$$F(E) \times \mathrm{Gl}(n) \to F(E)$$

is locally, hence globally, a smooth action.

By Lemma 11.1.3, the fiber $F(E_x)$ over $x \in M$ of the frame bundle of E is exactly the set of linear isomorphisms of the "standard fiber" $\mathbb{R}^n$ to the specific fiber E_x.

The construction for recovering a vector space V from its frame manifold $F(V)$ globalizes in a natural fashion to give a canonical way of recovering the vector bundle E from its associated frame bundle $F(E)$. The diagonal action of $\mathrm{Gl}(n)$ given by (11.2) extends by the same formula to a left action

$$\mathrm{Gl}(n) \times (F(E) \times \mathbb{R}^n) \to F(E) \times \mathbb{R}^n$$

and we let $F(E) \times_{\mathrm{Gl}(n)} \mathbb{R}^n$ denote the corresponding quotient space (with the quotient topology). The bundle projection $p : F(E) \to M$ passes to a well-defined map

$$\overline{p} : F(E) \times_{\mathrm{Gl}(n)} \mathbb{R}^n \to M$$

and the map

$$F(E) \times \mathbb{R}^n \to E,$$

again defined as in (11.1), passes to a well-defined map

$$\psi : F(E) \times_{\mathrm{Gl}(n)} \mathbb{R}^n \to E.$$

These maps are smooth and the diagram

$$\begin{array}{ccc} F(E) \times_{\mathrm{Gl}(n)} \mathbb{R}^n & \xrightarrow{\psi} & E \\ {\scriptstyle \overline{p}}\downarrow & & \downarrow{\scriptstyle \pi} \\ M & \xrightarrow[\mathrm{id}]{} & M \end{array}$$

commutes. It has also been arranged, by the very definitions, that ψ restricts to ψ_x on the vector space

$$\overline{p}^{-1}(x) = F(E_x) \times_{\mathrm{Gl}(n)} \mathbb{R}^n \subset F(E) \times_{\mathrm{Gl}(n)} \mathbb{R}^n,$$

carrying it isomorphically onto the vector space E_x, $\forall x \in M$. The inverse j_x of ψ_x, although defined by a choice of frame $\mathfrak{v} \in F(E_x)$, is independent of that choice (Lemma 11.1.4), $\forall x \in M$, and these fit together to give a global inverse

$$j : E \to F(E) \times_{\mathrm{Gl}(n)} \mathbb{R}^n.$$

In a local trivialization of E, one can choose n linearly independent smooth sections, using these to define j. It follows that j is locally, hence globally, smooth and ψ is a diffeomorphism. We summarize this discussion in a theorem.

Theorem 11.1.5. *The structure $\overline{p} : F(E) \times_{\mathrm{Gl}(n)} \mathbb{R}^n \to M$ is a vector bundle, canonically isomorphic to $\pi : E \to M$. The association $E \leftrightarrow F(E)$ is a one-to-one correspondence between the set of vector bundles over M and the set of associated frame bundles.*

Exercise 11.1.6. Let $\rho : \mathrm{Gl}(n) \times \mathbb{R}^n \to \mathbb{R}^n$ be the smooth action defined by

$$\rho(A, \vec{a}) = (A^{-1})^{\mathrm{T}} \vec{a}.$$

Given an n-plane bundle E over M, define $E^* = F(E) \times_\rho \mathbb{R}^n$ by analogy with the definition of $F(E) \times_{\mathrm{Gl}(n)} \mathbb{R}^n$ and prove that E^* is an n-plane bundle over M with each fiber E^*_x canonically isomorphic to the dual space of E_x. Again, it will be helpful to consider first the case $F(V) \times_\rho \mathbb{R}^n = V^*$, then the case of a product bundle, and finally the general case. This is exactly the dual bundle E^* introduced in Section 6.1

11.2. Principal G-Bundles

The frame bundle $F(E)$ is an example of a *principal G-bundle* where $G = \mathrm{Gl}(n)$. We will give the general definition after the following.

Definition 11.2.1. Let M and N be manifolds, together with smooth actions from the right

$$M \times G \to M,$$
$$N \times G \to N,$$

each written as $(x, g) \mapsto x \cdot g$. A smooth map $f : M \to N$ is *G-equivariant* (or *equivariant*, if the G-actions are understood from the context) if

$$f(x \cdot g) = f(x) \cdot g, \quad \forall x \in M, \quad \forall g \in G.$$

Definition 11.2.2. Let M and P be smooth manifolds, G a Lie group, and $p : P \to M$ a smooth map. Suppose that there is an open cover $\{U_\alpha\}_{\alpha \in \mathfrak{A}}$ of M and, $\forall \alpha \in \mathfrak{A}$, a commutative diagram

$$\begin{array}{ccc} p^{-1}(U_\alpha) & \xrightarrow{\psi_\alpha} & U_\alpha \times G \\ {\scriptstyle p}\downarrow & & \downarrow{\scriptstyle p_1} \\ U_\alpha & \xrightarrow[\mathrm{id}]{} & U_\alpha \end{array}$$

where ψ_α is a diffeomorphism. Suppose also that there is a smooth right action $P \times G \to P$, simply transitive on each fiber $p^{-1}(x)$. Finally, assume that ψ_α is equivariant with respect to this action and the right G-action

$$(U_\alpha \times G) \times G \to U_\alpha \times G,$$
$$((x, g), h) \mapsto (x, gh),$$

$\forall \alpha \in \mathfrak{A}$. Then, $p : P \to M$, together with this G-action, is called a principal G-bundle over M. The group G is called the *structure group* of the bundle.

Example 11.2.3. Let $p : P \to M$ be a principal G-bundle, $\sigma : M \to P$ a smooth map such that $p \circ \sigma = \mathrm{id}_M$. As in the case of vector bundles, σ is called a *section* of the principal bundle. While a vector bundle always admits sections (*e.g.*, the 0 section) this generally fails for principal bundles. Indeed, if σ is a section, define

$$\varphi : M \times G \to P$$

by

$$\varphi(x, g) = \sigma(x) \cdot g,$$

obtaining a diffeomorphism such that the diagram

$$\begin{array}{ccc} M \times G & \xrightarrow{\varphi} & P \\ {\scriptstyle p_1}\downarrow & & \downarrow{\scriptstyle p} \\ M & \xrightarrow[\text{id}]{} & M \end{array}$$

commutes. As in the case of vector bundles, φ is called a *trivialization* of the principal bundle P. Thus, the existence of a section is equivalent to triviality of a principal G-bundle.

Example 11.2.4. Suppose that the n-plane bundle $\pi : E \to M$ is given an explicit $\mathrm{O}(n)$-reduction. Equivalently, there is a positive definite inner product $\langle \cdot , \cdot \rangle_x$ on E_x that varies smoothly with $x \in M$. That is, $\langle s_1(x), s_2(x) \rangle_x$ is a smooth function of x, $\forall\, s_1, s_2 \in \Gamma(E)$ (*cf.* Exercise 3.4.16 for the case $E = T(M)$). In fact, this smoothly varying inner product can be thought of as a smooth "field" of inner products, often called a Riemannian metric on the bundle E. One can then define $\mathrm{O}(E) \subset F(E)$ to be the subset of frames that are orthonormal with respect to this smooth field of inner products and restrict p to a projection $p : \mathrm{O}(E) \to M$. If $U \subseteq M$ is an open, locally trivializing neighborhood for E, let $(s_1, \dots, s_n)$ be a smooth frame field on U (*i.e.*, a smooth section of $F(E|U)$). By an application of the Gram–Schmidt process to this frame field, we can assume that it is everywhere orthonormal, hence is a smooth section σ of $\mathrm{O}(E|U)$. All of $\mathrm{O}(E|U)$ can be swept out by applying right actions of $\mathrm{O}(n)$ to σ and one obtains a local trivialization

$$\begin{array}{ccc} U \times \mathrm{O}(n) & \xrightarrow{\theta} & \mathrm{O}(E|U) \\ {\scriptstyle p}\downarrow & & \downarrow{\scriptstyle p_1} \\ U & \xrightarrow[\text{id}]{} & U \end{array}$$

by setting $\theta(x, A) = \sigma(x) \cdot A$, $\forall\, x \in U$, $\forall\, A \in \mathrm{O}(n)$. Thus, $\mathrm{O}(E) \subset F(E)$ can be thought of as a subbundle that is invariant under the right action of $\mathrm{O}(n) \subset \mathrm{Gl}(n)$. In fact, $\mathrm{O}(n)$ is simply transitive on the fibers and $\mathrm{O}(E)$ is a principal $\mathrm{O}(n)$-bundle. The standard application of partitions of unity shows that there are infinitely many choices of Riemannian metrics on E and corresponding orthonormal frame bundles.

Example 11.2.5. Every n-plane bundle $\pi : E \to M$ admits $\mathrm{O}(n)$–reductions, but we know that it may or may not admit a $\mathrm{Gl}(k, n-k)$-reduction. In fact, such a reduction is equivalently a k-plane subbundle $E_k \subseteq E$ (Example 3.4.20). Given such a subbundle, let $F_k(E) \subseteq F(E)$ consist of the frames of E whose first k entries form a frame of E_k. As in the previous example, this forms a locally trivial subbundle of $F(E)$ that is invariant under the right action of $\mathrm{Gl}(k, n-k)$, a simply transitive action on each fiber. This is a principal $\mathrm{Gl}(k, n-k)$-bundle.

Example 11.2.6. An $\mathrm{O}(n, n-k)$-reduction of E corresponds to an *indefinite* inner product $\langle \cdot , \cdot \rangle_x$ on E_x (*cf.* Example 3.4.19) that varies smoothly with $x \in M$. Such a reduction may or may not exist but, if it does, one obtains a principal $\mathrm{O}(k, n-k)$-bundle of frames that are "orthonormal" with respect to this indefinite metric. The reader can supply the details.

Example 11.2.7. Let G be a Lie group, $H \subseteq G$ a closed subgroup. Then the quotient projection $p : G \to G/H$ is a principal H-bundle over the homogeneous space

G/H. Indeed, the local triviality is by Exercise 5.4.8 and the required property of the right H-action is obvious.

Example 11.2.8. The case of a principal bundle with *discrete* structure group G is noteworthy. A discrete group is simply an abstract group with the discrete topology (each point is an open set). This is a topological group and, if G is at most countably infinite, it can also be viewed as a Lie group of dimension 0. A principal G-bundle $p : P \to M$ is exactly a *regular covering space* (Definition 1.7.13) if G is a discrete group. Indeed, the local triviality of the bundle guarantees that each point $x \in M$ has an evenly covered neighborhood $U \subseteq M$. It can be seen that the group of homeomorphisms $g : P \to P$ such that $p \circ g = p$ is exactly the group G, acting (from the right) on the total space P of the bundle. This is the group of covering transformations and, by the definition of principal bundles, it permutes the fiber simply transitively.

We will want an appropriate definition of isomorphism for principal G–bundles over M. As in the case of vector bundles, it is useful to give an apparently weaker definition than one really wants and then prove that the stronger property holds.

Definition 11.2.9. If $p : P \to M$ and $p' : P' \to M$ are principle G-bundles, an isomorphism $\varphi : P \to P'$ is a smooth, G-equivariant map such that the diagram

$$\begin{array}{ccc} P & \xrightarrow{\varphi} & P' \\ {\scriptstyle p}\downarrow & & \downarrow{\scriptstyle p'} \\ M & \xrightarrow[\text{id}]{} & M \end{array}$$

commutes.

Exercise 11.2.10. If $\varphi : P \to P'$ is an isomorphism of principal G–bundles, prove that φ is a diffeomorphism and that φ^{-1} is also an isomorphism. Thus, isomorphism is an equivalence relation.

Associated to a principal G-bundle over M and any smooth left action of G on a manifold F, there is a smooth bundle over M with fiber F. This associated bundle is constructed using exactly the same technique whereby we recovered a vector bundle E from its associated frame bundle.

Definition 11.2.11. A (locally trivial) bundle over M with fiber a manifold F is a smooth map $\pi : E \to M$ with the following property. For each $x \in M$, there is an open neighborhood U of x in M and a commutative diagram

$$\begin{array}{ccc} \pi^{-1}(U) & \xrightarrow{\varphi} & U \times F \\ {\scriptstyle \pi}\downarrow & & \downarrow{\scriptstyle p_1} \\ U & \xrightarrow[\text{id}]{} & U \end{array}$$

in which φ is a diffeomorphism and p_1 is projection onto the first factor.

Exercise 11.2.12. Let $p : P \to M$ be a principal G-bundle and let F be a manifold. If $\rho : G \times F \to F$ is a smooth action, define the "diagonal" action of G on $P \times F$ in analogy with the definition you gave in Exercise 11.1.6 and denote the quotient space by $P \times_\rho F$. Show that the bundle projection p passes to a well-defined map $\pi : P \times_\rho F \to M$, and that this is a smooth, locally trivial bundle with fiber F. As usual, it will be helpful to consider first the case in which M reduces to a single point (*i.e.*, exhibit a canonical identification $G \times_\rho F = F$), then the case in which $P = M \times G$ is a trivial bundle, and finally the general case.

Definition 11.2.13. The locally trivial F-bundle constructed in Exercise 11.2.12 is called the F-bundle associated to the principal G-bundle by the group action ρ.

Normally, when the action and the principal bundle are understood from context, one refers simply to the "associated F-bundle".

Exercise 11.2.14. If $p : P \to M$ and $p' : P' \to M$ are isomorphic principal G–bundles over M and $\rho : G \times F \to F$ is a smooth action, prove that the associated bundles fit into a commutative diagram

$$\begin{array}{ccc} P \times_\rho F & \xrightarrow{f} & P' \times_\rho F \\ {\scriptstyle \pi}\downarrow & & \downarrow{\scriptstyle \pi'} \\ M & \xrightarrow[\text{id}]{} & M \end{array}$$

in which f is a diffeomorphism. One says that f is an isomorphism of these locally trivial F-bundles.

Exercise 11.2.15. If E is the associated F-bundle as above, show how each element $g \in P$ can be interpreted as an inclusion map $g : F \hookrightarrow E$ of the fiber over $p(g)$.

Exercise 11.2.16. Given a vector bundle E over M, show how to obtain the various associated tensor bundles, constructed in Section 7.4, as locally trivial bundles associated to the frame bundle $F(E)$.

11.3. Cocycles and Reductions

The notion of a G-cocycle γ was defined in Section 3.4 (Definition 3.4.10). There we considered subgroups $G \subseteq \mathrm{Gl}(n)$, but the definition works equally well for arbitrary Lie groups G. The notion of equivalence of $\mathrm{Gl}(n)$-cocycles (Definition 3.4.3) extends without change to G-cocycles. As in the comment following Definition 3.4.10, the set of equivalence classes $[\gamma]$ of G-cocycles is denoted by $H^1(M; G)$, the cohomology notation being motivated by analogies with Čech cohomology.

Given a principal G-bundle, a family of local trivializations

$$\begin{array}{ccc} p^{-1}(U_\alpha) & \xrightarrow{\psi_\alpha} & U_\alpha \times G \\ {\scriptstyle p}\downarrow & & \downarrow{\scriptstyle p_1} \\ U_\alpha & \xrightarrow[\text{id}]{} & U_\alpha, \end{array}$$

where $\{U_\alpha\}_{\alpha\in\mathfrak{A}}$ covers M, gives rise to a G-cocycle $\gamma = \{\gamma_{\alpha\beta}\}_{\alpha,\beta\in\mathfrak{A}}$ in a fairly obvious way. Indeed, over each nonempty intersection $U_\alpha \cap U_\beta$, the trivializations give a transition function $\psi_\alpha \circ \psi_\beta^{-1}$ of the form

$$\psi_\alpha \circ \psi_\beta^{-1}(x, g) = (x, \psi_{\alpha\beta}(x, g)).$$

If we set $\gamma_{\alpha\beta}(x) = \psi_{\alpha\beta}(x, e)$, then the G-equivariance of ψ_α and ψ_β implies that

$$\psi_\alpha \circ \psi_\beta^{-1}(x, g) = (x, \gamma_{\alpha\beta}(x)g).$$

The cocycle property for $\gamma = \{\gamma_{\alpha\beta}\}_{\alpha,\beta\in\mathfrak{A}}$ is obvious, as is the fact that the principal G-bundle $p : P \to M$ determines γ up to equivalence.

Exercise 11.3.1. Mimic the procedure in Section 3.4 for constructing n-plane bundles from $\mathrm{Gl}(n)$ to show how to construct a principal G-bundle from a G-cocycle. Show that the set of isomorphism classes of such bundles over M is canonically identified with $H^1(M;G)$.

By a common abuse of terminology, we often refer to $H^1(M;G)$ simply as the set of principal G-bundles over M.

If $H \subseteq G$ is a properly imbedded Lie subgroup, we can view H-cocycles on M as G-cocycles. If two such cocycles are equivalent as H-cocycles, they are also equivalent as G-cocycles, so we obtain a natural map of sets

$$\lambda : H^1(M;H) \to H^1(M;G).$$

It is possible that two H-cocycles not be equivalent, but become equivalent when viewed as G-cocycles, and so λ is not generally injective. It is not generally surjective either, since there is no reason *a priori* that a G-cocycle should be equivalent to some H-cocycle.

Definition 11.3.2. If $[\gamma] \in H^1(M;G)$, $[\eta] \in H^1(M;H)$ and $[\gamma] = \lambda[\eta]$, then the principal H-bundle $[\eta]$ is said to be an H-reduction of the principal G-bundle $[\gamma]$.

The following exercise lays bare the geometric meaning of this definition.

Exercise 11.3.3. Let $[\gamma] = \lambda[\eta]$ as in the above definition and let

$$q : Q \to M$$
$$p : P \to M$$

be, respectively, the principal H-bundle with cocycle η and the principal G-bundle with cocycle γ. Let $\rho : H \times G \to G$ be the canonical action of H on G by left multiplication and build the associated G-bundle

$$\pi : Q \times_\rho G \to M,$$

as in Exercise 11.2.12. Note that G has a natural right action on the total space $Q \times_\rho G$ of this bundle and prove that, equipped with this action, the associated G-bundle is a principal G-bundle isomorphic to $p : P \to M$. Use this to produce a commutative diagram

$$\begin{array}{ccc} Q & \xrightarrow{\;i\;} & P \\ {\scriptstyle q}\downarrow & & \downarrow{\scriptstyle p} \\ M & \xrightarrow[\mathrm{id}]{} & M \end{array}$$

in which i is a proper imbedding of Q as a submanifold of P that is invariant under the right action of H. Conversely, given such an imbedding of a principal H-bundle, show that the class $[\gamma]$ has the form $\lambda[\eta]$.

The infinitesimal G-structures of Section 3.4 (see Definition 3.4.12) are now seen to be G-reductions of the frame bundle $F(T(M))$.

Example 11.3.4. In Example 11.2.4, we saw that a Riemannian metric on an n-plane bundle E led to an $\mathrm{O}(n)$-reduction $\mathrm{O}(E) \subset F(E)$ exactly as in Exercise 11.3.3. Conversely, suppose that $Q \subset F(E)$ is an $\mathrm{O}(n)$-reduction as in that exercise. We will recover canonically a Riemannian metric on E giving this reduction. Indeed, let $x \in M$ and $\mathfrak{v} \in Q_x$, write $\mathfrak{v} = (v_1, \ldots, v_n)$ as a frame, and define a positive definite inner product $\langle \cdot, \cdot \rangle_{\mathfrak{v}}$ on E_x by requiring that

$$\langle v_i, v_j \rangle_{\mathfrak{v}} = \delta_{ij}, \quad 1 \le i, j \le n.$$

If $\mathfrak{v}' = (v_1', \ldots, v_n') \in Q_x$ is another choice, there is unique $A \in \mathrm{O}(n)$ such that $\mathfrak{v}' = \mathfrak{v} \cdot A$ and it is a routine computation to check that

$$\langle v_i', v_j' \rangle_{\mathfrak{v}} = \delta_{ij}, \quad 1 \le i, j \le n.$$

This shows independence of the choice of frame. In order to see that this inner product varies smoothly in a neighborhood of x, let $U \subset M$ be such a neighborhood, small enough that there is a smooth section $\sigma : U \to Q$. Then, for arbitrary $v, w \in \Gamma(E|U)$, the expression

$$\langle v(y), w(y) \rangle_{\sigma(y)}$$

depends smoothly on y and the assertion follows. The converse of Example 11.2.6 can be verified analogously.

Exercise 11.3.5. Check the converse of Example 11.2.5. That is, given a $\mathrm{Gl}(k, n-k)$-reduction, produce the k-plane subbundle of E giving rise to that reduction.

Exercise 11.3.6. With the hypotheses and notations of Exercise 11.3.3, assume further that

$$\mu : G \times F \to F$$

is a smooth action on the manifold F. Let μ denote also the restricted action

$$\mu : H \times F \to F$$

and exhibit an isomorphism

$$f : Q \times_\mu F \to P \times_\mu F$$

of associated bundles.

Example 11.3.7. If the n-plane bundle admits a (pseudo-) Riemannian metric, let $\mathrm{O}(E) \subset F(E)$ denote the orthonormal frame bundle. Let O stand for the respective groups $\mathrm{O}(n)$ or $\mathrm{O}(k, n-k)$ according to whether the metric is Riemannian or pseudo-Riemannian. Let

$$\rho : \mathrm{O} \times \mathbb{R}^n \to \mathbb{R}^n$$

be the standard left action. By Exercise 11.3.6, we then see that

$$\mathrm{O}(E) \times_\rho \mathbb{R}^n = F(E) \times_{\mathrm{Gl}(n)} \mathbb{R}^n = E,$$

canonically. This shows that E can be assembled using an O-cocycle. Generally, Exercise 11.3.6 shows that H-reductions of principal G-bundles allow us to assemble the associated bundles using an H-cocycle. This is the origin of the terminology H-reduction.

11.4. Frame Bundles and the Equations of Structure

In Section 10.6, we showed that a connection ∇ on an n-manifold M, together with a choice of frame field $\sigma = (X_1, \dots, X_n)$ on a trivializing neighborhood $U \subset M$ for $T(M)$, gives rise to a pair of equations relating certain differential forms (equations (10.9) and (10.10)) called the Cartan structure equations. These equations depend on the choice of local section σ of $F(M) = F(T(M))$, but it turns out that they can be lifted to the total space $F(M)$ to be globally defined independently of choices. The pullback by σ of these globally defined equations gives back the local equations. To emphasize the provisional nature of the local structure equations, we will write them with tildas over the forms as follows

$$d\widetilde{\theta} = -\widetilde{\omega} \wedge \widetilde{\theta} + \widetilde{\tau}, \tag{11.4}$$

$$d\widetilde{\omega} = -\widetilde{\omega} \wedge \widetilde{\omega} + \widetilde{\Omega}, \tag{11.5}$$

reserving the forms without tildas to denote the canonical global forms to be produced on the manifold $F(M)$.

If ∇ is the Levi-Civita connection of a Riemannian metric, the structure equations can be lifted to the total space $\mathrm{O}(M) = \mathrm{O}(T(M))$ of the reduced orthonormal frame bundle. In the case of a pseudo-Riemannian metric, there is also a unique Levi-Civita connection ∇ (just mimic Definition 10.2.10 and Exercise 10.2.12) and the structure equations will again lift globally to the total space $\mathrm{O}_k(M)$ of the orthonormal frame bundle. We will be particularly interested in both of these cases and so, in what follows, $p : P \to M$ will denote either the full frame bundle or the orthonormal frame bundle associated either to a Riemannian or pseudo-Riemannian metric. Similarly, G will denote any one of $\mathrm{Gl}(n)$, $\mathrm{O}(n)$ or $\mathrm{O}(k, n-k)$, as appropriate.

Our main application of this theory will be to prove the following basic result characterizing flatness of Riemannian and pseudo-Riemannian manifolds. (*cf.* Theorem 10.6.7).

Theorem 11.4.1. *Let M be a (pseudo-) Riemannian manifold, ∇ the Levi-Civita connection, and R the curvature tensor of ∇. The following are equivalent.*

(1) $R \equiv 0$.
(2) *There is a smooth atlas in which the metric coefficients are everywhere $g_{ij} \equiv \pm\delta_{ij}$, the negative sign occurring exactly for*
$$k + 1 \leq i = j \leq n.$$
(3) *There is a smooth atlas in which the Christoffel symbols are everywhere $\Gamma^k_{ij} \equiv 0$.*
(4) *Each $x \in M$ has a neighborhood U_x such that the holonomy of ∇ around each loop $\sigma \in \Omega(U_x, x)$ is the identity transformation.*
(5) *The Riemannian (respectively, pseudo-Riemannian) metric, as an infinitesimal $\mathrm{O}(n)$-structure (respectively, $\mathrm{O}(k, n-k)$-structure), is integrable.*

A (pseudo-) Riemannian manifold in which one, hence all, of these holds is said to be flat.

Exercise 11.4.2. As a review of the structure equations and for later use, check that, when ∇ is Levi-Civita for a Riemannian metric, then the $L(\mathrm{Gl}(n))$-valued forms $\widetilde{\omega}$ and $\widetilde{\Omega}$ actually take values in the subalgebra $L(\mathrm{O}(n))$. Similarly, if ∇ is

Levi-Civita for a pseudo-Riemannian metric, show that these forms are $L(\mathrm{O}(k,n-k))$-valued.

We will lift equations (11.4) and (11.5), as promised, by finding $\mathbb{R}^n$-valued forms θ and τ on P and $L(G)$-valued forms ω and Ω on P satisfying

$$d\theta = -\omega \wedge \theta + \tau, \tag{11.6}$$

$$d\omega = -\omega \wedge \omega + \Omega. \tag{11.7}$$

Furthermore, the choice of frame $(X_1, \dots, X_n)$ on U is a choice of smooth section $\sigma : U \to P|U$ and we will prove that

$$\begin{aligned}\widetilde{\theta} &= \sigma^*(\theta),\\ \widetilde{\tau} &= \sigma^*(\tau),\\ \widetilde{\omega} &= \sigma^*(\omega),\\ \widetilde{\Omega} &= \sigma^*(\Omega).\end{aligned}$$

Let $\zeta \in P$, $x = p(\zeta)$, and set $P_x = p^{-1}(x)$. The *vertical space* at $\zeta \in P$ will be

$$V_\zeta = T_\zeta(P_x) \subset T_\zeta(P).$$

Definition 11.4.3. The vertical subbundle $V \subset T(P)$ is $V = \bigcup_{\zeta\in P} V_\zeta$. The elements $X \in \Gamma(V)$ are called the vertical fields on P.

Each $E \in L(G)$ can be viewed as a vertical field on P. Indeed, E is an $n \times n$ matrix and e^{tE} is the one-parameter subgroup of G generated by E. Using the right action $P \times G \to P$, we obtain, for each $\zeta \in P$, a curve $\zeta \cdot e^{tE}$ which is at ζ at time $t = 0$. This curve lies in the fiber of P through ζ, so the corresponding infinitesimal curve is

$$\zeta \cdot E = \langle \zeta \cdot e^{tE} \rangle_{t=0} \in V_\zeta.$$

As ζ varies over P, this defines a vertical field $\overline{E}$ on P. The mapping $E \mapsto \overline{E}$ is a canonical linear injection $L(G) \hookrightarrow \Gamma(V)$.

If $U \subseteq M$ is an open, trivializing neighborhood for P, the trivializations $P|U \cong U \times G$ are in one-to-one correspondence with the sections $\sigma \in \Gamma(P|U)$. Indeed, $\sigma(x) \cdot B \leftrightarrow (x, B)$. Fix the choice of σ. Since $B \in G \subseteq \mathrm{Gl}(n)$ is a nonsingular $n \times n$ matrix and, as remarked above, each $E \in L(G)$ is an $n \times n$ matrix, the value of the vertical field $\overline{E}$ at $\zeta = (x, B) \in P|U$ is $\zeta \cdot E = (x, BE)$, where BE is the matrix product. This is because E, as a left-invariant vector field on G, has value at $B \in G$ given by BE. Thus, this way of viewing a matrix $E \in L(G)$ as a vertical field $\overline{E}$ on P is quite analogous to the way that E is viewed as a left-invariant field on G.

The right action $P \times G \to P$ induces a linear action

$$\mathfrak{X}(P) \times G \to \mathfrak{X}(P)$$

via the differential. If $X \in \mathfrak{X}(P)$ and $B \in G$, it will be natural to denote this action by $X \mapsto X \cdot B$. Consider also the automorphism $\mathrm{Ad}(B) : G \to G$, called the adjoint action and defined by $\mathrm{Ad}(B)(A) = B^{-1}AB$ (*cf.* Exercise 5.3.15). The differential $\mathrm{Ad}(B)_* : \mathfrak{X}(G) \to \mathfrak{X}(G)$ restricts to an automorphism of $L(G)$ where it will also be called $\mathrm{Ad}(B)$ and written $\mathrm{Ad}(B)(E) = B^{-1}EB$. The following is practically immediate.

Lemma 11.4.4. *Under the inclusion $L(G) \hookrightarrow \Gamma(V)$, $\mathrm{Ad}(B)(E) \mapsto \overline{E}\cdot B$, $\forall B \in G$, $\forall E \in L(G)$. In particular, $L(G) \subset \Gamma(V)$ is invariant under the right action of G.*

From now on, we denote $\overline{E}$ by E, identifying $L(G)$ as a vector subspace of $\Gamma(V)$. We will also write EB for the right translate $\overline{E}\cdot B$ of this vector field by $B \in G$.

Remark. Any basis of $L(G) \subset \Gamma(V)$ gives a trivialization of V. If desired, it is possible to specify a canonical choice of this basis. In the case that $G = \mathrm{Gl}(n)$, this will be $\{E^i_j\}_{1\le i,j\le n}$, where E^i_j is the $n \times n$ matrix having (i,j)th entry 1 and all remaining entries 0. The Lie algebra of $\mathrm{O}(k, n-k)$ consists of all matrices

$$\begin{bmatrix} A & B \\ B^{\mathrm{T}} & C \end{bmatrix},$$

where A is a skew symmetric, $k \times k$ matrix and C is $(n-k) \times (n-k)$ and skew symmetric, as is easily checked. For $1 \le i < j \le k$ and $k+1 \le i < j \le n$, set $A^i_j = E^i_j - E^j_i$. For all other $i < j$, set $A^i_j = E^i_j + E^j_i$. Then, $\{A^i_j\}_{1\le i<j\le n}$ is a canonical basis of $L(\mathrm{O}(k, n-k))$. The case in which $G = \mathrm{O}(n)$ is just the special case in which $k = n$, so the canonical basis is given by $A^i_j = E^i_j - E^j_i$, $1 \le i < j \le n$.

We are going to show that the connection ∇ defines a direct sum decomposition

$$T(P) = V \oplus H,$$

and a canonical choice of basis $\{E^i\}_{1\le i\le n}$ of $\Gamma(H)$. We call the bundle H the *horizontal* subbundle defined by ∇

Fix $\zeta_0 \in P$, $x_0 = p(\zeta_0) \in M$, and let $s : [0,\epsilon) \to M$ be a smooth curve with $s(0) = x_0$. The frame $\zeta_0 = (v_1, \dots, v_n)$ can be parallel transported along s. That is, each v_i is parallel transported and these remain linearly independent. In fact, in the case that ∇ is Levi-Civita for a (possibly indefinite) metric, the fact that ∇ respects the metric implies that the orthonormal frame parallel translates to orthonormal frames. Therefore, this parallel transport can be interpreted as a lift $s^\flat = (X^\flat_{1\,s(t)}, \dots, X^\flat_{n\,s(t)})$, $0 \le t < \epsilon$, of s to P starting at ζ_0. That is, the diagram

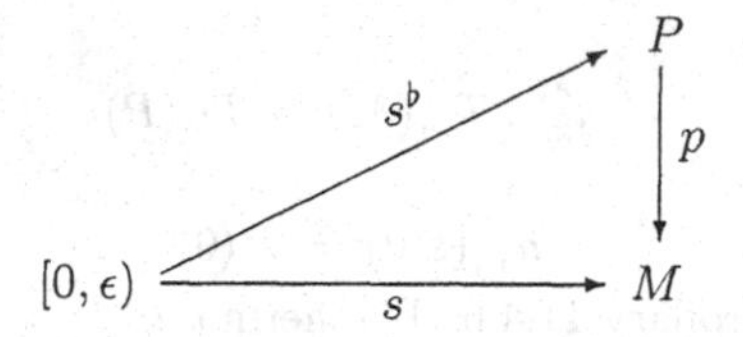

commutes. We think of $s^\flat$ as a "horizontal" lift and say that the initial velocity $\dot{s}^\flat(0) \in T_{\zeta_0}(P)$ is horizontal. If we can show that the set of all vectors in $T_{\zeta_0}(P)$ that can be obtained in this way is a vector subspace of dimension n, this will be our candidate for H_{ζ_0}. Similarly, $\dot{s}^\flat(t) \in H_{s^\flat(t)}$, justifying the term "horizontal lift".

A section $\sigma \in \Gamma(P|U)$, where U is a suitably small neighborhood of x_0, gives another way to lift s. If $\sigma = (X_1, \dots, X_n)$ and $\sigma(x_0) = \zeta_0$, set

$$s^\sharp(t) = (X_{1\,s(t)}, \dots, X_{n\,s(t)}), \quad 0 \le t < \epsilon.$$

Then,

$$s^\sharp(0) = \zeta_0$$

and the diagram

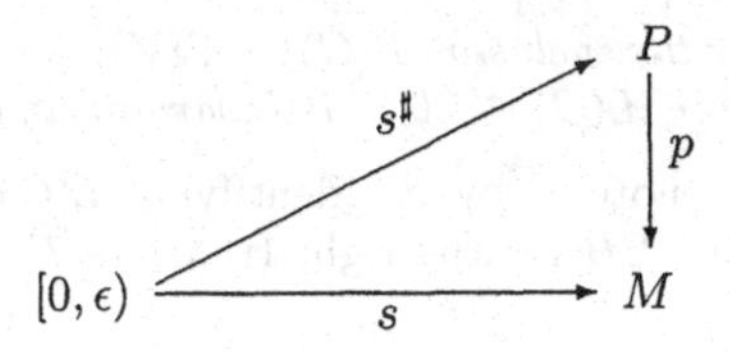

commutes.

Since $s^\flat(t)$ and $s^\sharp(t)$ lie in the fiber $P_{s(t)}$, there is a unique element $A(t) \in G$ such that

(11.8) $$s^\sharp(t) = s^\flat(t)A(t), \quad 0 \le t < \epsilon.$$

Clearly, $A : [0, \epsilon) \to G$ is smooth and $A(0) = I$, the identity matrix.

Finally, we use the local frame $\sigma = (X_1, \dots, X_n)$ to define the forms $\widetilde{\theta}$, $\widetilde{\tau}$, $\widetilde{\omega}$, and $\widetilde{\Omega}$ on U satisfying equation (11.4) and (11.5).

Lemma 11.4.5 (Key Lemma). $\dot{A}(0) = \widetilde{\omega}(\dot{s}(0)) \in L(G)$.

Before proving Lemma 11.4.5, we deduce some important consequences.

Corollary 11.4.6. *If $\sigma \in \Gamma(P|U)$ with $\sigma(x_0) = \zeta_0 \in P_{x_0}$, if*

$$s : [0, \epsilon) \to U$$

is smooth with $s(0) = x_0$, and if

$$s^\flat : [0, \epsilon) \to P|U$$

is the horizontal lift with $s^\flat(0) = \zeta_0$, then

$$\dot{s}^\flat(0) = -\zeta_0 \cdot \widetilde{\omega}(\dot{s}(0)) + \sigma_{*x_0}(\dot{s}(0)).$$

Proof. Indeed, differentiate $s^\sharp(t) = s^\flat(t)A(t)$ at 0, obtaining

$$\begin{aligned} \dot{s}^\sharp(0) &= \dot{s}^\flat(0) \cdot A(0) + s^\flat(0) \cdot \dot{A}(0) \\ &= \dot{s}^\flat(0) + \zeta_0 \cdot \widetilde{\omega}(\dot{s}(0)). \end{aligned}$$

But $\dot{s}^\sharp(0) = \sigma_{*x_0}(\dot{s}(0))$. □

We define

$$h_{x_0}^{\zeta_0} : T_{x_0}(M) \to T_{\zeta_0}(P)$$

by

$$h_{x_0}^{\zeta_0}(\dot{s}(0)) = \dot{s}^\flat(0),$$

this being linear by Corollary 11.4.6. Furthermore,

$$p_{*\zeta_0} \circ h_{x_0}^{\zeta_0} = \mathrm{id}_{T_{x_0}(M)},$$

since $-\zeta_0 \cdot \widetilde{\omega}(\dot{s}(0))$, being vertical, is annihilated by $p_{*\zeta_0}$, and

$$p_{*\zeta_0}(\sigma_{*x_0}(\dot{s}(0))) = \dot{s}(0).$$

In particular, $h_{x_0}^{\zeta_0}$ is injective.

Definition 11.4.7. The horizontal space at $\zeta_0 \in P$ is the n-dimensional vector space

$$H_{\zeta_0} = \mathrm{im}(h_{x_0}^{\zeta_0}).$$

The horizontal subbundle is

$$H = \bigcup_{\zeta \in P} H_\zeta.$$

Exercise 11.4.8. Prove that the horizontal bundle H really is a vector bundle and that $T(P) = V \oplus H$, each summand being an invariant subbundle under the right action of G.

In the following proof and subsequently, we will use the summation convention without further comment.

Proof of the key lemma. We take U to be a coordinate neighborhood and write $s(t) = (x^1(t), \dots, x^n(t))$. Write

$$A(t) = \left[a_k^j(t)\right],$$

where $a_k^j(0) = \delta_k^j$. The smooth frame $\sigma = (X_1, \dots, X_n)$ can also be written as a matrix valued function. Writing $X_{k\,s(t)} = u_k^i(t)\xi_{i\,s(t)}$ as a column vector, we get

$$s^\sharp(t) = \left[u_k^i(t)\right].$$

Similarly, write

$$s^\flat(t) = (X^\flat_{1\,s(t)}, \dots, X^\flat_{n\,s(t)}) = \left[v_j^i(t)\right].$$

In this notation, equation (11.8) becomes

$$v_j^i(t)a_k^j(t) = u_k^i(t). \tag{11.9}$$

Using the covariant derivative to express the fact that $s^\flat$ is a parallel frame along s gives

$$0 = \frac{\nabla X_j^\flat}{dt} = (\dot{v}_j^i + v_j^\alpha \dot{x}^\beta \Gamma^i_{\beta\alpha})\xi_{i\,s(t)},$$

so

$$\dot{v}_j^i(t) = -v_j^\alpha(t)\dot{x}^\beta(t)\Gamma_{\beta\alpha}(s(t)). \tag{11.10}$$

Differentiate equation (11.9) and get

$$\dot{v}_j^i(t)a_k^j(t) + v_j^i(t)\dot{a}_k^j(t) = \dot{u}_k^i(t). \tag{11.11}$$

In equation (11.11), set $t = 0$, note that

$$\left[v_j^i(0)\right] = s^\flat(0) = s^\sharp(0) = \left[u_j^i(0)\right],$$

and get

$$\begin{aligned}\dot{u}_k^i(0) &= \dot{v}_j^i(0)a_k^j(0) + v_j^i(0)\dot{a}_k^j(0)\\ &= \dot{v}_k^i(0) + u_j^i(0)\dot{a}_k^j(0).\end{aligned}$$

Use equation (11.10) at $t = 0$ to conclude that

$$\dot{u}_k^i(0) + u_k^\alpha(0)\dot{x}^\beta(0)\Gamma^i_{\beta\alpha}(x_0) = u_j^i(0)\dot{a}_k^j(0). \tag{11.12}$$

By the definition of the local connection form $\widetilde{\omega}$, we obtain

$$\begin{aligned}\widetilde{\omega}_j^i(\dot{s}(0))u_i^k(0)\xi_{k\,x_0} &= \widetilde{\omega}_j^i(\dot{s}(0))X_{i\,x_0}\\ &= \nabla_{\dot{s}(0)}X_j\\ &= \nabla_{\dot{s}(0)}(u_j^\alpha \xi_\alpha)\\ &= \dot{u}_j^\alpha(0)\xi_{\alpha\,x_0} + u_j^\alpha(0)\nabla_{\dot{x}^\beta \xi_{\beta\,x_0}}(\xi_\alpha)\\ &= \left(\dot{u}_j^k(0) + u_j^\alpha(0)\dot{x}^\beta(0)\Gamma^k_{\beta\alpha}(x_0)\right)\xi_{k\,x_0}.\end{aligned}$$

Setting the coefficients of $\xi_{k\,x_0}$ equal and applying equation (11.12), we obtain

$$u_i^k(0)\widetilde{\omega}_j^i(\dot{s}(0))) = \dot{u}_j^k(0) + u_j^\alpha(0)\dot{x}^\beta(0)\Gamma_{\beta\alpha}^k(x_0)$$
$$= u_i^k(0)\dot{a}_j^i(0).$$

In terms of matrix products, this says

$$\left[u_i^k(0)\right]\left[\widetilde{\omega}_j^i(\dot{s}(0))\right] = \left[u_i^k(0)\right]\left[\dot{a}_j^i(0)\right].$$

Since the matrix $\left[u_i^k(0)\right]$ is nonsingular, we conclude that

$$\dot{A}(0) = \widetilde{\omega}(\dot{s}(0)) \in L(G).$$

□

Definition 11.4.9. The tautological horizontal frame $(E_\zeta^1, \dots, E_\zeta^n)$, at $\zeta \in P$, is the unique n-frame of vectors in H_ζ such that

$$\zeta = (p_{*\zeta}(E_\zeta^1), \dots, p_{*\zeta}(E_\zeta^n)).$$

Equivalently, in terms of a local section $\sigma = (X_1, \dots, X_n)$ of P,

$$E_{\sigma(x)}^j = h_x^{\sigma(x)}(X_{j\,x}), \quad 1 \le j \le n.$$

As ζ varies over P, the reader can check that E_ζ^j varies smoothly. That is, $E^j \in \Gamma(H)$ and the tautological frame field $(E^1, \dots, E^n)$ is a canonical trivialization of H. This, together with the remark following Lemma 11.4.4, gives the following.

Lemma 11.4.10. *Given the connection ∇, the manifold P is canonically parallelizable.*

Indeed, for the case $P = F(M)$, the canonical basis of $\mathfrak{X}(P)$ is

$$\{E^i\}_{i=1}^n \cup \{E_j^i\}_{1 \le i,j \le n}$$

and, for the case $P = \mathrm{O}(M)$ or $\mathrm{O}_k(M)$, it is

$$\{E^i\}_{i=1}^n \cup \{A_j^i\}_{1 \le i < j \le n}.$$

Definition 11.4.11. The canonical coframe field for ∇ on P is the $\mathbb{R}^n$-valued 1-form θ such that

$$\theta | V \equiv 0;$$

$$\theta(a_i E^i) = \begin{bmatrix} a_1 \\ \vdots \\ a_n \end{bmatrix}.$$

Recall that, for each $\zeta \in P$, the vertical space is canonically $V_\zeta = L(G)$. This identification is understood in the following definition.

Definition 11.4.12. The connection form of ∇ on P is the $L(G)$-valued 1-form ω such that

$$\omega | H \equiv 0;$$
$$\omega_\zeta | V_\zeta = \mathrm{id}, \quad \forall \zeta \in P.$$

Remark. If we write

$$\theta = \begin{bmatrix} \theta^1 \\ \vdots \\ \theta^n \end{bmatrix},$$

$$\omega = \begin{bmatrix} \omega_1^1 & \omega_2^1 & \cdots & \omega_n^1 \\ \omega_1^2 & \omega_2^2 & \cdots & \omega_n^2 \\ \vdots & \vdots & & \vdots \\ \omega_1^n & \omega_2^n & \cdots & \omega_n^n \end{bmatrix},$$

then, in the case that $G = \mathrm{Gl}(n)$, the set

$$\{\theta^i\}_{i=1}^n \cup \{\omega_j^i\}_{1\le i,j\le n}$$

of 1-forms is dual to the canonical basis

$$\{E^i\}_{i=1}^n \cup \{E_j^i\}_{1\le i,j\le n}.$$

For $G = \mathrm{O}(n)$ or $\mathrm{O}(k, n-k)$, the dual to the canonical framing of P is

$$\{\theta^i\}_{i=1}^n \cup \{\omega_j^i\}_{1\le i<j\le n}.$$

Exercise 11.4.13. For $B \in G$, let $R_B : P \to P$ denote the right action of B and check that $R_B^*(\omega) = \mathrm{Ad}(B)\circ\omega$. That is, if $\zeta \in P$ and $v \in T_\zeta(P)$, then

$$(R_B^*(\omega))_\zeta(v) = \omega_{\zeta\cdot B}(v\cdot B) = \mathrm{Ad}(B)(\omega_\zeta(v)).$$

Remark. One can use this to generalize the notion of a connection to principal G-bundles $p : P \to M$, where G is a general Lie group. A connection on P will be an $L(G)$-valued 1-form ω on P such that

(1) if $\zeta \in P$ and $L(G) = V_\zeta \subset T_\zeta(P)$ is the vertical space at ζ, then $\omega_\zeta = \mathrm{id} : V_\zeta \to L(G)$;
(2) $R_b^*(\omega) = \mathrm{Ad}(b)\circ\omega$, $\forall b \in G$.

By the first of these conditions, $H = \ker(\omega)$ is an n-plane subbundle of $T(P)$ complementary to V. That is, $T(P) = V \oplus H$, and one calls H the "horizontal distribution". By the second condition, the horizontal distribution is invariant under the right action $P \times G \to P$. Piecewise smooth curves $s : [a,b] \to M$ have "horizontal lifts" $s^\flat : [a,b] \to P$, uniquely determined by the initial point $s^\flat(a) = \zeta \in P_{s(a)}$ and the requirement that $\dot{s}^\flat(t) \in H_{s^\flat(t)}$, $a \le t \le b$. This assertion is proven by the basic theorem of O.D.E. The horizontal lift is interpreted as "parallel transport" of ζ along s. Furthermore, if F is some manifold and $G \times F \to F$ is a smooth left action, there is an associated bundle $\pi : P \times_G F \to M$, with fibers diffeomorphic to F, and the connection on P defines a notion of parallel transport (or horizontal lifting) in this bundle along curves $s : [a,b] \to M$. Using the notation $[\zeta, x] \in P \times_G F$ for the equivalence class of $(\zeta, x) \in P \times F$, we define the parallel transport of $[\zeta, x] \in \pi^{-1}(s(a))$ to be $\tilde{s}(t) = [s^\flat(t), x]$, where $s^\flat(a) = \zeta$.

We return to the frame bundle $P = F(M), \mathrm{O}(M)$ or $\mathrm{O}_k(M)$ and the connection ∇. The following definitions are dictated by equations (11.6) and (11.7).

Definition 11.4.14. The torsion form of ∇ on P is the $\mathbb{R}^n$-valued 2-form τ on P given by

$$\tau = d\theta + \omega \wedge \theta.$$

The connection is said to be *symmetric* or *torsion free* precisely if $\tau \equiv 0$.

Definition 11.4.15. The curvature form of ∇ on P is the $L(G)$-valued 2-form Ω on P given by

$$\Omega = d\omega + \omega \wedge \omega.$$

The connection is said to be *flat* precisely if $\Omega \equiv 0$.

Lemma 11.4.16. *Let $\sigma = (X_1, \dots, X_n) \in \Gamma(P|U)$ and let $\widetilde{\theta}$, $\widetilde{\omega}$, $\widetilde{\tau}$, $\widetilde{\Omega}$ be the associated forms on U as in equations* (11.4) *and* (11.5). *Then,*

$$\begin{aligned} \widetilde{\theta} &= \sigma^*(\theta), \\ \widetilde{\omega} &= \sigma^*(\omega), \\ \widetilde{\tau} &= \sigma^*(\tau), \\ \widetilde{\Omega} &= \sigma^*(\Omega). \end{aligned}$$

Proof. If we verify the first two equations, the remaining two follow from the equations of structure. We use the notation in the remark following Definition 11.4.12.

For the first equation, we use $\theta|V \equiv 0$. Thus, for arbitrary $x \in U$ and $1 \leq j \leq n$,

$$\begin{aligned} \sigma_x^*(\theta^i_{\sigma(x)})(X_{j\,x}) &= \theta^i_{\sigma(x)}(\sigma_{*x}(X_{j\,x})) \\ &= \theta^i_{\sigma(x)}(h_x^{\sigma(x)}(X_{j\,x}) + \sigma(x) \cdot \widetilde{\omega}(X_{j\,x})) \\ &= \theta^i_{\sigma(x)}(h_x^{\sigma(x)}(X_{j\,x})) \\ &= \theta^i_{\sigma(x)}(E^j_{\sigma(x)}) \\ &= \delta^i_j \\ &= \widetilde{\theta}^i_x(X_{j\,x}). \end{aligned}$$

That is, $\sigma^*(\theta^i) = \widetilde{\theta}^i$, $1 \leq i \leq n$, so $\sigma^*(\theta) = \widetilde{\theta}$.

For the second equation, we use $\omega^\ell_k|H \equiv 0$ and consider the case of a general connection ($P = F(M)$). Thus, for arbitrary $x \in U$ and $1 \leq j \leq n$,

$$\begin{aligned} \sigma_x^*(\omega^\ell_{k\,\sigma(x)})(X_{j\,x}) &= \omega^\ell_{k\,\sigma(x)}(\sigma_{*\,x}(X_{j\,x})) \\ &= \omega^\ell_{k\,\sigma(x)}(h_x^{\sigma(x)}(X_{j\,x}) + \sigma(x) \cdot \widetilde{\omega}(X_{j\,x})) \\ &= \omega^\ell_{k\,\sigma(x)}(\sigma(x) \cdot \widetilde{\omega}(X_{j\,x})) \\ &= \omega^\ell_{k\,\sigma(x)}(\widetilde{\omega}^r_q(X_{j\,x}) E^r_{q\,\sigma(x)}) \\ &= \widetilde{\omega}^r_{q\,x}(X_{j\,x}) \omega^\ell_{k\,\sigma(x)}(E^r_{q\,\sigma(x)}) \\ &= \widetilde{\omega}^r_{q\,x}(X_{j\,x}) \delta^{\ell\,k}_{r\,q} \\ &= \widetilde{\omega}^\ell_{k\,x}(X_{j\,x}). \end{aligned}$$

For the case $P = \mathrm{O}(M)$ or $\mathrm{O}_k(M)$, replace $E^r_{q\,\sigma(x)}$ in the above with $A^r_{q\,\sigma(x)}$ and sum only over the indices $1 \leq r < q \leq n$, again obtaining

$$\sigma_x^*(\omega^\ell_{k\,\sigma(x)})(X_{j\,x}) = \widetilde{\omega}^\ell_{k\,x}(X_{j\,x}).$$

Since $x \in U$ and $1 \leq j \leq n$ are arbitrary, this gives $\sigma^*(\omega^\ell_k) = \widetilde{\omega}^\ell_k$, so $\sigma^*(\omega) = \widetilde{\omega}$. □

Lemma 11.4.17. *The following are equivalent for the connection ∇:*

(1) $\tau \equiv 0$;

(2) $\tau|(H \oplus H) \equiv 0$;
(3) ∇ *is symmetric.*

Proof. (1) $\Rightarrow$ (2) is immediate. We prove (2) $\Leftrightarrow$ (3). Let $\zeta_0 \in P$, set $x_0 = p(\zeta_0) \in M$, and let $U \subseteq M$ be an open neighborhood of x_0 such that $P|U$ is trivial. We can choose the section $\sigma \in \Gamma(P|U)$ so that $\sigma(x_0) = \zeta_0$ *and* so that $\sigma_{*x_0}(T_{x_0}(M)) = H_{\zeta_0}$. For this choice, we have

$$\tau_{\zeta_0}|(H_{\zeta_0} \times H_{\zeta_0}) \equiv 0 \Leftrightarrow 0 = \sigma^*_{x_0}(\tau_{\zeta_0}) = \widetilde{\tau}_{x_0}$$
$$\Leftrightarrow \text{the torsion tensor } T_{x_0} = 0.$$

Since $x_0 \in M$ is arbitrary, the assertion follows. In order to prove (3) $\Rightarrow$ (1), let ζ_0, x_0, and U be as above, and assume that ∇ is symmetric. Given any direct sum decomposition

$$T_{\zeta_0}(P) = V_{\zeta_0} \oplus \widetilde{H}_{\zeta_0},$$

one can choose $\sigma \in \Gamma(P|U)$ such that $\sigma_{*x_0}(T_{x_0}(M)) = \widetilde{H}_{\zeta_0}$. Then, the fact that $\sigma^*_{x_0}(\tau_{\zeta_0}) = \widetilde{\tau}_{x_0} = 0$ implies that

$$\tau_{\zeta_0}|(\widetilde{H}_{\zeta_0} \times \widetilde{H}_{\zeta_0}) \equiv 0.$$

Given $v \in V_{\zeta_0}$ and $w \in H_{\zeta_0}$, choose $\widetilde{H}_{\zeta_0}$, as above, and $u \in H_{\zeta_0}$ such that $v+u, w \in \widetilde{H}_{\zeta_0}$. This is clearly possible. Then

$$0 = \tau_{\zeta_0}(v+u, w) = \tau_{\zeta_0}(v, w) + \tau_{\zeta_0}(u, w) = \tau_{\zeta_0}(v, w).$$

This proves that

$$\tau_{\zeta_0}|(V_{\zeta_0} \times H_{\zeta_0}) \equiv 0.$$

By antisymmetry, we also have

$$\tau_{\zeta_0}|(H_{\zeta_0} \times V_{\zeta_0}) \equiv 0.$$

In order to prove that $\tau_{\zeta_0} = 0$, it remains to prove that

$$\tau_{\zeta_0}|(V_{\zeta_0} \times V_{\zeta_0}) \equiv 0.$$

Given $v, w \in V_{\zeta_0}$, let $u_1, u_2 \in H_{\zeta_0}$ be linearly independent. Then there is a complementary space $\widetilde{H}_{\zeta_0}$, as above, such that $v + u_1, w + u_2 \in \widetilde{H}_{\zeta_0}$. Thus,

$$0 = \tau_{\zeta_0}(v + u_1, w + u_2) = \tau_{\zeta_0}(v, w).$$

We have proven that $\tau_{\zeta_0} = 0$ for arbitrary $\zeta_0 \in P$. □

The following is proven by exactly the same argument.

Lemma 11.4.18. *The following are equivalent for the connection* ∇:
(1) $\Omega \equiv 0$;
(2) $\Omega|(H \oplus H) \equiv 0$;
(3) *The curvature tensor* $R \equiv 0$.

We can now deduce the following result.

Theorem 11.4.19. *The following properties of* ∇ *are equivalent*:
(1) *The n-plane distribution H on P is integrable;*
(2) $\Omega \equiv 0$;
(3) $R \equiv 0$;
(4) *for each $x \in M$, there is an open neighborhood U of x on which the holonomy group of ∇ is $H_x(U) = \{\text{id}\}$.*

Furthermore, if ∇ is symmetric, these properties are equivalent to

(5) $[E^i, E^j] \equiv 0$, $1 \le i, j \le n$.

Proof. We prove (1) $\Leftrightarrow$ (2) $\Leftrightarrow$ (3). The distribution H is exactly $\ker(\omega)$. Thus,

$$\begin{aligned}\Omega^i_j(E^k, E^\ell) &= d\omega^i_j(E^k, E^\ell) + \omega^i_r \wedge \omega^r_j(E^k, E^\ell) \\ &= E^k(\omega^i_j(E^\ell)) - E^\ell(\omega^i_j(E^k)) - \omega^i_j([E^k, E^\ell]) \\ &= -\omega^i_j([E^k, E^\ell]).\end{aligned}$$

Therefore, integrability of H is equivalent to $\Omega|(H \oplus H) \equiv 0$, which is equivalent to $\Omega \equiv 0$ and to $R \equiv 0$ by Lemma 11.4.18.

We prove that (1) $\Rightarrow$ (4). Thus, assume H to be integrable and choose $x \in M$, $\zeta \in p^{-1}(x)$, and let L be the leaf through ζ. Since $p_{*\zeta}$ carries $T_\zeta(L) = H_\zeta$ isomorphically onto $T_x(M)$, there is an open neighborhood $W \subseteq L$ of ζ carried diffeomorphically by p onto an open neighborhood $U \subseteq M$ of x. Let $s : [a, b] \to U$ be a piecewise smooth loop at x. Let $h_s : T_x(M) \to T_x(M)$ be the holonomy transformation around s. The horizontal lift $s^\flat : [a, b] \to P$ with initial point $s^\flat(a) = \zeta = (v_1, \dots, v_n)$ has terminal point $s^\flat(b) = (h_s(v_1), \dots, h_s(v_n))$. The curve $s^\flat$ must lie in W, hence it must be be the loop in W carried by p back to s. That is, $s^\flat(b) = s^\flat(a) = \zeta$ and $h_s(v_i) = v_i$, $1 \le i \le n$. Since the v_i's form a basis of $T_x(M)$, $h_s = \mathrm{id}_{T_x(M)}$. Since $s \in \Omega(M, x)$ was arbitrary, the holonomy group $H_x(U)$ of ∇ in U is trivial.

Conversely, supposing that (4) holds, we deduce (1). Let $(N, x^1, \dots, x^n)$ be a coordinate chart in M and set $W = p^{-1}(N)$. It will be enough to show that $H|W$ is integrable. Let $Z_i \in \Gamma(H|W)$ be the unique horizontal field that is p-related to ξ_i, $1 \le i \le n$. These fields span $H|W$ and their brackets $[Z_i, Z_j]$ are p-related to $[\xi_i, \xi_j] \equiv 0$. This implies that $[Z_i, Z_j] \in \Gamma(V|W)$, $1 \le i, j \le n$, but we will show the stronger fact that $[Z_i, Z_j] \equiv 0$, $1 \le i, j \le n$, so $H|W$ is integrable.

Let Ψ^i be the local flow in N generated by ξ_i, $1 \le i \le n$. These flows lift to local flows $\Psi^{i\flat}$ on $p^{-1}(N)$ with infinitesimal generator $Z_i \in \Gamma(H|W)$, $1 \le i \le n$. Let $x \in N$ and let $U \subseteq N$ be an open neighborhood of x so small that $H_x(U) = \{\mathrm{id}\}$. Let s^i_x be the flow line of Ψ^i in N from x to $y = \Psi^i_{t_1}(x)$. Similarly, s^j_y is the flow line of Ψ^j in N from y to $z = \Psi^j_{t_2}(y)$, $-s^i_z$ is the flow line of $(\Psi^i)^{-1}$ in N from z to $w = \Psi^i_{-t_1}$, and $-s^j_w$ is similarly defined from w to $x' = \Psi^j_{-t_2}(w)$. For sufficiently small values of t_1 and t_2, the p.r. curve

$$s = s^i_x + s^j_y - s^i_z - s^j_w$$

stays in U. Since these local flows commute, $x' = x$ and s must be a *loop* at x. Since the holonomy group $H_x(U)$ is trivial, the horizontal lift $s^\flat$ to any $\zeta \in p^{-1}(x)$ is a *loop* at ζ, and this implies that

$$\Psi^{i\flat}_{t_1}(\Psi^{j\flat}_{t_2}(\zeta)) = \Psi^{j\flat}_{t_2}(\Psi^{i\flat}_{t_1}(\zeta)),$$

for all small values of t_1 and t_2. Since $x \in N$ and $\zeta \in p^{-1}(x)$ are arbitrary, it follows that $[Z_i, Z_j] \equiv 0$, $1 \le i, j \le n$.

We turn to property (5). If $[E^k, E^\ell] \equiv 0$, $1 \le k, \ell \le n$, it is clear that H is integrable. To complete the proof of the theorem, we assume that ∇ is symmetric and prove that the integrability of H implies that $[E^k, E^\ell] \equiv 0$. Indeed,

$$\begin{aligned}
0 &= \tau^i(E^k, E^\ell) \\
&= d\theta^i(E^k, E^\ell) + \omega^i_j \wedge \theta^j(E^k, E^\ell) \\
&= E^k(\theta^i(E^\ell)) - E^\ell(\theta^i(E^k)) - \theta^i([E^k, E^\ell]) \\
&= -\theta^i([E^k, E^\ell]).
\end{aligned}$$

This implies that $[E^k, E^\ell] \in \Gamma(V)$, hence, when H is integrable, that this bracket vanishes identically, $1 \le k, \ell \le n$. □

Exercise 11.4.20. Deduce Theorem 11.4.1 From Theorem 11.4.19.

APPENDIX A

Construction of the Universal Covering

In this brief appendix, we give the construction of universal covering spaces, proving Theorem 1.7.29. The idea is that the path-lifting property of covering spaces suggests a way to form a covering space that has the path-lifting property *tautologically*. This is somehow the most natural covering space and the universal property will itself be a tautology.

Fix a path-connected, locally simply connected, pointed space (X, x_0) and let $\mathcal{P}(X, x_0)$ denote the set of paths $\sigma : [0,1] \to X$ with $\sigma(0) = x_0$. Let $\widetilde{X} = \mathcal{P}(X, x_0)/\sim_\delta$, the set of homotopy classes $[\sigma]$ mod the endpoints of paths $\sigma \in \mathcal{P}(X, x_0)$. Let $\widetilde{x}_0 = [x_0]$ and define

$$p : (\widetilde{X}, \widetilde{x}_0) \to (X, x_0)$$

by $p([\sigma]) = \sigma(1)$. We will put a topology on $\widetilde{X}$ relative to which p becomes a covering map.

Since X is locally simply connected, there is a base $\mathcal{B}$ of the topology of X consisting of simply connected open sets. For each $x \in X$, each basic neighborhood $U \in \mathcal{B}$ of x, and each $[\sigma] \in p^{-1}(x)$, define

$$\widetilde{U}_{[\sigma]} = \{[\tau] \in \widetilde{X} \mid \tau = \sigma \cdot \alpha \text{ where } \operatorname{im} \alpha \subset U \text{ and } \alpha(0) = x\}.$$

Call such subsets of $\widetilde{X}$ basic neighborhoods and let $\widetilde{\mathcal{B}}$ be the family of all basic neighborhoods.

Lemma A.1. *$\widetilde{\mathcal{B}}$ is the base of a topology on $\widetilde{X}$.*

Proof. We only need to show that the intersection of two basic neighborhoods is the union of basic neighborhoods. Accordingly, let

$$[\tau] \in \widetilde{U}_{[\sigma]} \cap \widetilde{V}_{[\sigma']}$$

and write

$$[\tau] = [\sigma \cdot \alpha],$$
$$[\tau] = [\sigma' \cdot \alpha'],$$

where $\alpha(0) = \sigma(1)$, $\alpha'(0) = \sigma'(1)$, $\operatorname{im} \alpha \subset U$ and $\operatorname{im} \alpha' \subset V$. Let $W \subseteq U \cap V$ be a simply connected neighborhood of $\tau(1)$. It is then evident that

$$\widetilde{W}_{[\tau]} \subseteq \widetilde{U}_{[\sigma]} \cap \widetilde{V}_{[\sigma']}.$$

Since $[\tau]$ was an arbitrary point of $\widetilde{U}_{[\sigma]} \cap \widetilde{V}_{[\sigma']}$, it follows that this intersection is a union of elements of $\widetilde{\mathcal{B}}$. □

As yet, we have not used the fact that $\mathcal{B}$ consists of simply connected neighborhoods. This is required for the following.

Lemma A.2. *The function $p : \widetilde{X} \to X$ is a covering map.*

Proof. If $U \in \mathcal{B}$, it is clear that $p^{-1}(U)$ is the union of all basic neighborhoods of the form $\widetilde{U}_{[\sigma]}$, where $\sigma \in \mathcal{P}(X, x_0)$ ranges over all paths with $\sigma(1) \in U$. Thus, p is continuous.

We claim that, for arbitrary $\widetilde{U}_{[\sigma]}$, the map $p : \widetilde{U}_{[\sigma]} \to U$ is bijective. Indeed, surjectivity is trivial (X is path-connected), so we prove injectivity. Suppose α and β are paths in U originating at $\sigma(1)$ such that

$$p([\sigma \cdot \alpha]) = p([\sigma \cdot \beta]).$$

Thus, α and β are paths in U with the same initial points and the same terminal points, so the simple connectivity of U implies that $\alpha \sim_\partial \beta$ and $[\sigma \cdot \alpha] = [\sigma \cdot \beta]$.

We have proven that each element $\widetilde{U}_{[\sigma]}$ of $\widetilde{\mathcal{B}}$ is carried by p one-to-one onto an element U of $\mathcal{B}$. So p is an open map and $p|\widetilde{U}_{[\sigma]}$ is a homeomorphism onto U.

It only remains for us to show that

$$\widetilde{U}_{[\sigma]} \cap \widetilde{U}_{[\tau]} \neq \emptyset \Rightarrow \widetilde{U}_{[\sigma]} = \widetilde{U}_{[\tau]}.$$

Indeed, for arbitrary $[\sigma \cdot \gamma] \in \widetilde{U}_{[\sigma]}$, we will show that this element belongs also to $\widetilde{U}_{[\tau]}$. The reverse inclusion follows by the same proof, proving equality of sets. By assumption, there are paths α and β in U such that

$$\begin{aligned} \alpha(0) &= \sigma(1), \\ \beta(0) &= \tau(1), \\ [\sigma \cdot \alpha] &= [\tau \cdot \beta]. \end{aligned}$$

The last equality implies that $\alpha(1) = \beta(1)$, and so we can define a path $\gamma' = (\beta \cdot \alpha^{-1}) \cdot \gamma$. Then

$$\tau \cdot \gamma' \sim_\partial (\tau \cdot \beta) \cdot (\alpha^{-1} \cdot \gamma) \sim_\partial (\sigma \cdot \alpha) \cdot (\alpha^{-1} \cdot \gamma) \sim_\partial \sigma \cdot \gamma.$$

Thus $[\sigma \cdot \gamma] = [\tau \cdot \gamma'] \in \widetilde{U}_{[\tau]}$. □

Examining this proof, the reader should see that the requirement that X be locally simply connected can be weakened to "semi-locally simply connected" (*cf.* the remark following Theorem 1.7.29).

Observe that the path-lifting property, valid for all covering spaces, is transparent for this one. Indeed, if $\sigma \in \mathcal{P}(X, x_0)$ and $t \in [0, 1]$, define

$$\sigma_t(s) = \sigma(ts), \quad 0 \leq s \leq 1.$$

Then $\widetilde{\sigma}(t) = [\sigma_t]$, $0 \leq t \leq 1$, defines a continuous path $\widetilde{\sigma}$ in $\widetilde{X}$, issuing from $\widetilde{x}_0$, and

$$p(\widetilde{\sigma}(t)) = p([\sigma_t]) = \sigma(t), \quad 0 \leq t \leq 1.$$

This is what we meant by the introductory remark that $p : \widetilde{X} \to X$ "has the path-lifting property tautologically". Similarly, since this lift of σ terminates at the point $\widetilde{\sigma}(1) = [\sigma]$, and since $[\sigma]$ is an arbitrary point of $\widetilde{X}$, we see that $\widetilde{X}$ is path-connected. Finally, if σ is a loop at x_0, $\widetilde{\sigma}$ will be a loop at $\widetilde{x}_0$ if and only if $[\sigma] = [x_0]$. Thus, suppose $\gamma : [0, 1] \to \widetilde{X}$ is a loop at $\widetilde{x}_0$. Then $p \circ \gamma$ is a loop in X at x_0 and γ is, by definition, the lift of $p \circ \gamma$. By the preceding remarks, $p \circ \gamma \sim_\partial x_0$ and the homotopy lifting property then implies that $\gamma \sim_\partial \widetilde{x}_0$. These remarks prove the following.

Lemma A.3. *The space $\widetilde{X}$ is simply connected.*

The following completes the proof of Theorem 1.7.29.

Theorem A.4. *The function* $p : \widetilde{X} \to X$ *is a universal covering map. Furthermore, a covering space* $\widehat{p} : \widehat{X} \to X$ *is isomorphic to the universal covering if and only if* $\widehat{X}$ *is simply connected.*

Proof. Let $\widehat{p} : \widehat{X} \to X$ be a covering space with $\widehat{X}$ connected. Choose a basepoint $\widehat{x}_0 \in \widehat{p}^{-1}(x_0)$. We define (the unique) map f making the following diagram commutative:

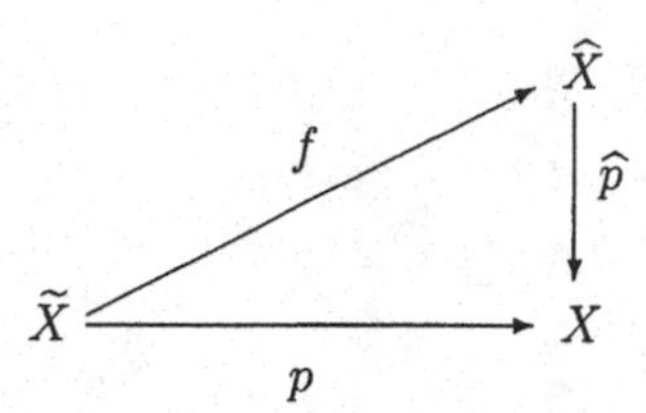

If $[\sigma] \in \widetilde{X}$, we let $\widehat{\sigma}$ denote the unique lift of the path σ to a path in $\widehat{X}$ starting at $\widehat{x}_0$. We then define $f([\sigma]) = \widehat{\sigma}(1)$, a function such that $\widehat{p} \circ f = p$. As V ranges over $\mathcal{B}$, the connected components of $\widehat{p}^{-1}(V)$ range over a base $\widehat{\mathcal{B}}$ for the topology of $\widehat{X}$. Evidently, if $\widehat{V}$ is a component of $\widehat{p}^{-1}(V)$, where $V \in \mathcal{B}$, the components of $f^{-1}(\widehat{V})$ are among the components of $p^{-1}(V)$ and it follows that f is continuous. Thus, $p : \widetilde{X} \to X$ is the universal cover. If $\widehat{p}$ is also universal, the uniqueness theorem (Lemma 1.7.26), together with Lemma A.3, implies that $\widehat{X}$ is simply connected. Conversely, suppose that $\widehat{X}$ is simply connected. By Exercise 1.7.24, f is a covering map and, by Exercise 1.7.28, $\widehat{X}$ will be evenly covered by f. Since $\widetilde{X}$ is path-connected, it follows that f is a homeomorphism, proving that $\widehat{p} : \widehat{X} \to X$ is also the universal cover. $\square$

The following concludes the proof of Theorem 13.22.

Theorem A.4. *Let $p : \tilde{X} \to X$ be a universal covering map, [illegible] connected manifold [illegible] X [illegible] universal covering [illegible] and [illegible] $\tilde{X}$ is simply connected.*

Proof. [illegible] $\tilde{X}$ [illegible] space [illegible] connected. Choose [illegible] point [illegible] the following diagram commutes:

[illegible] the [illegible] of the path [illegible] in $\tilde{X}$ [illegible] at $\tilde{x}_0$. [illegible] $= \tilde{\sigma}(1)$ [illegible] $= \tilde{x}$. As [illegible] the connected components of [illegible] over [illegible] of X [illegible] component [illegible] where $V \in$ [illegible] the components [illegible] are the components [illegible] follows that [illegible] $\tilde{X} \to$ [illegible] universal [illegible] (Lemma [illegible]) together with [illegible] A.3 [illegible] that [illegible] is simply connected. [illegible] X [illegible] covering [illegible] Since [illegible] is also [illegible] ☐

APPENDIX B

The Inverse Function Theorem

The following simple lemma will be used in the proof of Theorem 2.4.1 and in that of Theorem 2.8.4.

Lemma B.1 (Contraction mapping lemma). *Let X be a complete metric space with metric ρ and let $T : X \to X$ be a mapping. If there is a constant $c \in (0, 1)$ such that*

$$\rho(T(x), T(y)) \leq c\rho(x, y), \ \ \forall\, x, y \in X,$$

then T has a unique fixed point $x_0 \in X$. Furthermore, for each $x \in X$,

$$\lim_{n\to\infty} T^n(x) = x_0.$$

Proof. Let $x \in X$ and remark that $1 + c + c^2 + \cdots = C < \infty$, so

$$\rho(T^n(x), T^{n+k}(x)) \leq c^n \sum_{i=1}^{k} \rho(T^{i-1}(x), T^i(x)) \leq c^n C \rho(x, T(x)),$$

implying that $\{T^n(x)\}_{n=1}^{\infty}$ is a Cauchy sequence. Our hypothesis also implies that T is continuous. By completeness, set

$$x_0 = \lim_{n\to\infty} T^n(x).$$

Then,

$$\begin{aligned} T(x_0) &= T(\lim_{n\to\infty} T^n(x)) \\ &= \lim_{n\to\infty} T(T^n(x)) \ \ \text{(by continuity)} \\ &= \lim_{n\to\infty} T^{n+1}(x) \\ &= x_0. \end{aligned}$$

Since T fixes x_0 and strictly reduces distances, the fact that x_0 is the only fixed point is clear, as is the remaining assertion of the lemma. □

Definition B.2. A mapping $T : X \to X$, as above, is called a contraction mapping.

Another useful tool in the proof of the inverse function theorem is the mean value theorem from multivariable calculus. Let U and V be open subsets of $\mathbb{R}^n$, $\Phi : U \to V$ a map of class C^k for some $1 \leq k \leq \infty$. If the line segment $\{tq + (1 - t)p\}_{0 \leq t \leq 1}$ is contained in U, then the mean value theorem asserts that

$$\Phi(q) - \Phi(p) = \left(\int_0^1 J\Phi(tq + (1-t)p)\, dt\right) \cdot (q - p). \tag{B.1}$$

Turning to the proof of Theorem 2.4.1, we let $\Phi : U \to V$ be a C^k map between open subsets of $\mathbb{R}^n$ and assume that $J\Phi(p)$ is nonsingular, for some $p \in U$. We must find an open neighborhood W_p of p in U that Φ carries C^k-diffeomorphically onto an open neighborhood of $\Phi(p)$. By appropriate changes of coordinates, we lose no generality in assuming that $p = 0 = \Phi(p)$ and that $J\Phi(0) = I_n$. Thus, the associated mapping $\Psi : U \to \mathbb{R}^n$, defined by

$$\Psi(x) = x - \Phi(x),$$

satisfies $\Psi(0) = 0$ and $J\Psi(0) = 0$. Finally, for each positive real number η, let B_η denote the closed ball in $\mathbb{R}^n$ of radius η and centered at 0.

Since $J\Phi(x)$ is continuous in x, we can choose $\eta > 0$ so small that $J\Phi(x)$ is nonsingular, $\forall\, x \in B_\eta$. We fix this condition and, in fact, the following.

Lemma B.3. *There is a value of $\eta > 0$ so small that, if $x_1, x_2 \in B_\eta$, then*

$$\|\Psi(x_1) - \Psi(x_2)\| \leq \|x_1 - x_2\|/2,$$
$$\|\Phi(x_1) - \Phi(x_2)\| \geq \|x_1 - x_2\|/2.$$

Proof. Since Ψ is of class C^1 and $J\Psi(0) = 0$, an application of equation (B.1) to the mapping Ψ implies that, for $\eta > 0$ sufficiently small,

$$\|\Psi(x_1) - \Psi(x_2)\| \leq \|x_1 - x_2\|/2.$$

Since $\Phi(x) = x - \Psi(x)$, it follows that

$$\|\Phi(x_1) - \Phi(x_2)\| \geq \|x_1 - x_2\| - \|\Psi(x_1) - \Psi(x_2)\| \geq \|x_1 - x_2\|/2.$$

□

Corollary B.4. *For each $y \in B_{\eta/2}$, there is a unique $x \in B_\eta$ such that $\Phi(x) = y$.*

Proof. Define T_y on B_η by $T_y(z) = y + \Psi(z)$. By the first inequality in Lemma B.3, it is clear that $T_y(B_\eta) \subseteq B_\eta$. By this same inequality,

$$\|T_y(x_1) - T_y(x_2)\| = \|\Psi(x_1) - \Psi(x_2)\| \leq \|x_1 - x_2\|/2,$$

so T_y is a contraction mapping on the complete metric space B_η. Let $x \in B_\eta$ be the unique fixed point and remark that

$$x = T_y(x) = y + \Psi(x) = y + x - \Phi(x)$$

is satisfied if and only if $\Phi(x) = y$. □

Let $Z = \operatorname{int}(B_{\eta/2})$ and $W = \Phi^{-1}(Z)$. These are open neighborhoods of 0 in V and U, respectively.

Corollary B.5. *The mapping $\Phi : W \to Z$ is a homeomorphism.*

Proof. We have shown that Φ maps W one-to-one onto Z, so it remains to be proven that Φ^{-1} is continuous. But the equations $\Phi(x_1) = y_1$ and $\Phi(x_2) = y_2$ and the second inequality in Lemma B.3 imply that

$$\|\Phi^{-1}(y_1) - \Phi^{-1}(y_2)\| = \|x_1 - x_2\| \leq 2\|y_1 - y_2\|,$$

proving the assertion. □

Lemma B.6. *The map $\Phi^{-1}$is differentiable at each point of Z and*

$$J(\Phi^{-1}) = (J\Phi)^{-1} \circ \Phi^{-1}.$$

Proof. Let $b = \Phi(a) \in Z$, $a \in W$. By differentiability at a, we can write

$$\Phi(x) - \Phi(a) = J\Phi(a) \cdot (x - a) + \|x - a\|\vec{\epsilon}(x, a),$$

where

$$\lim_{x \to a} \vec{\epsilon}(x, a) = 0.$$

Since $J\Phi(a)$ is nonsingular, we can write

$$x - a = J\Phi(a)^{-1} \cdot (\Phi(x) - \Phi(a)) - \|x - a\| J\Phi(a)^{-1} \cdot \vec{\epsilon}(x, a).$$

Writing $x = \Phi^{-1}(y)$, we obtain

$$\begin{aligned}\Phi^{-1}(y) - \Phi^{-1}(b) &= J\Phi(a)^{-1} \cdot (y - b) - \|x - a\| J\Phi(a)^{-1} \cdot \vec{\epsilon}(x, a) \\ &= J\Phi(a)^{-1} \cdot (y - b) - \|y - b\|\vec{\delta}(y, b),\end{aligned}$$

where

$$\vec{\delta}(y, b) = \frac{\|x - a\|}{\|y - b\|} J\Phi(a)^{-1} \cdot \vec{\epsilon}(x, a).$$

But $y \to b$ if and only if $x \to a$ and we have the inequality

$$\frac{\|x - a\|}{\|y - b\|} \le 2$$

by Lemma B.3. That is,

$$\lim_{y \to b} \vec{\delta}(y, b) = 0$$

and Φ^{-1} is differentiable at $b \in Z$ with

$$J\Phi^{-1}(b) = J\Phi(a)^{-1} = J\Phi(\Phi^{-1}(b))^{-1}.$$

Since $b \in Z$ is arbitrary, all assertions follow. □

Corollary B.7. *The map Φ^{-1} is of class C^k on Z.*

Proof. Since Φ is of class C^k, the entries in the matrix $(J\Phi)^{-1}$ are functions of class C^{k-1} and Corollary B.5, together with Lemma B.6, implies that $J(\Phi^{-1})$ is continuous. That is, Φ^{-1} is of class C^1. If $k = 1$, we are done. Otherwise, feeding this new fact back into Lemma B.6 implies that Φ^{-1} is of class C^2. Continuing in this way (forever, if $k = \infty$), we complete the proof. □

We have proven the C^k version of Theorem 2.4.1, $1 \le k \le \infty$.

Remark. It is not hard to adapt the above proof to work for mappings

$$F : U \to \mathbf{F},$$

where $U \subseteq \mathbf{E}$ is open and $\mathbf{E}, \mathbf{F}$ are Banach spaces over $\mathbb{R}$. One says that F is differentiable at $p \in U$ if there exists a bounded linear transformation

$$JF(p) : \mathbf{E} \to \mathbf{F}$$

(the *Jacobian* of F at p) such that

$$\lim_{x \to p} \frac{F(x) - F(p) - JF(p) \cdot (x - p)}{\|x - p\|} = 0.$$

As usual, one shows that such a linear transformation is unique and that its existence implies the continuity of F at p. If this condition holds for all $p \in U$, we obtain a map

$$JF : U \to \mathcal{L}(\mathbf{E}, \mathbf{F}),$$

where $\mathcal{L}(\mathbf{E},\mathbf{F})$ denotes the Banach space of bounded linear transformations from $\mathbf{E}$ to $\mathbf{F}$. If the map JF is continuous, we say that F is of class C^1 on U. As usual, one obtains the chain rule

$$J(F \circ G) = JF \circ JG$$

for C^1 functions, as well as the fact that a bounded linear transformation is its own Jacobian. Inductively, one defines F to be of class C^k on U, $k \geq 1$, if JF is defined and of class C^{k-1} on U. If F is of class C^k on U, $\forall\, k \geq 1$, then F is of class C^∞ on U. If F is a C^k mapping of U, one-to-one onto an open subset $V \subseteq \mathbf{F}$, $k \geq 1$, and if $F^{-1} : V \to U$ is also of class C^k, then F is said to be a C^k diffeomorphism of U onto V. In our proof of the inverse function theorem, we chose $\eta > 0$ so small that $J\Phi(x)$ is invertible, $\forall\, x \in B_\eta$. The usual determinant argument for this is unavailable in infinite dimensions, but it remains elementary that the subset of elements in $\mathcal{L}(\mathbf{E},\mathbf{F})$ with bounded inverses is open (*cf.* [**24**, pp. 71–72]). Finally, the mean value theorem (equation (B.1)) is completely elementary for general Banach spaces (*cf.* [**24**, p. 107]), so the proof that we have given for the finite dimensional case of the inverse function theorem goes through unchanged.

Theorem B.8. *Let* $\mathbf{F}$ *and* $\mathbf{E}$ *be Banach spaces,* $U \subseteq \mathbf{E}$ *an open subset, and let* $F : U \to \mathbf{F}$ *be of class* C^k *on* U, $1 \leq k \leq \infty$. *If* $p \in U$ *and* $JF(p)$ *is an isomorphism of Banach spaces, then there is an open neighborhood* W *of* p *in* U *that is carried* C^k *diffeomorphically by* F *onto an open neighborhood* $F(W)$ *of* $F(p)$ *in* $\mathbf{F}$. *The Jacobian of* F^{-1} *at* $F(x)$ *is the inverse of* $JF(x)$, $\forall\, x \in W$.

Here, of course, by an "isomorphism of Banach spaces" we mean a bounded linear transformation with bounded inverse. The final statement of Theorem B.8 is just the equation

$$J(F^{-1}) = (JF)^{-1} \circ F^{-1}$$

(Lemma B.6).

There is a corresponding version of the implicit function theorem (Corollary 2.4.11) for Banach spaces. This will give a remarkably elegant way of proving the smooth dependence on initial conditions in the fundamental theorem of O.D.E. In order to state this implicit function theorem, we will need some notation.

Let $\mathbf{E}$, $\mathbf{F}$, and $\mathbf{H}$ be Banach spaces, $U \subseteq \mathbf{E}$ and $V \subseteq \mathbf{F}$ open subsets, and let $F : U \times V \to \mathbf{H}$ be of class C^k on $U \times V$, $1 \leq k \leq \infty$. Denoting the variables in U and V by x and y, respectively, let $(p,q) \in U \times V$ and form the "partial Jacobian" $J_yF(p,q) \in \mathcal{L}(\mathbf{F},\mathbf{H})$ (respectively, $J_xF(p,q) \in \mathcal{L}(\mathbf{E},\mathbf{H})$) by holding x (respectively, y) fixed and treating F as a function of the remaining variable.

Theorem B.9. *Let* $F : U \times V \to \mathbf{H}$ *be of class* C^k *as above, let* $(p,q) \in U \times V$, $F(p,q) = c \in \mathbf{H}$, *and assume that* $J_yF(p,q) : \mathbf{F} \to \mathbf{H}$ *is an isomorphism of Banach spaces. Then there exists an open neighborhood* W *of* p *in* U *and a unique* C^k *map* $\varphi : W \to V$ *such that* $\varphi(p) = q$ *and, on* W,

$$F(x, \varphi(x)) \equiv c.$$

Proof. Let

$$G : U \times V \to \mathbf{E} \times \mathbf{H}$$

be defined by the formula

$$G(x,y) = (x, F(x,y)),$$

a C^k map with Jacobian

$$JG(p,q) = \begin{bmatrix} \mathrm{id}_{\mathbf{E}} & 0 \\ * & J_y F(p,q) \end{bmatrix}.$$

This is an isomorphism of the Banach space $\mathbf{E} \times \mathbf{F}$ onto $\mathbf{E} \times \mathbf{H}$, so the inverse function theorem provides an open neighborhood of (p,q) in $U \times V$ that is carried by G diffeomorphically onto an open neighborhood of (p,c) in $\mathbf{E} \times \mathbf{H}$. Since the Banach space $\mathbf{E} \times \mathbf{F}$ has the Cartesian product topology, this neighborhood of (p,q) can be taken to be of the form $W \times W'$, where W is an open neighborhood of p and W' an open neighborhood of q. On $G(W \times W')$, the inverse transformation has a formula

$$G^{-1}(x,z) = (x, H(x,z))$$

for a unique C^k map $H : G(W \times W') \to W'$. Thus, the desired map φ has the formula $\varphi(x) = H(x,c)$. Indeed,

$$(x, F(x, \varphi(x))) = G(x, \varphi(x)) = G(x, H(x,c)) = (x,c).$$

Since G is a diffeomorphism, φ is unique. □

APPENDIX C

Ordinary Differential Equations

We prove Theorem 2.8.4. Remark that the general system (time dependent with parameters $z = (z^1, \ldots, z^r) \in V \subseteq \mathbb{R}^r$)

$$\frac{dx^i}{dt} = f^i(t, z, x^1(t,z), \ldots, x^n(t,z)), \quad 1 \leq i \leq n, \quad -\epsilon < t < \epsilon,$$

on an open subset $U \subseteq \mathbb{R}^n$, can be viewed as an autonomous system without parameters on the open subset $(-\epsilon, \epsilon) \times V \times U \subseteq \mathbb{R}^{n+r+1}$, by adjoining the equations

$$\frac{dt}{dt} = 1,$$
$$\frac{dz^1}{dt} = 0,$$
$$\vdots$$
$$\frac{dz^r}{dt} = 0.$$

Consequently, we formulate the proof for the autonomous case without parameters on an open subset $U \subseteq \mathbb{R}^n$:

$$(*) \qquad \frac{dx^i}{dt} = f^i(x^1(t), \ldots, x^n(t)), \quad 1 \leq i \leq n.$$

We will assume that $1 \leq k \leq \infty$ and that $f^i \in C^k(U)$, $1 \leq i \leq n$, and prove that the solution defines a local flow of class C^k. This will involve an induction on k in which the remark in the previous paragraph becomes crucial.

No generality will be lost by taking $U = \mathbb{R}^n$ and assuming that the vector field $X = (f^1, \ldots, f^n)$ is compactly supported. Indeed, we are proving a local theorem near $x_0 \in U$, so X can be damped off to 0 outside of a relatively compact region $W \subset \overline{W} \subset U$, containing a given closed ball $B_\eta(x_0)$, then extended by 0 to all of $\mathbb{R}^n$. In this way, we will be considering a complete vector field X on $\mathbb{R}^n$ and will find a uniform parameter interval $(-c, c)$ on which the solution curves are defined for all choices of initial condition $x \in \mathbb{R}^n$. Restricting to $x \in B_\eta(x_0)$ and taking $c > 0$ smaller, if necessary, we see that integral curves to X, starting in $B_\eta(x_0)$ and parametrized on $(-c, c)$, must stay in the region W where X has not been altered.

C.1. Existence and uniqueness of solutions

Since we will not be thinking of the vector field X as a differential operator, we will write $X(x)$ for X_x. A curve $s : (-\delta, \epsilon) \to \mathbb{R}^n$ is integral to X if and only if

$$s(t) = s(0) + \int_0^t X(s(u))\, du, \quad -\delta < t < \epsilon.$$

This formula suggests a mapping of a certain complete metric space into itself which will turn out to be a contraction mapping (Definition B.2).

Let K be a Lipschitz constant for X. This exists because X is C^1 and compactly supported. Let $0 < c < 1/K$.

In what follows, $E(\mathbb{R}^n)$ will denote the Banach space of all continuous paths

$$s : [-c, c] \to \mathbb{R}^n,$$

with the sup norm.

Lemma C.1.1. *For each $a \in \mathbb{R}^n$, the transformation*

$$T_a : E(\mathbb{R}^n) \to E(\mathbb{R}^n),$$

$$T_a(s)(t) = a + \int_0^t X(s(u))\, du, \quad -c \le t \le c,$$

is a contraction mapping.

Proof. If $s_1, s_2 \in E(\mathbb{R}^n)$, then

$$\begin{aligned}
\|T_a(s_1)(t) - T_a(s_2)(t)\| &= \left\| \int_0^t (X(s_1(u)) - X(s_2(u))\, du \right\| \\
&\le c \sup_{-c \le u \le c} \|X(s_1(u)) - X(s_2(u))\| \\
&\le cK \sup_{-c \le u \le c} \|s_1(u) - s_2(u)\|.
\end{aligned}$$

That is,

$$\|T_a(s_1) - T_a(s_2)\| \le cK\|s_1 - s_2\|$$

and $0 < cK < 1$, so T_a is a contraction mapping. □

By the contraction mapping lemma, it follows that there is a unique curve $s \in E(\mathbb{R}^n)$ with $T_a(s) = s$. As remarked above, this says, equivalently, that there is a unique solution $s(t) = (x^1(t), \dots, x^n(t))$ to $(*)$, parametrized on $[-c, c]$ and having initial condition $s(0) = a$.

Lemma C.1.2. *The solution curve $s : [-c, c] \to \mathbb{R}^n$ is of class C^{k+1}.*

Proof. Indeed, the equation

$$s(t) = a + \int_0^t X(s(u))\, du,$$

together with the fact that X is C^k, implies that, if s is of class C^j, some $0 \le j \le k$, then s is actually of class C^{j+1}. But $s \in E(\mathbb{R}^n)$ is of class C^0, by definition, hence induction on j gives the assertion. □

Remark. If the vector field X is only required to be Lipschitz, the above argument goes through to provide a unique C^1 solution parametrized on $[-c, c]$.

We have not quite proven the uniqueness of solutions as formulated in Theorem 2.8.4. Suppose that s_1 and s_2 are two solutions, parametrized on respective closed, nondegenerate intervals J_1 and J_2 about 0, and both satisfying the initial condition $s_1(0) = s_2(0) = a$. Let $J \subseteq J_1 \cap J_2$ be the largest closed, nondegenerate subinterval about 0 on which $s_1 = s_2$. By what has just been proven, J is not empty. If $J \neq J_1 \cap J_2$, then one endpoint $\overline{a}$ of J lies in both J_1 and J_2 and $\overline{a} = s_1(\tau) = s_2(\tau)$. Then $\sigma_i(t) = s_i(\tau + t)$ is a solution of $(*)$ with $\sigma_i(0) = \overline{a}$,

$i = 1, 2$, so it follows from what has just been proven that s_1 and s_2 agree on a larger subinterval than J after all. That is, $J = J_1 \cap J_2$ and we have proven the existence and uniqueness part of Theorem 2.8.4.

Define

$$\Phi : [-c, c] \times \mathbb{R}^n \to \mathbb{R}^n$$

such that, for each $x \in \mathbb{R}^n$ and $-c \leq t \leq c$, $\Phi(t, x)$ describes the unique integral curve to X with $\Phi(0, x) = x$. Denote this integral curve by s_x and define

$$\varphi : \mathbb{R}^n \to E(\mathbb{R}^n)$$

by $\varphi(x) = s_x$.

Lemma C.1.3. *The map φ admits a Lipschitz constant B. That is,*

$$\|\varphi(x) - \varphi(y)\| \leq B\|x - y\|, \ \forall\, x, y \in \mathbb{R}^n.$$

In particular, φ is continuous.

Proof. Let $x, y \in W$ and remark that

$$\|s_x - T_y(s_x)\| = \|T_x(s_x) - T_y(s_x)\| = \|x - y\|.$$

Using notation established above, set $\epsilon = cK \in (0, 1)$ and write

$$\|s_x - T_y^q(s_x)\| \leq \sum_{j=1}^{q} \|T_y^{j-1}(s_x) - T_y^j(s_x)\| \leq \sum_{j=1}^{q} \epsilon^{j-1}\|x - y\|.$$

Since $T_y^q(s_x) \to s_y$ in $E(\mathbb{R}^n)$ as $q \to \infty$ and

$$\sum_{j=1}^{\infty} \epsilon^{j-1} = B < \infty,$$

thc assertion is established. □

Corollary C.1.4. *The map*

$$\Phi : (-c, c) \times \mathbb{R}^n \to \mathbb{R}^n$$

is a local C^0 flow.

We emphasize that, because of our simplifying assumption that X is compactly supported, the parameter interval $(-c, c)$ is uniform for all initial values $x \in \mathbb{R}^n$. The proof of Lemma 4.1.10 is applicable, therefore, and gives

Corollary C.1.5. *The compactly supported vector field X generates a unique C^0 flow*

$$\Phi : \mathbb{R} \times \mathbb{R}^n \to \mathbb{R}^n.$$

The hardest part of the proof of Theorem 2.8.4 is to show that this flow is of class C^k. It turns out that, if we can prove it to be C^1, a rather ingenious recursive argument yields an inductive proof that Φ is of class C^k.

In order to prove that Φ is C^1, we will verify the hypotheses of the implicit function theorem (Theorem B.9) for the map

$$F : \mathbb{R}^n \times E(\mathbb{R}^n) \to E(\mathbb{R}^n),$$

defined by

$$F(x, s) = x - s + \int_0^* X \circ s.$$

This notation is understood to define a continuous curve in $\mathbb{R}^n$ by the formula

$$F(x,s)(t) = x - s(t) + \int_0^t X(s(u))\,du, \quad -c \le t \le c.$$

One has $F(x,s) = 0$ if and only if $s = s_x$ is the integral curve to X with initial value $s_x(0) = x$. The implicit function theorem will guarantee that the map $\varphi(x) = s_x$, satisfying $F(x, \varphi(x)) = 0$, is of class C^1 (in fact, of class C^k). It is not obvious that this implies C^1 smoothness for the flow Φ itself, but this will be the case.

Before giving the details, we make a small digression.

C.2. A digression concerning Banach spaces

Let $G : \mathbf{F} \to \mathbf{H}$ be a C^k map of Banach spaces, $1 \le k \le \infty$, and let $E(\mathbf{F})$ and $E(\mathbf{H})$ denote the associated Banach spaces of continuous paths, parametrized on a fixed, bounded, nondegenerate interval $[-c, c]$. We define a map

$$F : E(\mathbf{F}) \to E(\mathbf{H})$$

by the formula

$$F(s) = \int_0^* G \circ s.$$

We are going to prove that F is also of class C^k.

First, remark that

$$\int_0^* JG \circ s \in E(\mathfrak{L}(\mathbf{F}, \mathbf{H}))$$

can be interpreted as an element of $\mathfrak{L}(E(\mathbf{F}), E(\mathbf{H}))$, $\forall\, s \in E(\mathbf{F})$. Indeed, if $\sigma \in E(\mathbf{F})$, then

$$\int_0^t JG(s(u)) \cdot \sigma(u)\,du \in \mathbf{H}, \quad -c \le t \le c,$$

so $\int_0^* JG \circ s$ can be viewed as a transformation sending $\sigma \in E(\mathbf{F})$ to

$$\int_0^* (JG \circ s) \cdot \sigma \in E(\mathbf{H}).$$

This is evidently a linear transformation of $E(\mathbf{F})$ into $E(\mathbf{H})$ and it is bounded because $JG(s(u))$ is uniformly bounded, $-c \le u \le c$.

Lemma C.2.1. *The map F is of class at least C^1 and, as s ranges over $E(\mathbf{F})$,*

$$JF(s) = \int_0^* JG \circ s.$$

Proof. Let $s_0 \in E(\mathbf{F})$. Then

$$\lim_{s \to s_0} \frac{F(s) - F(s_0) - \int_0^* (JG \circ s_0) \cdot (s - s_0)}{\|s - s_0\|}$$
$$= \lim_{s \to s_0} \int_0^* \frac{G \circ s - G \circ s_0 - (JG \circ s_0) \cdot (s - s_0)}{\|s - s_0\|} = 0$$

since the integrand converges to 0 uniformly on $[-c, c]$. Thus, $\int_0^* JG \circ s_0$ satisfies the definition of $JF(s_0)$, $\forall\, s_0 \in E(\mathbf{F})$. To see that JF is continuous, write

$$\|JF(s_1) - JF(s_2)\| = \left\| \int_0^* (JG \circ s_1 - JG \circ s_2) \right\| \le c\|JG \circ s_1 - JG \circ s_2\|$$

and appeal to the continuity of JG. □

Suppose, now, that it has been proven that F is of class C^r, some $1 \leq r < k$, and that $J^r F(s) = \int_0^* J^r G \circ s$, $\forall\, s \in E(\mathbf{F})$. The lemma gives the case $r = 1$ and it also provides the inductive step: $J^r F$ is of class C^1 and

$$J^{r+1}F(s) = \int_0^* J^{r+1}G \circ s, \quad \forall\, s \in E(\mathbf{F}).$$

Corollary C.2.2. *If $G : \mathbf{F} \to \mathbf{H}$ is of class C^k, $1 \leq k \leq \infty$, then F is also of class C^k and*

$$JF(s) = \int_0^* JG \circ s, \quad \forall\, s \in E(\mathbf{F}).$$

C.3. Smooth dependence on initial conditions

We return to the map F of $\mathbb{R}^n \times E(\mathbb{R}^n)$ into $E(\mathbb{R}^n)$ given by the formula

$$F(x, s) = x - s + \int_0^* X \circ s.$$

From the previous section, this has the same smoothness class C^k as the vector field X and

$$J_s F(x, s) = -\operatorname{id}_{E(\mathbb{R}^n)} + \int_0^* JX \circ s.$$

Let $\|JX\| = M$. For the following lemma, we take $c > 0$ smaller, if necessary, so that $c < 1/M$.

Lemma C.3.1. *For each $(x, s) \in \mathbb{R}^n \times E(\mathbb{R}^n)$, the bounded linear operator $J_s F(x, s)$ has a bounded inverse.*

Proof. Indeed, for arbitrary $\sigma \in E(\mathbb{R}^n)$,

$$\left\| \int_0^t JX(s(u)) \cdot \sigma(u)\, du \right\| \leq cM\|\sigma\|, \quad -c \leq t \leq c.$$

It follows that the operator $L = \int_0^* JX \circ s$ has norm $\|L\| \leq cM < 1$, hence that

$$R = \sum_{j=0}^{\infty} (-1)^{j+1} L^j$$

converges. That is, $R \in \mathcal{L}(E(\mathbb{R}^n), E(\mathbb{R}^n))$ and

$$R \circ (L - \operatorname{id}_{E(\mathbb{R}^n)}) = (L - \operatorname{id}_{E(\mathbb{R}^n)}) \circ R = \operatorname{id}_{E(\mathbb{R}^n)}$$

as asserted. □

Thus, the hypothesis of Theorem B.9 is verified.

Corollary C.3.2. *The map $\varphi : \mathbb{R}^n \to E(\mathbb{R}^n)$, defined by $\varphi(x) = s_x$, is smooth of class C^k.*

In fact, we only need to know that φ is of class C^1.

Corollary C.3.3. *The global flow Φ on $\mathbb{R}^n$ is smooth of class at least C^1.*

Proof. By Corollary C.3.2, φ is smooth of class at least C^1, so

$$\varphi(x+h) - \varphi(x) = J\varphi(x) \cdot h + \|h\|\delta(h),$$

where

$$\lim_{h\to 0} \delta(h) = 0 \text{ in } E(\mathbb{R}^n).$$

That is,

$$\lim_{h\to 0} \delta(h)(t) = 0 \text{ uniformly on } [-c, c].$$

Thus, from

$$\begin{aligned}\Phi(t, x+h) - \Phi(t, x) &= \varphi(x+h)(t) - \varphi(x)(t) \\ &= (J\varphi(x) \cdot h)(t) + \|h\|\delta(h)(t)\end{aligned}$$

we deduce that the partial Jacobian

$$J_x\Phi(t, x) \cdot h = (J\varphi(x) \cdot h)(t)$$

exists and, for each $h \in \mathbb{R}^n$, is continuous in (t, x). By successively substituting $h = e_i$, $1 \le i \le n$, we conclude that all entries of the matrix $J_x\Phi(t, x)$ are continuous functions of $(t, x) \in [-c, c] \times \mathbb{R}^n$. Since the flow lines are integral to X, we also see that

$$\frac{\partial \Phi(t, x)}{\partial t} = X(\Phi(t, x))$$

is continuous in (t, x). Thus, all entries of $J\Phi(t, x)$ are continuous on $[-c, c] \times \mathbb{R}^n$ and Φ is of class C^1 there. The parameter interval $[-c, c]$ of this local C^1 flow being uniform for all initial values $x \in \mathbb{R}^n$, we obtain the unique global C^1 flow

$$\Phi : \mathbb{R} \times \mathbb{R}^n \to \mathbb{R}^n$$

generated by X. □

Remark. In order to prove Lemma C.3.1, we had to allow c to be chosen small enough. Our final conclusion, however, was that the *global* flow Φ is C^1. We are going to prove inductively that, for $1 \le q \le k$, Φ is of class C^q (the case $q = 1$ is Corollary C.3.3). At each step of the induction, c may have to be chosen smaller. Nonetheless, at each step the conclusion is about the global flow, so the induction works even for the case $k = \infty$.

Let $2 \le j \le k$ and assume that it has been shown that the flow $\Phi(t, x)$ is of class C^{j-1}. We will show that $\partial\Phi(t, x)/\partial t$ and $J_x\Phi(t, x)$ are both of class C^{j-1} in (t, x), concluding that $\Phi(t, x)$ is of class C^j.

Lemma C.3.4. *The expression $\partial\Phi(t, x)/\partial t$ is of class C^{j-1} in (t, x).*

Proof. Indeed,

$$\partial\Phi(t, x)/\partial t = X(\Phi(t, x))$$

and X is of class C^k, Φ of class C^{j-1}, and $j \le k$. □

Lemma C.3.5. *The expression $J_x\Phi(t, x)$ is of class C^{j-1} in (t, x).*

Proof. Since we know that Φ is of class at least C^1, we can differentiate

$$\Phi(t, x) = x + \int_0^t X(\Phi(u, x))\, du$$

under the integral sign with respect to the variables x, obtaining

$$J_x\Phi(t,x) = \mathrm{id}_{\mathbb{R}^n} + \int_0^t JX(\Phi(u,x))J_x\Phi(u,x)\,du.$$

Then,

$$\frac{\partial}{\partial t}J_x(\Phi(t,x)) = JX(\Phi(t,x))J_x(\Phi(t,x)),$$

and this can be interpreted as a time dependent system of O.D.E. with parameters x. The unknown functions are the entries of the matrix $J_x(\Phi(t,x))$, and so the system is linear. The coefficients are entries of the matrix $JX(\Phi(t,x))$, hence are of class C^{j-1} in (t,x). Thus, this system is of class C^{j-1}. As remarked at the beginning of this appendix, time dependent systems with parameters are equivalent to autonomous systems without parameters, so the inductive hypothesis guarantees that the entries of $J_x(\Phi(t,x))$ are smooth of class C^{j-1}. □

The proof of Theorem 2.8.4 is complete.

Remark. It would have been possible to formulate and prove the O.D.E. theorem entirely in the context of Banach spaces (see, *e.g.*, Lang [**24**, pp. 132–145], who credits the idea of using the implicit function theorem to Pugh and Robbin), but we have avoided significant technical details by resisting the temptation to do so. On the other hand, judicious use of the implicit function theorem in infinite dimensions does seem to simplify the proof of the finite dimensional theorem.

C.4. The Linear Case

In the proof of Theorem 10.1.13, we appealed to the theorem that a linear system of O.D.E. has solutions that are defined on the largest parameter interval (b,c) on which the system itself is defined. Here is the formal theorem.

Theorem C.4.1. *Let $A:(b,c)\to\mathfrak{M}(n)$ be smooth of class C^k. Then the system*

$$\dot{s}(t) = A(t)\cdot s(t), \quad b<t<c,$$

with initial condition $s(t_0)=a$ has solution $s(t)$ defined for $b<t<c$.

Proof. Let $(b',c')\subseteq(b,c)$ be the maximal open interval about t_0 on which the solution $s(t)$ is defined. If $c'<c$, we deduce a contradiction as follows. Fix a basis $\{w_1,\dots,w_n\}$ of $\mathbb{R}^n$ and, for $1\le i\le n$, let $\sigma_i(t)$ be the unique solution of

$$\dot{\sigma}_i(t) = A(t)\cdot\sigma_i(t),$$
$$\sigma_i(c') = w_i,$$

defined on some interval $(c'-\epsilon,c'+\epsilon)\subseteq(b,c)$. Choose $t_*\in(c'-\epsilon,c')$, so close to c' that $\{\sigma_1(t_*),\dots,\sigma_n(t_*)\}$ is also a basis of $\mathbb{R}^n$. This is possible by the continuity of the solutions σ_i. Thus, there are constants $c_1,\dots,c_n\in\mathbb{R}$ such that

$$\sum_{i=1}^n c_i\sigma_i(t_*) = a.$$

By the linearity of the system, it is clear that

$$\sigma(t) = \sum_{i=1}^n c_i\sigma_i, \quad c'-\epsilon<t<c'+\epsilon,$$

is also a solution and that $\sigma(t_*) = s(t_*)$. By uniqueness of solutions, σ and s agree on $(c' - \epsilon, c')$, so σ extends the solution s to the larger interval $(b', c' + \epsilon)$. This contradicts the maximality of (b', c'), proving that $c' = c$. Similarly, $b' = b$. □

APPENDIX D

The de Rham Cohomology Theorem

In this appendix, our goal is to prove the de Rham theorem for Čech and singular cohomology. In order to avoid any possible confusions between cohomology theories, we will denote the de Rham cohomology by $H^*_{\mathrm{DR}}(M)$. Our proof will show that the graded algebra structures are isomorphic. In particular, this will show that the de Rham cohomology algebra $H^*_{\mathrm{DR}}(M)$ is a purely topological invariant of the manifold M (Theorem 8.9.6).

D.1. Čech cohomology

This section consists largely of definitions and statements of basic properties. Few proofs will be given because they are mostly routine (and tedious) computations.

The first step is to define the cohomology of an open cover $\mathfrak{U} = \{U_\alpha\}_{\alpha\in\mathfrak{A}}$ of the space X with coefficients in a commutative ring R with unity.

Definition D.1.1. If $p \geq 0$ is an integer, a (Čech) p-simplex of $\mathfrak{U}$ is an ordered $(p+1)$-tuple $(U_{\alpha_0}, U_{\alpha_1}, \dots, U_{\alpha_p})$ of elements of $\mathfrak{U}$ such that $U_{\alpha_0} \cap \cdots \cap U_{\alpha_p}$ is nonempty.

Definition D.1.2. If $p \geq 0$ is an integer, an R-valued (Čech) p-cochain on $\mathfrak{U}$ is a function φ that, to each p-simplex $(U_{\alpha_0}, U_{\alpha_1}, \dots, U_{\alpha_p})$, assigns an element

$$\varphi(U_{\alpha_0}, U_{\alpha_1}, \dots, U_{\alpha_p}) = \varphi_{\alpha_0\alpha_1\cdots\alpha_p} \in R.$$

The set of all R-valued p-cochains on $\mathfrak{U}$ is denoted by $\check{C}^p(\mathfrak{U}; R)$.

Evidently, the operations of "simplexwise addition" of cochains and "simplexwise scalar multiplication" make $\check{C}^p(\mathfrak{U}; R)$ into an R-module. More precisely,

$$(\varphi + \psi)_{\alpha_0\alpha_1\cdots\alpha_p} = \varphi_{\alpha_0\alpha_1\cdots\alpha_p} + \psi_{\alpha_0\alpha_1\cdots\alpha_p}$$

and

$$(a\varphi)_{\alpha_0\alpha_1\cdots\alpha_p} = a(\varphi_{\alpha_0\alpha_1\cdots\alpha_p}),$$

$\forall\, a \in R$ and $\forall\, \varphi, \psi \in \check{C}^p(\mathfrak{U}; R)$.

Definition D.1.3. The (Čech) coboundary operator

$$\delta : \check{C}^p(\mathfrak{U}; R) \to \check{C}^{p+1}(\mathfrak{U}; R)$$

is the R-linear map defined by the formula

$$\delta(\varphi)_{\alpha_0\alpha_1\cdots\alpha_{p+1}} = \sum_{i=0}^{p+1} (-1)^i \varphi_{\alpha_0\cdots\widehat{\alpha}_i\cdots\alpha_{p+1}}.$$

Thus, we obtain a sequence

$$\xrightarrow{\delta} \check{C}^p(\mathfrak{U};R) \xrightarrow{\delta} \check{C}^{p+1}(\mathfrak{U};R) \xrightarrow{\delta} \check{C}^{p+2}(\mathfrak{U};R) \xrightarrow{\delta} \cdots .$$

The following is a routine computation.

Lemma D.1.4. *The sequence of coboundary operators satisfies* $\delta^2 = 0$.

Definition D.1.5. The module of (Čech) p-cocycles is

$$\check{Z}^p(\mathfrak{U};R) = \ker(\delta) \cap \check{C}^p(\mathfrak{U};R)$$

and the module of (Čech) p-coboundaries is

$$\check{B}^p(\mathfrak{U};R) = \operatorname{im}(\delta) \cap \check{C}^p(\mathfrak{U};R).$$

As usual, $\check{B}^p(\mathfrak{U};R) \subseteq \check{Z}^p(\mathfrak{U};R)$.

Definition D.1.6. The pth Čech cohomology of $\mathfrak{U}$, with coefficients in the ring R, is

$$\check{H}^p(\mathfrak{U};R) = \check{B}^p(\mathfrak{U};R)/\check{Z}^p(\mathfrak{U};R).$$

We have defined $\check{H}^*(\mathfrak{U};R)$ as a graded R-module. It is made into a graded algebra by the "cup product".

Definition D.1.7. If $\varphi \in \check{C}^p(\mathfrak{U};R)$ and $\psi \in \check{C}^q(\mathfrak{U};R)$, the cup product $\varphi\psi \in \check{C}^{p+q}(\mathfrak{U};R)$ of these cochains is defined by

$$(\varphi\psi)_{\alpha_0\cdots\alpha_{p+q}} = \varphi_{\alpha_0\cdots\alpha_p}\psi_{\alpha_p\cdots\alpha_{p+q}}.$$

This makes $\check{C}^*(\mathfrak{U};R)$ into a graded algebra. Another straightforward computation proves the following.

Lemma D.1.8. *If* $\varphi \in \check{C}^p(\mathfrak{U};R)$ *and* $\psi \in \check{C}^q(\mathfrak{U};R)$, *then*

$$\delta(\varphi\psi) = (\delta\varphi)\psi + (-1)^p\varphi(\delta\psi).$$

This is formally the same as the formula for the exterior derivative of the wedge product of a p-form and a q-form. As in that case, we get the following consequence.

Lemma D.1.9. *The graded module* $\check{Z}^*(\mathfrak{U};R)$ *of Čech cocycles is a graded subalgebra of* $\check{C}^*(\mathfrak{U};R)$ *and* $\check{B}^*(\mathfrak{U};R) \subseteq \check{Z}^*(\mathfrak{U};R)$ *is a 2-sided ideal. Consequently, cup product is well defined on* $\check{H}^*(\mathfrak{U};R)$, *making that graded* R*-module into a graded algebra over* R.

Finally, we will define the Čech cohomology algebra of the space X to be the "direct limit"

$$\check{H}^*(X;R) = \varinjlim \check{H}^*(\mathfrak{U};R)$$

over finer and finer open covers. We make this precise.

Let $\{V_\alpha^*\}_{\alpha\in\mathfrak{A}}$ be a family of graded R-algebras, indexed by a *partially ordered set* $\mathfrak{A}$. That is, there is a partial ordering $\alpha \preceq \beta$ on $\mathfrak{A}$ such that, whenever $\alpha, \beta \in \mathfrak{A}$, $\exists\gamma \in \mathfrak{A}$ with $\alpha \preceq \gamma$ and $\beta \preceq \gamma$. Assume also that, whenever $\alpha \preceq \beta$, there is given a homomorphism $\varphi_\alpha^\beta : V_\alpha^* \to V_\beta^*$ of graded algebras and that, whenever $\alpha \preceq \beta \preceq \gamma$, then $\varphi_\beta^\gamma \circ \varphi_\alpha^\beta = \varphi_\alpha^\gamma$.

We say that $\{V_\alpha^*, \varphi_\alpha^\beta\}_{\alpha,\beta\in\mathfrak{A}}$ is a *directed system of graded* R*-algebras*.

On the disjoint union

$$\mathcal{V}^* = \coprod_{\alpha \in \mathfrak{A}} V^*_\alpha,$$

define the equivalence relation $\sim$ generated by

$$v \sim \varphi^\beta_\alpha(v),$$

where $v \in V^*_\alpha$ and $\alpha \preceq \beta$. Then $\mathcal{V}^*/\!\sim$ has a natural graded R-algebra structure. Indeed, scalar multiplication $a[v] = [av]$ is clearly well defined. As for addition, if $[v], [w] \in \mathcal{V}^*/\!\sim$ are represented by $v \in V^*_\alpha$ and $w \in V^*_\beta$, find $\gamma \in \mathfrak{A}$, $\alpha \preceq \gamma$, $\beta \preceq \gamma$, and set

$$[v] + [w] = [\varphi^\gamma_\alpha(v) + \varphi^\gamma_\beta(w)].$$

It is trivial to check that this is well defined and that these operations make $\mathcal{V}^*/\!\sim$ into a graded R-module. Similarly, the algebra multiplication passes to a well defined multiplication making $\mathcal{V}^*/\!\sim$ into a graded R-algebra.

Definition D.1.10. In the above situation, we set

$$\varinjlim V^*_\alpha = \mathcal{V}^*/\!\sim$$

and call this graded R-algebra the direct limit of the directed system of graded R-algebras.

Example D.1.11. For a differentiable manifold M, let $\mathfrak{U}_x$ denote the set of open neighborhoods U of $x \in M$. This is a directed system under the partial order $U \preceq V \Leftrightarrow U \supseteq V$. The graded algebras $\{A^*(U)\}_{U \in \mathfrak{U}_x}$ form a directed system under the restriction homomorphisms

$$\rho^V_U(\omega) = \omega | V,$$

where $\omega \in A^*(U)$ and $U \supseteq V$. Then

$$\mathcal{A}^*_x = \varinjlim A^*(U)$$

is just the graded algebra of *germs* at $x \in M$ of smooth forms.

Let $\mathfrak{O}(X)$ denote the set of open covers of X. This is partially ordered by:

$$\mathfrak{U} \preceq \mathcal{V} \Leftrightarrow \mathcal{V} \text{ is a refinement of } \mathfrak{U}.$$

Since any two open covers of X have a common refinement, this makes $\mathfrak{O}(X)$ into a directed system. If $\mathfrak{U} = \{U_\alpha\}_{\alpha \in \mathfrak{A}}$, $\mathcal{V} = \{V_\beta\}_{\beta \in \mathfrak{B}}$, and $\mathfrak{U} \preceq \mathcal{V}$, then there is a choice function $i : \mathfrak{B} \to \mathfrak{A}$ such that $V_\beta \subseteq U_{i(\beta)}$, $\forall\, \beta \in \mathfrak{B}$. This induces a homomorphism

$$i^\# : \check{C}^*(\mathfrak{U}; R) \to \check{C}^*(\mathcal{V}; R)$$

of graded algebras, where

$$i^\#(\varphi)_{\beta_0\beta_1\cdots\beta_p} = \varphi_{i(\beta_0)i(\beta_1)\cdots i(\beta_p)}.$$

The following is trivial.

Lemma D.1.12. *The homomorphism $i^\# : \check{C}^*(\mathfrak{U}; R) \to \check{C}^*(\mathcal{V}; R)$ satisfies $i^\# \circ \delta = \delta \circ i^\#$.*

We cannot use $i^\#$ as a homomorphism $\varphi^{\mathcal{V}}_{\mathfrak{U}} : \check{C}^*(\mathfrak{U}; R) \to \check{C}^*(\mathcal{V}; R)$ for a directed system of algebras. The problem is that $i^\#$ depends on arbitrary choices, so we could never guarantee that $\varphi^{\mathcal{W}}_{\mathcal{V}} \circ \varphi^{\mathcal{V}}_{\mathfrak{U}} = \varphi^{\mathcal{W}}_{\mathfrak{U}}$. But the above lemma implies that $i^\#$ induces $i^* : \check{H}^*(\mathfrak{U}; R) \to \check{H}^*(\mathcal{V}; R)$ and it turns out that, at the level of cohomology, the arbitrariness disappears.

Definition D.1.13. If $\mathfrak{U} \preceq \mathcal{V}$ in $\mathfrak{O}(X)$, if $i, j : \mathfrak{B} \to \mathfrak{A}$ are two choice functions, as above, and if $p \in \mathbb{Z}$, define

$$S : \check{C}^p(\mathfrak{U}; R) \to \check{C}^{p-1}(\mathcal{V}; R)$$

by the formula

$$S(\varphi)_{\beta_0\beta_1\cdots\beta_{p-1}} = \sum_{\ell=0}^{p-1}(-1)^\ell \varphi_{i(\beta_0)\cdots i(\beta_\ell)j(\beta_\ell)\cdots j(\beta_{p-1})}.$$

As usual, if $p - 1 < 0$, we understand that $\check{C}^{p-1}(\mathfrak{U}; R) = 0$ and $S = 0$. The following is checked by a routine (if somewhat tedious) computation, left to the reader.

Lemma D.1.14. *If i, j, and S are as above, then*

$$S \circ \delta + \delta \circ S = j^\# - i^\#.$$

Consequently, $i^* = j^* : \check{H}^*(\mathfrak{U}; R) \to \check{H}^*(\mathcal{V}; R)$.

By this lemma, whenever $\mathfrak{U} \preceq \mathcal{V}$ in $\mathfrak{O}(X)$, we define a homomorphism

$$\varphi_{\mathfrak{U}}^{\mathcal{V}} = i^* : \check{H}^*(\mathfrak{U}; R) \to \check{H}^*(\mathcal{V}; R)$$

of graded algebras that is independent of the (allowed) choice of $i : \mathfrak{B} \to \mathfrak{A}$.

Lemma D.1.15. *If $\mathfrak{U} \preceq \mathcal{V} \preceq \mathcal{W}$ in $\mathfrak{O}(X)$, then $\varphi_{\mathcal{V}}^{\mathcal{W}} \circ \varphi_{\mathfrak{U}}^{\mathcal{V}} = \varphi_{\mathfrak{U}}^{\mathcal{W}}$.*

Proof. Indeed, set

$$\mathfrak{U} = \{U_\alpha\}_{\alpha\in\mathfrak{A}},$$
$$\mathcal{V} = \{V_\beta\}_{\beta\in\mathfrak{B}},$$
$$\mathcal{W} = \{W_\gamma\}_{\gamma\in\mathfrak{C}},$$

and let $i : \mathfrak{B} \to \mathfrak{A}$, $j : \mathfrak{C} \to \mathfrak{B}$ be suitable choice functions. Then $i \circ j : \mathfrak{C} \to \mathfrak{A}$ is an allowed choice function relative to the refinement $\mathfrak{U} \preceq \mathcal{W}$. But

$$\varphi_{\mathfrak{U}}^{\mathcal{W}} = (i \circ j)^* = j^* \circ i^* = \varphi_{\mathcal{V}}^{\mathcal{W}} \circ \varphi_{\mathfrak{U}}^{\mathcal{V}}.$$

□

Thus, we get a directed system $\{\check{H}^*(\mathfrak{U}; R), \varphi_{\mathfrak{U}}^{\mathcal{V}}\}_{\mathfrak{U},\mathcal{V}\in\mathfrak{O}(X)}$ of graded algebras over R.

Definition D.1.16. The Čech cohomology of the space X with coefficients in R is the direct limit

$$\check{H}^*(X; R) = \varinjlim \check{H}^*(\mathfrak{U}; R),$$

taken over the directed system $\mathfrak{O}(X)$.

Let $f : X \to Y$ be a continuous map between spaces. Given $\mathfrak{U} \in \mathfrak{O}(Y)$, define $f^{-1}(\mathfrak{U}) \in \mathfrak{O}(X)$ by the usual pullback construction. If we are given a p-simplex $(f^{-1}(U_{\alpha_0}), \dots, f^{-1}(U_{\alpha_p}))$ of $f^{-1}(\mathfrak{U})$, then it is clear that $(U_{\alpha_0}, \dots, U_{\alpha_p})$ is a p-simplex of $\mathfrak{U}$. Consequently, each Čech cochain $\theta \in \check{C}^p(\mathfrak{U}; R)$ has a natural pullback $f^\#(\theta) \in \check{C}^p(f^{-1}(\mathfrak{U}); R)$. This defines a homomorphism

$$f^\# : \check{C}^*(\mathfrak{U}; R) \to \check{C}^*(f^{-1}(\mathfrak{U}); R)$$

of graded algebras.

Lemma D.1.17. *If $f : X \to Y$, as above, then $f^{\#} \circ \delta = \delta \circ f^{\#}$ and there is canonically defined an induced homomorphism*

$$f^* : \check{H}^*(Y;R) \to \check{H}^*(X;R)$$

of graded algebras over R. This makes Čech cohomology into a contravariant functor on the category of topological spaces and continuous maps.

Definition D.1.18. If $\mathfrak{A}$ is a directed system, a cofinal subsystem $\mathfrak{C} \subseteq \mathfrak{A}$ is a directed subsystem with the property that, whenever $\alpha \in \mathfrak{A}$, $\exists \gamma \in \mathfrak{C}$ such that $\alpha \preceq \gamma$.

Finally, the proof of the following lemma is a straightforward application of definitions.

Lemma D.1.19. *Let $\{V^*_\alpha, \varphi^\beta_\alpha\}_{\alpha,\beta\in\mathfrak{A}}$ be a directed system of graded R-algebras. If $\mathfrak{C} \subseteq \mathfrak{A}$ is a cofinal subset, then there is a canonical isomorphism*

$$\varinjlim V^*_\alpha = \varinjlim V^*_\gamma$$

of graded R-algebras, where the first limit is taken over all $\alpha \in \mathfrak{A}$ and the second is taken over all $\gamma \in \mathfrak{C}$.

By Corollary 10.5.8, the family of simple covers (Definition 8.5.5) of a differentiable manifold is a cofinal subset of $\mathfrak{O}(M)$.

Corollary D.1.20. *The Čech cohomology $\check{H}^*(M;R)$ of a differentiable manifold can be computed by taking the limit only over the directed system of simple covers.*

D.2. The de Rham–Čech complex

The proof we will give of the de Rham–Čech theorem is essentially that of André Weil [48]. The main step is to prove the following.

Theorem D.2.1. *If $\mathfrak{U}$ is a simple cover of M, there is a canonical isomorphism*

$$\psi_{\mathfrak{U}} : \check{H}^*(\mathfrak{U};\mathbb{R}) \to H^*_{\mathrm{DR}}(M)$$

of graded algebras and, if $\mathfrak{U} \preceq \mathfrak{V}$, where both are simple covers, then the diagram

$$\begin{array}{ccc} \check{H}^*(\mathfrak{U};\mathbb{R}) & \xrightarrow{\varphi^{\mathfrak{V}}_{\mathfrak{U}}} & \check{H}^*(\mathfrak{V};\mathbb{R}) \\ & {\scriptstyle\psi_{\mathfrak{U}}}\searrow \quad \swarrow{\scriptstyle\psi_{\mathfrak{V}}} & \\ & H^*_{\mathrm{DR}}(M) & \end{array}$$

is commutative.

The equivalence of de Rham theory and Čech theory follows easily. Indeed, the Čech cohomology can be computed by passing to the limit over the simple covers only (Corollary D.1.20), so the isomorphisms $\psi_{\mathfrak{U}}$ induce a well-defined homomorphism

$$\psi : \check{H}^*(M;\mathbb{R}) \to H^*_{\mathrm{DR}}(M).$$

The fact that each $\psi_{\mathfrak{U}}$ is an isomorphism implies the same for ψ.

Theorem D.2.2 (de Rham). *There is a canonical isomorphism*

$$H^*_{\mathrm{DR}}(M) = \check{H}^*(M;\mathbb{R})$$

of graded $\mathbb{R}$-algebras.

For use in the following section, we record the following corollary, implicit in the above argument.

Corollary D.2.3. *If $\mathfrak{U}$ is a simple cover, the natural homomorphism of $\check{H}^*(\mathfrak{U};\mathbb{R})$ into the limit $\check{H}^*(M;\mathbb{R})$ is an isomorphism of graded algebras.*

In order to prove Theorem D.2.1, we will build an enormous cochain complex of graded algebras that includes both $(A^*(M), d)$ and $(\check{C}^*(\mathfrak{U};\mathbb{R}), \delta)$ as subcomplexes. If $\mathfrak{U}$ is simple, we will prove that the inclusions of these subcomplexes induce isomorphisms in cohomology.

Fix an open cover $\mathfrak{U} = \{U_\alpha\}_{\alpha\in\mathfrak{A}}$ of M. For the following definitions, it is not necessary that $\mathfrak{U}$ be simple.

Definition D.2.4. A Čech p-cochain on $\mathfrak{U}$ with values in A^q is a function φ that, to each p-simplex $(U_{\alpha_0}, U_{\alpha_1}, \dots, U_{\alpha_p})$ of $\mathfrak{U}$ assigns

$$\varphi_{\alpha_0\alpha_1\cdots\alpha_p} \in A^q(U_{\alpha_0}\cap U_{\alpha_1}\cap\cdots\cap U_{\alpha_p}).$$

The set of all such cochains will be denoted $E^{p,q}(\mathfrak{U}) = \check{C}^p(\mathfrak{U}; A^q)$.

Although the coefficient ring $A^q(U_{\alpha_0}\cap U_{\alpha_1}\cap\cdots\cap U_{\alpha_p})$ changes with each simplex, one can still add cochains simplexwise and multiply them by real scalars. These operations make $E^{p,q}(\mathfrak{U}) = \check{C}^p(\mathfrak{U}; A^q)$ into a real vector space.

There is also a *bigraded* multiplication

$$E^{p,q}(\mathfrak{U}) \otimes E^{r,s}(\mathfrak{U}) \to E^{p+r,q+s}(\mathfrak{U}).$$

In defining this and other operations, we make the notation less bewildering to the eye by abusing it (the notation, that is). Whenever respective forms have been defined on respective open sets with common, nonempty intersection, addition of such forms and exterior products of such forms are understood to be defined on their common domain. For instance, if $(U_{\alpha_0}, U_{\alpha_1}, U_{\alpha_2})$ is a 2-simplex of $\mathfrak{U}$ and $\omega_{\alpha_i\alpha_j} \in A^q(U_{\alpha_i}\cap U_{\alpha_j})$, then

$$\omega_{\alpha_1\alpha_2} - \omega_{\alpha_0\alpha_2} + \omega_{\alpha_0\alpha_1} \in A^q(U_{\alpha_0}\cap U_{\alpha_1}\cap U_{\alpha_2}).$$

Similarly, if $\varphi_{\alpha_0\alpha_1} \in A^q(U_{\alpha_0}\cap U_{\alpha_1})$ and $\psi_{\alpha_1\alpha_2} \in A^s(U_{\alpha_1}\cap U_{\alpha_2})$, then

$$\varphi_{\alpha_0\alpha_1}\wedge\psi_{\alpha_1\alpha_2} \in A^{q+s}(U_{\alpha_0}\cap U_{\alpha_1}\cap U_{\alpha_2}).$$

With these conventions understood, we define the bigraded multiplication as follows. If $\varphi \in E^{p,q}(\mathfrak{U})$ and $\psi \in E^{r,s}(\mathfrak{U})$, then $\varphi\psi \in E^{p+r,q+s}(\mathfrak{U})$ is defined on a $(p+r)$-simplex $(U_{\alpha_0}, \dots, U_{\alpha_p}, \dots, U_{\alpha_{p+r}})$ by

$$(\varphi\psi)_{\alpha_0\cdots\alpha_{p+r}} = (-1)^{qr}\varphi_{\alpha_0\cdots\alpha_p}\wedge\psi_{\alpha_p\cdots\alpha_{p+r}} \in A^{q+s}(U_{\alpha_0}\cap\cdots\cap U_{\alpha_{p+r}}).$$

Often we suppress explicit reference to the simplex on which this formula is being evaluated and write

$$\varphi\psi = (-1)^{qr}\varphi\wedge\psi.$$

We say that $E^{**}(\mathfrak{U}) = \{E^{p,q}(\mathfrak{U})\}_{p,q=0}^\infty$ is a bigraded algebra under this multiplication. Note how this operation combines the cup product from Čech theory with the exterior multiplication from de Rham theory. The strange sign $(-1)^{qr}$ will be needed in the proof of Lemma D.2.8.

Definition D.2.5. The de Rham operator $\varepsilon : E^{p,q}(\mathfrak{U}) \to E^{p,q+1}(\mathfrak{U})$ is defined by setting

$$(\varepsilon\varphi)_{\alpha_0\alpha_1\cdots\alpha_p} = (-1)^p d(\varphi_{\alpha_0\alpha_1\cdots\alpha_p}) \in A^{q+1}(U_{\alpha_0} \cap U_{\alpha_1} \cap \cdots \cap U_{\alpha_p}),$$

for arbitrary $\varphi \in \check{C}^p(\mathfrak{U}; A^q)$ and for every p-simplex $(U_{\alpha_0}, U_{\alpha_1}, \ldots, U_{\alpha_p})$ of $\mathfrak{U}$.

Definition D.2.6. The Čech operator $\delta : E^{p,q}(\mathfrak{U}) \to E^{p+1,q}(\mathfrak{U})$ is defined by setting

$$(\delta\varphi)_{\alpha_0\alpha_1\cdots\alpha_{p+1}} = \sum_{i=0}^{p+1}(-1)^i \varphi_{\alpha_0\cdots\widehat{\alpha}_i\cdots\alpha_{p+1}} \in A^q(U_{\alpha_0} \cap U_{\alpha_1} \cap \cdots \cap U_{\alpha_{p+1}}),$$

$\forall \varphi \in \check{C}^p(\mathfrak{U}; A^q)$ and for every p-simplex $(U_{\alpha_0}, U_{\alpha_1}, \ldots, U_{\alpha_p})$ of $\mathfrak{U}$.

The following are evident:

- $\varepsilon^2 = 0$,
- $\delta^2 = 0$,
- $\varepsilon \circ \delta = -\delta \circ \varepsilon$.

Remark that the sign $(-1)^p$ in the definition of ε is responsible for the anticommutativity of ε and δ.

$$\begin{array}{ccccccc}
\vdots & & \vdots & & \vdots & & \\
\uparrow{\scriptstyle\varepsilon} & & \uparrow{\scriptstyle\varepsilon} & & \uparrow{\scriptstyle\varepsilon} & & \\
E^{0,2}(\mathfrak{U}) & \xrightarrow{\delta} & E^{1,2}(\mathfrak{U}) & \xrightarrow{\delta} & E^{2,2}(\mathfrak{U}) & \xrightarrow{\delta} & \cdots \\
\uparrow{\scriptstyle\varepsilon} & & \uparrow{\scriptstyle\varepsilon} & & \uparrow{\scriptstyle\varepsilon} & & \\
E^{0,1}(\mathfrak{U}) & \xrightarrow{\delta} & E^{1,1}(\mathfrak{U}) & \xrightarrow{\delta} & E^{2,1}(\mathfrak{U}) & \xrightarrow{\delta} & \cdots \\
\uparrow{\scriptstyle\varepsilon} & & \uparrow{\scriptstyle\varepsilon} & & \uparrow{\scriptstyle\varepsilon} & & \\
E^{0,0}(\mathfrak{U}) & \xrightarrow{\delta} & E^{1,0}(\mathfrak{U}) & \xrightarrow{\delta} & E^{2,0}(\mathfrak{U}) & \xrightarrow{\delta} & \cdots
\end{array}$$

Figure D.2.1. The de Rham–Čech complex

Definition D.2.7. For each integer $m \geq 0$, $E^m(\mathfrak{U}) = \bigoplus_{p+q=m} E^{p,q}(\mathfrak{U})$ and the total differential operator $D : E^m(\mathfrak{U}) \to E^{m+1}(\mathfrak{U})$ is $D = \varepsilon + \delta$.

It is a good idea to picture $E^{**}(\mathfrak{U})$ laid out as a first quadrant array in the (p, q)-plane, having $E^{p,q}(\mathfrak{U})$ at the point (p, q) of the integer lattice as in Figure D.2.1, with the de Rham operators ε as vertical arrows and the Čech operators δ as horizontal arrows. This array is called the de Rham–Čech complex. The total degree of an element $\varphi \in E^{p,q}(\mathfrak{U})$ is $p+q$ and $E^m(\mathfrak{U})$ is spanned by the elements of total degree m. One can view $E^m(\mathfrak{U})$ in this diagram as lying along the diagonal $p+q=m$. If $\varphi \in E^{p,q}(\mathfrak{U})$, where $p+q=m$, then

$$D(\varphi) = \varepsilon(\varphi) + \delta(\varphi) \in E^{p,q+1}(\mathfrak{U}) \oplus E^{p+1,q}(\mathfrak{U}) \subset E^{m+1}(\mathfrak{U}).$$

Lemma D.2.8. *The pair $(E^*(\mathfrak{U}), D)$ is a cochain complex in which $E^*(\mathfrak{U})$ is a graded algebra over $\mathbb{R}$ and*

$$D(\varphi\psi) = D(\varphi)\psi + (-1)^m \varphi D(\psi),$$

where $\varphi \in E^m(\mathfrak{U})$.

Proof. Indeed, $E^*(\mathfrak{U}) = \{E^m(\mathfrak{U})\}_{m=0}^{\infty}$ is a graded vector space and it is clear that the bigraded multiplication in $E^{**}(\mathfrak{U})$ induces a graded algebra structure on $E^*(\mathfrak{U})$. Because of the anticommutativity of ε and δ,

$$D^2 = \varepsilon^2 + \varepsilon \circ \delta + \delta \circ \varepsilon + \delta^2 = 0.$$

Finally, it is only necessary to verify the Leibnitz formula for $\varphi \in E^{p,q}(\mathfrak{U})$ and $\psi \in E^{r,s}(\mathfrak{U})$, $p+q = m$. Suppress reference to the $(p+r)$-simplex $(U_{\alpha_0}, \dots, U_{\alpha_{p+r}})$ and compute

$$\begin{aligned}
\varepsilon(\varphi\psi) &= (-1)^{p+r} d((-1)^{rq} \varphi \wedge \psi) \\
&= (-1)^{p+r+rq}(d(\varphi) \wedge \psi + (-1)^q \varphi \wedge d(\psi)) \\
&= (-1)^{p+r+rq+p+(q+1)r}(\varepsilon(\varphi)\psi) + (-1)^{p+r+rq+q+r+rq}(\varphi\varepsilon(\psi)) \\
&= \varepsilon(\varphi)\psi + (-1)^{p+q}\varphi\varepsilon(\psi).
\end{aligned}$$

That is,

$$\varepsilon(\varphi\psi) = \varepsilon(\varphi)\psi + (-1)^m \varphi\varepsilon(\psi). \tag{D.1}$$

Similarly, suppress reference to the $(p+r+1)$-simplex $(U_{\alpha_0}, \dots, U_{\alpha_{p+r+1}})$ and compute

$$\begin{aligned}
\delta(\varphi\psi) &= (-1)^{rq}(\delta(\varphi) \wedge \psi + (-1)^p \varphi \wedge \delta(\psi)) \\
&= (-1)^{rq}(-1)^{rq}\delta(\varphi)\psi + (-1)^{p+rq}(-1)^{q(r+1)}\varphi\delta(\psi) \\
&= \delta(\varphi)\psi + (-1)^{p+q}\varphi\delta(\psi).
\end{aligned}$$

That is,

$$\delta(\varphi\psi) = \delta(\varphi)\psi + (-1)^m \varphi\delta(\psi). \tag{D.2}$$

By adding equation (D.1) and equation (D.2), we obtain the desired Leibnitz rule for D. □

Lemma D.2.9. *There are canonical inclusions*

$$(A^*(M), d) \xhookrightarrow{i} (E^*(\mathfrak{U}), D),$$

$$(\check{C}^*(\mathfrak{U}; \mathbb{R}), \delta) \xhookrightarrow{j} (E^*(\mathfrak{U}), D)$$

of subcomplexes, respecting the graded algebra structures.

Proof. Indeed, if $\omega \in A^q(M)$, $i(\omega) \in \check{C}^0(\mathfrak{U}; A^q) = E^{0,q}(\mathfrak{U})$ assigns to each 0-simplex (U_{α_0}) the element $i(\omega)_{\alpha_0} = \omega | U_{\alpha_0}$. Since

$$(\delta(i(\omega)))_{\alpha_0\alpha_1} = \omega | U_{\alpha_0} \cap U_{\alpha_1} - \omega | U_{\alpha_0} \cap U_{\alpha_1} = 0,$$

we see that

$$D(i(\omega)) = \varepsilon(i(\omega)) = i(d\omega).$$

Likewise, if $\varphi \in \check{C}^p(\mathfrak{U}; \mathbb{R})$, $j(\varphi) \in \check{C}^p(\mathfrak{U}; A^0)$ assigns to each p-simplex $(U_{\alpha_0}, \dots, U_{\alpha_p})$ the 0-form on the open set $U_{\alpha_0} \cap \dots \cap U_{\alpha_p}$ that is the constant function $\varphi_{\alpha_0 \cdots \alpha_p} \in \mathbb{R}$.

Clearly, $\varepsilon(\varphi)_{\alpha_0\cdots\alpha_p} = 0$, and so j also commutes with the coboundary operators of these complexes. It is clear from these definitions that

$$i(\omega \wedge \eta) = i(\omega)i(\eta),$$

whenever $\omega \in A^q(M)$ and $\eta \in A^s(M)$, and that

$$j(\varphi\psi) = j(\varphi)j(\psi),$$

whenever $\varphi \in \check{C}^p(\mathfrak{U};\mathbb{R})$ and $\psi \in \check{C}^r(\mathfrak{U};\mathbb{R})$. □

Corollary D.2.10. *There are canonical homomorphisms*

$$i^* : H^*_{\mathrm{DR}}(M) \to H^*(E^*(\mathfrak{U}), D)$$

and

$$j^* : \check{H}^*(\mathfrak{U};\mathbb{R}) \to H^*(E^*(\mathfrak{U}), D)$$

of graded algebras.

We augment the rows of the diagram in Figure D.2.1 by i. That is, the new rows are

$$A^q(M) \xrightarrow{i} E^{0,q}(\mathfrak{U}) \xrightarrow{\delta} E^{1,q}(\mathfrak{U}) \xrightarrow{\delta} \cdots .$$

Similarly, we augment the columns by j:

$$\check{C}^p(\mathfrak{U};\mathbb{R}) \xrightarrow{j} E^{p,0}(\mathfrak{U}) \xrightarrow{\varepsilon} E^{p,1}(\mathfrak{U}) \xrightarrow{\varepsilon} \cdots .$$

Lemma D.2.11. *The augmented diagram has exact rows.*

Proof. If $\omega \in A^q(M)$, we have seen in the proof of Lemma D.2.9 that $\delta(i(\omega)) = 0$. Conversely, if $\varphi \in \check{C}^0(\mathfrak{U};A^q)$ and $\delta(\varphi) = 0$, then the forms $\varphi_{\alpha_0} \in A^q(U_{\alpha_0})$ and $\varphi_{\alpha_1} \in A^q(U_{\alpha_1})$ must agree on $U_{\alpha_0} \cap U_{\alpha_1}$, if this intersection is nonempty. Hence, the forms $\varphi_\alpha \in A^q(U_\alpha)$ piece together smoothly to give a form $\omega \in A^q(M)$ such that $i(\omega) = \varphi$. This proves exactness at $E^{0,q}(\mathfrak{U})$.

We prove exactness at $E^{p,q}(\mathfrak{U}) = \check{C}^p(\mathfrak{U};A^q)$ when $p \geq 1$. Let $\{\lambda_\alpha\}_{\alpha\in\mathfrak{A}}$ be a smooth partition of unity subordinate to $\mathfrak{U}$. Define

$$\Lambda : E^{p,q}(\mathfrak{U}) \to E^{p-1,q}(\mathfrak{U})$$

as follows. Given $\varphi \in \check{C}^p(\mathfrak{U};A^q) = E^{p,q}(\mathfrak{U})$, define $\Lambda(\varphi) \in \check{C}^{p-1}(\mathfrak{U};A^q)$ to be the element, the value of which on the $(p-1)$-simplex $(U_{\alpha_0},\dots,U_{\alpha_{p-1}})$ is

$$\Lambda(\varphi)_{\alpha_0\cdots\alpha_{p-1}} = \sum_\alpha \lambda_\alpha \varphi_{\alpha\alpha_0\cdots\alpha_{p-1}},$$

where each term of this locally finite sum is interpreted, in the obvious way, as a q-form on $U_{\alpha_0} \cap \cdots \cap U_{\alpha_{p-1}}$. If $\delta(\varphi) = 0$, the reader can check that $\varphi = \delta(\Lambda(\varphi))$, proving exactness at $E^{p,q}(\mathfrak{U})$. □

Lemma D.2.12. *If the cover $\mathfrak{U}$ is simple, the augmented diagram has exact columns.*

Proof. If $\zeta \in \check{C}^p(\mathfrak{U};\mathbb{R})$, we have seen in the proof of Lemma D.2.9 that $\varepsilon(j(\zeta)) = 0$. Conversely, it is clear that, if $\varphi \in \check{C}^p(\mathfrak{U},A^0)$ and $\varepsilon(\varphi) = 0$, then, $d(\varphi_{\alpha_0\cdots\alpha_p}) = 0$, for each p-simplex $(U_{\alpha_0},\dots,U_{\alpha_p})$. The fact that $U_{\alpha_0} \cap \cdots \cap U_{\alpha_p}$ is connected implies that $\varphi_{\alpha_0\cdots\alpha_p} \in A^0(U_{\alpha_0} \cap \cdots \cap U_{\alpha_p})$ is constant. Thus, we can define $\zeta \in \check{C}^p(\mathfrak{U};\mathbb{R})$ by

$$\zeta_{\alpha_0\cdots\alpha_p} = \text{ the constant } \varphi_{\alpha_0\cdots\alpha_p}$$

and $j(\zeta) = \varphi$. This proves exactness at $E^{p,0}$.

If $q \geq 1$, exactness at $E^{p,q}(\mathfrak{U})$ follows from the Poincaré lemma and the fact that $U_{\alpha_0} \cap \cdots \cap U_{\alpha_p}$, if not empty, has the de Rham cohomology of $\mathbb{R}^n$. □

Lemma D.2.13. *If the cover $\mathfrak{U}$ is simple, the homomorphisms i^* and j^* are surjective.*

Proof. Let $\zeta \in E^m(\mathfrak{U})$ be a D-cocycle. We show that the element $[\zeta] \in H^*(E^*(\mathfrak{U}), D)$ is in the image both of i^* and j^*. If $m = 0$, then $\varepsilon\zeta = 0 = \delta\zeta$ and the assertions follow from Lemmas D.2.11 and D.2.12, respectively. Assume, therefore, that $m \geq 1$.

The cocycle ζ lies along the diagonal $p + q = m$, so we write

$$\zeta = \sum_{p=0}^{m} \zeta_p,$$

where $\zeta_p \in E^{p,m-p}(\mathfrak{U})$, $0 \leq p \leq m$. The equation $D(\zeta) = 0$ implies that $\varepsilon(\zeta_0) = 0$, so Lemma D.2.12 allows us to find $\xi_0 \in E^{0,m-1}(\mathfrak{U})$ with $\varepsilon(\xi_0) = \zeta_0$. Then $\zeta - D(\xi_0)$ has 0 as its component in $E^{0,m}(\mathfrak{U})$ and is D-cohomologous to ζ. Suppose, inductively, that $\xi_k \in E^{m-1}(\mathfrak{U})$ has been found so that $\zeta - D(\xi_k)$ has 0 as its component in $E^{p,m-p}(\mathfrak{U})$, $0 \leq p \leq k$. Since this is still a D-cocycle, exactness of the column $p = k+1$ allows us to find $\theta \in E^{k+1,m-k-2}(\mathfrak{U})$ such that $\zeta - D(\xi_k) - D(\theta)$ has 0 as its component in $E^{p,m-p}(\mathfrak{U})$, $0 \leq p \leq k+1$. We take $\xi_{k+1} = \xi_k + \theta$. By finite induction, there is $\xi_m \in E^{m-1}(\mathfrak{U})$ such that $\zeta - D(\xi_m)$ is concentrated in $E^{m,0}(\mathfrak{U})$. Thus, without loss of generality, we assume that $\zeta \in E^{m,0}(\mathfrak{U}) = \check{C}^m(\mathfrak{U}; A^0)$. The fact that this is a D-cocycle implies that $\varepsilon(\zeta) = 0 = \delta(\zeta)$. Thus, there is $c \in \check{C}^m(\mathfrak{U}; \mathbb{R})$ such that $j(c) = \zeta$ and $j(\delta(c)) = \delta(\zeta) = 0$. Since j is one-to-one, $\delta(c) = 0$, so $[c] \in \check{H}^m(\mathfrak{U}; \mathbb{R})$ and $j^*[c] = [\zeta]$.

An entirely parallel argument, using Lemma D.2.11 (exactness of the rows), shows that we can assume that ζ is concentrated in $E^{0,m}(\mathfrak{U})$ and that there is $\omega \in A^m(M)$ such that $d\omega = 0$ and $i^*[\omega] = [\zeta]$. □

Lemma D.2.14. *If the cover $\mathfrak{U}$ is simple, the homomorphisms i^* and j^* are injective.*

Proof. Suppose that $\omega \in A^m(M)$ has $d\omega = 0$ and $i^*[\omega] = 0$. If $m = 0$, then ω is a locally constant function and $i(\omega)$ vanishes on every 0-simplex (U_{α_0}). That is, $\omega \equiv 0$, so $[\omega] = 0 \in H^0_{\mathrm{DR}}(M)$. If $m \geq 1$, then $i(\omega) \in E^{0,m}(\mathfrak{U})$ is of the form $i(\omega) = D(\xi)$, $\xi \in E^{m-1}(\mathfrak{U})$. Write

$$\xi = \sum_{p=0}^{m-1} \xi_p,$$

where $\xi_p \in E^{p,m-1-p}(\mathfrak{U})$, $0 \leq p \leq m-1$. Since the component of $i(\omega)$ in $E^{m,0}(\mathfrak{U})$ is 0, $\delta(\xi_{m-1}) = 0$ and Lemma D.2.11 implies that $\xi_{m-1} = \delta(\theta)$, $\theta \in E^{m-2,0}(\mathfrak{U})$. Thus, $\xi' = \xi - D(\theta)$ has component 0 in $E^{m-1,0}(\mathfrak{U})$ and $D(\xi') = i(\omega)$. Again using Lemma D.2.11 and finite induction, we see that no generality is lost in assuming that ξ is concentrated in $E^{0,m-1}(\mathfrak{U})$ and that $\delta(\xi) = 0$. By one more appeal to Lemma D.2.11, we find a unique $\eta \in A^{m-1}(M)$ such that $i(\eta) = \xi$. Then, $i(d\eta) = \varepsilon(i(\eta)) = i(\omega)$ and, i being injective, $\omega = d\eta$. That is, $[\omega] = 0$ as desired.

A completely parallel argument, using Lemma D.2.12, proves that j^* is injective. □

If the cover $\mathfrak{U}$ is simple, we define

$$\psi_{\mathfrak{U}} = (i^*)^{-1} \circ j^* : \check{H}^*(\mathfrak{U};\mathbb{R}) \to H^*_{\mathrm{DR}}(M),$$

an isomorphism of graded $\mathbb{R}$-algebras by Lemmas D.2.13 and D.2.14. The following completes the proof of Theorem D.2.1.

Lemma D.2.15. *If $\mathfrak{U} \preceq \mathfrak{V}$ are simple covers of M, then the diagram*

$$\begin{array}{ccc} \check{H}^*(\mathfrak{U};\mathbb{R}) & \xrightarrow{\varphi_{\mathfrak{U}}^{\mathfrak{V}}} & \check{H}^*(\mathfrak{V};\mathbb{R}) \\ & \searrow^{\psi_{\mathfrak{U}}} \quad {}^{\psi_{\mathfrak{V}}}\swarrow & \\ & H^*_{\mathrm{DR}}(M) & \end{array}$$

is commutative.

Proof. Indeed, if $\mathfrak{U} = \{U_\alpha\}_{\alpha\in\mathfrak{A}}$ and $\mathfrak{V} = \{V_\beta\}_{\beta\in\mathfrak{B}}$, recall that $\varphi_{\mathfrak{U}}^{\mathfrak{V}}$ is induced by a choice function $\ell : \mathfrak{B} \to \mathfrak{A}$ such that $V_\beta \subseteq U_{\ell(\beta)}$, $\forall \beta \in \mathfrak{B}$. The same choice function ℓ defines a homomorphism

$$\varphi_{\mathfrak{U}}^{\mathfrak{V}} : H^*(E^*(\mathfrak{U}), D) \to H^*(E^*(\mathfrak{V}), D)$$

and the diagram

$$\begin{array}{ccccc} H^*(M) & \xrightarrow{i^*} & H^*(E^*(\mathfrak{U}), D) & \xleftarrow{j^*} & \check{H}^*(\mathfrak{U};\mathbb{R}) \\ \mathrm{id}\downarrow & & \varphi_{\mathfrak{U}}^{\mathfrak{V}}\downarrow & & \varphi_{\mathfrak{U}}^{\mathfrak{V}}\downarrow \\ H^*_{\mathrm{DR}}(M) & \xrightarrow{i^*} & H^*(E^*(\mathfrak{V}), D) & \xleftarrow{j^*} & \check{H}^*(\mathfrak{V};\mathbb{R}) \end{array}$$

commutes. □

Remark. In fact, the de Rham–Čech isomorphism

$$\psi : \check{H}^*(M;\mathbb{R}) \to H^*_{\mathrm{DR}}(M)$$

is *functorial.* That is, whenever $f : M \to N$ is a smooth map between manifolds, the diagram

$$\begin{array}{ccc} \check{H}^*(N;\mathbb{R}) & \xrightarrow{f^*} & \check{H}^*(M;\mathbb{R}) \\ \psi\downarrow & & \downarrow\psi \\ H^*_{\mathrm{DR}}(N) & \xrightarrow[f^*]{} & H^*_{\mathrm{DR}}(M) \end{array}$$

is commutative. That is, on smooth manifolds, Čech theory and de Rham theory are equivalent as *functors.* Checking this functoriality is straightforward.

D.3. Singular Cohomology

We will define the graded singular cohomology algebra $H^*(M;\mathbb{R})$ and prove the following.

Theorem D.3.1. *There is a canonical isomorphism*

$$\check{H}^*(M;\mathbb{R}) = H^*(M;\mathbb{R})$$

of graded R-algebras.

We note that with little extra effort, the proof of this theorem can be carried out with $\mathbb{R}$ replaced by an arbitrary commutative ring with unity. It can also be generalized to a larger class of topological spaces than manifolds.

The proof of Theorem D.3.1 will be analogous to that of Theorem D.2.2. Coupled with Theorem D.2.2, this will prove the de Rham theorem for singular cohomology.

Theorem D.3.2 (de Rham). *There is a canonical isomorphism*

$$H^*_{\mathrm{DR}}(M) = H^*(M;\mathbb{R})$$

of graded $\mathbb{R}$*-algebras.*

Again, these isomorphisms are easily checked to be functorial.

The singular cohomology should be defined via duality at the chain level. Recall (Definition 8.2.11) that $C_p(M;\mathbb{R})$ denotes the real vector space with basis the set $\Delta_p(M)$ of smooth (respectively, continuous) singular p-simplices in M. This is called the space of singular p-chains. Set

$$C^p(M;\mathbb{R}) = \mathrm{Hom}_{\mathbb{R}}(C_p(M;\mathbb{R}),\mathbb{R}),$$

the space of singular p-cochains. The boundary operator

$$\partial : C_{p+1}(M;\mathbb{R}) \to C_p(M;\mathbb{R})$$

has adjoint

$$\partial^* : C^p(M;\mathbb{R}) \to C^{p+1}(M;\mathbb{R}),$$

called the *singular coboundary* operator and the identity $\partial^2 = 0$ dualizes to $\partial^{*2} = 0$. In the standard fashion, this gives rise to the vector spaces $Z^*(M;\mathbb{R})$ and $B^*(M;\mathbb{R})$, called the spaces of singular cocycles and singular coboundaries, respectively. The singular cohomology theory is then the quotient

$$H^*(M;\mathbb{R}) = Z^*(M;\mathbb{R})/B^*(M;\mathbb{R}).$$

The above construction is *functorial.* That is, smooth (respectively, continuous) maps $f : N \to M$ induce graded, $\mathbb{R}$-linear maps

$$f_\# : C_*(N;\mathbb{R}) \to C_*(M;\mathbb{R}),$$
$$f^\# : C^*(M;\mathbb{R}) \to C^*(N;\mathbb{R}).$$

These commute, respectively, with ∂ and ∂^*, hence pass to graded, $\mathbb{R}$-linear maps, f_* and f^* on homology and cohomology, respectively. The usual functorial properties are satisfied, making singular homology into a covariant functor, singular cohomology into a contravariant functor. For homology, this is the content of Exercise 8.2.24. Dualizing this gives the corresponding result for cohomology.

It remains that we define the graded algebra structure on $H^*(M;\mathbb{R})$. Multiplication in this algebra is called the singular cup product and will be denoted by a dot "$\cdot$" to distinguish it from the Čech cup product. For this, let $n = p+q$, $p, q \geq 0$, and consider the maps

$$\sigma_p : \Delta_p \to \Delta_n,$$
$$\sigma^q : \Delta_q \to \Delta_n,$$

defined by

$$\sigma_p(x^1,\dots,x^p) = (x^1,\dots,x^p,0,\dots,0),$$
$$\sigma^q(x^1,\dots,x^q) = (0,\dots,0,x^1,\dots,x^q).$$

One calls σ_p the front p-face operator and σ^q the back q-face operator. Given $\varphi \in C^p(M;\mathbb{R})$ and $\psi \in C^q(M;\mathbb{R})$, the cup product $\varphi \cdot \psi \in C^n(M;\mathbb{R})$ will be completely determined by its values on the set $\Delta_n(M)$ of singular n-simplices on M. This is because this set is a basis for the vector space $C_n(M;R)$. If $s \in \Delta_n(M)$, define

$$\varphi \cdot \psi(s) = \varphi(s \circ \sigma_p)\psi(s \circ \sigma^q).$$

A little combinatorics gives the following expected relation.

Lemma D.3.3. *If $\varphi \in C^p(M;\mathbb{R})$ and $\psi \in C^q(M;\mathbb{R})$, then*

$$\partial^*(\varphi \cdot \psi) = (\partial^*\varphi) \cdot \psi + (-1)^p \varphi \cdot (\partial^*\psi).$$

As usual, we get the following consequence.

Lemma D.3.4. *The graded vector space $Z^*(M;\mathbb{R})$ of singular cocycles is a graded subalgebra of $C^*(M;\mathbb{R})$ and $B^*(M;\mathbb{R}) \subseteq Z^*(M;\mathbb{R})$ is a 2-sided ideal. Consequently, cup product is well defined on $H^*(M;\mathbb{R})$, making that graded vector space into a graded algebra over $\mathbb{R}$. Finally, if $f : N \to M$ is smooth, the induced map*

$$f^* : H^*(M;\mathbb{R}) \to H^*(N;\mathbb{R})$$

is a homomorphism of graded algebras.

Remark. In the above discussion, we allowed $\Delta_q(M)$ to be either the set of smooth singular simplices in M or the set of continuous ones. This yields two possibly different singular cohomologies, the smooth and the continuous. The proof that we will give of Theorem D.3.1 works equally well in either case, hence both theories, being canonically isomorphic to Čech cohomology, are identical. Also, since continuous maps between manifolds are homotopic to smooth ones (Subsection 3.8.B), the homotopy invariance of singular theory (not proven here, but *cf.* [**13**]) implies that these theories are canonically isomorphic as *functors*.

Corollary D.3.5. *The singular cohomology algebra, computed by using smooth singular simplices, is canonically and functorially equivalent to that computed by using continuous singular simplices.*

Let $\mathfrak{U} = \{U_\alpha\}_{\alpha \in \mathfrak{A}}$ be an open cover. We are going to mimic the construction of the de Rham–Čech complex to produce a singular-Čech complex $E^{**}(\mathfrak{U})$. Ultimately, we will need $\mathfrak{U}$ to be simple. The proof of Theorem D.3.1 will then proceed almost exactly as that of Theorem D.2.1.

Definition D.3.6. A Čech p-cochain on $\mathfrak{U}$ with values in C^q is a function φ that, to each p-simplex $(U_{\alpha_0}, U_{\alpha_1}, \dots, U_{\alpha_p})$ of $\mathfrak{U}$ assigns

$$\varphi_{\alpha_0\alpha_1\dots\alpha_p} \in C^q(U_{\alpha_0} \cap U_{\alpha_1} \cap \cdots \cap U_{\alpha_p};\mathbb{R}).$$

The set of all such cochains will be denoted $E^{p,q}(\mathfrak{U}) = \check{C}^p(\mathfrak{U};C^q)$.

Once again, this is naturally a real vector space. We have simply replaced $A^q(U_{\alpha_0} \cap U_{\alpha_1} \cap \cdots \cap U_{\alpha_p})$ with $C^q(U_{\alpha_0} \cap U_{\alpha_1} \cap \cdots \cap U_{\alpha_p};\mathbb{R})$ in the earlier definition. Similarly, we get a bigraded multiplication on the resulting double complex and, setting $\varepsilon = (-1)^p\partial^* : E^{p+q}(\mathfrak{U}) \to E^{p,q+1}$ and defining δ in complete analogy with Definition D.2.6, we complete the definition of the double complex. The total differential is $D = \varepsilon + \delta$.

The first significant difference between the current and former construction is that there is no natural inclusion of $(C^*(M;\mathbb{R}),\partial^*)$ into $(E^*(\mathfrak{U}),D)$. If $\varphi \in$

$C^q(M;\mathbb{R})$, $U_\alpha \in \mathfrak{U}$, and if $i_\alpha : U_\alpha \hookrightarrow M$ is the inclusion, then we can define the restriction of φ to U_α by

$$\varphi|U_\alpha = i_\alpha^\#(\varphi).$$

This defines a homomorphism

$$i : (C^q(M;\mathbb{R}), \partial^*) \to (\check{C}^0(\mathfrak{U}; C^q), \varepsilon),$$
$$i(\varphi)_\alpha = \varphi|U_\alpha$$

of cochain complexes. However, it is quite possible that $\varphi \neq 0$, but that $\varphi|U_\alpha = 0$, for every $\alpha \in \mathfrak{A}$, and so i will not be injective. The solution to this is the delicate subdivision process for singular homology (see commentary following Proposition 8.5.13) and cohomology that allows computation of these theories using only "$\mathfrak{U}$-small" singular simplices (Definition 8.5.11). The sketch for homology accompanying Definitions 8.5.11 and 8.5.12 dualizes to a similar procedure for cohomology and the references for details are the same. We note that the definition of cup product also works for this $\mathfrak{U}$-small theory and record the following.

Theorem D.3.7. *The inclusions $\Delta_q^{\mathfrak{U}}(M) \hookrightarrow \Delta_q(M)$, $q \geq 0$, induce canonical isomorphisms of graded vector spaces*

$$H_*^{\mathfrak{U}}(M;\mathbb{R}) = H_*(M;\mathbb{R}),$$
$$H^*_{\mathfrak{U}}(M;\mathbb{R}) = H^*(M;\mathbb{R}).$$

In the case of cohomology, this is an isomorphism of graded algebras.

One now notes that

$$i : (C^q_{\mathfrak{U}}(M;\mathbb{R}), \partial^*) \to (\check{C}^0(\mathfrak{U}; C^q), \varepsilon)$$

is injective.

Lemma D.3.8. *There are canonical inclusions*

$$(C^*_{\mathfrak{U}}(M;\mathbb{R}), \partial^*) \overset{i}{\hookrightarrow} (E^*(\mathfrak{U}), D),$$
$$(\check{C}^*(\mathfrak{U};\mathbb{R}), \delta) \overset{j}{\hookrightarrow} (E^*(\mathfrak{U}), D)$$

of subcomplexes, respecting the graded algebra structures.

Proof. Indeed, for each Čech 1-simplex $(U_{\alpha_0}, U_{\alpha_1})$ and each cochain $\varphi \in C^q_{\mathfrak{U}}(M;\mathbb{R})$, we see that

$$(\delta i(\varphi))_{\alpha_0\alpha_1} = \varphi|U_{\alpha_0} \cap U_{\alpha_1} - \varphi|U_{\alpha_0} \cap U_{\alpha_1} = 0$$

for the above definition of i. Thus,

$$D(i(\varphi)) = \varepsilon(i(\varphi)) = i(\partial^*\varphi).$$

For the definition of j, note that, for each Čech p-simplex, $(U_{\alpha_0}, \dots, U_{\alpha_p})$, $C^0(U_{\alpha_0} \cap \cdots \cap U_{\alpha_p})$ is just the set of arbitrary $\mathbb{R}$-valued functions on the open set $U_{\alpha_0} \cap \cdots \cap U_{\alpha_p}$ (the singular 0-simplices in a space are just the points). A Čech 0-cochain $\psi \in \check{C}^0(\mathfrak{U};\mathbb{R})$ assigns to each p-simplex $(U_{\alpha_0}, \dots, U_{\alpha_p})$ an element $\psi_{\alpha_0\cdots\alpha_p} \in \mathbb{R}$. We define $j(\psi) \in \check{C}^p(\mathfrak{U}; C^0)$ by letting $j(\psi)_{\alpha_0\cdots\alpha_p}$ denote the constant function on $U_{\alpha_0} \cap \cdots \cap U_{\alpha_p}$ with value $\psi_{\alpha_0\cdots\alpha_p}$. Evidently, $j(\psi) = 0$ if and only if $\psi = 0$, and so j is injective, By the definition of the singular coboundary operator, the coboundary of a constant 0-cochain vanishes, so

$$D(j(\psi)) = \delta(j(\psi)) = j(\delta\psi).$$

The fact that i and j respect the graded algebra structures can be checked by the reader. □

Again we augment the rows of $E^{**}(\mathfrak{U})$ by i, obtaining the sequence

$$C^q(M;\mathbb{R}) \xhookrightarrow{i} E^{0,q}(\mathfrak{U}) \xrightarrow{\delta} E^{1,q}(\mathfrak{U}) \xrightarrow{\delta} \cdots,$$

and we augment the columns by j, obtaining the sequence

$$\check{C}^p(\mathfrak{U};\mathbb{R}) \xhookrightarrow{j} E^{p,0}(\mathfrak{U}) \xrightarrow{\varepsilon} E^{p,1}(\mathfrak{U}) \xrightarrow{\varepsilon} \cdots.$$

Lemma D.3.9. *If the cover $\mathfrak{U}$ is locally finite (in particular, if $\mathfrak{U}$ is simple) the augmented diagram has exact rows.*

Proof. We have seen that $\delta \circ i = 0$. Conversely, if $\psi \in \check{C}^0(\mathfrak{U}; C^q)$ and $\delta\psi = 0$, we see that, for each 1-simplex $(U_{\alpha_0}, U_{\alpha_1})$,

$$\psi_{\alpha_0}|U_{\alpha_0} \cap U_{\alpha_1} = \psi_{\alpha_1}|U_{\alpha_0} \cap U_{\alpha_1},$$

and so the ψ_αs patch together to define $\psi' \in C^q_{\mathfrak{U}}(M)$ such that $i(\psi') = \psi$. This proves exactness at $E^{0,q}(\mathfrak{U})$.

We prove exactness at $E^{p,q}(\mathfrak{U}) = \check{C}^p(\mathfrak{U}; C^q)$ when $p \geq 1$. For this, we construct $\Lambda : E^{p,q}(\mathfrak{U}) \to E^{p-1,q}(\mathfrak{U})$, $p \geq 1$, such that

$$\delta\varphi = 0 \Rightarrow \varphi = \delta\Lambda(\varphi).$$

Since $\delta^2 = 0$, this will prove the lemma.

Let $\nu : \Delta^{\mathfrak{U}}_q(M) \to \mathbb{Z}^+$ be defined by

$$\nu(\sigma) = \text{ cardinality of } \{U_\alpha \in \mathfrak{U} \mid \sigma(\Delta_q) \subset U_\alpha\}.$$

The hypothesis that $\mathfrak{U}$ is locally finite guarantees that $\nu(\sigma)$ is finite. If $U_\alpha \in \mathfrak{U}$ and $(U_{\alpha_0}, \dots, U_{\alpha_{p-1}})$ is a Čech $(p-1)$-simplex, define

$$i^{\alpha_0\cdots\alpha_{p-1}}_\alpha : C^q(U_\alpha \cap U_{\alpha_0} \cap \cdots \cap U_{\alpha_{p-1}}) \to C^q(U_{\alpha_0} \cap \cdots \cap U_{\alpha_{p-1}})$$

by

$$i^{\alpha_0\cdots\alpha_{p-1}}_\alpha(\varphi)(\sigma) = \begin{cases} \varphi(\sigma), & \text{if } \sigma(\Delta_q) \subset U_\alpha \cap U_{\alpha_0} \cap \cdots \cap U_{\alpha_{p-1}}, \\ 0 & \text{otherwise.} \end{cases}$$

Finally, for each $\varphi \in E^{p,q}(\mathfrak{U})$, define $\Lambda(\varphi) \in E^{p-1,q}(\mathfrak{U})$ by

$$\Lambda(\varphi)_{\alpha_0\cdots\alpha_{p-1}} = \frac{1}{\nu} \sum_{\alpha\in\mathfrak{A}} i^{\alpha_0\cdots\alpha_{p-1}}_\alpha(\varphi_{\alpha\alpha_0\cdots\alpha_{p-1}}).$$

A direct computation proves the desired identity. □

Lemma D.3.10. *If the cover $\mathfrak{U}$ is simple, the augmented diagram has exact columns.*

Proof. We have seen that $\varepsilon \circ j = 0$. Conversely, if $\psi \in \check{C}^p(\mathfrak{U}; C^0)$ and $\varepsilon(\psi) = 0$, the definition of the singular coboundary operator implies that, for each Čech p-simplex $(U_{\alpha_0}, \dots, U_{\alpha_p})$, $\psi_{\alpha_0\cdots\alpha_p}$ is constant on each component of $U_{\alpha_0} \cap \cdots \cap U_{\alpha_p}$. Since the cover is simple, this open set is connected and $\psi_{\alpha_0\cdots\alpha_p}$ is constant. That is, $\psi \in \operatorname{im} j$.

Finally, exactness at $E^{p,q}(\mathfrak{U})$, $q \geq 1$, is proven by dualizing the construction in Example 8.2.17, proving that the singular cohomology of a contractible space is the same as that of a point. Since the cover is simple, the open set $U_{\alpha_0} \cap \cdots \cap U_{\alpha_p}$ is contractible, for each Čech p-simplex $(U_{\alpha_0}, \dots, U_{\alpha_p})$, and the desired exactness follows. □

At this point, the arguments of the previous section can be reproduced, practically word for word, to prove the following.

Theorem D.3.11. *If $\mathfrak{U}$ is a simple cover, there is a canonical isomorphism*

$$\check{H}^*(\mathfrak{U};\mathbb{R}) = H^*_{\mathfrak{U}}(M;\mathbb{R})$$

of graded algebras.

By Corollary D.2.3, $\check{H}^*(\mathfrak{U};\mathbb{R}) = \check{H}^*(M;\mathbb{R})$ and, by Theorem D.3.7, $H^*_{\mathfrak{U}}(M;\mathbb{R}) = H^*(M;\mathbb{R})$, completing the proof of Theorem D.3.1, hence of Theorem D.3.2.

Remark. The advantage of this proof of the de Rham theorem is that it demonstrates the equivalence, at the level of cohomology, of the singular cup product with exterior multiplication of forms. The disadvantage is that it completely disguises the fact that the equivalence of cohomology classes of forms with those of singular cocycles is achieved by integration of forms on (smooth) singular cycles.

Bibliography

[1] F. Adams, *Vector fields on spheres*, Ann. Math. **75** (1962), 603–632.

[2] L. Auslander and R. E. MacKenzie, *Introduction to Differentiable Manifolds*, McGraw–Hill, New York, NY, 1963.

[3] W. Boothby, *An Introduction to Differentiable Manifolds and Differential Geometry*, Academic Press, New York, NY, 1975.

[4] R. Bott and J. Milnor, *On the parallelizability of the spheres*, Bull. Amer. Math. Soc. **64** (1958), 87–89.

[5] R. Bott and L. Tu, *Differential Forms in Algebraic Topology*, Springer-Verlag, New York, NY, 1982.

[6] A. Candel and L. Conlon, *Theory of foliations, i.*, American Mathematical Society, Providence, RI, 1999.

[7] S. K. Donaldson, *An application of gauge theory to the topology of 4-manifolds*, J. Diff. Geo. **18** (1983), 269–316.

[8] J. Dugundji, *Topology*, Allyn and Bacon, Boston, MA, 1966.

[9] B. Eckmann, *Gruppentheoretischer Beweis des Satzes von Hurwitz-Radon über die Komposition quadratischer Formen*, Comm. Math. Helv. **15** (1942), 358–366.

[10] S. Eilenberg and N. Steenrod, *Foundations of Algebraic Topology*, Princeton University Press, Princeton, NJ, 1952.

[11] D. Freed and K. Uhlenbach, *Instantons and Four-Manifolds*, Springer-Verlag, New York, NY, 1984.

[12] R. Gompf, *Three exotic $\mathbb{R}^4$'s and other anomalies*, J. Diff. Geo. **18** (1983), 317–328.

[13] M. Greenberg and H. Harper, *Lectures on Algebraic Topology*, The Benjamin Cummings Publishing Co., Reading, MA, 1981.

[14] S. Helgason, *Differential Geometry and Symmetric Spaces*, Academic Press, New York, NY, 1962.

[15] N. J. Hicks, *Notes on Differential Geometry*, D. Van Nostrand, New York, NY, 1965.

[16] M. W. Hirsch, *Differential Topology*, Springer-Verlag, New York, NY, 1976.

[17] D. Hoffman and W. H. Meeks III, *A complete, embedded minimal surface with genus one, three ends and finite total curvature*, J. Diff. Geo. **21** (1985), 109–127.

[18] ———, *Embedded minimal surfaces of finite topology*, Ann. Math. **131** (1990), 1–34.

[19] ———, *Minimal surfaces based on the catenoid*, Amer. Math Monthly **97** (1990), 702–730.

[20] W. Hurewicz and H. Wallman, *Dimension Theory*, Princeton Univ. Press, Princeton, NJ, 1948.

[21] M. Kervaire and J. Milnor, *Groups of homotopy spheres, I*, Ann. Math. **77** (1963), 505–537.

[22] S. Kobayashi and K. Nomizu, *The Foundations of Differential Geometry, I*, Wiley (Interscience), New York, NY, 1963.

[23] ———, *The Foundations of Differential Geometry, II*, Wiley (Interscience), New York, NY, 1969.

[24] S. Lang, *Real Analysis*, 2nd edition, Addison-Wesley, Reading, MA, 1983.

[25] E. L. Lima, *Commuting vector fields on S^3*, Ann. Math. **81** (1965), 70–81.

[26] W. S. Massey, *Algebraic Topology: An Introduction*, Harcourt, Brace, and World, Inc., New York, NY, 1967.

[27] J. Milnor, *Some consequences of a theorem of Bott*, Ann. Math. **68** (1958), 444–449.

[28] ———, *Morse Theory*, Princeton University Press, Princeton, NJ, 1963.

[29] ———, *Problem list*, Seattle Topology Conference, 1963.

[30] ———, *Topology from a Differentiable Viewpoint*, University of Virginia Press, VA, 1965.

[31] E. Moise, *Geometric Topology in Dimensions* 2 *and* 3, Springer-Verlag, New York, NY, 1977.

[32] S. B. Myers and N. E. Steenrod, *The group of isometries of a Riemannian manifold*, Ann. Math. **40** (1939), 400–416.

[33] W. F. Newns and A. G. Walker, *Tangent planes to a differentiable manifold*, J. London Math. Soc. **31** (1956), 400–407.

[34] B. O'Neill, *Elementary Differential Geometry*, Academic Press, New York, NY, 1966.

[35] H. Poincaré, *Analysis situs*, Journal de l'École Polytechnique **1** (1895), 1–121.

[36] T. Radó, *Über den Begriff der Riemannsche Fläche*, Acta Univ. Szeged **2** (1924–26), 101–121.

[37] G. Reeb, *Sur Certaines Propriétés Topologiques des Variétés Feuilletées*, Hermann, Paris, 1952.

[38] S. Smale, *Generalized Poincaré's conjecture in dimensions greater than four*, Ann. of Math. **74** (1961), 391–406.

[39] E. Spanier, *Algebraic Topology*, McGraw–Hill, Inc., New York, NY, 1966.

[40] M. Spivak, *Differential Geometry, Volume II*, Publish or Perish Publishing Co., Boston, MA, 1970.

[41] ———, *Differential Geometry, Volume I, Second Edition*, Publish or Perish, Inc., Houston, Texas, 1979.

[42] J. Stallings, *The piecewise linear structure of euclidean space*, Proc. Camb. Phil. Soc. **58** (1962), 481–488.

[43] N. Steenrod, *The Topology of Fiber Bundles*, Princeton University Press, Princeton, NJ, 1951.

[44] J. J. Stoker, *Differential Geometry*, Wiley (Interscience), New York, NY, 1969.

[45] D. Tischler, *On fibering certain foliated manifolds over S^1*, Topology **9** (1970), 153–154.

[46] J. W. Vick, *Homology Theory*, Springer-Verlag, New York, NY, 1982.

[47] S. Halperin W. Greub and R. Vanstone, *Connections, Curvature and Cohomology, Volume II*, Academic Press, New York, NY, 1972.

[48] G. Wallet, *Nullité de l'invariant de Godbillon–Vey d'un tore*, C. R. Acad Sci. Paris, Série A **283** (1976), 821–823.

[49] F. Warner, *Foundations of Differentiable Manifolds and Lie Groups*, Springer-Verlag, New York, NY, 1983.

[50] H. Whitney, *The self intersections of a smooth n-manifold in 2n-space*, Ann. Math. **45** (1944), 220–246.

Index

Printed in the United States of America